D1068593

Vacuum Structure in Intense Fields

NATO ASI Series

Advanced Science Institutes Series

A series presenting the results of activities sponsored by the NATO Science Committee, which aims at the dissemination of advanced scientific and technological knowledge, with a view to strengthening links between scientific communities.

The series is published by an international board of publishers in conjunction with the NATO Scientific Affairs Division

A	**Life Sciences**	Plenum Publishing Corporation
B	**Physics**	New York and London
C	**Mathematical and Physical Sciences**	Kluwer Academic Publishers
D	**Behavioral and Social Sciences**	Dordrecht, Boston, and London
E	**Applied Sciences**	
F	**Computer and Systems Sciences**	Springer-Verlag
G	**Ecological Sciences**	Berlin, Heidelberg, New York, London,
H	**Cell Biology**	Paris, Tokyo, Hong Kong, and Barcelona
I	**Global Environmental Change**	

Recent Volumes in this Series

Series B: Physics

Vacuum Structure in Intense Fields

Edited by

H. M. Fried

Brown University
Providence, Rhode Island

and

Berndt Müller

Duke University
Durham, North Carolina

Plenum Press
New York and London
Published in cooperation with NATO Scientific Affairs Division

Proceedings of a NATO Advanced Study Institute on
Vacuum Structure in Intense Fields,
held July 31–August 10, 1990,
in Cargèse, Corsica, France

Library of Congress Cataloging-in-Publication Data

NATO Advanced Study Institute on Vacuum Structure in Intense Fields
  (1990 : Cargèse, France)
    Vacuum structure in intense fields / edited by H.M. Fried and
Berndt Muller.
        p.   cm. -- (NATO ASI series. Series B. Physics ; vol. 255)
    "Published in cooperation with NATO Scientific Affairs Division."
    "Proceedings of a NATO Advanced Study Institute on Vacuum
Structure in Intense Fields, held July 31-August 10, 1990, in
Cargèse, Corsica, France."
    Includes bibliographical references and index.
    ISBN 0-306-43910-7
    1. Quantum electrodynamics--Congresses. 2. Heavy ion collisions-
-Congresses. 3. Quantum field theory--Congresses.
4. Electromagnetic fields--Congresses.   I. Fried, H. M. (Herbert
Martin) II. Müller, Berndt. III. North Atlantic Treaty
Organization. Scientific Affairs Division. IV. Title: Intense
fields. V. Series.
QC793.3.Q35N38 1990
537.6'7--dc20   91-15446
                                                         CIP

© 1991 Plenum Press, New York
A Division of Plenum Publishing Corporation
233 Spring Street, New York, N.Y. 10013

Printed in the United States of America

PREFACE

This Advanced Study Institute (ASI) brought together two distinct "schools of approach" to Quantum Electrodynamics (QED) in the presence of intense, external, electromagnetic fields, in an effort to lay a joint foundation for a needed theoretical explanation of the sharp e^+e^- "resonances" observed in the scattering of very heavy ions.

These (GSI/Darmstadt) experiments, whose history, latest reconfirmations, and most recent data were presented in three opening sessions (Bokemeyer, Koenig), show a smooth background of positron (e^+) production, as a function of e^+ kinetic energy. Superimposed upon this background are four very sharp peaks, of narrow widths (≤ 30 KeV) and of clear experimental significance (~ 5 standard deviations). Most of these peaks correspond to sharp, essentially back-to-back electron-positron emission in the ions' center of mass.

Following the approach of "supercritical" potential theory (SPT), where the total ionic charge unit Z satisfies $Z > 137$, it has been possible to provide a detailed and apparently correct understanding of the smooth e^+e^- background; a coherent description of different facets of this approach, emphasizing the nature of the charged, supercritical vacuum, was described by the authors responsible for the invention of SPT (Greiner, Müller, Rafelski). In addition, predictions for related phenomena were outlined by other lecturers using the SPT approach (Bawin, Soff, Sørensen).

SPT has had little success, however, in understanding the sharp peak structure, and this remains the goal of those following the approach of quantum field theory (QFT) in the nonperturbative domain of strongly-coupled fields. Lectures were given on possible directions within QFT, by considering the possibility of a new phase of QED in intense fields (Chodos by analytic methods, Dagotto by numerical computation); and by considering nonperturbative, "loop bremsstrahlung" approximations to an ionic scattering amplitude (Fried). Other lecturers presented more general QFT discussions (Consoli, Cottingham, Minakata, Savvidy). In addition, brief contributions were given by a handful of students and visitors whose current research work is relevant to the topic and themes of this ASI; these contributions have been intermixed in this Volume, as appropriate, with those of the main lecturers, as they were in the sessions of the ASI.

During the two-week period of this institute, lecturers of both the SPT and QFT persuasions were deliberately intermixed, so that all participants were exposed to and informed by all presentations. For the readers of this Volume, however, who may have neither the patience nor the opportunity to sample all of this material at one sitting, it was felt that an arrangement of these various topics by general subject would be the most efficient.

Finally, it may be emphasized that, in this book, one touches upon but one aspect of the many phenomena expected to be of high physical interest when intense external fields are present. It is hoped that the ideas and techniques discussed here will find application to a wide variety of problems whose common theme is Physics in the presence of intense external fields.

H.M. Fried
B. Müller

CONTENTS

QUANTUM ELECTRODYNAMICS IN INTENSE FIELDS

RELATED QUANTUM FIELD THEORY/PARTICLE THEORY CONSIDERATIONS

SUMMARY LECTURE

VACUUM STRUCTURE - An Essay

Johann Rafelski

University of Arizona
Department of Physics
Tucson, AZ 85721

Introduction: In this essay I review in qualitative terms the present understanding of the vacuum structure and how it interplays with the general understanding of the physical laws. I discuss in turn: (The) Quantum Vacuum, The Electric Vacuum, The Confining Strong Interaction Vacuum, The Known and Unknown Higgs Vacuum, The Gravitating Vacuum and present at the end Historical Highlights

(THE) QUANTUM VACUUM

In an encyclopedia you will probably read that the vacuum is a space devoid of matter. And it has no structure. When you finish reading this essay, a totally opposite picture should emerge in your mind: the surprising development of physics research of the last 25 years is the recognition that the vacuum has structure and that it plays a key role in understanding the laws of physics. The same set of laws of physics operating in a different vacuum describes very different phenomena. The vacuum actually provides us with a background in which the world is embedded, in which the world of physical phenomena occurs. We now consider the vacuum much like a medium in classical physics. Vacuum, the "nothing", actually is "something".

We should remind ourselves that the concept of vacuum was actually born long before the birth of modern physics. The vacuum was conceived in order to be able to speak of the absence of matter in the old Greek understanding of nature. Greek philosophers needed the vacuum for the atoms, which they thought were the substance that makes up matter, to move about in and to rearrange into different forms of matter. And when the vacuum was resurrected in post-Newtonian era, it had to be the space devoid of matter in which material bodies or planets or stars could move; the greek concept of atoms had been lost temporarily. Today in almost everyone's eyes, space devoid of matter is the space from which all visible and real matter has been removed. Normally the vacuum is taken to be a region of space in which nothing is to be found, nothing of any material. This archaic concept derives, of course, from the past era of classical physics, but quantum physics and also relativity are changing our understanding of the "nothing".

Maybe we should first try to 'define' i.e. narrow the meaning of the word 'vacuum'. We have to define it in a way that takes into account the concepts of modern physics, for example that matter is only a certain form of energy, and if we talk about taking matter out of space to produce a vacuum, we also have to remind ourselves that, if we leave behind any form of

Vacuum Structure in Intense Fields, Edited by
H.M. Fried and B. Muller, Plenum Press, New York, 1991

energy, that energy appears to us as matter. An electromagnetic field filling an empty space devoid of matter, could, in certain cases, convert to matter, and so we must immediately extend the definition **Vacuum** = space devoid of matter to "**space devoid of matter and fields**". But there is a second aspect that is very important to the modern understanding of the vacuum that derives from need for quantum physics: the processes in which energy materializes to matter do not happen on a macroscopic scale. They happen on a microscopic, sub-atomic scale and those processes are determined by **quantum physics**.

The key point has to do with the **uncertainty relation** of quantum mechanics. The quantum mechanical motion may be seen as certain average over all accessible paths of action. Therefore we cannot make a firm prediction which of the paths will be taken, except in the classical limit in which path of least action dominates the physical motion. It is crucial to appreciate that on the atomic scale we lose a certainty about knowing things. We can only express a likelihood of what will happen, and only in the classical limit does this likelihood assume a dimension of certainty. There is thus an inherent uncertainty in the statement that a region of space is devoid of fields and matter. Naively, and somewhat incorrectly this is referred to as the uncertainty relation between energy and time. This uncertainty relation tells us that it is not possible to measure energy absolutely accurately in a finite time. It would take an infinite time to measure energy precisely.

So the influence of quantum mechanics is particularly strongly felt if one considers a region devoid of matter for a very short period of time. Even if we were to make every effort to take all forms of energy from this region, we still could find a fluctuation in energy. We recognize that the vacuum now is very much more difficult to define, and one must leave out all concepts which are classical in nature, classical in the sense that they refer to certainty, and not to probability. Comparing two quantum states, you would say that the one which is lower in energy is the "true" vacuum state and it can be observed for arbitrarily long time. We thus see that combining these two very important points - the fact that energy and matter are one and the same and that the absence of energy cannot be assured on a short time scale - we are led to a new definition of the 'local' vacuum: **(a region) of space-time is in its vacuum state when it has the lowest possible energy**.

I have implicitly assumed in the above definition that regions of space-time may be associated with different vacua. But it is possible that even after taking out all forms of energy, we really still encounter locally a "false" vacuum, a state higher in energy than the truly lowest energy state. There must be a barrier between the two states, and even if the one state has a higher energy than the other one, the probability for it to decay into the second one in a finite time may be very, very small. Both may correspond to "local" minima of energy, the 'true' vacuum corresponding to the lowest minimum and the 'false' vacuum corresponding to one of the other minima. This is reminiscent of different aggregate states of matter. So we actually arrive at the remarkable observation that the **vacuum is actually something complex and structured** and possibly differs from one region of space-time to another.

One should realize here that if we were to induce a change of the vacuum from one state to another, it would be a catastrophic change in all likelihood, because many properties of physical systems would change. This is again analogous to situation with ordinary matter. What can happen is that the vacuum state which has higher energy content, even though it may appear to be more ordered and simple, it may be unstable, and this false vacuum may disintegrate with a big bang.

Some current research literature addresses this nightmare of theoretical physicists! Fortunately, considering the various circumstances accurately, the conclusion is that there is not yet an immediate danger. We do not have to worry very much about destroying the vacuum because the experimental means we have to disturb the vacuum are all inferior to what Nature already often tried, and nevertheless we are here. But soon the conditions that we make in our big particle accelerators are going to exceed what the cosmic radiation has often tried out, and it is conceivable that these reactions could trigger the disintegration of our vacuum should it not be the most stable state of the Universe.

Do we have any reason to believe that the vacuum could now be in a state which could undergo an explosive decay? Why should we believe that we are perhaps not in the ground state, in the basic state of the vacuum? We know little about the instant the Universe was created. But what happened even a tiny instant later, we think we know, especially the fact that some time after the moment of creation, the Universe was very hot. The hot Universe has cooled down since, as the expansion did work against the attractive gravitational force. Now, it is quite well-known that if cooling occurs rapidly, non-adiabatically, then you often do not come to form the state of lowest energy; often you form the state of higher entropy.

During a non-adiabatic expansion, sometimes you get stuck in a super-cooled phase and precisely that could have happened to the vacuum in which we live. Should several false vacuum states exist, it would actually be quite likely that ours is not the true one! From this discussion arises a further practical amendment to the definition of the 'vacuum': **Our vacuum state is the state of lowest energy which can be reached given the evolutionary boundary conditions of our physical system.**

THE ELECTRIC VACUUM

In order to begin discussing in more detail the aspects of the vacuum which we have just mentioned in the introduction, one should certainly take as an example one in which we know some of the interactions from experience. Without doubt the interaction which we know best is the electromagnetism. So I think we best begin our discussion considering the electromagnetic structure of the vacuum. When we speak of electromagnetic interactions, we must consider at least two different types of particles. One might be the atomic nuclei (which carry positive electrical charge) and the others would be the electrons (which are negatively charged). Then, of course, we must have the positrons, the antiparticles of the electrons, which are also positively charged. An atom, as we all know, is a bound state of an electron with the atomic nucleus, and all the atomic properties follow from this. All the chemical molecular properties of matter also derive from the electromagnetic forces among these particles whose effects operate in accord with the principles of quantum mechanics to produce the detailed structure of the matter of our everyday experience. One important aspect of this in our context is that, for example, we can measure and also very accurately calculate the binding energies of the electrons in atoms and molecules. The important detail is that actually this binding is a little bit more than a first impression because, as we will now describe, **the presence of the small localized nuclear charge induces a dielectric polarization of the vacuum** in such a fashion as to make the binding of the electrons slightly stronger than we expect while ignoring the role of the vacuum.

To understand this effect, I have to introduce Dirac's "sea". Soon after quantum mechanics was discovered, one of the essential developments needed was to reconcile quantum mechanics and relativity. Dirac at that time proposed an equation, which describes the quantum mechanical motion of relativistic electrons. However, the Dirac equation exhibits very special properties not found previously in the study of the motion of non-relativistic particles. There are bound states of electrons which nearly corresponded to the previously established non-relativistic states, there also occurred new states which had substantially lower, in fact, negative energy eigenvalue. Dirac interpreted these states later as belonging to antiparticles. Essentially, he postulated the existence of antimatter. It turns out that these negative energies also exist in relativistic classical mechanics, because the energy of a particle is given by the expression $E = mc^2\sqrt{1 - (v/c)^2}$, where m is the mass, v is the velocity of the particle

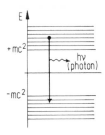

Fig. 2.1. Within the framework of quantum mechanics an electron carrying positive energy would fall down into a lower and lower state under the emission of electromagnetic radiation.

and c the speed of light. In the square root we can choose the positive or the negative sign, so we have positive and negative energy also in classical mechanics. But this is not yet important because in classical mechanics a particle can change the energy only continuously. Therefore, in a world where all the particles are in positive energy states, they will always stay in positive energy states, because they cannot gradually go to the negative energy states, as there is an energy gap between these states, $2mc^2$. So we have the freedom of talking about positive or negative energy particles in classical mechanics. There could be two 'worlds' symmetric to each other. When we write down the quantum Dirac equation, we find the relation relativistic quantum mechanics establishes between these two worlds. It is impossible to disregard the negative energies, because we must have a 'complete' set of states of the quantum mechanical Hamiltonian in order to be sure that the hermitian Hamiltonian, which gives us real eigenenergies, is also selfadjoint. This means that it describes in principle completely the time evolution of the physical system under consideration. In quantum mechanics we have, in particular, jumps between different states; transitions between states occur when they are separated by a finite energy interval and there is an interaction due to the electric charge. As a result, if we have only positive energy electrons in the beginning, an electron in one of these states could make a transition, a jump, into one of the negative energy states emitting a photon (see Figure 2.1). This electron could then fall into even lower energy states and in the end it would be giving off an infinite amount of energy. This causes a big conceptual problem.

In order to rule out the existence of such a perpetuum mobile, Dirac had to invent the antiparticle. The way he did it was to introduce the Dirac sea. He filled all the negative energy states with electrons! This resolves the perpetuum mobile problem since an electron, due to the Pauli exclusion principle, cannot coexist with another electron in the same quantum state. It is important to note that there is no classical mechanics analogue to this mechanism. So now an electron cannot make a transition to the filled Dirac sea. Should there be a hole in the sea of negative energy states, it is now interpreted as an antiparticle (see Figure 2.2): the filling of the hole generates radiation energy in excess of $2mc^2$. The antimatter is invented. But the price we now pay for this ingenious mechanism is high: the sea of negative energy electrons implies that the vacuum state has a very negative energy, indeed infinitely negative. While this is only philosophically a difficulty, since we can measure only energy differences, a more significant problem also posed is that an infinite number of electrons in the negative energy states, this Dirac sea, carry an infinite charge.

This observation makes a further refinement of the description necessary (see Figure 2.3), namely we will have to ensure that there is complete symmetry between the vacua of electrons and positrons, such that the two charges cancel, while the energies, as the formalism easily shows, add up! The particle - antiparticle symmetry is introduced in response to the need to have a charge neutral vacuum - it is referred commonly as 'charge conjugation

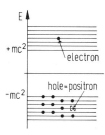

Fig. 2.2. Dirac's picture of the vacuum: The Dirac Sea.

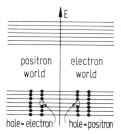

Fig. 2.3. Symmetric 'Dirac Sea' of electrons and positrons.

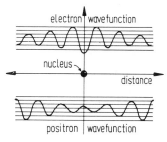

Fig. 2.4. An atomic nucleus attracts electrons with positive energy (positrons with negative energy) and repels electrons with negative energy (positrons with positive energy).

symmetry'. The small print here is that the state in which we live surely is not symmetric as we do not find antimatter in the Universe! This may require more thinking about the initial conditions we had in the Universe.

It would seem that this beautiful academic gedanken-experiment construct is condemned from the outset. However, there is an amazingly simple and convincing experimental confirmation of the picture. The presence of the sea of electrons and positrons lends itself to polarization by any applied field, in particular the one generated by the atomic nucleus. The positive nuclear charge will attract the sea electrons ever so slightly, while repelling the sea positrons. Indeed, there is nothing particular about a nucleus, any electric field will do just that: it will polarize the vacuum. We have exactly the same situation as we have in a macroscopic medium. But the vacuum polarization is somewhat different from polarization of matter, because the electrons in the Dirac sea have negative energy. And that implies that if a force attracts the sea electron, they like to move away since these states are indeed positron states (see Figure 2.4). If one works it out, one arrives at the result that the observable polarization charge density in the vacuum due to an applied electrical field will actually strengthen the measured electrical field rather than weaken it. This is analogous to what one normally would expect, especially if one thinks of a behavior of a polarizable medium, any electrical field will be asymptotically weakened by the polarization charge generated, but as we approach the source, the electrical field increases .

It turns out that this is one of the very important aspects of the understanding of the vacuum and consequently each interaction must be carefully studied to identify the sign of the polarization: in quantum electrodynamics the attraction is enhanced, in some other theories, in particular in 'non-abelian' gauge theories we cannot distinguish between polarization and self-interaction phenomena, and the sign changes. When the required calculations are done, the first result one finds out [aside of some complications (symmetry breaking divergences) indicating that our theories are yet not fully consistent] is that there is indeed still an **infinite contribution to the displacement charge**. This, however, does not disturb the theorist, because he realizes that the definition of the unit of charge is subject to an arbitrary re-normalization. All charges we observe are charges which we measure. Electromagnetic charges are measured by finding the coefficient of the $1/r^2$ Coulomb force at large distances. And consequently if a vacuum acts back on the charges at short distances in such a way as to change the value of a charge which we observe asymptotically, that can be absorbed into the definition of charge in the process of re-normalization, it is the re−normalized charge we observe.

This can be implemented consistently only if the effect is the same for all particles, irrespective of their nature. That indeed will be true **if the re-normalization constant is not the property of the particle, but rather a general property of the vacuum**: in the same vacuum the response to any particle of unit charge must be the same. Actually one can prove a theorem based on the formalism developed which shows that this is true, relating this phenomenon to the fact that the electromagnetic fields are gauge invariant. Thus we learn that the observed electrical charge is derived from a bare electrical charge which is renormalized by the properties of the vacuum, characterized by gauge invariance, to the actual observed physical value.

There is another short range effect of the polarization phenomenon which implies a modification of the inverse square law of the force between two charges. The change in the space distribution of charge, the polarization cloud, has thus an observable effect. It is very useful first to know the range of this displacement charge. Then we can identify physical systems that have the size of this displacement charge as their typical dimension. Clearly, it cannot be true that the vacuum polarization effect is visible for macroscopic distances, because then we couldn't have experimentally verified the Coulomb law of electric forces in the first place. So it must be something which has a short range, a microscopic range. Consider the origin of the displacement charge: the electrical field of the nucleus is pushing away the negative energy electrons and attracting the negative energy positrons. In different terms this corresponds to the creation of one or more virtual electron−positron pairs for a very short time or over a very short distance. And the energy it takes to do that is of the order of twice the rest energy of the electron. And the momentum that such a pair would have to carry would be of magnitude $m \cdot c$.

The uncertainty relation now tells us that the product of the uncertainties in momentum and position is of the order of Planck's constant $\hbar$ and we find that the typical length over which this will happen is $\lambda_c/2\pi = \hbar/mc$. This distance, the so-called electron Compton wave length, is about a hundred times larger than the size of the nucleus and it is about 200 times smaller than the size of an atom. So we need a physical system in which we can do an experiment and which is 200 times smaller than an atom. There are two quantities we can consider which determine the size of an atom: the strength of the electrical charge and the mass of the electron. The first alternative leads one to consider the innermost bound electron in a heavy atom, say lead: the disadvantage is that it comes with 82 electrons. In principle this would be a very nice case, but it is a bit obscured by the other electromagnetic phenomena stemming from the numerous charged bodies nearby. Another alternative is offered by the surprising and totally not understood fact that there is a heavy electron which is called the muon. Our ignorance of the situation is summarized by giving the fact a magic name: MUON (from the greek letter μ). But for the purpose of the present discussion we simply accept that an electron has a heavy companion, with a (mean) lifetime of $2.2\mu\text{sec}$, called the muon, which has a 208 times greater mass than the electron. The size of a muonic atom is about two hundred times smaller than the usual one, just as required to sample the charge distribution of the vacuum polarization.

The next question is: precisely how large polarization effect do we expect? The displacement charge contained in the vacuum polarization can be calculated to be about one thousandth of the source charge, and the effect on the energy of a given state in high Z or muonic atoms will be of the same magnitude. In order to measure it precisely, one usually would like to exploit a difference. The useful and distinctive feature is that the deviation from the Coulomb law of interaction with the nucleus will be seen only at short distances. Therefore, only those states which have a capacity to probe the short distances will be measurably affected; one thus exploits the splitting in the energy of the different quantum states, of which one is well-overlapping the nucleus and the other not. The observed splitting turns out to be just what theory predicts, of the order of one thousandth of the binding energy, and in muonic atoms in particular, the vacuum polarization is the dominant contribution to the additional energy displacements. One can hence say that the vacuum polarization phenomenon has been experimentally established with precision of about 0.1%. It has been confirmed by experimental tests to such a degree of precision that we can convincingly argue that our understanding of the vacuum role in the interaction between electrically charged particles is very well-established indeed.

Another aspect of electromagnetic interactions is light. Let us now take a good look here at a light wave travelling through the vacuum. In the absence of other electromagnetic fields, a photon will travel freely through the vacuum. That is just how we define the velocity of light. We measure it as the speed by which light propagates in the vacuum surrounding us. Experiments show that this velocity is independent of the frequency and wave length of light. This also means that photons which make up the lightwaves are massless. Therefore, we can observe photons which come from distant stars many light years away. But the fact that we have measured the velocity of light to have the value of about 300 000 kilometers per second does not account for all possible vacuum effects on light propagation. The uncertainty relation permits any photon to become an electron-positron pair for some time, and then to recombine and become a photon again, all this is already incorporated into the physical properties of the photon as we know it. But due to this vacuum polarization effects, two beams of light will also interact with each other in an observable way! The virtual scattering of a light quantum from a vacuum electron-positron pair makes possible for a second photon to scatter from the first photon: the second photon must arrive just when the first photon is an electron-positron pair. But an experiment to measure light-light scattering in empty space is not simple, and has not been successfully carried through. The difficulty is the smallness of the charge distribution that the photon induces in the vacuum. Despite the smallness of this effect nobody doubts in the reality of this phenomenon. **A philosophical aspect to remember is that a new interaction (photon – photon scattering, photon – field scattering, photon slitting etc.) arises due to the properties of the vacuum.**

The next question that arises in your mind is now probably: can light velocity be exceeded? This most fundamental question is a difficult one. There is a maximum velocity, we think to be the light velocity. This is a consequence of space symmetry and **the equivalence of all inertial observers.** Now we have to consider the role of the vacuum: what we know today is in particular that if one has a vacuum in which there is a strong electric field or a strong magnetic field, the velocity of light can change: the velocity of light may be less than the one we normally measure in 'empty space'. So within the 'class' of equivalent inertial observers the answer is 'no'. But to 'find' the 'yes' answer, all we need to do is to make a small 'mistake' by sneaking in through some back door an effectively 'non-equivalent' observer and occasionally we hear about such 'exciting' new theory. But is the light velocity now the "vacuum" velocity? And, of course, we do not know if there are new worlds out there, with their own class of equivalent observers, ...Clearly, this transcends the contents of our present discussion, and in particular of any accessible experimental information.

We have learned that the electric vacuum can have a structure, and we have seen that some of the effects are indeed such that experiments prove that this is indeed so. Let us now consider if aside of its energy content the vacuum can acquire any other properties. Consider as an example, two conducting plates. The quantum vacuum between them is never empty. There is always a certain probability that it contains particles or energy quanta. Now we know that in a vacuum between conducting plates, the energies of the states of these quanta will be different from those in a free, large volume. The quanta which can virtually appear between the plates must satisfy the constraint that all electric fields on the surface of the conductor must vanish, because of the currents which are induced in the conductor (see Figure 2.5). Two conducting plates thus define the specific frequencies of electromagnetic waves which can be virtually excited. Even though we cannot radiate this virtual radiation, its energy depends on the distance between the plates, and if we sum over all these possible excitations, then we find that the energy of the vacuum must depend on the separation between the two plates. Since the energy changes as we move the plates closer, there is a vacuum force, a vacuum pressure if one wishes to see it that way, which depends on the distance between the plates. This force called the Casimir effect, has actually been measured in experiments and we have seen that a vacuum can act on a macroscopic object which modifies its properties. Even if it is only the effect of the boundary on the vacuum.

A much more profound next issue is the question if and how the vacuum can change its structure under appropriate conditions. The vacuum may have different types of structure, and it may be caught up in one structure, and possibly if conditions changed, it could change into another structure characterized by well distinguishable macroscopic properties. Observation of such phenomena would constitute the ultimate proof that our understanding of the vacuum, at least in principle is not Ptolemean (which means that we are making a fundamental mistake in a basic notion, leading on to more and more complex descriptions that, in principle, are accurate but in practice prevent any further significant progress in un-

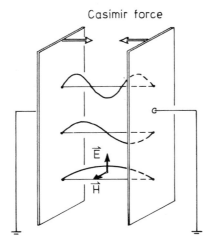

Casimir force

Fig. 2.5. Casimir Effect: Virtual states between two conducting plates

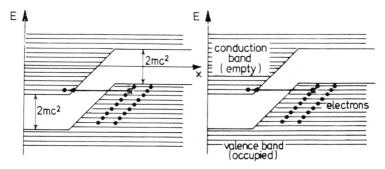

Fig. 2.6. A very strong electrical field is capable of producing electron-positron pairs from the vacuum. States of positive and negative energy play the role of valence and conduction band.

derstanding). The two structures must be distinguished by some property, and if we discuss electric, or electromagnetic, vacuum states, one of the most important properties certainly is charge.

We have seen that in the Dirac sea picture the charge of the vacuum is zero. But this had to do with the fact that the vacuum was only polarized by a relatively small charge and was symmetric initially between electrons and positrons. In order to generate a real charge in the vacuum, real vacuum polarization, rather than just a displacement current (charge) we must exploit the fact that we have already broken the particle-antiparticle symmetry between electrons and positrons by taking an external source, the nuclear charge. Therefore in principle there is no reason to believe that the charge of the vacuum is zero. Indeed, we have to carefully verify that our tacit assumption is satisfied for even 'weak' nuclear charges.

We can view the states of electrons and positrons as being analogous to the conduction and valence bands of insulators (see Figure 2.6). As soon as the potential of the electrical field can bridge the gap between the electron and positron states, this will create an electron-positron pair. When we apply a very strong electric field to insulators, these insulators can spark occasionally. This is due to some remaining residual electrical conductivity in the insulating material, which can lead to an avalanche of ionization. The threshold that is necessary to create an electron-positron pair in the vacuum plays a similar role to the threshold that must be overcome in an insulator to make a spark. We would expect the vacuum to spark not only in the vicinity of atomic nuclei, but also if we were to apply a sufficiently strong (constant homogeneous) electrical field to the vacuum. But it turns out that, if one calculates the probability for such a pair production to occur, one finds that in normal macroscopic electrical fields, even the strongest ones that can be produced, this probability is so small that we would not be able to observe it in a lifetime. The essential point, as we have already seen, is the strength of the electrical field at subatomic distances. We have identified during the discussion of the vacuum polarization the magnitude of the characteristic length. So now we have to manage to change the electrical field from basically zero to the required value over a similar distance. This is indeed impossible with macroscopic fields. But it is possible with the electrical field of a large nucleus with $Z > 137$ (see Figure 2.7).

When the electrical field is made sufficiently strong so that the **vacuum polarization becomes real**, then, due to charge conservation, during the process of change of the vacuum from a neutral to a charged state, some charge must be emitted. And if in the vicinity of the electrical field of the nucleus, the vacuum carries the charge of an electron, the emitted particle must surely be a positron. This is, of course, in accord with the principle of charge conservation: the vacuum charge will always be balanced by the process of emission of the oppositely charged particle. In essence, the vacuum region surrounding the nucleus is actually negatively charged after the positron has been emitted.

Greiner and his school of theoretical physics in Frankfurt have devoted considerable attention to these phenomena over the past 20 years. They have in particular shown that these positrons have a well defined energy. The nuclear electrical field applied determines how large this energy will be. This prediction can be tested in experiments which employ different electrical field strengths. But nuclei with a sufficient number of protons to make a field which is strong enough to produce a real vacuum polarization do not exist: it is necessary to collide two very heavy nuclei, like uranium nuclei, at one another so that they come very close together. A small area of space then contains a total of 184 protons, or any other similar number selected by the experimentalists. During the time that the nuclei spend together, one may hope to see the transition from the neutral vacuum to a charged vacuum. For the past 15 years such experiments were only possible at the GSI laboratory in Darmstadt, W. Germany, where the high Z ions could be accelerated to above their Coulomb barrier (see Figure 2.8).

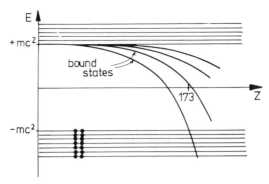

Fig. 2.7. The vacuum polarization becomes real when a bound electron state dives into the sea of negative energy states.

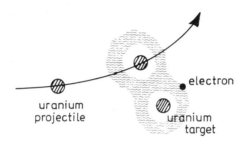

Fig. 2.8. During the collision of two very heavy atomic nuclei (e.g. uranium with uranium) a bound state of electrons is bound simultaniously to both nuclei in a 'quasi-molecular' state.

It can be argued that there is not enough time to observe the vacuum decay if the two nuclei simply fly past each other. If the time during which the vacuum becomes overcritical is short, this is the time available for the vacuum decay and hence the energy of the emitted positron is broadened by the uncertainty relationship and one cannot distinguish positrons emitted by the decaying vacuum from the many positrons produced by other more mundane processes. This phenomenon can be counterbalanced to some extent by judicious choice of scattering parameters permitting the nuclei to glue together in a grazing collision. Such experiments have been performed for quite some time now and the sharp positron lines have indeed been found in the spectra, the energy of which coincides more or less with the energy calculated for the positrons in the vacuum. However, further experiments are needed to ascertain the delicate nature of the observed positron lines. In particular, the lines observed are not showing the systematic characteristic which is expected. Thus either they are due to an experimental error, or a new, previously unseen complex particle, or possibly more involved structure of the vacuum than initially proposed: it seems that we can safely exclude them as possible positron lines accompanying the formation of the naive charged vacuum introduced above. This subject matter is currently under much scrutiny, and we will have to await further reports before the last word can be said about this matter.

An interesting point to keep in mind is that within the presently known zoo of elementary particles the electron is the best candidate for study of the vacuum structure and sparking of the vacuum. In principle we need particles to fit the Compton wave length of the electrical field generated by the colliding nuclei. Probably the best case would be a heavy electron about five to ten times lighter than a muon. The muon is too heavy – its mass is larger than the depth of the potential of heavy nuclei. A particle with greater than an electron's mass up to about 30 times an electron's mass would fit better into the nuclear potential well, and probe the vacuum in a much more profound way.

Of course, we so far only considered the structure of the vacuum as far as the observed part of the electromagnetic interaction is concerned. As everybody knows, in nature there exist other interactions. For example, gravitational binding is essential for the creation of planetary systems and so-called weak interactions are essential for the natural radioactivity of atoms...and strong interactions bind atomic nuclei together. We must now explore, how the understanding of the vacuum we have just begun fits into the greater picture of the diverse interactions governing our world.

THE CONFINING STRONG INTERACTION VACUUM

Let us now consider how the world would look like if light-light scattering would be not an induced phenomenon, but a basic property to start with. Note that there should be more than one kind of photon: we would need photon-like particles that come in several different types (charges), not necessarily particles and antiparticles, but particles with a new type of internal property other than just the polarization. Effectively we need a new kind of charge commonly referred to as 'color' to give to these new photons called gluons. The regular photon does not have an electrical charge otherwise our vacuum would be very opaque. Even the extremely weak effect of photon-photon scattering which we have just discussed implies that there is some kind of a photon-photon effective interaction. What would happen if this interaction were attractive? Photons could cluster. Now, photons are Bosons and like to be in the same quantum state, a fact which makes lasers possible. If there were to be some kind of significant attractive interaction between them, they would like to clump together, like matter, but unstable against collapse; and ever more photons would like to be in the clump.

So if photons have charged brothers and sisters, gluons (see Figure 3.1), the vacuum structure must be rather complex. Actually this gluon structure of the vacuum is believed to be the explanation of why the constituents of elementary particles, called **quarks**, cannot travel freely through the vacuum in which we live. Normally we do not see the gluonic structure of the vacuum as gluons interact neither with photons nor with electrons, because they are not electrically charged. They also do not interact directly with the constituents of atomic nuclei – protons and neutrons – although these are constructed from quarks which interact with gluons. But the nucleons are thought to be neutral with regard to the new charge and hence do not interact much with the clumped vacuum. Nevertheless the vacuum prevents the constituents from separating from each other as individual quarks feel the charge of the gluonic vacuum structure. A nucleon is like a bubble containing the quarks which abhor the vacuum (see Figure 3.2). What is probably happening is that quarks try to get out of their nucleon bubbles all the time, but they scatter from the clumps of gluons in the surrounding condensed gluon vacuum, and are thus prevented from becoming free and independent particles. Before we go on with this discussion, we should say that there is yet **no direct evidence** for this picture of the 'subatomic' world. But it works well as a conceptual foundation of many models of strong interactions.

There is something we should understand better. We argue that around the bubbles is gluon vacuum, and to keep the bubbles together it must effectively exert some inward directed pressure on the quark bubble (see Figure 3.3). There is actually experimental proof that the vacuum can exert a pressure. We discussed already in the last section the Casimir effect of the vacuum of quantum electrodynamics. This effect is not directly related to the vacuum of the strong interaction and really is caused by the altered space boundaries. But since the issue is a matter of principle and not one of detail, this will clearly disperse any doubts about the ability of the vacuum to press quark bubbles together.

The vacuum can exert pressure on the quarks in dependence on the size of the bubbles, now professionally called 'bags' in which the quarks are confined. We have to realize that the region in which the quarks can exist is free of the vacuum structure which, as we discussed,

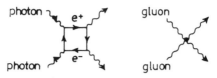

Fig. 3.1 Photon-photon vacuum interaction and gluon-gluon fundamental interaction.

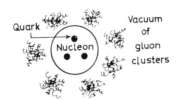

Fig. 3.2. Nucleons, i.e. protons and neutrons, are like bubbles in the vacuum made out of clumps of gluons. The building stones of the nucleons, the quarks, can only be found inside the bubble because the gluonic vacuum is opaque and impenetrable for them.

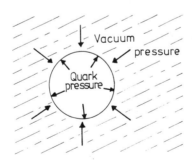

Fig. 3.3. The bubbles in which quarks are found are held together by the pressure of the gluon vacuum.

is probably some complicated gluon cluster state that repels the quarks. The confining constraints at the boundary will have the consequence that the energy states of virtual particles are changed, as in the Casimir effect, and consequently there is a pressure acting on the surface of this region. If we were to remove the quarks from such a region of space, then we would expect gluonic clusters to fall into this region and occupy it, as they have a natural tendency to be everywhere, i.e., we expect that the structured vacuum has a tendency to enter the region which is occupied by quarks. Thus quarks do prevent this from happening. No empty bags, that is holes in the vacuum, appear possible.

Numerous experiments show that individual quarks are permanently confined to the interiors of the elementary particles, and the question is asked: how can it be that we do not observe any single quark? Our present answer is: because the vacuum is structured - in order to remove a quark which carries the strong interaction charge, this structure has to be displaced, and the displacement of the vacuum structure costs a lot of energy. But let me caution that in principle free quarks are not impossible. What would be required is that the range of the glue interaction be finite, as is the case with the weak interaction that we will discuss in the next chapter. If that were the case, then even the clustered gluon vacuum would permit the existence of a very large region of space filled with only a single quark because the field around the quark extends only over a finite region and therefore can displace the structure of the vacuum only in a finite volume. So free quarks are possible only if the gluon range is finite. It is somewhat similar with the basic law of electrodynamics: If the inverse square law of a force between two particles would be attenuated, this would have grave consequences for the observed charge inside a sphere. If you make the sphere larger, the observed charge would be reduced by attenuation. The same is true here if the range of the gluon force is attenuated, then we can have single quarks, because only a finite amount of energy is necessary in order to disturb the structure of the vacuum. Said differently, **colored** (i.e. not neutral) states of quarks and gluons are a priori not impossible, but must be very heavy as the vacuum structure is altered over significant volumes. But it is more natural in the picture of the strongly interacting vacuum we have developed to expect that quarks are permanently confined.

In order to test these possibilities it is certainly important to look whether one can find single quarks in nature. People have looked for free quarks in many, many experiments, and I even remember that somebody has claimed to have seen individual quarks. But this discovery remained unconfirmed. Most of these experiments really try to find a very special property that one assigns to quarks, namely that their charge is **not** a multiple of the unit of the electron's charge. This is indeed a special property, special in the sense that it is different from what we find looking at electrons and protons. But it is not special at all in the sense that charge differences between different quarks are always integer, which is a nice and theoretically a very much intriguing and appreciated feature. If quark search experiments do not succeed, then our picture of interactions is actually much more beautiful and theoretically more internally consistent. This is so, simply because the interactions between quarks very much resemble the interaction between electrically charged matter that one almost has the impression that they are of the same kind, that one could see a certain unification of the fundamental interactions emerging even at this deeper level.

The strongly interacting, confining vacuum is a beautiful example of the principle that the laws of physics we uncover have to be supplemented by a knowledge of the vacuum state. As we have seen, the effective properties of interactions like the electro-magnetic interaction or the gluon interaction depend on the vacuum state. Because the vacuum state of gluons is quite different from that of the electromagnetic interaction, the properties of this interaction are also quite different. The theory of quarks, gluons and the structured vacuum of what is commonly

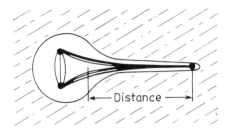

Fig. 3.4. The energy required to pull a quark out of the nucleon bubble increases in proportion with the distance. In order to completely isolate the quark an infinite amount of energy would be necessary.

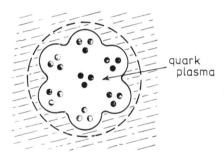

quark
plasma

Fig. 3.5. If one could compress an atomic nucleus hard enough, the individual bubbles would fuse into one single giant bubble in which the quarks can move freely.

called "quantum chromodynamics", the quantum theory of color interactions between quarks and gluons. This theory really depends in an extreme sense on the understanding of the fundamental structure of the gluon vacuum.

Regrettably, I think we know nothing really about the complex structure of the vacuum in mathematical detail. We realize that because gluons are (color) charged and because of the observed absence, or at least, the very likely absence of color charges at low mass scale the vacuum should be structured. On the other hand we have not been able to mathematically derive the properties of this structure and to understand the details of the new vacuum state starting from first principles. But 'theoretical' difficulties are not critical for the validity of a qualitative idea. The history of science shows many instances where this situation has occurred. I am thinking for example about the theory of superconductivity. Superconductivity was actually discovered accidentally, as are many important phenomena, and it took some forty years to unravel the structure of the superconducting state, and the knowledge just got a big jolt with the recent experimental advent of high temperature superconductors. And the physics that went into understanding it is in many respects similar to what we are discussing in the context of the strongly interacting vacuum. There are many different ways and methods to address this problem: we will have to find a practical approach to understand the confining vacuum and with it the theory of strong interactions which not only is responsible for the structure of nucleons, but also for the nuclear forces which bind them together.

The clustered gluon vacuum of quantum chromodynamics is a theoretical challenge – and clearly it would be quite interesting and helpful if we were able to establish a path to experimental investigation of its properties. We would like to recreate in a small region of space-time conditions akin to the early Universe, in which the glue was evaporated, perhaps better said, melted globally. We would like to have such a system with "global" (though still microscopic) melted vacuum. Such a global state must be large as compared to the typical structural scale of the perturbative vacuum inside hadrons. If one looks at an atomic nucleus with a magnifying glass, it would probably look something like an assembly of bubbles in the glue liquid, each bubble being like...hole in Swiss cheese. Therefore what one can think about would be to simply compress a nucleus, compress it so much that the bubbles, the holes in Swiss cheese, start to overlap, and finally form one huge bubble. Such a classical picture is of course not correct when one tries to understand properties of compressed nuclei, this picture is overly simplified, but we do not understand too well anyway how the nucleus is actually built up of quarks and gluons, so maybe such a classical picture is not so bad after all as a starting point. Some ten years ago we realized that the path to an extended local change of the vacuum is in compressing the nuclei. The perturbative vacuum in which inside nucleons quarks can exist, will probably allow quarks and gluons, also those which are evaporated from the vacuum structure, to move almost freely in a relatively speaking 'large' volume, but they will remain confined by the surrounding normal vacuum.

What one might do to press the nucleons in nuclei together is to bang two nuclei into each other. Such an experiment may not be easy as the nucleons may not want to do just what we want them to do. They may disperse without much compression, or actually compress each other, but for a much too short time for the melting to occur. Certainly, we will try it out. We will have to use very powerful accelerators and sophisticated experimental techniques. And it is not to be expected that all nuclear collisions will lead to the new form of vacuum. Indeed it is likely that many other things will happen as well. But sometimes the giant bubble of empty "perturbative" vacuum we want to look for will be formed, hopefully. Because of the inherent uncertainties in this approach, it would be difficult to obtain the required enormous sums of money for such an experiment. But fortunately these very powerful accelerators already exist, and they are already being used for such experiments.

We now consider briefly what happens when we collide large atomic nuclei at a sufficiently high speed so that the individual nucleons cannot move away from the collision zone when these nuclei collide. A significant compression of the nuclei must follow, and this compression will bring the individual nucleons much closer together than they otherwise would be in an individual nucleus. When they are very close together and even overlap, we can view their interior as being one large bubble. This bubble is filled with the different quarks which, of course, are brought in by the nucleons of the incoming nucleus. However, these quarks will no longer be in the lowest state which is allowed physical laws. Because if you compress something suddenly, you cannot avoid heating it up.

A "softer" approach to melt the nuclear structure involves the annihilation of anti-nuclear matter on nuclear matter. Currently we have at our disposal only antiproton beams. In such an annihilation process the quarks from the projectile antiproton and from some deep nucleon in the nucleus annihilate into individual gluons. These may, in turn, short linteract with the surrounding nuclear matter and the annihilation energy is shared with the surrounding nuclear matter, melting the vacuum locally.

The whole scenario that we are developing here, namely that we have a region of space which is large on a microscopic scale, which is filled with particles that have a high temperature and which have a charge, is very similar to the electron-ion plasma formed when atoms dissociate into electrons and ions at high temperature or high density; the light electrons carry a negative charge and the heavy ions carry a positive charge. The amounts of both must be the same in order for the plasma to be neutral. Quark-gluon plasma, which we want to create in nuclear collisions or antimatter annihilation, all the carriers of strong 'color' charge have a similar mass, perhaps even a zero mass. As consequence this plasma is more homogeneous and the components will be more closely coupled and will come to equilibrium much more easily.

One of the first physical questions which must be addressed, is about the heat needed to melt the gluon cluster vacuum: the so called "latent" heat. We easily can make an order of magnitude estimate of this quantity. I think that today we have only very vague ideas about the precise number, and probably any figure that is between five times smaller than one GeV per unit hadronic volume and five times larger has been given in the literature. We have here a situation which is very different from the one, for example, in the experiment on vacuum polarization which we discussed in the previous section. There we had a precise idea and prediction of an effect. Here, we do not know if we have not goofed on our basic theoretical, conceptual understanding, quite aside of the rather loose prediction about even the most essential of the quantities in question.

However, the collisions between nuclei may proceed so fast that the region of space which has been converted into what we call quark-gluon plasma will not only be small but also short-lived. We have to think hard about how one can observe the properties of the melted vacuum. We are looking for messengers from the new phase of matter which will uniquely and unequivocally tell us that we have created the required conditions and that our ideas were right in the first place. For this purpose we need particles which interact fairly strongly to be copiously produced, but not too strongly, or else they will not get out. Also, these particles ought to be actually made inside the new phase, not to be brought in with the colliding matter, which could falsify the conclusions. Furthermore, our messengers must carry to us information how this phase looks and disclose the secrets of its structure.

Among the species which are not present in normal nuclear matter there is in particular a third, fairly common kind of quark, the "strange" quark, which can be easily created in collisions of elementary particles. The mass of these strange quarks is quite comparable to the typical energies of quarks in what we have considered to be the quark-gluon plasma. So when quarks or gluons in this plasma collide, they will certainly produce strange quarks. The speed at which these quarks will be produced is well-determined by the known strength of interaction between gluons and these strange quarks. So we can compute the rate of production of such strange quarks in the quark-gluon plasma. They will not be easily absorbed again, as this would require that two: strange particle and antiparticle meet again. Hence, they will not disappear once they have been created. Thus the essential point is that they can reach the observer after the plasma cools down or dissociates, although not individually, but built into other particles, such as for example, kaons (K), lambdas (Λ) or omegas (Ω). It turns out in a detailed study that strange quarks and strange anti-quarks in plasma are mainly produced by gluons, which are not so abundant in normal nuclei. Thus overabundant strangeness provides also an indirect evidence for the presence of gluons in the plasma.

Once these experiments have been carried out and we have arrived at a better understanding of the transition between normal nuclear matter and quark matter, which is accompanied by a change in the strongly interacting vacuum, then we can probably better understand the structure of the very dense, compact stars which are called neutron stars. They are really nothing but giant nuclei having a radius of several kilometers. As soon as quarks were invented, or very soon thereafter, it was realized that the interiors of these neutron stars could consist of deconfined quark matter. The structure of the interior of these stars (nuclei vs quark) influence the understanding of the basic properties, such as mass, radius, and brightness of such objects.

THE KNOWN AND UNKNOWN HIGGS VACUUM

I believe that a third stage of the discovery of the vacuum will come when we collide quarks and leptons or quarks and quarks at ultrahigh energies. If we do not aim high as to have a unified theory of quarks and leptons, but first consider a theory which describes the unification of the electromagnetic and weak interactions, which in turn provides for a slight generalization of the electromagnetic vacuum, then there is already such a model. Before entering totally into the unknown, let us first look at the key features of this rather successful, but clearly yet incomplete understanding aimed at the electro-weak phenomena.

Interestingly, this is presently the one case of a so-called fundamental theory, in which the key properties derive from properties of the vacuum, in what I would like to call a 'toy model' of vacuum structure. This model is amazingly successful. It permits us on one hand to describe the wealth of electro-weak phenomena, often called 'weak interactions' and on the other hand it gives us clues how the more involved problem, the treatment of the entire quark and lepton world may be attempted. The weak and the electromagnetic interactions form an ideal playground to build practical experience with models of the vacuum structure, and to learn in a simple way how to express the possibility that different properties of physical laws can arise in different physical vacua.

Let me recall a few facts about the weak interaction, which is mediated by particles similar to photons or gluons, which we call the intermediate bosons W and Z. These particles have been recently discovered at the antiproton $Sp\bar{p}S$ collider at CERN, and are routinely produced in the LEP (CERN), SLC (Stanford) and Tevatron (Fermilab) facilities. But they differ from both photons and gluons in that they are relatively heavy, nearly 100 times as massive as a proton, and still viewed to be elementary. Now, it is not so easy to understand why they are heavy, because the theory which describes these weak interactions is almost identical in structure to the theory that describes the interactions transmitted by photons and gluons and hence these bosons should be perhaps also massless. So when developing a theoretical framework, theoreticians came to think that maybe the **mass of the intermediate bosons is a property of the vacuum!**

We have seen that we could have quarks in the 'perturbative' QCD vacuum and no quarks in the other 'true' vacuum. Now we consider a still more complex structure. We want a particle to have very large mass in 'true' vacuum and to be perhaps even massless in the

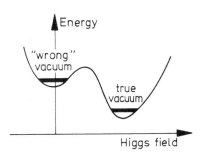

Fig. 4.1. The Higgs field gives the intermediate boson a mass.

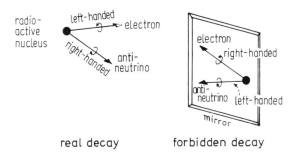

real decay   forbidden decay

Fig. 4.2. The weak interaction violates the symmetry of reflexion: the reflected image turns the relationship between the direction of motion (arrow) and spin (ring) of the particles around.

'false' vacuum. Let me recall the well known fact that mass of the particle that mediates an interaction determines the interaction range. This fact derives basically from the same principles which we invoked to argue that the vacuum is structured within the realm of quantum mechanics. When a charged particle emits a photon, the emission process takes some energy, and therefore the photon can only exist for a short time, since the uncertainty relation allows the energy conservation to be violated only for a short time. Now, if the particle that is emitted is very massive, this means that it can only exist for a much shorter time than a photon, and therefore it can travel only a very small distance, and this limits the range of the interaction. What we really mean to say is that the vacuum responds in this way to the presence of several particles, limiting their mutual range of interaction.

Returning now to the concrete question about the large mass of W and Z, the mediators of weak interactions: we wish the vacuum to generate the mass, which requires a macroscopic presence of some mass field commonly called 'Higgs'. The vacuum can be characterized by which value the 'Higgs' field assumed. In a specific toy model, what we do is to actually express the mass as being proportional to the value of the Higgs field which in the true vacuum is large (see Figure 4.1). It is most convenient to think, that there is only one fundamental mass in a theory, and that other masses derive from it in some way via constants which are dimensionless. In particular, if this original, fundamental mass is provided by the true vacuum value of the Higgs field, then the masses of the particles in the theory are generated by a very ingenious series of couplings between the Higgs field and these particles. And I should say that, of course, this sounds very ad hoc at first sight. But there are experimental predictions of such a scheme which can be very specific and some predictions have been extremely successful. However, these successes really do not prove the last point, all we know is that it is correct to imagine that there is a vacuum mass field.

Next, in order to be able to address more details of this situation we must first look at certain symmetries of space and how the vacuum structure relates to them. Naively, it shouldn't matter if we describe the laws of physics in a coordinate system which has, e.g. right-handed coordinates. We must realize here that left and right-handed are really just names of things, and an arbitrary choice to some extent, because who knows what is the actual left and right hand. Indeed the weak interactions are not symmetric against the exchange of left- and right-handed coordinate systems, as was surprisingly discovered some 35 years ago (see Figure 4.2) - the radioactive decay of nuclei does not look the same if we look at it in a mirror. What I am tacitly introducing here is the fact that each massless particle which has an intrinsic spin, has also another reference frame invariant property, the helicity. A particle is moving in the direction of its spin or against it. We have found that all neutrinos emerging from a weak interaction process are left-handed and all antineutrinos are right-handed.

This of course does not preclude the existence of the other neutrino handedness, but it requires an asymmetry in the sense that in the 'true' vacuum the other handedness of neutrinos are very heavy. But the issue before us is that the observed interactions break the symmetry, and that the easy way to interpret this would be to say that the vacuum- state properties are the source of this symmetry breaking. Perhaps in the beginning there were two degenerate vacua, the one which was left-handed, and the other which was right-handed. Nature chooses between the two possibilities in a process we call spontaneous symmetry breaking. A nice and simple analogy to recall here is the ferromagnetism. In disordered phase the elementary magnets, the magnetic moments of the lattice sites do not have a preferred orientation. But as temperature drops below the Curie point, a macroscopic structure evolves with the orientation of the magnet being depended in some accidental way on the history of the ferromagnet. When we look at it microscopically, in every ferromagnet there are various

$$\begin{pmatrix} \nu_e \\ e^- \end{pmatrix} \begin{pmatrix} \nu_\mu \\ \mu^- \end{pmatrix} \begin{pmatrix} \nu_\tau \\ \tau^- \end{pmatrix} \cdots$$

$$\begin{pmatrix} u \\ d \end{pmatrix} \begin{pmatrix} c \\ s \end{pmatrix} \begin{pmatrix} t \\ b \end{pmatrix} \cdots$$

Fig. 4.3. The known species of quarks known to us can be divided in pairs into groups ("families"). Each family of quarks corresponds to a family of leptons, i.e. particles related to the electron.

regions with different orientations of the magnetic field. Pushing this analogy, it could be that the vacuum prefers left and right-handed interactions in different parts of the Universe. While such a possibility cannot a priori be excluded and would indeed be a nice path to effectively restore the right-left handed symmetry in the Universe, I would caution against too much enthusiasm here in view of lack of any experimental evidence. Indeed the uniformity for of the black body background radiation suggests that the Universe was highly uniform until very late in its evolution, which in turn suggests that all interactions between particles were uniformly and universally maintained in the Universe.

Another path to resolve the right–left asymmetry in the vacuum is to allow the mass Higgs field which gives different masses to the right and left-handed particles, for some particles at least. Naturally, this amounts to a postponement of the resolution of this riddle to the understanding of the Higgs, but is it not what we always do, giving names to all those phenomena we can not yet comprehend? But now we have several interesting alternatives, falsifiable by experimental approaches. What we would like to do experimentally is to melt the weak vacuum in order to restore the symmetry. However, the scale for melting the vacuum may be forbidding: in strong interactions it is one GeV, one proton mass as we have discussed. In weak interactions our scale will be generated by the characteristic mass associated with the weak interaction, which is about one hundred times the proton mass if we assume that the Higgs is responsible for its own scale. We should anticipate that for the weak vacuum that the change of its properties will be a hundred times more difficult to achieve than for the strong, though it is not necessarily so. We may not be able to separate the weak interaction vacuum from an even more complex issue: the possibility that quarks and leptons are somehow related to each other, and hence as we melt the massive Higgs field, all particle masses vanish and we unify them into a basic element of matter.

Quarks, as far as their interactions are concerned, come in different kinds, three different colors and several flavour doublets (u,d) , (c,s), (..t?,b) as we call them, and leptons come first in a doublet of (charge, neutral), and also in different flavors (electron, muon, tau) (see Figure 4.3). Very recent experiments involving the decays of the Z - bosons have shown that it is very likely that no further particle families with such 'small' masses exist. So there is the riddle of three flavors. The hidden and perhaps profound relationship between generations of quarks and leptons, which we know very little about at present is still to be discovered. We should remember that there is little experimental guidance in building such structures, and hence chances of finding the right approach are exceedingly small today. On the other hand, there are very surprising and little understood symmetries that connect quarks and leptons on one side, and there are amazing symmetry breakings: aside from the parity breaking we discussed above there is the CP symmetry breaking: it is the product of the charge reflection and parity reflection operators which is not respected by the unknown vacuum!

It would be very surprising if this relationship were an accident. The simplest point

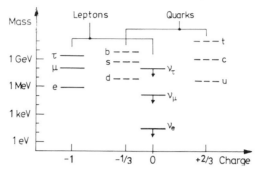

Fig. 4.4. Quarks and leptons can be ordered in such a way that they differ by one unit of charge each as shown here horizontally. Vertically their mass is shown - each mark on the scale denotes one order of magnitude.

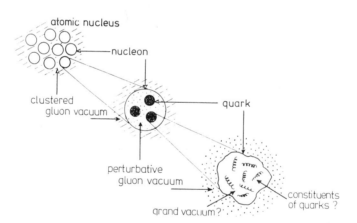

Fig. 4.5. It is assumed today that the quarks also have an inner structure which is determined by the vacuum of the unified interaction. The structural principle of the nucleons built out of quarks would then repeat itself at a lower level.

showing some intrinsic connection between all particles are the charges of quarks and electrons. We know that charges of quarks can be +2/3 or -1/3 of one elementary charge, so that they can differ just by one electron charge. And the charges of leptons, like electrons and neutrinos, the latter one being neutral, also differ by one unit of the electron charge (see Figure 4.4). It would be very surprising if this fact that the difference in the charges is the same for quarks and electrons would be accidental. Thus there is some yet not understood regularity which connects the particle families. I further recall that the sum of the charges of all quarks and leptons is the same, a fact very necessary for diverse theoretical considerations. So actually one would like to view quarks and leptons as being a reflection of a certain substructure which resides in them.

So possibly in order to get at the vacuum of weak interactions, we will have to connect the quark and lepton sectors. Then, we would come to the possibility that quarks and leptons can annihilate each other, as they are really two different facets of the common elementary object, but one is probably the matter and the other antimatter. This implies that matter as we know it is potentially unstable: if we had a piece of the vacuum in which electrons and quarks are particle-antiparticle to each other, protons and electrons could disintegrate! After some extensive searches for the so called proton decay, which was not observed, we have today a lower limit on the lifetime of the proton in our vacuum, perhaps as long as 10^{35}y. Our ignorance of why such a decay or annihilation reaction of quarks on electrons is impossible is clothed into the statement that the baryon number is conserved. Baryon number is a quantum number given to each proton, a quark carries by postulate just 1/3 of it. We do not know of any fundamental reason why baryon number should or should not be conserved. In principle, at least, a quark could annihilate with an electron. This has never been observed, as we have said, but it could happen even in our present vacuum on a very, very small scale. But it also may be that the structure of the vacuum forbids this process to occur, we really have no way to judge.

Maybe we again have the same picture which we had with quarks being constituents of the protons, that inside quarks there is a new kind of vacuum to be found, a vacuum amenable to conversion of quarks and leptons into each other (see Figure 4.5). If we want to probe such new processes, then we will probably have to go to much higher energies than we can today. While we probably do not yet have accelerators to carry out such experiments, we might at least look at the sky and in particular ask whether there are particles occurring in nature which carry the united vacuum on a relatively large scale with them and thus could stimulate the various reactions, possibly by some kind of a catalyst? I really mean a catalyst for the conversion of normal matter into radiation!. And it has been speculated recently that magnetic monopoles may be such a catalyst.

Unlike electric elementary charges, the magnetic monopoles may not be pointlike objects. It has been suggested that they may have an internal charge and that they may carry with them part of a different vacuum, which belongs to the interaction that possibly connects quarks and leptons. These magnetic monopoles arise within the scope of a proposed unified theory of all known gauge interactions. These monopoles carry with them this vacuum which comes from the early stages of the Universe, much the same way the charged vacuum in heavy ion collisions is carrying locally by the atomic nuclei. But the magnetic poles are expected to be very massive, so heavy that we have no prospect of making them with accelerators in the foreseeable future. But they might have been produced in the early Universe, when there was plenty of energy around. It is possible that they survived the evolution of the Universe and could be part of the cosmic particle flux.

When monopoles impact on the Earth, they very likely would fall right through its surface to the center. Another point is that, of course, the Universe is very large now, and if there were only 'few' monopoles left from the beginning of time, so we can wait long before one comes our way, in particular considering that they could be stopped (since they interact with matter) in gravitational traps such as centers of galaxies, stars or even planets, a fact almost unavoidable should monopoles be very heavy. Therefore, it might be quite futile if we just simply wait for a decay of a proton or the passage of a monopole through an apparatus.

Whenever such magnetic monopoles encounter a quark, energy is set free. This is possibly neither a particle annihilation nor particle decay process, it is like a reaction process in which a quark is converted into electrons in the unknown Vacuum. But if we believe that quarks can be transformed into leptons in the remnant of the vacuum, the monopole, then the energy released will just be the difference between the masses of these particles. We can calculate this from conservation principles, but something that one has to realize in that context is that at the moment we really do not understand where from the masses of particles arise. And similarly we do not have a theory to compute the reaction rates in this new unified vacuum of quarks and leptons. But maybe in the centers of galaxies there is another source of energy, the magnetic monopole burning up baryons.

In much of our discussion in the previous sections we have been considering how in consequence of vacuum structure in strong fields energy is converted into matter, that is particles. We now see that it is really the reverse process which is truly the greatest challenge, at the same time as it is perhaps the most fascinating issue in science. The dream of every science fiction author and I suppose of many of the readers of this essay is to know a practical way to convert matter into radiation on demand, into energy! What keeps us at present from using the energy that is contained in the mass of matter surrounding us is that the proton is a very stable form of matter. It does not decay, thereby releasing its energy. We only know that the proton lives on the average for much longer than the age of the Universe. But we know this only to be true in our vacuum, which developed during the expansion and the cooling of the Universe, and it certainly is possible, and even suggestive considering matter-antimatter asymmetry that the proton would not be stable in a different vacuum. At least in principle there is no reason to believe that this is impossible, as long as there is no fundamental understanding of the baryon number conservation, and what makes me believe this is the fact that we only observe matter in the Universe. No antimatter has ever been seen. The same forces that lead to the formation of matter without the antimatter may be harnessed to convert matter back into energy.

Sometime, somewhere in the Universe matter has been made without antimatter. Or, antimatter is around us and we just do not recognize it as antimatter. Or matter and antimatter originally were equally abundant, but somehow antimatter has been destroyed or converted into matter. As there is far more radiation in the Universe than matter, and we know that in the early Universe a considerable fraction of this radiation must have existed in the form of equal amounts of matter and antimatter, the asymmetry between matter and antimatter was not so great as it seems, at least as far as the underlying mechanism is concerned.

THE GRAVITY VACUUM

If we can understand in terms of the vacuum structure the coupling of gravitation to matter, we could possibly be able to turn it on and off. This coupling is in eyes of theoretical physicists most arbitrary, as a dimensioned object called 'G' is describing its strength. Here it is important to remember that this constant relates the inertial measured mass to the gravitational force acting between massive bodies. The proportionality constant between the two masses is contained in the gravitational constant. There is a profound difference between inertial mass and gravitational mass, even though these two are believed to be strictly proportional to each other according to the famous Einstein's equivalence principle. This constant

G is known by intricate analysis of various natural phenomena to be independent of time to a very high degree of accuracy and it is believed that it remains uninfluenced by the expansion of the Universe. Thus a way to influence (the vacuum which affects) the gravitational constant is not on the horizon. This is in particular so since there is a well-known phenomenon that there is at present no understanding of quantum gravity: that is understanding in principle how microscopic quantum objects gravitate and respond to gravitation. The difficulty is that gravity always is a global theory, describing in principle the entire space manifold, while quantum physics is a microscopic theory, concerned with the small environment of a point in space time.

The helpful hint is that gravitational constant carries a dimension. Therefore it could indeed be that this constant has something to do with the structure of the vacuum, as has been proposed in 'Higgs' gravity models. That brings up another dimensioned constant in this field the so called cosmological constant which Einstein introduced into his equations of General Relativity, and later again discarded because it was experimentally determined to be nearly zero. This cosmological constant essentially measures the amount of gravitating energy contained in the vacuum. Since we discussed that there are many different types of vacua, all of which have a different structure, in many cases a complicated structure, it is hard to understand why the gravitational effect of the vacuum is so small. A vacuum doesn't have much of an inertia. But it should have a gravitating mass. For some reason there is practically no gravitation of our vacuum. The common way of hand-waving this problem away is to say that we renormalize the vacuum not to gravitate. But that's not true. We cannot do this because the Universe develops with time. When we renormalize this constant away in the very early Universe, we probably would have to struggle in order to have this constant still zero today, since the early Universe was probably a melted vacuum. When the change to the frozen vacuum of today occurred, then this constant would acquire a nonvanishing value. So there is an intrinsic difficulty.

If today the vacuum has almost no gravitational effect, then if at some time earlier in the Universe the vacuum had a different structure, then certainly this vacuum must have had a strong gravitating influence. Without doubt it would have strongly influenced the expansion of the Universe. If one wants to compute the time sequence of the Universe, one, of course, usually proceeds backwards. One starts from what one sees today and takes the established laws which govern the evolution of gravitating bodies - the Universe is such a body in principle - and computes backwards in time. The Universe is expanding and has been expanding for a long time. So in the past the Universe was smaller and its energy density was higher than today. At some point in the past we arrive at an energy density in the Universe that would have been sufficiently high to permit us to imagine that the vacuum of the Universe had melted. This would happen for the strong vacuum at about one GeV/fm^3 And this melting may have reoccurred several times at earlier points in the evolution of the Universe, we just know too little about the 'unknown' vacuum.

Let us look at this last transition. At that point, when we go backwards, we must assume that the unit of volume of the vacuum would suddenly gravitate with a mass which corresponds to one GeV. That would seem to make the expansion of the Universe much more difficult, but when one works it out this phenomenon favors the expansion!. The reason for this is a somewhat curious feature of Einstein's theory of gravitation. As we discussed, a vacuum not only can have a nonvanishing energy, but also a pressure. Now in the absence of a constraint, be it high density or temperature, the melted vacuum would go over into the frozen vacuum that has lower energy. The region of space filled by the melted vacuum thus tends to become smaller. In other words, the melted vacuum is associated with a negative pressure, if the pressure of the frozen vacuum is normalized to be zero. Now in Einstein's theory not only mass gravitates, but so does pressure. Under normal conditions the gravitating effect of a pressure is imperceptibly small, but the vacuum pressure is large, so large that it overwhelms the gravity of the mass contained in the vacuum. The net gravitating effect of the melted vacuum is therefore repulsive.

Let us briefly look at the evolution of the Universe forward through the transition between vacua. The Universe must have been filled with the melted vacuum, with many particles contained in it which more than balanced the negative pressure of the melted vacuum. Now as the Universe expanded and cooled down, at some point in time, the positive pressure of the particles was no longer large enough to balance that vacuum pressure. At that moment the transition to the frozen vacuum should have taken place, confining the quarks to the interior of microscopic regions of melted vacuum called nucleons and mesons. However, if the expansion proceeded sufficiently fast, then the transition to the frozen vacuum could not occur immediately but only sometime later, and the Universe would spend some period of its evolution in the 'wrong' vacuum state, with its negative pressure no longer balanced by the particles contained in it. This supercooled, melted vacuum existed for a certain period of time, until it began to be converted into the frozen vacuum in a process very much like an explosion. The frozen vacuum started in small bubbles which grew at the speed of light. In the conversion process, the latent heat of the melted vacuum was transformed into real thermal energy, and the Universe was heated up again. But the really important period was, of course, the one in which the Universe contained the supercooled melted vacuum. The gravitational action of its negative pressure actually pressed the Universe apart. This works as a kind of antigravity. The expansion during that phase can be calculated to be exponential and in analogy to the economical troubles this has been termed .... inflation.

The very rapid growth of the size of the Universe suggest a smoothing of the small inhomogeneities in the distribution of matter, which are generated by the tendency of gravitation to coagulate the mass into clumps. And the Universe appears today to have been very uniform through the late stages at which atoms formed and radiation and matter decouple. This has been observed very recently by determining the structure of the black body background radiation, originating from the period that radiation and pressure decouple, when the temperature in the Universe became so low, that excitations of atoms became difficult. This blackbody spectrum turns out to be extremely precisely following the expectation for a entirely homogenous Universe. Thus somehow no structure has arisen through the very late stages, and then it must have come about much more rapidly than we can today understand in terms of dynamics of gravitating bodies.

Let me now change the subject somewhat: the black body background radiation is like an echo of the birth of the Universe. It is actually a very remarkable thing, as it points in another way to the need for the structured vacuum. We recall that in order to have waves propagating through the vacuum, more than a century ago many physicists were demanding some form of ether. In order to settle the issue, Michelson set out to prove that the velocity of light is the same in all directions of space, and independent of the motion of the observer. Such an experiment does not prove or disprove the existence of an ether, as Einstein very clearly pointed out. Einstein was very careful to point out that the laws of physics, which are relativistically invariant, make observation of a relativistically invariant ether impossible. However, what he didn't consider at that time was that even when the laws of physics are relativistically invariant, the state of the Universe can depend on the initial condition, on how the Universe started. And where it started. The frame of reference which is defined by the expansion of the Universe is a preferred reference frame in many ways. I can determine now an absolute velocity of the Earth with respect to the frame of reference defined by the rest-frame in which the black body radiation is homogenous. The background radiation provides me with an absolute frame of reference in which the Universe started. It does not, of course, provide a medium in which electromagnetic waves travel, but I can tell you in principle and also in practice what velocity the Earth has with respect to this ether. And all our spaceships at any place in the Universe can always tell at what velocity with respect to the ether, and therefore to the earth they are moving. The crucial point is that we have an absolute frame of reference provided for us from the beginning of the Universe by the frame in which the inflation of the vacuum occurred.

Acknowledgement: Much of the understanding presented in this manuscript arises from work done in collaboration with W. Greiner and B. Müller. I thank I. Bialynicki-Birula and M. Danos for the thorough reading of the manuscript.

HISTORICAL HIGHLIGHTS

ca. 500 B.C.
Parmenides (founder of the Eleatic School of philosophy) teaches that the "void" is unnecessary for a description of the world.

ca. 450 B.C.
Empedocles describes experiments with the so called "klepshydra" which is used to demonstrate that nature does not allow the creation of a macroscopic vacuum ("horror vacui").

ca. 400 B.C.
Peak of the atomic teachings under Democrit who stated that all material is built up out of indivisible atoms moving about in the micro-vacuum.

ca. 350 B.C.
Aristotle supports the theory of "horror vacui". The entire space is filled with the four elements (fire, earth, water, air) and with 'ether'. The terms 'ether' and 'vacuum' can be largely viewed as synonymous for the following 2,200 years.

1643/44
Torricelli's barometer experiments are carried out.

1650
Otto von Guericke invents the air pump and demonstrates the pressure difference between air and vacuum.

1687
Newton describes the classical conception of absolute space in "Principia Mathematica Philosophiae Naturalis".

ca. 1850
Robert Boyle shows sound cannot diffuse in the vacuum, whereas light can pass through unimpeded.

1873
Maxwell develops the uniform theory of electromagnetism and predicts electromagnetic waves propagating in the ether.

1887
The Michelson-Morley experiment shows that the vacuum is not filled with a material ether influencing the propagation of light.

1905 - 15
Einstein develops the special and general relativity theory: principles of equivalence of iner-

tial observers, the equivalence of inertial and heavy mass. Curved space-time (vacuum) is introduced.

1925 - 35
Development of quantum mechanics. Heisenberg formulates the uncertainty relation. Heisenberg, Pauli, Dirac and others develop the quantum theory of the electromagnetic vacuum and antimatter is predicted. Prediction of the vacuum polarization and the light-on-light scattering.

1947
Lamb and Rutherford discover experimentally the splitting of the states in the L-shell of the hydrogen atom and thus prove the existence of virtual particles in the vacuum.

1946 - 50
Feynman, Schwinger and Tomonaga develop quantum electrodynamics in its modern form and introduce the renormalization of the vacuum state.

1949
H.C. Casimir shows that the zero point energy of the vacuum is variable and that this phenomenon can be measured as it is associated with vacuum pressure.

1960 - 70
First vacuum structure models are introduced in particle physics. Formulation of the unified theory of electromagnetic and weak interaction by Glashow, Weinberg and Salam and introduction of the intermediate bosons W and Z.

1968 to date
Study of Vacuum in strong fields by Greiner and his Frankfurt school are followed by experimental efforts at GSI to detect the vacuum decay.

1973 to date
Development of Quantum Chromodynamics of quarks and gluons and the understanding of the vacuum of strong interactions.

1990
COBE (Cosmic Background Explorer) determines the uniformity of the background black body radiation and measures the absolute velocity of the Earth with respect to the frame of reference of last inflationary expansion of the Universe.

INVESTIGATION OF (e+e−) CORRELATIONS IN HEAVY ION COLLISIONS WITH THE DOUBLE ORANGE SPECTROMETER

W. Koenig

Gesellschaft für Schwerionenforschung
D-6100 Darmstadt, Fed. Rep. of Germany

1. INTRODUCTION

About 10 years ago positron spectra obtained from heavy-ion (HI) collisions of U+U and U+Cm ions at the Coulomb barrier showed a narrow ($\sim$ 80 keV) unexpected line. In subsequent investigations, performed mainly by two groups at GSI, the ORANGE and the EPOS collaboration, positron lines were observed for a variety of HI systems including the so-called subcritical systems ($Z_{target} + Z_{proj} < 173$) and, thus, excluding spontaneous positron emission as the origin of these lines. Within the experimental resolution of several 10 keV the line energies seemed to be independent of the HI system. This triggered the idea that the "monoenergetic" positrons could originate from the two-body decay of a neutral particle into an (e+e−) pair. A first pioneering experiment of the EPOS collaboration investigating (e+e− coincidences gave indeed a result consistent with the assumption of a two-body decay. Since then various systems were investigated, and more complex patterns of angular and energy correlations of the (e+e−) pair were observed. This report describes the status of the experiments performed by the ORANGE collaboration. The results obtained by the EPOS collaboration are summarized in the contribution of H. Bokemeyer to this conference. In the following solely experimental results will be presented and discussed. No attempt will be made to describe the various theoretical models and speculations which can be found elsewhere [1].

2. THE ORANGE SET UP

An iron-free Orange β-spectrometer (diameter $\sim 1m$) focuses e+ emitted in the backward hemisphere ($\theta_{e+} = 110° - 142°$) onto a focal cylinder centered along the beam axis (left part of Fig. 1). A second identical Orange spectrometer detects e− emitted in the forward hemisphere ($\theta_{e+} = 35° - 70°$) in coincidence with the e+ (right part of Fig. 1). The focal cylinder of each spectrometer is covered by 60 high-resolution Si detectors. The energy resolution of the sum-energy $E_\Sigma = E_{e+} + E_{e-}$ amounts to $\sim$10 keV (FWHM). At a constant field a momentum bin $\Delta p/p = 24\%$, corresponding to $\Delta E = 100$ keV at E$\sim$ 300 keV, is analyzed simultaneously for each lepton. A larger energy range of typically 200 keV to 500 keV is scanned by sweeping the spectrometers with their momentum bins differing by $\sim 10\%$, in order to account for the Doppler shift due to the motion of the c.m.-system. In the particular case of U+Ta collisions, the difference energy $E_\Delta = E_{e+} - E_{e-}$ between the two leptons was scanned additionally in order to study a larger range of energy correlations. A single detector accepts a momentum bin of $\Delta p/p \sim 2\%$ corresponding to $\Delta E = 10$ keV at E$\sim$ 300 keV. This fits the energy resolution of the detectors of $\Delta E \sim 6$ keV (e+) and ~ 8 keV (e−), respectively. The requirement that the energy measured by the detector pulse height corresponds to the momentum given by the

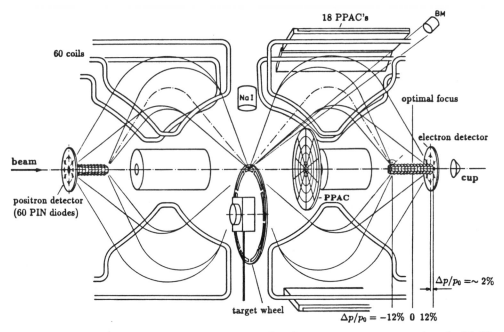

Fig. 1. Experimental setup consisting of two iron free Orange spectrometers equipped with 60 Si detectors each. The forward spectrometer is surrounded by 18 position-sensitive heavy-ion detectors (PPAC) and contains a further heavy-ion detector in its center. A mini ionization chamber is used to monitor beam and target quality, and a NaI detector measures high-energetic γ-rays.

spectrometer setting and the detector position allows an unique identification of e^+ and e^- originating from the target, and suppresses background very efficiently [2].

A central position-sensitive HI-detector measures the polar angles in between $14° \leq \theta_{ion} \leq 35°$ with an accuracy of $\sim 1°$. All azimuthal angles ϕ_{ion} are covered with a resolution of $60°$. This detector defines the lower limit of accepted e^- polar angles θ_{e^-}. A set of 18 large-area detectors, surrounding the forward spectrometer, detect HI scattered within $40° \leq \theta_{ion} \leq 70°$ with an accuracy of $\sim 0.5°$. Further detectors are used to monitor the beam energy and target quality as well as the emission of high-energetic γ-rays.

The design described above was chosen, in particular, for identifying back-to-back emission of monoenergetic (e^+e^-) pairs. The opening angle of the (e^+e^-) pair is measured within a range of $\theta_{e^+e^-} = 40° - 180°$ in the lab. system. Since the focal cylinders are subdivided into 6 columns, each containing 10 detectors, the total azimuthal angular range of $\phi = 0° - 360°$ is divided into 6 bins with $\Delta\phi = 60°$. Thus, the total range of opening angles is divided into four bins centered at $\theta_{e^+e^-} = 70°, 90°, 130°$, and $180°$ with a FWHM of $\sim \pm20°$ and a solid-angle ratio of $1 : 2 : 2 : 1$, respectively. Fig. 2 shows the distribution of opening angles in the lab. system obtained by a Monte-Carlo simulation of a two body decay. The mean velocity and velocity distribution of the emitter is chosen in such a way, that previously measured e^+ -line widths are reproduced [1,5]. The distribution labelled a) in the left part of Fig. 2 shows the distribution including only the kinematical broadening due to the motion of the emitter, whereas curve b) includes multiple scattering inside the high-Z target (U, $400\mu g/cm^2$). Obviously, the measured distribution does not depend significantly on the choice of parameters for the velocity distribution of the emitter, but is governed by multiple scattering. The right part of Fig. 2 shows the opening angle distribution taking into account the resolution of the setup. For comparison, the dashed line shows the distribution of an E0 or E2 internal-pair

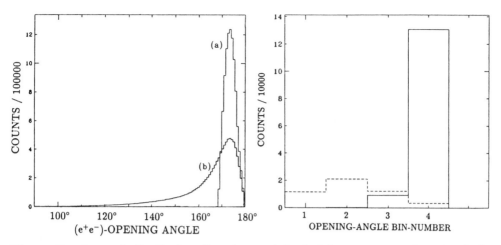

Fig. 2. Opening-angle distribution of an e^+, e^- pair in the lab. system assuming a two body decay of an emitter moving with c.m. velocity.

left: a) transformation in the lab. system only

b) including multiple scattering in the target (U, $400\mu gcm^2$).

right: the same as in the left part, but including the resolution of the exp. setup (full line).

dashed line: distribution of an E0 or E2 IPC process multiplied by a factor of 10.

conversion process (IPC). Relative to a two-body decay, the IPC intensity is multiplied by a factor of 10. In case of a back-to-back decay, practically all ($\sim$ 95 %) (e$^+$e$^-$) pairs are comprised in the 180° bin of the experimental set up. The Monte-Carlo simulations give a detection probability of $\sim$ 30% for back-to-back emitted (e$^+$e$^-$) pairs with respect to the detection of the positron. The suppression of internal pair creation (IPC) with respect to an isotropic back-to-back emission in the c.m. system has been estimated to be as large as a factor $\sim 10^{-3}$ (E0 and E2 transitions).

Since the electrons are detected in the forward, the positrons in the backward hemisphere relative to the c.m.-velocity a two-body decay of an emitter with $v \sim v_{cm}$ yields in the lab. system negative energy differences $E_\Delta \sim -40$ keV of the pair. On the other hand, IPC as well as uncorrelated (e$^+$e$^-$) emission is dominated by pairs with positive energy differences.

3. RESULTS

Using the setup described above a series of e$^+$e$^-$-coincidence measurements was carried out to investigate the collision systems U+Pb at a beam energy of (E/A)= 5.9 MeV/u, U+U at (E/A)= 5.8, 5.9, 6.0 MeV/u, and recently, U+Ta at (E/A)=6.3 MeV/u. In the following the three systems are discussed separately.

3.1 U + Pb

The full-line histogram in Fig.3 shows the e$^+$e$^-$-sum-energy spectrum obtained from U+Pb collisions for opening angles $\theta_{e^+e^-} = 180°\pm20°$ and for $E_{e^+}^{cm} = E_{e^-}^{cm}\pm17\%$. In the left part of Fig. 3 (e$^+$e$^-$) pairs were measured in coincidence with all heavy ions detected ($29° \leq \theta_{ion}^{cm} \leq 151°$), in the right part the HI-scattering angles were restricted to the angular range $53.5° \leq \theta_{ion}^{cm} \leq 151°$ (impact parameter (b) = 2 fm - 16.5 fm). In the latter case, the cut-off angle $\theta_{ion}^{cm} \leq 53.5°$ corresponds to the maximum scattering angle of projectiles in the lab. system. Only target recoils from peripheral collisions are scattered to larger lab. angles. Two narrow sum lines were observed with energies centered at $E_\Sigma = (575 \pm 6)$ keV and (787 ± 8) keV. Their widths of (15 ± 4) keV (FWHM) are consistent with the expected experimental resolution. The corresponding

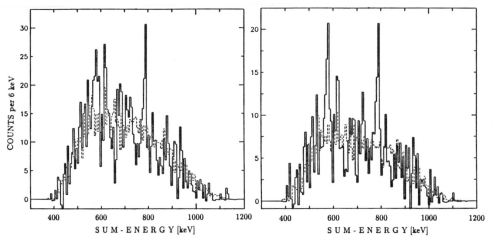

Fig. 3. e^+e^--sum-energy spectra obtained for U+Pb collisions at 5.9 MeV/u and for $E_{e^+}^{cm} \sim E_{e^-}^{cm}$. The full-line histograms correspond to the opening-angle bin of $180° \pm 20°$, while the dashed-line histograms display the remaining opening-angle range of observation. The latter were adjusted in height to fit the total intensity of the spectrum obtained for the 180° bin.
left: coincidences with all HI detected (b = 2 fm - 32 fm).
right: central, quasielastic HI collisions only (b = 2 fm - 16.5 fm).

spectra, gained for opening angles not containing the 180° bin, show no structures (dashed-line histogram). The latter were adjusted in height in order to fit the spectra obtained for the 180° bin. The statistical significance of the lines amounts to ~ 4 and ~ 5 standard deviations, respectively. The line characteristic is consistent with a two-body decay of an emitter being essentially at rest in the c.m. system of the colliding ions ($\beta_{cm} \leq 0.05c$). Under this assumption, the differential cross sections of the lines, $(d\sigma/d\Omega_{ion}^{cm})$, were found to be ~ 0.07 and ~ 0.11 μb/sr, respectively, in agreement with our previous results [3].

Comparing the left and right part of Fig. 3, the peak-to-background ratio is enhanced for more central, quasi-elastic collisions. This indicates a steeper impact parameter dependence of the line production mechanism as compared to the usual dynamically induced pair production.

Stimulated by an indication of a γ-line observed at an energy corresponding to the e^+e^--sum energy of 787 keV (Fig. 4), Monte Carlo studies were performed to investigate the question of whether or not internal pair creation (IPC) from one nucleus can lead to lines as narrow as observed in the measured sum spectra. Assuming an E1 transition and requiring an emission within the 180° bin of the spectrometer a narrow line (30 keV) can be obtained. For the sum over the other opening angle bins only a broad distribution (FWHM=150 keV) is obtained which might be hidden in the continuous background distribution (Fig. 5). However, Monte Carlo simulations show that the angular resolution of the Double Orange spectrometer is sufficient to perform an event-by-event Doppler-shift correction assuming emission from the projectile or recoil nucleus. In that case, IPC should result in a narrow peak (FWHM = 30 keV), even for all opening angles besides the 180° bin. According to the line intensity observed in the spectrum obtained for the 180° bin (Fig. 3) an - at least - 5 standard deviation effect should be visible in the Doppler corrected spectrum. This is in clear disagreement with the experimental results (Fig. 6).

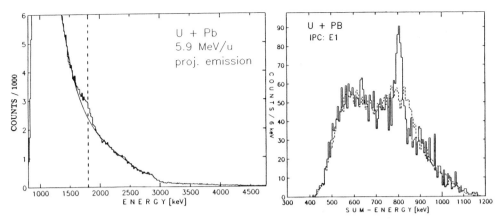

Fig. 4. The γ-ray spectrum obtained for U+Pb collisions at 5.9 MeV/u under the same kinematical constraints as the e^+e^--sum-energy spectrum shown in Fig. 3 . The γ-ray energy is event-by-event corrected for Doppler shift, assuming emission from the projectile-like scattered ion. The dashed line specifies the γ-ray energy which corresponds to the (e^+e^-)-sum-energy line at 790 keV, shown in Fig. 3. The full line displays the result of a fit to the continuous distribution outside the energy region around 1.8 MeV.

Fig. 5. (e^+e^-)-sum-energy spectra for the 180° bin of the spectrometer (full line) and the remaining opening angles $\theta_{e^+e^-} = 40° - 170°$ obtained by a Monte-Carlo simulation. An E1-IPC-process with a transition energy of 1.8 MeV is assumed, superimposed on the continuous distribution produced by uncorrelated (e^+e^-)-emission. The kinematical constraints correspond to those of Fig. 3 .

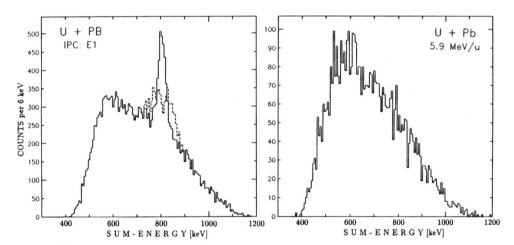

Fig. 6. (e^+e^-)-sum-energy spectra obtained by an event-by-event Doppler-shift correction assuming emission from the projectile-like scattered ions. All (e^+e^-)-opening angles apart from the 180° bin $(\theta_{e^+e^-} = 40° - 170°)$ are taken into analysis.

left: Monte-Carlo simulation assuming an E1-IPC transition.

The dashed line shows the uncorrected distribution.

right: Measured data.

3.2 U + U

The investigation of the collision system U+U at three beam energies with comparable statistics revealed a more complex picture as that observed in U+Pb collisions. At 5.9 MeV/u, we found again a sum line at ~815 keV, which exhibits a 180° characteristic as reported [3], even though its intensity is reduced appr. by a factor of two. At the other two beam energies (5.8 and 6.0 MeV/u), the intensity of the line is reduced below the level of statistical significance. The shift in the line energy of ~25 keV, when going from U+Pb to U+U collisions, points to a dependence of the line energy on the heavy-ion system. In this context, the effect of the difference in target thickness [Pb (~900 μg/cm^2), U (~400 μg/cm^2)], apart from the negligible difference in the energy loss of the leptons, is presently not clear.

In addition, a sum line at E_Σ ~555 keV is also seen, predominantly at 5.8 MeV/u (Fig. 7) with an statistical significance of ~ 4σ. Another line is indicated at E_Σ ~630 keV (Fig. 7, right part). They appear, however, at all opening angles $\theta_{e^+e^-}$ of observation, in disagreement with the scenario of a back-to-back decay. As shown in Fig. 8, the contribution from the 180° bin is negligible.

These features underline that rather complex production mechanisms must be involved in the observed phenomenon, and that not all lines found have to be necessarily of the same nature. None of the lines observed fit the typical signature investigated so far within the IPC scenario. In case of an emission from the fast moving projectile or recoil ions and opening angles around 90° a Doppler broadening of the sum-energy lines of ~ 100 - 150 keV is expected (cf. Fig. 5, dashed line). The IPC scenario fails, furthermore, to reproduce narrow single positron lines. Nevertheless, it should be mentioned that the theoretical description of the IPC process is currently not complete. In general the spin of excited nuclei is aligned perpendicular to the HI-scattering plane and, thus, the transition favors certain spin substates. For this case the opening-angle distribution is unknown. Furthermore, the Doppler broadening is strongly affected by the emission angle with respect to the HI-scattering plane. In case of the U + U system no experimental evidence was found for an in-plane / out-of-plane anisotropy. In addition, calculations assuming such an anisotropy for the emissions of (e$^+$e$^-$) pairs from the scattered nuclei did not yield quantitative agreement with the measured data.

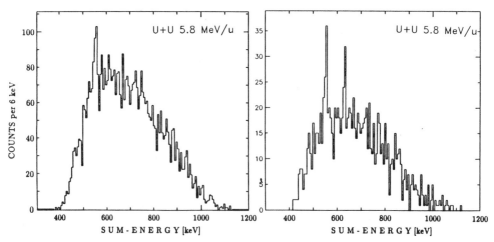

Fig. 7. (e$^+$e$^-$)-sum-energy spectrum obtained for U+U collisions at 5.8 MeV/u, for the entire angular range of opening angles $\theta_{e^+e^-}$.

left: all (e$^+$e$^-$) pairs investigated without any coincidence with the heavy ions required (i.e. no cut in the off-line analysis applied).

right: (e$^+$e$^-$) pairs with $E_{e^+}^{cm} \sim E_{e^-}^{cm}$ measured in coincidence with heavy ions scattered into an angular range of $\theta_{lab} = 20° - 35°$ and the complementary range $\theta_{lab} = 70° - 55°$.

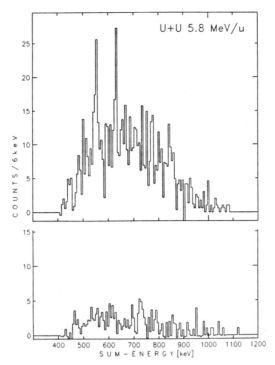

Fig. 8. (e^+e^-)-sum-energy spectrum obtained for U+U collisions at 5.8 MeV/u and for $E_{e^+}^{cm} \sim E_{e^-}^{cm}$.

upper part: All opening angles besides the 180° bin.   lower part: 180° bin only.

3.3 U + Ta

The system U+Ta was investigated in order to perform a direct comparison with data obtained by the EPOS collaboration. At the beam energy of 6.3 MeV/u two lines at (625 ± 8) keV and (805 ± 8) keV were reported[4,5]. We also found two lines at (635 ± 6) and (805 ± 8) keV in agreement with the EPOS results. Furthermore, our setup allowed for the first time to investigate e^+e^--pair emission for deep inelastic reactions. At 6.3 MeV/u the grazing angle amounts to $\sim 19°$ for the Ta-like recoil ion. This is well within the the range of $14° - 35°$ of the forward heavy-ion detector. By triggering on deep inelastic binary collisions with mass drift towards symmetry and large Q-values ($|Q| > 200$ MeV) as well as on fission products of U-like ions, a pronounced peak at 635 keV (statistical significance: 6.5σ) was observed (Fig. 9). The strong alignment of the fission products within the reaction plane shows that the fission events are correlated with large transfer of angular momentum in the first (binary) step of the reaction[6]. As compared to elastic scattering, for which the same line was observed, the emission probability per collision is increased by a factor of 10. The main feature of this trigger on deep inelastic events is probably the large reaction time expected for large mass drifts towards symmetry and large transfer of angular momentum. At incident beam energies still close to the Coulomb barrier the radial kinetic energy at the contact point of the two nuclei is small. Thus, immediate heating ($\tau \sim 10^{-21}$s) of the nuclei by converting this energy into internal excitation is reduced as compared to higher incident beam energies.

In case of the line at 635 keV the momentum correlation of the e^+ and e^- again is in disagreement with a two-body decay. The line intensity is most pronounced at $\theta_{e^+e^-} = 90°$, but a significant contribution ($\sim 17\%$) arises also from the 180° bin. It is comparable to the solid angle fraction of this bin (16.7%). In addition, the difference-energy distribution is very

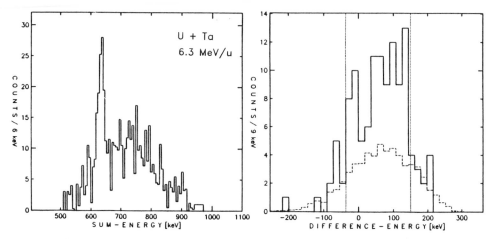

Fig. 9. (e^+e^-)-energy spectra obtained for deep inelastic U+Ta collisions at 6.3 MeV/u.

left: (e^+e^-)-sum-energy spectrum obtained for difference energies within the window shown below.

right: (e^+e^-)-difference-energy spectra obtained for sum-energy windows at the line position (full line) and on each side of the peak (dashed line).

broad and (taken out the Doppler shift of $\sim$ 40 keV) centered at a positive value of $\sim$ 100 keV (see Fig. 9).

In case of deep inelastic U+Ta collisions the large spread of masses, atomic numbers, excitation energies, and angular momenta results in an extremely small intensity contribution obtained from a specific nuclear transition. Thus, it is rather unlikely that the observed line is related to IPC.

4. DISCUSSION AND SPECULATIONS

For two systems (U + U and U + Ta) a narrow sum-energy line was observed even at opening angles smaller than 90°. This represents strong evidence for a process in which a third heavy partner is involved, taking momentum but nearly no energy. Because this third partner must be essentially at rest, it is probably a target atom. Furthermore the sum-energy lines are observed in coincidence with two scattered heavy ions, thus, a two-step process is required. Fig. 10 shows the momentum transfer to a third heavy partner obtained from the measured angular distributions for the systems U + U and U + Ta. Although the error bars are quite large, the distribution seems to peak around a momentum transfer of (800-1000) keV/c. This value is comparable or even larger than the mean momentum of the emitted leptons. The mean momentum transfer corresponds to a characteristic radius which is of the order of the compton wavelength of the electron. This result is in agreement with both, an IPC process as well as the speculation that an extended neutral particle (consisting of charged constituents) is destroyed by a secondary scattering process on the high-Z target atoms. However, a deexcitation process of a nucleus at rest via IPC requires the excitation of this nucleus with a reasonable probability such that a two step process can take place within the target with reasonable probabilities. A production of such a nucleus via e.g. neutron capture or a (n,n') reaction is rather unlikely since fast neutrons are emitted in the first step (several MeV) and the cross sections for the second step are of the order of the size of the nucleus. Furthermore, in such a secondary reaction additional levels should be excited and the corresponding γ-rays should be observed. In case of the U + Ta collisions, a large high resolution (4 keV) Ge(i) detector was used to monitor the γ-ray spectra. No γ-rays were observed in the MeV-range which could be attributed to such a process i.e. a narrow line without Doppler broadening in coincidence with scattered heavy

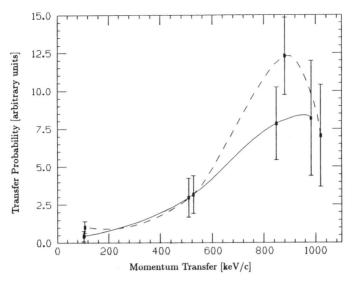

Fig. 10. Distribution of momentum transfers to a third heavy partner obtained from the opening angle distribution measured for two sum-energy lines. Full line: 635 keV line observed for U + Ta collisions. Dashed line: 555 keV line observed for U + U collisions.

ions. On the other hand, although the measured data are still consistent with the idea that an extended neutral particle might be produced, no definite proof for such a scenario is given by the current experiments. Data with better statistics and higher resolution with respect to the emission angles of the leptons are necessary to shed more light on the observed phenomena. This is in particular needed, since the experiments are not guided by theoretical models or predictions.

5.(e^+e^-) SCATTERING EXPERIMENTS

In addition to the heavy-ion experiments, a series of e^+e^- -(Bhabha-) scattering experiments was carried out in collaboration with the ILL in Grenoble during the last years. Bhabha scattering provides a model-independent test for the existence of neutral particles which have an e^+e^- decay branch. In the first experiments the excitation function was measured with a statistical accuracy of 0.2 %, giving no evidence for a resonance with an invariant mass around 1.8 MeV/c^2. These results have established a lower limit of 3.5×10^{-13} s for the particle lifetime which, however, is still away from the upper bound of $\sim 10^{-10}$ s, set by the heavy-ion experiments [7]. In a follow-up experiment, the sensitivity was improved for long-lived particles by employing an active-shadow technique. For lifetimes shorter than 7.5×10^{-12} s ($\Gamma_{e^+e^-} < 8.8\times10^{-5}$ eV), the results rule out the existence of a neutral particle [8].

References

1. B. Müller in Atomic Physics of highly ionized atoms, R. Marrus ed., Plenum Press N.Y. (1989)

   c.f. other contributions to these proceedings

2. W. Koenig, F. Bosch, P. Kienle, C. Kozhuharov, H. Tsertos, E. Berdermann, S. Huchler, W. Wagner, Z. Phys. **A328**, 129 (1987).

3. W. Koenig, E. Berdermann, F. Bosch, S. Huchler, P. Kienle, C. Kozhuharov, A. Schröter, S. Schuhbeck, H. Tsertos, Phys. Lett. **B 218**, 12 (1989).

4. P. Salabura, H. Backe, K. Bethge, H. Bokemeyer, T.E. Cowan, H. Folger, J.S. Greenberg, K. Sakaguchi, D. Schwalm, J. Schweppe, K.E. Stiebing, Phys. Lett. **B 245**, 153(1990).

5. H. Bokemeyer, contribution to these proceedings

6. Y. Civelekoglu,P. Glässel,D. v. Harrach, R. Männer, H. J. Specht, J. B. Wilhelmy, GSI annual report 1979, GSI 80-3 (1980), page 30 and GSI annual report 1978 GSI 79-11 (1979), page 31

7. H. Tsertos, C. Kozhuharov, P. Armbruster, P. Kienle, B. Krusche, K. Schreckenbach, Phys. Rev. D 40, 1397 (1989).

8. S.M. Judge, B. Krusche, K. Schreckenbach, H. Tsertos, P. Kienle, Phys. Rev. Lett. 65, 972 (1990).

POSITRON LINE EMISSION IN VERY HEAVY ION COLLISIONS:
RESULTS FROM THE EPOS COLLABORATION

H. Bokemeyer

Gesellschaft für Schwerionenforschung
D-6100 Darmstadt, Fed. Rep. of Germany

I. INTRODUCTION

The detection of narrow lines in positron singles- and electron positron sum-energy spectra measured after heavy-ion collisions at the UNILAC accelerator of GSI, Darmstadt has generated increasing interest in the past years. Whereas the line phenomenon itself appears to be experimentally well established, no conceivable explanation is given until today although many different attempts have been made. They reach from conventional processes to newly proposed mechanisms, including spontaneous positron emission in long-living quasiatomic heavy-ion formations as well as the production of light neutral fundamental particles. This article summarizes some of the main features of the positron lines as they have been obtained by the EPOS collaboration [SalB90]. No attempt is made to describe the experiment itself in full detail; instead we refer to [BokS89,Bok87,CowG87,Bok90]. Two other groups, the ORANGE [KoeB87,KoeB89] and the TORI [RheB89] collaborations are also studying these subjects. For the results of the ORANGE group see the contribution of W. Koenig to this conference. Comprehensive collections of theoretical aspects have been given by various other lectures during this school and can be found e.g. in [Gre88,FacT89] and in a recent review [Mül89].

II. MAIN FEATURES OF THE EPOS SET-UP

The EPOS transport system installed at the UNILAC utilizes a solenoidal magnetic field in order to selectively transport electrons and positrons away from the highly radioactive target area to special set-ups of high-resolution Si(Li)-detectors placed along the solenoid axis at distances > 1 m. The main elements of EPOS are shown in fig. 1. The set-up and its general performance concerning both singles and electron positron coincidence experiments has already been comprehensively discribed in [Bok87, CowG87] and shall not be repeated here. Only the following additional features installed for the most recent measurements will shortly be described: (i) The solid angle covered by the two parallel plate detectors used to detect the two scattered ions in coincidence with positrons and electrons has been enlarged: each detector subtends an angular range of $20° < \theta_{ion} < 70°$ and $-60° < \phi_{ion} < 60°$ with respect to the beam axis. (ii) The intrinsic energy resolution of the Si(Li) electron detector device was improved to $\delta E_e < 10$ keV. A time-resolution of $\delta t \lesssim 5$ ns has been achieved both for the electron and the positron detectors ($E_e = 350$ keV). The solid angle for the leptons is rotational symmetric around the solenoid axis, however, the polar angular range accepted is different for positrons and electrons (see fig. 2a). (iii) In the most recent experiment forward (F) and backward (B) emission of the leptons with respect to the beam direction could be directly distinguished by means of a mechanical sub-

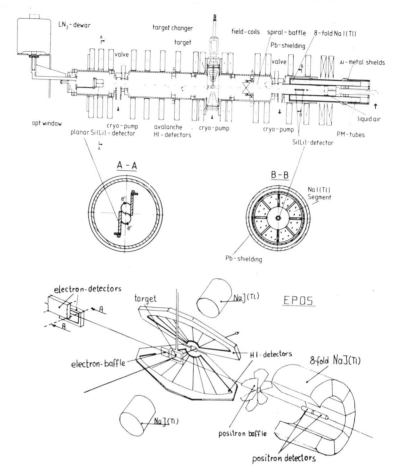

Fig. 1. Schematic view (top) and perspective drawing (bottom) of the EPOS spectrometer in a version optimized for electron-positron correlation experiments. The middle panel shows cross sections through the electron (left) and positron (right) detection area.

division of the detectors. This is based on the fact that, independent from ϑ_e and the lepton energy, any lepton emitted e.g. in the forward hemisphere (with respect to the beam direction) on its spiral path always returns to the axis from a backward direction and vice versa (see figs. 2b and 2c). In order to avoid the detection of low-energy δ-electrons, the detector geometry of the electron detectors has to be more complicated than for the positron detector. As a consequence the amount of separation between forward and backward emission is energy dependent and is optimized for an electron energy range of 300 keV to 400 keV. These and all other properties of the EPOS system for the detection of electrons and positrons have been studied extensively with Monte Carlo simulation techniques [Cow88,Sal90].

III. TOPOLOGY OF POSITRON SINGLES AND ELECTRON POSITRON COINCIDENCE LINES

At present three groupes of lines are identified in the positron singles spectra as well as in the sum-energy spectra of coincident electron-positron emission, in the latter at roughly twice the singles energies. The individual lines of the groups are at essentially equal energies in each of the collision systems all exceeding a combined charge $Z_{ua} \simeq 163$. To start with, fig. 3 presents a compilation of the positron singles lines. The lines were found to appear preferentially at beam energies around the

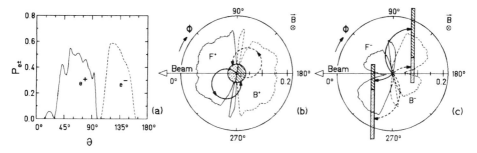

Fig. 2. (a) Transport probability for positrons and electrons with a kinetic energy of 350 keV as a function of the lepton emission angle θ with respect to the solenoid axis.
(b) Polar diagram of the positron transport probability as a function of azimuthal angle ϕ with respect to the solenoid axis. To distinguish between forward (F) and backward (B) (dashed) emission with respect to the beam direction the positron detector is subdivided into two halfes as indicated (View from target to the positron detector).
(c) As (b) but for electrons (view from the electron detectors to the target). The individual electron detectors are subdivided parallel to the solenoid axis as indicated by dashed and dotted areas. From the trajectories shown it is obvious that always the upper part of each detector corresponds to forward emission and vice versa.

Coulomb barrier and for particular regions of ion scattering-angles (see sect. V.) which generally belong to elastic or quasielastic collisions ($\Delta A \lesssim 5\,amu$, $\Delta Q \lesssim 20\,MeV$) . Consequently the given spectra are individually selected for such preferential conditions. Clearly the lines group around three mean energies of $\sim 320\,keV$, $\sim 360\,keV$, and $\sim 430\,keV$ (table 1). No spectra are shown in the cases, where the respective line has not been found in the particular experiment performed at a limited beam energy range. A subset of these line spectra has been published previously [CowB85]. Note, that for the EPOS device, if isotropic emission and negligible intrinsic width can be assumed, the laboratory system (LS) line energies are practically identical with the intrinsic emission energies and the LS widths of the positron singles lines reflect the velocity of the emitting system. Typically the observed widths are of $\Delta E \simeq 80\,keV$ corresponding to emitter velocity values of $v_{em} \simeq 0.05$ c, i.e. to values in the order of the heavy-ion center-of-mass (CM) velocity [CowB85].

Table 1. Positron singles line energies for various heavy collision systems

collision system	mean line energies [keV]		
$^{238}U + ^{248}Cm$	328 ± 9	-	440 ± 12
$^{232}Th + ^{248}Cm$	326 ± 10	358 ± 9	427 ± 11
$^{238}U + ^{238}U$	313 ± 8	348 ± 9	-
$^{232}Th + ^{238}U$	-	358 ± 11	445 ± 9
$^{238}U + ^{232}Th$	323 ± 10	-	439 ± 13
$^{232}Th + ^{232}Th$	318 ± 9	-	446 ± 10
$^{232}Th + ^{181}Ta$	-	376 ± 8	-

The continous part of the spectra is well described by a superposition of nuclear internal pair conversion (IPC) and quasiatomic positron production. The first contribution is derived from the simultaneously measured γ-ray spectra incorporating an E1/E2 multipolarity mixing with the E1 component dominant for $E_\gamma \gtrsim 1.6\,MeV$; the experimentally determined mixing ratio proved to be globally valid for these deformed nuclei at collision energies around the Coulomb barrier energy [Sak90]. The quasiatomic rate is calculated from theory [ReiM81] with a scaling factor $f_{QA} = 1.00 \pm 0.05$ obtained from a simultanous fit to the total spectra of the above collision systems [Sak90]. The calculated spectra contain all instrumental details including detector resolution and are normalized to the rate of scattered particles.

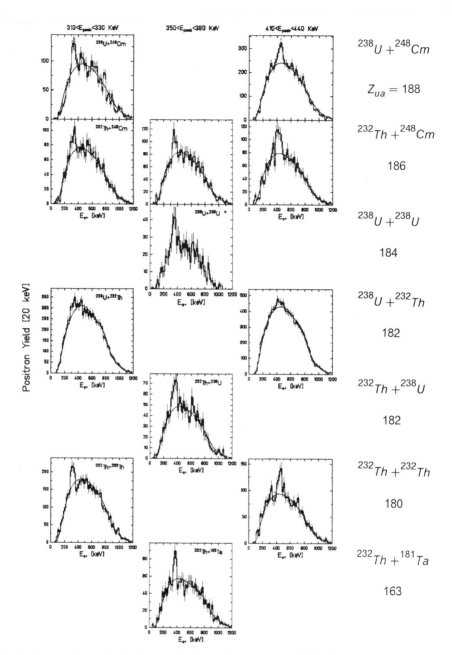

Fig. 3. The positron singles line family for various systems ($163 \leq Z_{ua} \leq 186$) as observed with the EPOS spectrometer. The spectra are ordered columnwise for increasing line energies with mean values around ~ 320 keV, ~ 360 keV, and ~ 430 keV. Each row stems from the collision system indicated, but the individual spectra are differently selected for beam energies and for windows in the heavy-ion scattering angles θ positioned symmetric around $\theta_1 - \theta_2 = 0^0$.

The family of sharp sum-energy lines detected in the spectra of the energy sum, $E_\Sigma = E_+ + E_-$, of the lepton pairs is displayed in fig. 4 (a 760-keV sum-energy line found in a pioneering experiment for $^{238}U + ^{232}Th$ collisions [CowB86] has been omitted). The continous part of the sum-energy spectra is again well described by a superposition of conversion processes from nuclear transitions and of quasiatomic electron-positron emission. The calculations correctly incorporate the high multiplicities of the individual processes and are normalized to the total positron yield with normalization factors close to 1. The characteristic line parameters are summarized in table 2. The narrow sum-energy lines again form an ensemble of three lines with mean energies around ~ 610 keV, ~ 750 keV and ~ 810 keV analogous to the positron singles-lines. The sum-energy line widths, ΔE_Σ, however, amount to values of $\lesssim 40$ keV, which is much less than what is expected from the individual Doppler broadening of the lepton energies unless a mutual cancellation of the energy shifts of the leptons of a pair is considered. If existing such a mechanism also implies maximally Doppler broadened structures in the associated spectra of the difference of the lepton energies, $E_\Lambda = E_+ + E_-$ (analogous to the broadening of the singles lines), located symmetric around $E_\Lambda = 0$ (if no additional energy shifts different for electrons and positrons are to be considered). In fact, such a correlated pattern is generally observed in the experiment, but some of the details are not strictly fulfilled: e.g. the centroid and the width of the broad difference energy structures vary with a tendency to higher positron than electron energies such as expected from the interaction with the Coulomb field of the heavy ions. Nevertheless, the actual simultanous existence of narrow sum and broad difference-energy lines is sufficient to exclude conventional nuclear IPC as the origin of these lines. The kinematic behaviour of IPC is that of a three-body decay which opposite to the present experimental appearance results in continously distributed difference-energy spectra. This has been shown in a series of Monte Carlo (MC) calculations performed to study the instrumental response on known electron positron production processes [Cow88,Sal90]. Independent from these considerations conventional IPC is moreover excluded from the narrowness of the widths of the sum-energy lines, since in order to explicitly obtain a narrow sum-energy peak from IPC, the source has to move very slowly in the LS ($v < 0.01c$). Each positron event, however, is detected in coincidence with both scattered particles with high velocities ($0.04c \lesssim v \lesssim 0.12c$) belonging to (quasi)-elastic scattering with only little mass loss ($\Delta A < 5$ amu) . No lines were found in delayed coincidences with the two heavy scattering products. Moreover, no appropriate γ-ray lines (excluding also external pair conversion) have been found in the simultaneously measured γ-ray spectra. Also no respective electron conversion lines considering nuclear E0 transitions were detected in the cases the conversion electron spectra have been explicitly measured.

Table 2. **Properties of the sum-energy lines.**

collision system	E_{beam} [MeV/u]	$<E_\Sigma>$ [keV]	ΔE_Σ [keV]	$<E_\Lambda>$ [keV]	ΔE_Λ [keV]	counts
$^{238}U + {}^{232}Th$	5.86 to 5.90	608 ± 8	$25 \pm 3^{b)}$	~ -10	~ 170	75 ± 17
	$\sim 5.83^{a)}$	760 ± 20	≤ 80	$\sim +10$	~ 140	35 ± 9
	5.87 to 5.90	809 ± 8	$40 \pm 4^{b)}$	$\sim +30$	~ 130	105 ± 20
$^{238}U + {}^{181}Ta$	6.24 to 6.38	625 ± 8	$20 \pm 3^{c)}$	$\sim +30$	~ 230	41 ± 10
	5.93 to 6.13	748 ± 8	$33 \pm 5^{c)}$	$\sim +150$	~ 450	120 ± 25
	6.24 to 6.38	805 ± 8	$27 \pm 3^{c)}$	$\sim +220$	~ 280	49 ± 14
a): [CowB85]		b),c): intrinsic resolution ~ 13 keV ~ 10 keV, respectively				

The similarity in the appearance of singles and coincidence positron lines is striking. It is not only that the cross sections are of the same order ($d\sigma/d\Omega_{HI} \simeq 5$ μb/sr) but the mean energies of the ~ 610-keV and ~ 810-keV sum-energy lines also nicely match with the measured positron singles line energies at roughly half the sum-line energies, a correspondence which in particular holds if the observed asymmetries of the difference-energy structures are taken into account (the association of the 748-keV line in $^{228}U + {}^{232}Ta$ with a very broad and asymmetric structure in the difference-energy spectrum is still under investigation). Recently the direct correspondence between singles and coincidence lines was directly proved within the same set of data for the 625- and the 805-keV lines of $^{238}U + {}^{232}Th$. More arguments that the singles and the coincidence lines belong to the same origin are given below.

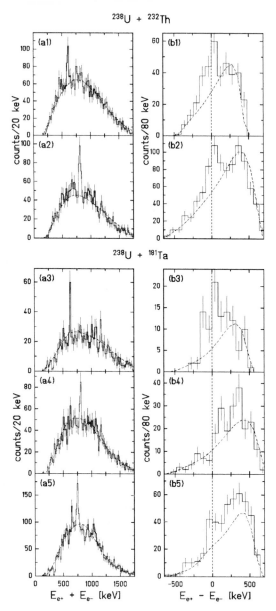

Fig. 4. The sum-energy line family for $^{238}U + ^{232}Th$ (up) and $^{238}U + ^{181}Ta$ (down) collisions. as observed with the EPOS spectrometer. The sum (left) and associated difference (right) energy spectra are integrated over the experimental ion angular range of $20° \lesssim \theta \lesssim 70°$ and are gated with lepton energy and time-of-flight windows as examplified in fig. 5.

IV. THE LEPTON ANGULAR DISTRIBUTION OF THE SUM-ENERGY LINES

In view of the far reaching consequences of the two-body decay hypothesis of a neutral object (see the contributions to this conference), originally introduced as the source of the narrow lines [e.g. CowB85, BalB85, Schäf89 also GraS89], an unambigous prove or disprove is desired. Independent tests are based on the assumed coupling of such objects or their components to known particles or refer to the inverse process, i.e. Bhabha scattering. As a matter of fact, these experiments all miss the particular circumference of the heavy-ion collision. The present results leave practically no room for the additional existence of fundamental or composed species with lifetimes $\lesssim 10^{-10}$ s and sizes which do not considerably exceed the Compton wavelength of the electron (see e.g. in: [Mül89, Schäf89, FacT89, Bok90, TseK89, ScheR90]); longer lifetimes are exluded by the fiducial volume of the heavy-ion experiments itself. A direct test of the hypothesis, however, needs the determination of the invariant mass of the line producing lepton pairs itself within the heavy ion experiment. Unfortunately, the measurement of the lepton emission-angles (or the opening angles) is hindered due to the strong angular scattering of the low-energetic ($\simeq 300$ keV) leptons in the targets (~ 300 $\mu g/cm^2$). Thus, only rough and indirect measurements of the lepton emission-angles have been performed so far.

IV.1 The Doppler shift method

The Doppler shift ΔE of the leptons is in first order proportional to $\cos \vartheta_{e,HI}$, $\vartheta_{e,HI}$ being the lepton emission-angle relative to the velocity vector of the emitting system. In the particular case of an object decaying at rest in the heavy-ion CM system this dependence results in the cancellation of the individual shifts of the back-to-back emitted leptons in the energy sum. The actual appearance of narrow lines in the sum energy together with the structures in the difference energy explicitly lead to the original proposal of a strongly simplified particle scenario, i.e. the two-body decay of a neutral object essentially at rest in the heavy-ion CM system.

IV.2 The time-of-flight method

The time-of-flight, t_e, of the leptons along their trajectories they need to reach the respective detector is determined by their velocity parallel to the solenoid axis. It is thus proportional to $1/\cos \vartheta_e$, with ϑ_e denoting the lepton emission angle with respect to the solenoid axis. The experimental time-resolution as compared to the time-of-flight range (~ 15 ns) is sufficient to subdivide the lepton angular ranges into two major parts, corresponding to perpendicular or parallel emission to the solenoid axis. Since the solid-angle, where electrons and positrons are accepted in the solenoid, are oriented opposite to each other (both form opposite cones perpendicular to the beam direction (see fig. 1 and 2(a)) this feature can be used to distinguish between an emission of the positron opposite or perpendicular to the direction of the electron of the pair.

This technique is demonstrated for the 608-keV and 809-keV sum-energy lines of $^{238}U + {}^{232}Th$ in fig. 5. The left part displays the result of a MC study where a small fraction of a pair decay of an object at rest in the CM system, contributing to only 3% of the singles positron yield, is admixed to the bulk of nuclear and quasiatomic electron-positron emission. In accordance with sect. IV.1 it becomes clear from this study, that already the restriction to events with lepton energies consistent with a back-to-back decay by means of a wedge-shaped window reduces the less correlated background contributions (fig. 5(a)): window W contains only 37% of the background but 100% of the two-body decay. The additional restriction of the time-of-flight window A by means of the different angular correlation further reduces the background by 55% whereas 81% of the back-to-back mode are contained in this window (fig. 5b).

The behaviour of the 809-keV sum-energy line obtained in the $^{238}U + {}^{232}Th$ measurement at beam energies within 5.87 and 5.90 MeV/u on these windows is studied in the middle column of fig. 5. The spectrum fig. 5(c) is a projection of the wedge shaped window W of fig. 5(a). The line at 809 keV containing (110 ± 33) events reaches a 3σ confidence level. If in addition the lepton time-of-flight window A of fig. 5(b) is applied to these data the spectrum fig. 5(d) results. There the 809-keV line with (90 ± 14) events now arrives at a 6σ level. Since this behaviour is fully adequate to the above MC study of back-to-back emitted pairs we conclude that the features of the 809-keV sum-energy line are consistent with a two-body decay scenario.

This, however, is different for the 608-keV line displayed in the right column (here also a slightly increased beam energy range from 5.86 to 5.90 MeV/u has been se-

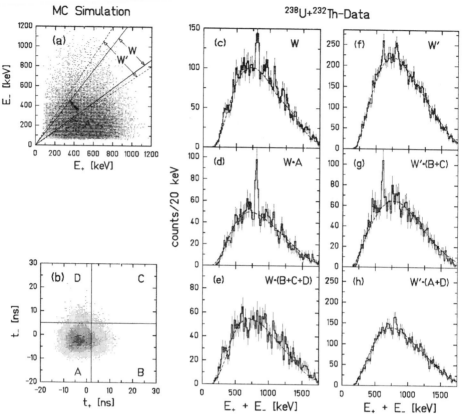

Fig. 5. Sum-energy spectra of coincident electron-positron emission in $^{238}U + {}^{232}Th$ colli-sions associated with a beam energy range of 5.86 to 5.90 MeV/u (middle column) and 5.86 to 5.90 MeV/u (right column). For the two-dimensional event distributions (left column) plotted as functions of the lepton energies and the lepton times-of-flight, respectively, see text. The sum-energy spectra are differently gated with the two-dimensional windows as indicated (see also text).

lected). The spectrum of fig. 5(f) again is exclusively associated to a wedge-shaped window W' being similar but slightly enlarged as compared to W ((106 ± 25) line events; 4σ). This correlation together with the narrow width of the line again indicates a mo-mentum correlated pair emission, but contrary to the 809-keV line, not the time-of-flight window A but windows B and C were found to accomodate the emission characteristic of the line events (fig. 5(g), (80 ± 13) events, 6σ). This latter behaviour of the 608-keV line is inconsistent with the idealised two-body decay scenario with the object at rest in the heavy-ion CM system. Comparable dependences as for the 608-keV and for the 809-keV lines have also been observed - with less statistical evidence - for the 625-keV and for the 805-keV sum-energy lines of $^{238}U + {}^{181}Ta$, respectively, although there the 625-keV line is associated with a larger time-of-flight window which still partly overlaps with the window A of fig. 5(b). In all cases the wedge-shaped windows have been ad-justed according the shifts of the structures in the associated difference-energy spectra (fig. 4).

IV.3 The hemisphere correlation

In the EPOS magnetic transport system a most direct access to the emission di-rection is given by the geometrical correspondence between the lepton azimuthal an-gle ϕ_e with respect to the solenoid axis and the place where its trajectory hits the detector. The use of this technique shortly described in sect. II implies, that the target and the detector system both are aligned along the rotational symmetry axis of the solenoid. The method fails if the place of emission of the lepton is considerably removed from this axis (in the order of mm) as in the case of a delayed emission from a fast moving nucleus. Obviously, this feature includes the possibility for a distinction between a preferentially back-to-back or otherwise correlated pair emission. A MC simulation of the realistic experimental situation in fact predicts a considerable separation power, R, with values of 0.8 for the case of back-to-back and 0.4 for uncorrelated emission; R is defined as the ratio between the number of events observed in detector combinations which belong to back-to-back emission and the number of totally observed electron-positron events. For the ^{90}Sr IPC source a value of R = (0.39 ± 0.01) has been measured which nicely confirms the calculations.

The technique has been utilized recently for the $^{238}U + ^{181}Ta$ experiments [Sal90]. The results still suffer from poor statistics. For the 625-keV and the 805-keV sum-energy line the ratios of R = (0.63 ± 0.20) and R = (0.74 ± 0.15), respectively, are consistent with the two-body decay hypothesis. The ratio R = (0.40 ± 0.10) obtained for the 748-keV sum-energy line is of better statistical relevance. Clearly R misses the value expected for the back-to-back scenario but it can also not be explained by processes like conventional nuclear IPC: the line is found to be associated with the emission of only the electron into the forward hemisphere irrespective from the emission direction of the positron. The details of the line appearence indicate a emission from a fast moving system off from the solenoid axis, which may confuse the hemisphere correlation technique. Including off-axis emission in the MC simulations, again any known IPC process is in conflict with the narrow width of the sum-energy line and other details of the experimental lepton angular correlation pattern.

V. THE ION ANGULAR DEPENDENCE OF THE LINES

Based on a reliable absolute knowledge of the background spectra (see sect. III) also the yield of both singles and coincidence lines could recently be derived as a function of the ion scattering-angle [Sak90]. Fig. 6 displays the impact parameter dependence of the production probability of the 748-keV sum-energy line of $^{238}U + ^{181}Ta$. This is considered the most favourable case for this analysis because there not only a large mass asymmetry of the system, sufficient to separate central from peripheral quasielastic collisions, but also a sufficient statistical quality of the line is attained. The pair emission probability is plotted as a function of the nuclear distance $R_{min} = a(1 + 1/ \sin \theta_{cm})$ with 2a equal to the nuclear distance of closest approach in a head-on collision. An exponential increase with a slope constant $2\alpha = (0.35 \pm 0.07)$ appears to characterize the data up to nuclear distances close to the nuclear interaction radius. Qualitatively, such exponential behaviour of the probability is well known from nucleon transfer and from quasiatomic processes. However, for heavy collision systems the dependence of the one-neutron transfer into the nuclear ground state with slope parameters typically of $2\alpha = 1.08$ ($^{238}U + ^{197}Au$, [WirB87]) is steeper, whereas the quasiatomic positron production results in a less steep increase ($2\alpha \simeq 0.13$, [ArmK79, Sak90]). Similar results have been obtained for the R_{min} dependence of the 625-keV and 805-keV sum-energy lines where slope parameters $2\alpha = 0.42 \pm 0.08$ and 0.21 ± 0.07, respectively, have been deduced. Taken together, a mean slope parameter of $2\alpha \simeq 0.33$, corresponding to a mean energy transfer of $\Delta E \simeq 3.8$ MeV, may qualitatively describe the situation for the three lines.

Comparable kinematic dependences have been observed for all of the investigated collision systems both for the positron singles as well as for the coincidence lines. But in most cases, due to the insufficient mass asymmetry, only LS ion angular dependences could be derived. All lines exploit angular distributions which essentially are symmetric around $\Delta\theta = \theta_1 - \theta_2 = 0°$, the principal mirror symmetry axis of the overall scattering distributions, with mean values $|\Delta\theta|$ around 10° to 20°. In fact, this circumstance has always been used to improve the peak-to-background ratio of the singles positron lines (sect. III, [CowB85]). We may generally assume that the central collisions play the dominant role in the line production process (as observed for the

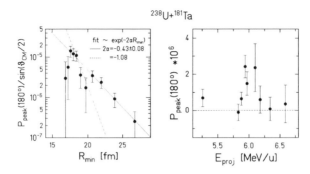

Fig. 6. Kinematic dependence of the 748-keV ($^{238}U + {}^{181}Ta$) sum-energy line. (Left): R_{min} dependence, of the line emission probability per scattered ion pairs (assuming intrinsic back-to-back emission). For the exponential fit see text. (Right): Observed yield of line events per scattered ion pair versus beam energy.

$^{238}U + {}^{181}Ta$ system). Then mean nuclear distances $\overline{R}_{min}$ of 18 to 21 *fm* with a mean width ΔR_{min} in the order of 1.2 *fm* can be deduced from the LS angular distributions.

Such a kinematic behaviour implies the existence of corresponding windows of favourable projectile energies, ΔE_{proj}, in the order of 0.2 to 0.3 *MeV/u* located around the Coulomb barrier energy. In fact, this is consistent with the excitation function measured for the 748-keV sum-energy line [BokS89] and with the circumstance that the 625-keV and 805-keV sum-energy lines in $^{238}U + {}^{181}Ta$ have only been found at beam energies between 6.2 and 6.4 *MeV/u*. Also the 608-keV and 809-keV sum-energy lines in $^{238}U + {}^{232}Th$ both display a clear threshold behaviour with the onset of the line production at beam energies > 5.85 *MeV/u* (no higher values than 5.9 *MeV/u* have been probed during this experiment).

VI. CONCLUDING REMARKS

At present, the ensemble of positron lines manifests itself as a globally valid (collision systems with a combined charge $163 \leq Z_{ua} \leq 188$) and a systematically ordered (set of ≥ 3 groups of lines) phenomenon. Singles and coincidence lines can be attributed to the same pair emission process, where the electrons and positrons essentially share the available energy. In detail the positron energies typically appear to be larger than the electron energies just as it can be anticipated from an interaction with a Coulomb-potential of a positive charge at the time of emission. Looking to the ion angular dependence the line production rates are strongly enhanced for closest collisions of heaviest nuclei although a genuine connection with any nuclear or quasiatomic process has not been proved and even appears to be excluded for known mechanisms. Nevertheless, such kinematic dependence strongly argues that the puzzling line production process is intimately related to particular features of heavy-ion collisions as e.g. the presence of exclusively strong, localized, and time varying electric fields.

The reproduction in follow-up experiments as well as the observed systematic dependences of the lines prove these as experimentally established. Detailed MC studies and the variant appearance and characteristics of the lines excludes instrumental effects as the source of the lines. The ORANGE collaboration reports on comparable results obtained in recent experiments with $^{238}U + {}^{238}U$, $^{238}U + {}^{208}Pb$ and $^{238}U + {}^{181}Ta$ collisions ([BerB90] and contribution of W. Koenig to this conference). Still, however, the positron line phenomena awaits a consistent and well accepted explanation. It thus remains a challenging problem for both theoreticians and experimentalists. After the pioneering phase of detecting the lines now systematic investigations of the production mechanism and the lepton emission characteristica are urgently needed. Since the cross sections were found to be small, the experimental conditions of the experiment - although already highly optimized - had to be further improved. A a total reconstruction of the EPOS device increases the electron positron coincidence detection efficiency by a factor ~ 5 and the installation of a separate ECR source raises the UNILAC beam intensity. Besides the aim to gain in the statistical significance, particular interest will then be devoted to the details of the lepton angular-distributions and the reaction parameters as well as to the question of the existence of the lines at lighter systems. We are confident, that this will shed light both into the decay as well as into the production process of the lines.

REFERENCES

[ArmK79] P. Armbruster and P. Kienle, Z. Physik A291 (1979)399

[BalB85] A.B. Balantekin, C. Bottcher, M.R. Strayer, and S.J. Lee, Phys. Rev. Lett. 55 (1985)461

[BerB90] E. Berdermann, F. Bosch, T. Ishikawa, P. Kienle, I. Koenig, W. Koenig, C. Kozhuharov, A. Schröter, and H. Tsertos, Ann. Rep. GSI-90-1, GSI Darmstadt (1990), p. 129

[Bok84] H. Bokemeyer in: Selected Topics in Nucl. Struct., J. Stachura, ed., Inst. of Nucl. Phys., Jag. Univ., Cracow, Polen, (1984), p.335 and GSI-Preprint 84-83, GSI, Darmstadt (1984)

[Bok86] H. Bokemeyer in: Selected Topics in Nucl. Structr., R. Broda and J. Stachura, eds., Inst. of Nucl. Phys., Jag. Univ., Cracow, Polen (1986)

[Bok87] H. Bokemeyer in: [Gre87], p.195

[Bok90] H. Bokemeyer, Habilitationsschrift, J.W. Goethe Univ., Frankfurt (1989) and GSI, Darmstadt (1990)

[SalB90] P. Salabura, H. Backe, K. Bethge, H. Bokemeyer, T.E. Cowan, H. Folger, J.S. Greenberg, K. Sakaguchi, D. Schwalm, J. Schweppe, and K.E. Stiebing, subm. to Phys. Lett. and
H. Bokemeyer, H. Folger, K. Bethge, D. Kraft, K. Sakaguchi, E. Stiebing, T. Cowan, J.S. Greenberg, J. Schweppe, H. Backe, D. Schwalm, P. Salabura, Ann. Rep. GSI-90-1, GSI Darmstadt (1990), p. 130

[BokS89] H. Bokemeyer, P. Salabura, D. Schwalm, and K.E. Stiebing, in: [FacT89], p. 77

[Cow88] T. Cowan, Ph.D.-thesis, Yale Univ., New Haven (1989)

[CowB85] T. Cowan, H. Backe, M. Begemann, K. Bethge, H. Bokemeyer, H. Folger, J.S. Greenberg, H. Grein, A. Gruppe, Y. Kido, M. Klüver, D. Schwalm, J. Schweppe, K.E. Stiebing, N. Trautmann and P. Vincent, Phys. Rev. Lett. 54 (1985) 1761

[CowB86] T. Cowan, H. Backe, K. Bethge, H. Bokemeyer, H. Folger, J.S. Greenberg, K. Sakaguchi, D. Schwalm, J. Schweppe, K.E. Stiebing and P. Vincent, Phys. Rev. Lett. 56 (1986) 444

[CowG87] T. Cowan and J.S. Greenberg in: [Gre87], p. 111

[FacT89] Test of Fundamental Laws in Physics, O. Fackler and J. Trân Thanh Vân, eds., Editions Frontières, Gif sur Yvette(1989) ISBN 2-86332-064-5

[GraS89] S. Graf, S. Schramm, J. Reinhardt, B. Müller, and W. Greiner, J. Phys. G15 (1989) 1467

[Gre87] Physics of Strong Fields, W. Greiner, ed., Plenum Press, New York (1987) ISBN 0-366-42577-7

[Kie89a] P. Kienle in: Selected Topics in Nucl. Structure, Inst. of Nucl. Phys., Jag. Univ., Cracow (1989), Poland

[Kie89b] P. Kienle in: [FacT89], p. 63

[KoeB87] W. Koenig, F. Bosch, P. Kienle, C. Kozhuharov, H. Tsertos, E. Berdermann, S. Huchler, W. Wagner, Z. Physik A328 (1987) 129

[KoeB89] W. Koenig, E. Berdermann, F. Bosch, S. Huchler, P. Kienle, C. Kozhuharov, A. Schröter, S. Schuhbeck, H. Tsertos, Phys. Lett. B218 (1989) 12

[Mül89] B. Müller in: At. Phys. of Highly Ionized Atoms, R. Marrus, ed., Plenum Press, New York (1989)

[ReiM81] J. Reinhardt, B. Müller, and W. Greiner, Phys. Rev. A24 (1981) 103

[RheB89] M. Rhein, B. Blank, E. Bozek, E. Ditzel, H. Friedemann, H. Jäger, E. Kankeleit, G. Klotz-Engmann, M. Krämer, V. Lips, C. Müntz, H. Oeschler, A. Piechaczek, I. Schall, C. Wille in: [TraF89], p. 195

[Sak90]   K. Sakaguchi, Ph.D.-thesis, J.W. Goethe Univ., Frankfurt (1989) and GSI-90-05, GSI Darmstadt (1990) ISSN 0171-4546

[Sal90]   P. Salabura, Ph.D.-thesis, Jag. Univ. Cracow (1989) and   GSI-90-06, GSI, Darmstadt (1990) ISSN 0171-4546

[SalB90] P. Salabura, H. Backe. K. Bethge, H. Bokemeyer, T.E. Cowan, H. Folger, J.S. Greenberg, K. Sakaguchi, D. Schwalm, J. Schweppe, and K.E. Stiebing, subm. to Phys. Lett. and H. Bokemeyer, H. Folger, K. Bethge, D. Kraft, K. Sakaguchi, E. Stiebing, T. Cowan, J.S. Greenberg, J. Schweppe, H. Backe, D. Schwalm, P. Salabura, Ann. Rep. GSI-90-1, GSI Darmstadt (1990), p. 130

[Schäf89] A. Schäfer, Journ. Phys. $\underline{G15}$ (1989) 371

[SenB87] P. Senger, H. Backe, M. Begemann-Blaich, H. Bokemeyer, P. Glässel, D.v. Harrach, M. Klüver, W. Konen, K. Poppensieker, K.E. Stiebing, J. Stroth, K. Wallenwein in: [Gre87], p. 423

[Tra87]   The Standard Model, J. Trân Thanh Vân, ed., Editions Frontières, Gif sur Yvette (1987) ISBN 2-86332-047-5

[TseK89] H. Tsertos, C. Kozhuharov, P. Armbruster, P. Kienle, B. Krusche, K. Schreckenbach, Phys. Rev. $\underline{D40}$ (1989) 1397

[WirB87] G. Wirth, W. Brüchle, M. Brügger, Fan Wo, K. Sümmerer, F. Funke, J.V. Kratz, M. Lerch, N. Trautmann, Phys. Lett. $\underline{B177}$ (1986)282

[ScheR90] A. Scherdin, J. Reinhardt, W. Greiner, and B. Mueller, Low Energy e^+e^- scattering, Inst. f. Theor. Physik, Univ. Frankfurt (1990) UFTP 248/1990, subm. to Rev. Prog. Phys

STRONG COULOMB COUPLING OF CHARGED PARTICLES

M. Bawin

Institut de Physique B5, Université de Liège
4000 Liège 1, Belgium

1. INTRODUCTION

According to our present understanding of Quantum Electrodynamics of strong fields, we expect spontaneous pair production of electron positron pairs to occur in the external Coulomb field of a nucleus of charge Z when Z exceeds a critical value $Z_c \approx 170$.[1] The corresponding critical value Z_c^0 above which spontaneous creation of pion pairs (π^-, π^+) occurs is $Z_c^0 \approx 10^4$, so that the interest for such phenomenon would seem quite academical. Nevertheless the study of scalar boson "condensation" is of theoretical interest for (at least) two reasons. First, boson condensation of charged scalar particles in a strong Coulomb field may give us some insight into the structure of the QCD vacuum and its gluon condensate.[2] Second, it provides us with a simple model for the "restructuring" of the vacuum in a high density medium[3] that may exist, for instance, in the collision of two heavy ions, inducing a state of matter under "extreme conditions". Sections 2-3 are devoted to the discussion of this subject within classical electromagnetic field theory. We shall discuss in Sec. 4 a related topic which is being currently investigated in great detail in lattice QED,[4] i.e. the much more complicated problem of two spin one-half charged particles in mutual strong Coulomb interaction. Our interest for such a problem stems from recent heavy-ion experiments at Darmstadt and the conjecture of a new phase of QED.[5] We shall see that a reasonable non perturbative approach to the relativistic two-body problem leads to an equation whose eigenvalue spectrum is very different from the spectrum of the Klein-Gordon (or Dirac) equation and does not lead to the existence of a critical coupling strength above which spontaneous pair production may occur. Thus, a simple extrapolation of external field results to the problem of two mutually interacting particles may be quite misleading, at least if one trusts some reasonable two-body relativistic wave equations.

Vacuum Structure in Intense Fields, Edited by
H.M. Fried and B. Muller, Plenum Press, New York, 1991

2. KLEIN-GORDON PARTICLES IN STRONG COULOMB FIELDS

Consider a pion of mass m in the external Coulomb field V(r) of an external source of charge Z. We shall always consider in these lectures that the Coulomb interaction has been regularized at the origin. Specifically, we shall choose :

$$V(r) = - \frac{\alpha Z}{r} \qquad (r > r_0) \qquad (1)$$

$$V(r) = - \frac{\alpha Z}{r_0} \qquad (r < r_0) \qquad (2)$$

(α is the fine structure constant).

In formulas (1) and (2) r_0 can be arbitrarily small but not zero. So a "point" nucleus could be for instance, a nucleus with $r_0 \approx 10^{-27}$ fm, rather than $r_0 = 0$. It is well known[6] that a regularization of the Coulomb problem for Klein-Gordon particles is needed for $\alpha Z > 1/2$. Formulas (1) and (2) provide a very simple and physically intuitive way of performing such a regularization. The ground-state radial Klein-Gordon (K.G.) equation to be solved is then :[7]

$$u'' + \left[(E - V)^2 \right] - m^2 u(r) = 0 \quad , \qquad (3)$$

where E is the pion energy. Its spectrum is given by :[7]

$$\left. \frac{W'_{K\mu}(\rho)}{W_{K\mu}(\rho)} \right|_{r = r_0} = \tilde{K} \cotan \tilde{K} r_0 \quad , \qquad (4)$$

where $W_{K\mu}(\rho)$ is the Whittaker function, and

$$K = \frac{E\alpha Z}{\left(m^2 - E^2 \right)^{1/2}} \quad , \qquad (5)$$

$$\mu = \left(\frac{1}{4} - \alpha^2 Z^2 \right)^{1/2} \quad , \qquad (6)$$

$$\rho = 2 \left(m^2 - E^2 \right)^{1/2} r \quad , \qquad (7)$$

$$\tilde{K} = \left[\left(E + \frac{\alpha Z}{r_0} \right)^2 - m^2 \right]^{1/2} \quad . \qquad (8)$$

The bound states spectrum coming from Eq. (4) with decreasing values of r_0 starting at $r_0 = 10^{-3}$ fm down to $r_0 = 10^{-27}$ fm is given in Fig. 1.

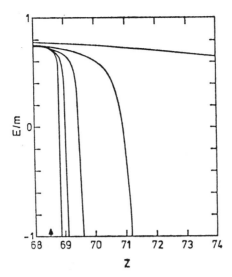

Fig. 1. *1s bound-state energy spectrum for a Klein-Gordon particle with interaction (1)-(2) as a function of the external charge Z. The different curves (from upper right to lower left) refer to a radius $r_0 = 10^{-3}, 10^{-9}, 10^{-15}, 10^{-21}$ and 10^{-27} fm, respectively, and to m = pion mass. Note that our results hold for any mass, provided that the radius r_0 is rescaled in such a way that mr_0 keeps a constant value.*

One can see that for a given value of r_0, there exists a critical Z-value such that $E = -m$ at $Z = Z_c$. It would be tempting to conclude that for $Z > Z_c$ spontaneous pair creation of (π-, π+) pairs then may take place, just as in the corresponding problem for Dirac particles. However, the mathematical structure of the K.G. equation is quite different from the Dirac equation, and its interpretation as a single particle equation for $Z > Z_c$ is beset with a number of difficulties, both mathematical and physical. Let us first observe that the conserved norm N associated with the Klein-Gordon equation (3) is :

$$N = \int_0^\infty (E - V) u^2 dr \quad .$$

$$(9)$$

For sufficiently small V, we can associate particle states with $N > 0$ solutions, while $N < 0$ solutions are anti-particle states, i.e. the K.G. equation is endowed with an indefinite metric. Now, if we compute the eigenvalue spectrum from (4), we find that complex eigenvalues occur for $Z > Z_c$.[8] Furthermore, the corresponding eigenstates necessarily have a zero norm,[9] which makes their physical interpretation rather difficult. Even worse, the set of eigenstates $\{u_E(r)\}$ no longer form a complete set. So-called associated eigenstates must be added to the true eigenstates in order to span the whole Hilbert space. Such associated eigenstates u_a satisfy the equation :[10]

$$\left[(E - V)^2 - m^2\right] u_a + u''_a = -2(E - V) u \quad ,$$

$$(10)$$

where u satisfies (3). Needless to say, the physical interpretation of u_a is still more problematic than the interpretation of $u(r)$ for $Z > Z_c$. Rather than trying to give a meaning to eigenstates of the Klein-Gordon equation for $Z > Z_c$, we shall instead interpret the occurrence of complex eigenvalues to signal the breakdown of the validity of the K.G. equation in the supercritical regime. We shall see below that this viewpoint is consistent with our solution of the problem of a fixed source in electromagnetic interaction with a classical scalar field. Before doing that, however, let us consider a related problem which can be solved within the single-particle approach to the K.G. equation. Suppose we were to allow nuclei to have an arbitrary small radius. Can these nuclei have an arbitrary charge ? This problem was first considered by Gärtner, Müller, Reinhardt and Greiner,[11] who showed that "point" nuclei with $Z > 137$ could not exist. Let us now show how this result is modified if there exist point scalar particles (π) or mass m greater than the electron mass m_e. Consider in Fig. 1 the curve relative to $r_0 = 10^{-27}$ fm. One can see that there exists a critical value $Z_{cr}^{(1)}$ such that $E_\pi = -m_e$, as one has $m_\pi > m_e$. For any value of $Z > Z_{cr}^{(1)}$, the "inverse β-decay" reaction :

$$N \rightarrow (N,\pi^-) + e^+ + \nu_e \qquad (11)$$

can now take place.

In reaction (11), N denotes the supercritical ($Z > 68$) point nucleus, (N,π^-) a π^- bound state in the Coulomb field of the nucleus. This instability sets in just before Z reaches $Z_{cr}^{(2)}$, the threshold value at which one has : $E = -m_\pi$. It follows that stable point nuclei with $Z > 68$ cannot exist if there exists a point scalar particle of mass greater than the electron mass.

3. CHARGED SOURCES IN CLASSICAL SCALAR ELECTRODYNAMICS

We saw in Sec. 2 that the K.G. equation for a particle in an external Coulomb field could no longer be interpreted as a single particle equation for $Z > Z_c$, i.e. when complex eigenvalues arise. Obviously, a field theoretic formulation of the problem is needed in the supercritical domain. As is often the case, one has a choice between finding approximate solutions to the Quantum Field Theoretic problem and solving exactly the corresponding classical problem. The Quantum problem was studied by Klein and Rafelski,[12] while Mandula[13] solved the classical field equations. Like Mandula, we shall then study the exact solution to the classical field equations, which are expected to be a good approximation to the Quantum Field Theoretic problem, as we are dealing with bosons.

The Lagrangian density $\mathcal{L}$ describing the interaction of a charged boson field ϕ with the electromagnetic field A^μ in the presence of an externally prescribed current j_μ^{ext} is given by[14]

$$\mathcal{L} = \left(\partial_\mu \phi - ieA_\mu \phi\right)\left(\partial^\mu \phi^* + ieA^\mu \phi^*\right) - m^2 \phi \phi^* - \frac{1}{4}\left(\partial_\mu A^\nu\right)\left(\partial^\mu A_\nu\right) - e j_\mu^{ext} A^\mu \ . \qquad (12)$$

From (12), one gets the following equations of motion (we take $\alpha \equiv e^2/4\pi = 1/137$) :

$$\left(\partial_\mu - ieA_\mu\right)^2 \phi + m^2 \phi = 0 \ , \tag{13}$$

$$\Box A_\mu - \partial_\mu \partial^\nu A_\nu + ie\left(\phi^* \overset{\leftrightarrow}{\partial}_\mu \phi - 2\, ieA_\mu \phi^* \phi\right) = e\, j_\mu^{ext} \ . \tag{14}$$

Taking now

$$j_\mu^{ext} = \rho(r)\delta_{\mu 0} \equiv \frac{eZ\delta\left(r - r_0\right)}{r^2}\, \delta_{\mu 0} \ , \tag{15}$$

where $\rho(r)$ is a static prescribed charged distribution corresponding to a total charge Ze distributed on a shell of radius r_0, and working in the radiation gauge

$$\vec{\nabla}\cdot\vec{A} = 0 \ , \tag{16}$$

one can see that Eqs. (13) and (14) have the trivial Coulomb solution :

$$A_0 = \frac{eZ}{4\pi r} \qquad (r > r_0)$$

$$\tag{17}$$

$$A_0 = \frac{eZ}{4\pi a} \qquad (r < r_0)$$

$$\phi = \vec{A} = 0 \ . \tag{18}$$

The energy density T_{00} corresponding to $\mathcal{L}$ from Eq. (12) is given by :

$$T_{00} = \left|\left(i\frac{\partial}{\partial t} - eA_0\right)\phi\right|^2 + \left|\vec{\nabla}\,\phi\right|^2 + m^2|\phi|^2 + \frac{1}{2}\left(\vec{\nabla} A_0\right)^2 \ . \tag{19}$$

We now wish to show that Eqs. (2) and (3) have solutions with $\phi \neq 0$ whose corresponding energy is lower than the energy corresponding to solutions (6) and (7) above some critical Z value.

As in Ref. 13 we look for spherically symmetric solutions of the form

$$\phi(r) = e^{i\omega t}\, \tilde{\phi}(r) \ , \tag{20}$$

$$A_0 \equiv A_0(r) \ , \tag{21}$$

$$\vec{A} = 0 \ . \tag{22}$$

Upon writing :

$$eA_0 + \omega = \frac{f}{r} \quad , \tag{22}$$

$$\tilde{\phi} = \frac{g}{\sqrt{2}er} \quad , \tag{23}$$

Eqs. (13), (14) and (19) become, respectively,

$$\frac{d^2g}{dr^2} + \frac{f^2 - m^2r^2}{r^2} g = 0 \quad , \tag{24}$$

$$\frac{d^2f}{dr^2} - \frac{g^2}{r^2} f = -\frac{\alpha Z}{r_0} \delta(r - r_0) \quad , \tag{25}$$

$$T_{00} = \frac{f^2g^2}{2e^2r^4} + \frac{1}{2e^2}\left[\left(\frac{g}{r}\right)'\right]^2 + \frac{m^2}{2e^2}\frac{g^2}{r^2} + \frac{1}{2e^2}\left[\left(\frac{f}{r}\right)'\right]^2 \quad . \tag{26}$$

The total energy E associated with T_{00} is of course given by

$$E = 4\pi \int_0^\infty T_{00} r^2 dr \quad . \tag{27}$$

Note that Eqs. (24) and (25) do not really contain two independent parameters m and r_0. The only (dimensionless) parameter is actually mr_0. This can be seen by using $x = mr$ as a new variable in Eqs. (24) and (25).

If we drop the nonlinear term g^2 in (25), then the radial Klein-Gordon equation is recovered from Eq. (24) :

$$\left[\frac{d^2}{dr^2} + (eA_0 - \omega)^2 - m^2\right] \tilde{g} = 0 \quad , \tag{28}$$

where A_0 is given by Eq. (17).

Guided by the corresponding analysis in the massless case[13] and our knowledge of the solutions to the linearized Klein-Gordon equation, we look for solutions to (24) and (25) satisfying the boundary conditions

$$f \underset{r \to \infty}{\simeq} -mr + \alpha Q \quad , \tag{29}$$

$$g \underset{r \to \infty}{\simeq} C \exp\left[-(8m\alpha Qr)^{1/2}\right] \quad , \tag{30}$$

$$f \underset{r \to 0}{\simeq} f_1 r \quad , \tag{31}$$

$$g \underset{r \to 0}{\simeq} g_1 r \ , \qquad\qquad (32)$$

where f_1, g_1, Q, and C are arbitrary constants to be determined, Q being the effective charge of the source. Equation (25) also requires

$$f'\big|_{r=r_0+\varepsilon} - f'\big|_{r=r_0-\varepsilon} = -\frac{\alpha Z}{r_0} \ . \qquad\qquad (33)$$

Continuity of f, g, g' together with (33) will determine the value of the four arbitrary constants.

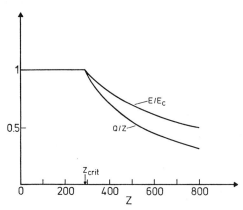

Fig. 2. Variation of the ratio Q/Z as a function of Z ; eQ is the effective charge of the source while eZ is the bare charge of the fixed source (i.e., in the absence of any scalar field). The other curve gives the ratio E/E_c as a function of Z ; E is the energy of the field while E_c is the energy associated with a pure Coulomb solution. Both curves show that for $Z > Z_{crit} = 290$ the partially screened solutions $(Q < Z)$ have lower energy than the Coulomb solution.

Our choice of $\omega = -m$ in (22) is determined by the fact that, as discussed in Sec. 1, we expect a breakdown of the validity of the one-particle interpretation of the K.G. equation for $Z > Z_{cr}$. For numerical convenience, we took $r_0 = 0.8$ fm, $Z_{cr} = 290$, $m = 139.6$ MeV. With these values of our parameters, Fig. 2 shows how the total energy E and total charge of the field vary with Z. One can see that Eqs. (24) and (25) have solutions with $g \neq 0$ that have lower energy than the Coulomb solution $(g = 0)$ for $Z > Z_{cr} = 290$.

The remarkable feature of these solutions is that they correspond to a partial screening of the external source, in contrast with the corresponding result for massless pions, where total screening occurs.[2] Note that our result is in qualitative agreement with that of Klein and Rafelski[12] who used a coherent-state approximation to the quantum-field-theoretic problem. Figure 3 illustrates the behavior of the solutions f and g for $Z > 524 > Z_{crit}$. We note that f is practically linear in a wide range of r. In Fig. 4 we illustrate the importance of various components to the field energy E by decomposing T_{00} in the following way :

$$T_{00} = T_{00}^{em} + T_{00}^{pion} + T_{00}^{int} \ , \tag{34}$$

where :

$$T_{00}^{pion} = \left[\left(\frac{g}{r} \right)' \right]^2 + 2 \, m^2 \frac{g^2}{r^2} \ , \tag{35}$$

$$T_{00}^{em} = \left[\left(\frac{f}{r} \right)' \right]^2 \ , \tag{36}$$

$$T_{00}^{int} = \frac{f^2 g^2}{r^4} - m^2 \frac{g^2}{r^2} \ . \tag{37}$$

This decomposition ensures that for $Z = 0$ T_{00}^{pion} is the free pion field energy density, while, for $g = 0$, T_{00}^{em} is the energy density of the free electromagnetic field. T_{00}^{int} then stands for the interaction energy density.

To conclude, we have found that classical massive scalar electrodynamics with a fixed external source has solutions corresponding to a partial screening of the source for values of the external charge larger than some critical value. These solutions have lower energy than the Coulomb solution, thus implying that, within classical electrodynamics, there exists an upper limit on the charge of stable external sources.

Our previous results have focused on the description of stable solutions to the equations of motion. Let us now study whether there exist solutions with energy smaller than the Coulomb solution for $Z < Z_c$. On very general ground, we expect these solutions to be unstable.[13] Our results[15] are shown in Figs. 5 and 6. They show that indeed partially screened solutions with energy lower than the Coulomb solution exist in the range $Z_0 < Z < Z_c$, where Z_0 is the Z-value at which the (linearized) K.G. equation has zero-energy solutions.

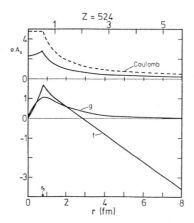

Fig. 3. Shape of the fields f and g in configuration space (lower part) for Z = 524 (> Z_{crit}). Their asymptotic behavior is as in Eqs. (29)-(32). The upper part gives the electric potential field A_0, which is related to f through relation (22) (solid curve). The electric potential field corresponding to the pure Coulomb solution [Eq. (17)] is also shown for comparison. The scale on the top gives the distance r in pion Compton wavelength units.

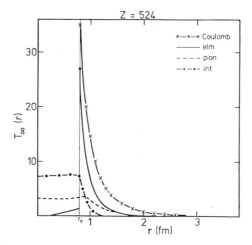

Fig. 4. Behavior of the various components to the energy density T_{00} [Eqs. (35)-(37)], as a function of r, for Z = 524. The energy density corresponding to the pure Coulomb solution [Eq. (17)] is shown for comparison. It vanishes for r < r_0.

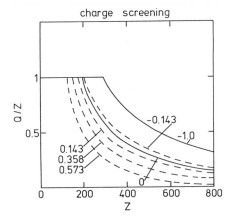

Such solutions are obtained with boundary conditions at infinity :[15]

$$f(r) \underset{r \to \infty}{\simeq} \omega r + \alpha Q \tag{38}$$

$$g(r) \underset{r \to \infty}{\simeq} C \exp\left[-\left(m^2 - \omega^2\right)^{1/2} r\right] . \tag{39}$$

These results suggest that even for $Z_0 < Z < Z_c$, the single-particle interpretation of the K.G. equation might be questionable.

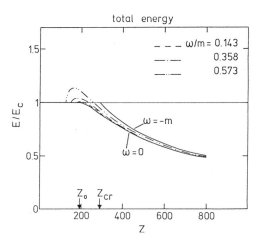

4. STRONG COULOMB COUPLING AND THE RELATIVISTIC TWO-BODY PROBLEM

We saw in Sec. 2 that the Klein-Gordon equation leads to complex energy eigen-value in sufficiently strong Coulomb fields. Our study of the problem of a fixed source in classical electromagnetic interaction with a charged scalar field has then led us to the following simple picture : the occurrence of complex eigenvalues is a signal that the ground state of the system no longer is the bare external source. In sufficiently strong Coulomb fields the ground state corresponds to an external source partially screened by the inter-acting scalar field. It would be of great interest to know whether the same picture also holds for a given particle of finite mass m, spin one-half and charge Z. Indeed, this would imply that single particle states of strongly charged particles would have an energy larger than some multiparticle "condensate" corresponding to the ground state of the system. One would therefore arrive at a picture of "effective confinement" of single particle states, which would be actually excited states of some multiparticle ground state.[16] We shall therefore now study the following problem : does a "reasonable" relativistic two-body wave equation for two spin one-half equal-mass particles in strong Coulomb interaction lead to an energy spectrum which indicates some instability of the system. For instance, we would expect the existence of negative total energy bound states to indicate that the two-body system becomes unstable with respect to spontaneous particle-antiparticle pair creation, whereas our experience with the Klein-Gordon equation would also lead to expect that the existence of complex eigenvalues would be associated with some kind of pair condensation. It is unfortunately impossible to give a clear-cut answer to this problem. Indeed, although there certainly exists a well defined framework for describing relativistic particles in external fields (Klein-Gordon, Dirac and higher spin equations), the situation is quite different for two interacting particles. Although one can in principle formally write a two-body Bethe-Salpeter (B-S) wave equation, its practical solution usually requires making drastic approximations on the form of the interaction kernel, implying loss of gauge invariance, and incorrect static or high-energy limit.[17,18]

In order to see this in some detail, consider the B-S equation for two charged spin one-half particles (with masses m_1 and m_2) in electromagnetic interaction :

$$\left(\gamma_1 p_1 + m_1\right)\left(\gamma_2 p_2 + m_2\right) \psi\left(p_1, p_2\right) = \int K\left(p_1, p_2, q_1, q_2\right) \psi\left(q_1, q_2\right) d^4 q_1 d^4 q_2 \ . \qquad (40)$$

In Eq. (40), the γ 's are the usual Dirac gamma-matrices, and p_i is a 4-vector momentum variable of particle i (i = 1,2), while $\left(\gamma_i p_i + m_i\right)^{-1}$ is its bare propagator. Let us then consider the set of "generalized ladder graphs" describing the electromagnetic interaction (photon exchange) between particles 1 and 2. Schematically, K can be written :

$$K = K_1 + K_2 + K_3 + ... \qquad (41)$$

with :

$$K_1 = \quad\quad (42)$$

$$K_2 = \quad\quad (43)$$

$$K_3 = \quad + \quad + \quad\quad (44)$$

K_1 corresponds to the exchange of one photon, K_2 is the two-photon crossed ladder graph, K_3 contains all three-photon irreducible ladder graphs, and so on.

Suppose we now ask the following question : how many K_i's should we keep in Eq. (40) in order that our B-S equation (40) gives us back the Dirac equation in the static limit $m_2 \to \infty$? The answer is : all of them ![17] The same result holds if one requires Eq. (40) to be gauge invariant and to have the correct high-energy (eikonal) limit.[17,18] In order to avoid problems related to the truncation of the ladder series (41), a great many authors have proposed alternatives to the B-S equation

4.1. Elementary introduction to Relativistic Quantum Constraint Dynamics

We shall now discuss an alternative approach to the relativistic two-body problem, known as Relativistic Quantum Constraint Dynamics. We shall first do so for the simplest case, i.e. two scalar particles with scalar interactions. Furthermore, in this lecture we shall only sketch the main features of this approach and we shall hardly be able to do justice to the mathematical and physical contents of Relativistic Quantum Constraint Dynamics (RQCD) for which we must refer the reader to the literature.[19] The essential idea of RQCD is to describe the interaction between particle 1 (mass m_1) and particle 2 (mass m_2) (both spinless) by means of two Klein-Gordon equations :

$$\left[p_1^2 - m_1^2 - S\left(X_1,X_2\right) \right] \psi \left(X_1,X_2\right) = 0 \ , \quad\quad (45)$$

$$\left[p_2^2 - m_2^2 - S\left(X_1,X_2\right) \right] \psi \left(X_1,X_2\right) = 0 \ . \quad\quad (46)$$

Equation (45) describes the scalar interaction of particle 1 in the field of particle 2, while Eq. (46) describes the interaction of particle 2 in the field of particle 1. $S(X_1,X_2)$ is some Lorentz-invariant function of the coordinates. Now the system (45)-(46) only makes sense if Eqs. (45) and (46) are compatible. This implies :

$$\left[H_1, H_2\right] = 0 \quad , \tag{47}$$

where we define :

$$H_i = p_i^2 - m_i^2 - S\left(X_1, X_2\right) \qquad (i = 1,2) \quad . \tag{48}$$

From condition (47), one finds that S must actually be a function of the invariant $\widetilde{X}^2$ with :

$$\widetilde{X}^2 \equiv X^2 + (X.P)^2/W^2 \quad , \tag{49}$$

where :

$$X = X_1 - X_2 \equiv \left(t_1 - t_2, \vec{r}\right) \quad , \tag{50}$$

$$P = p_1 + p_2 \quad , \tag{51}$$

$$W = \left(P^2\right)^{1/2} \quad . \tag{52}$$

In particular, one finds that $\widetilde{X}^2 = -\vec{r}^2$ in the centre-of-mass system $(\vec{P} = 0)$. The compatibility condition (47) thus determines the form of the Lorentz invariant upon which S may depend. Let us now introduce the additional variables :

$$p = \frac{E_2 p_1 - E_1 p_2}{2W} \quad , \tag{53}$$

$$E_i = \frac{1}{W}\left(p_i \cdot P\right) \qquad (i = 1,2) \quad . \tag{54}$$

Formulas (54) can also be written :

$$E_1 = \frac{W^2 - m_1^2 + m_2^2}{2W} \quad , \tag{55}$$

$$E_2 = \frac{W^2 - m_2^2 + m_1^2}{2W} \quad . \tag{56}$$

From (53), (51), (55), (56), one then finds :

$$p \cdot P = 0 \quad . \tag{57}$$

Formulas (51), (53) reduce to the usual formulas for total and relative momentum in the nonrelativistic limit. From (45), (46), (55), (56), (57), one then finds :

$$\left(H_1 - H_2\right)\psi \equiv 0 \quad . \tag{58}$$

On the other hand, (45) and (46) then lead to the same equation :

$$\left(p^2 - b^2(W) + S\right)\psi = 0 \ , \tag{59}$$

with :

$$b^2(W) = \frac{W^4 - 2W^2\left(m_1^2 + m_2^2\right) + \left(m_1^2 - m_2^2\right)^2}{4W^2} \ . \tag{60}$$

$b^2(W)$ is the invariant square of the relative 3-momentum of two free particles in the C.M. system.

Let us finally introduce :

$$\varepsilon_w = \frac{W^2 - m_1^2 - m_2^2}{2W} \ , \tag{61}$$

so that we have :

$$b^2 = \varepsilon_w^2 - m_w^2 \ , \tag{62}$$

with

$$m_w = \frac{m_1 m_2}{W} \ . \tag{63}$$

In the C.M. system, (59) then writes :

$$\left(\vec{p}^2 + m_w^2 + S\left(|\vec{r}|\right) - \varepsilon_w^2\right)\psi\left(\vec{r}\right) = 0 \ . \tag{64}$$

Note that, because of (57), we have a purely 3-dimensional wave equation, without relative time dependence. This is of course a major computational advantage of Eq. (64) over the B-S equation (40). Furthermore, Eq. (64) enjoys a property not shared by the B-S equation with a truncated ladder series in the interaction kernel : it reduces to the K.G. equation (with a scalar interaction) in the static limit (i.e. when either mass becomes infinite). This can be seen at once from (61) and (64). Equation (64) also has the correct eikonal limit in the high energy region.[17]

4.2. Two spin one-half particles in electromagnetic interaction

In order to extend our formalism to Dirac particles with vector (electromagnetic) interactions, all we have to do in principle is to replace (45) and (46) by their corresponding Dirac counterpart. Now, however, ψ is a 16-component object, so that actual computations are quite involved. Fortunately it turns out that, in the equal mass case, a singlet solution φ obeys the simple equation :[20]

$$\left[\vec{p}^{\,2} + m_w^2 - \left(\varepsilon_w - V\right)^2\right] \varphi = 0 \quad, \tag{65}$$

where V is the Coulomb interaction.

Actually, (65) could have been obtained from the free two-body wave equation by the simple substitution :

$$\varepsilon_w \;\to\; \varepsilon_w - V \tag{66}$$

in complete analogy with the Klein-Gordon equation. One can furthermore establish a connection between the B.S. equation (40) and Eq. (65).[21]

4.3. Strongly coupled Positronium (SCP)

We shall now study bound states solutions to (65) with V again given by : (α is the electromagnetic coupling constant)

$$V = -\frac{\alpha}{r} \qquad (r > r_0) \quad, \tag{67}$$

$$V = -\frac{\alpha}{r_0} \qquad (r < r_0) \quad, \tag{68}$$

where r_0 is an arbitrary cut-off radius.

As pointed out in the introduction, one can hope that strongly coupled QED may give us some insight into short distance QCD and the possible existence of a new phase of QED for sufficiently large values of α. Let us mention that, in spite of its simplicity, Eq. (65) used in the conjunction with $V = -\alpha/r$ yields the correct field-theoretic result for the positronium ground state up to order α^4 in a very straightforward way. This should be contrasted with quite involved standard calculations of positronium ground state based upon the B-S equation.

Solutions to Eq. (65) can be written down immediately as Eq. (65) is formally identical with the S-wave radial Klein-Gordon equation. One gets :[22]

$$\varphi(r) = \frac{A}{r} \, W_{k,\mu}(\rho) \, \frac{\sin \tilde{K} r_0}{W_{k,\mu}(\rho)\big|_{r = r_0}} \qquad (r > r_0) \quad, \tag{69}$$

where $W_{k,\mu}(\rho)$ is Whittaker's function,

$$\varphi(r) = \frac{A}{r} \, \sin \tilde{K} r \qquad (r < r_0) \quad, \tag{70}$$

with :

$$\mu = \left(\frac{1}{4} - \alpha^2\right)^{1/2} \tag{71}$$

$$k = \frac{\varepsilon_w \alpha}{\left(m_w^2 - \varepsilon_w^2\right)^{1/2}} \tag{72}$$

$$\rho = 2\,Kr \tag{73}$$

$$K = \left(m_w^2 - \varepsilon_w^2\right)^{1/2} \tag{74}$$

$$\tilde{K} = \left[m_w^2 - \left(\varepsilon_w + \frac{\alpha}{r_0}\right)^2\right]^{1/2} . \tag{75}$$

Bound state solutions are obtained by matching $r\,\phi(r)$ and its derivative at $r = r_0$:

$$\tilde{K} \cotan \tilde{K}\, r_0 = \left. \frac{\dfrac{d}{dr} W_{k,\mu}(\rho)}{W_{k,\mu}(\rho)} \right|_{r = r_0} . \tag{76}$$

Results are displayed in Fig. 7.

One can see that the energy of the $^1S_0$ state remains positive for all α-values, indicating <u>no instability</u> of the two-body system with respect to spontaneous pair creation. This should actually be not too surprising, as Eq. (65) contains a term $(2\,\varepsilon_w\,A)$ which becomes infinitely repulsive as W becomes arbitrarily small, as shown by formula (61). We can see that the existence of strongly attractive interactions in Eq. (65) must not be necessarily interpreted as a signal of instability with respect to spontaneous pair creation, in contrast with the corresponding situation for Klein-Gordon particles.

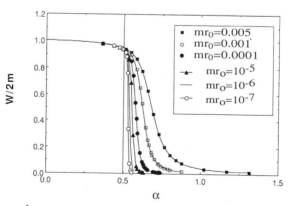

Fig. 7. Energy (W) of $^1S_0$ state of strongly coupled positronium as a function of α for different values of mr_0 .

Curves in Fig. 7 do however show a striking change in the energy scale as α goes from the region $\alpha < \alpha_{cr}$ to the region $\alpha > \alpha_{cr}$, where $\alpha_{cr} \simeq 1/2$ for $mr_0 = 10^{-7}$. As long as α is less than α_{cr}, the mass m of the particle provides the energy scale. For $\alpha > \alpha_{cr}$, there is a sharp variation of W with α. However, for some value of $\alpha = \alpha_0 > \alpha_{cr}$ (but still very close to α_{cr}), W now scales with $m^2 r_0$ and becomes arbitrarily small for any $\alpha > \alpha_0$.

One can illustrate this behavior by means of analytic expressions for W which approximate the exact solution in different α-regions.

(1) For $\alpha < 1/2$, $r_0 = 0$, Eq. (3) can be solved exactly with interaction (6). The ground state energy is given by :[22]

$$W_0^2 = 2m^2 \left\{ 1 + \left[1 + \frac{\alpha^2}{\left(\frac{1}{2} + \sqrt{\frac{1}{4} - \alpha^2} \right)^2} \right]^{-1/2} \right\} . \tag{77}$$

Formula (77) shows that $W_0 \sim 2m$ for $0 < \alpha < 1/2$.

(2) For α slightly larger than 1/2 ($\lambda \equiv \left(\alpha^2 - \frac{1}{4} \right)^{1/2}$ real and small), $\frac{W}{2m} \ll 1$, $m\, r_0 \ll 1$, one can show from (76) that the ground state energy W_0 is given by :

$$W_0 = 2\, \alpha\, m^2\, r_0\, e^{\frac{\pi - 2\gamma}{\lambda}} . \tag{78}$$

where γ is Euler's constant : $\gamma = 0.577....$

(3) For $\alpha \gg 1$, $\frac{W}{2m} \ll 1$, one finds[22] from (76) :

$$W_0 \cong \frac{2\, m^2\, r_0}{\alpha} . \tag{79}$$

A numerical study shows (79) to be already an excellent approximation for $\alpha > 1$. Figure 8 shows that $\dfrac{W}{2\, m^2\, r_0\, \alpha}$ is a bounded function of (α) as mr_0 becomes arbitrarily small. It shows that our formula (78) provides an upper bound for W_0 for all α.

Let us note that mr_0 can be considered a nonperturbative modification of the Coulomb interaction at small distance. This interpretation, however, is not compulsory : the occurrence of an arbitrary scale for $\alpha > 1/2$ can be regarded as a mathematical conse-quence of the self-adjointness of the eigenvalue problem defined by (65).[22]

To conclude, we found that strongly coupled "positronium" as described by RQCD is stable with respect to the spontaneous creation of particle-antiparticle pairs at any finite value of the coupling constant. The new feature, however, is that for any $\alpha > 1/2$, the ground state energy becomes arbitrarily small (but nonzero). As already emphasized, these results are in sharp contrast with corresponding results from the Dirac equation,

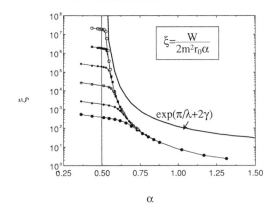

Fig. 8. Values of the ratio ξ (Eq. (78)) as a function of α for the state of strongly coupled positronium and for several values of mr_0 $(10^{-7}, 10^{-6}, 10^{-5}, 10^{-4}, 10^{-3}$ and 5×10^{-3}, respectively from top to bottom). The full line indicates the limiting value of ξ, given by Eq. (78).

even though Eq. (65) reduce to the Dirac equation when the mass of one of the particles becomes infinite. Thus the stability properties of a two-body system with equal masses may be quite different from the unequal mass case, at least if one trusts Eq. (65) in the nonperturbative regime.

ACKNOWLEDGMENTS

This work was supported by the National Fund for Scientific Research (Belgium).

REFERENCES

1. W. Greiner, B. Müller and J. Rafelski, "Quantum Electrodynamics of Strong Fields" Springer-Verlag, Berlin (1981).
2. B. Müller and J. Rafelski, Phys. Rev. D25:566 (1982).
3. A.B. Migdal, Rev. Mod. Phys. 50:107 (1978).
4. J.B. Kogut, E. Dagotto and A. Kocic, Nucl. Phys. B317:253 (1989).
5. D.G. Caldi and A. Chodos, Phys. Rev. D36:2876 (1987) ; D.G. Caldi, A. Chodos, K. Everding, D.A. Owen and S. Vafaeisefat, Phys. Rev. D39:1432 (1989) ; L.S. Celenza, V.K. Mishra, C.M. Shakin and K.F. Liu, Phys. Rev. Lett. 57:55 (1986) ; L.S. Celenza, C.R. Ji and C.M. Shakin, Phys. Rev. D36:2144 (1987) ; Y.J. Ng and Y. Kikuchi, Phys. Rev. D36:2880 (1987) ; Y. Kikuchi and Y.J. Ng, Phys. Rev. D38:3578 (1988).
6. K.M. Case, Phys. Rev. 80:797 (1950).
7. M. Bawin and J. Cugnon, Phys. Lett. 107B:257 (1981).
8. V.S. Popov, Sov. Phys. JETP 32:526 (1971).
9. B. Schroer and J.A. Swieca, Phys. Rev. D2:2938 (1970).
10. V.A. Rizov, H. Sazdjian and I.T. Todorov, Ann. Phys. (N.Y.) 165:59 (1985).
11. P. Gärtner, B. Müller, J. Reinhardt and W. Greiner, Phys. Lett. 95B:181 (1980).

12. A. Klein and J. Rafelski, Z. Phys. A284:71 (1978).

13. J.E. Mandula, Phys. Lett. 68B:495 (1977).

14. M. Bawin and J. Cugnon, Phys. Rev. D28:2091 (1983).

15. M. Bawin and J. Cugnon, Phys. Rev. D37:2344 (1988).

16. J. Rafelski, Phys. Lett. 79B:419 (1978).; A.J.G. Hey, D. Horn and J.E. Mandula, Phys. Lett. 80B:90 (1978).

17. I.T. Todorov, Quasi-Potential Approach to the Two-Body Problem in Quantum Field Theory, in "Properties of the Fundamental Interactions", A. Zichichi, Ed., Editrice Compositori, Bologna (1973) vol. 9, part C, p. 953.

18. R. Barbieri, M. Ciafaloni and p. Menotti, Nuovo Cim. 55A:701 (1968).

19. H.W. Crater and P. Van Alstine, Phys. Rev. D36:3007 (1987).

20. P. Van Alstine and H. Crater, Phys. Rev. D34:1932 (1986).

21. H. Sazdjian, Phys. Lett. 156B:381 (1985) ; J. Math. Phys. 28:2618 (1987).

22. M. Bawin and J. Cugnon, Phys. Lett. B: to be published.

QUANTUM ELECTRODYNAMICAL CORRECTIONS IN CRITICAL FIELDS

G. Soff

Gesellschaft für Schwerionenforschung (GSI), Planckstraße 1
Postfach 110 552, D-6100 Darmstadt, West Germany

1. INTRODUCTION

We investigate field-theoretical corrections, such as vacuum polarization and self energy to study their influence on strongly bound electrons in heavy and superheavy atoms. In critical fields ($Z \simeq 170$) for spontaneous e^+e^- pair creation the coupling constant of the external field $Z\alpha$ exceeds 1 thereby preventing the ordinary perturbative approach of quantum electrodynamical corrections which employs an expansion in $Z\alpha$. For heavy and superheavy elements radiative corrections have to be treated to all orders in $Z\alpha$. The Feynman diagrams for the lowest-order (a) vacuum polarization and (b) self energy are displayed in Fig. 1. The double lines indicate the exact propagators and wave functions in the Coulomb field of an extended nucleus. Fig. 2 shows an $Z\alpha$-expansion of the vacuum polarization graph. The dominant effect is provided by the Uehling contribution being visualized by the first diagram on the right hand side. It is linear in the external field and thus of order $\alpha(Z\alpha)$.

2. VACUUM POLARIZATION OF ORDER $\alpha(Z\alpha)^n$

The influence of the attractive Uehling potential on electronic binding energies for $Z > 100$ has been calculated already by Werner and Wheeler [1]. For the critical nuclear charge number Z_{cr} at which $E_{1s} = -mc^2$ the Uehling potential leads to an 1s-energy shift of $\Delta E_{vp}^{(n=1)} = -11.8$ keV [2], which decreases Z_{cr} by one third of a unit. The remaining vacuum-polarization part in lowest order of the fine-structure constant α but to all orders in $(Z\alpha)^n$ with $n \geq 3$ is more difficult to elaborate. First evaluations of this contribution were presented by Gyulassy [3-5] and by Rinker and Wilets [6-8]. These authors made use of the angular momentum decomposition of the electron propagator in spherical symmetric potentials that was developed by Wichmann and Kroll [9].

Vacuum Structure in Intense Fields, Edited by
H.M. Fried and B. Muller, Plenum Press, New York, 1991

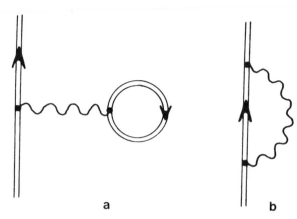

Figure 1. Feynman diagrams for the lowest-order (a) vacuum polarization and (b) self energy.

From bound state QED [10] the energy shift corresponding to the total vacuum polarization is given by

$$\Delta E = 4\pi i \alpha \int d(t_2 - t_1) \int d\vec{x}_2 \int d\vec{x}_1 \, \bar{\phi}_n(x_2)\gamma^\mu \phi_n(x_2) D_F(x_2 - x_1) \, Tr[\gamma_\mu S_F(x_1, x_1)] \tag{1}$$

where the photon propagator is

$$D_F(x_2 - x_1) = \frac{-i}{(2\pi)^4} \int d^4k \, \frac{e^{-ik(x_2-x_1)}}{k^2 + i\epsilon} \tag{2}$$

and the Feynman propagator for the electron can be represented by

$$S_F(x_2, x_1) = \frac{1}{2\pi i} \int_{C_F} dz \sum_n \frac{\phi_n(\vec{x}_2)\bar{\phi}_n(\vec{x}_1)}{E_n - z} \, e^{-iz(t_2-t_1)}$$

$$= \frac{1}{2\pi i} \int_{C_F} dz \, \mathcal{G}(\vec{x}_2, \vec{x}_1, z)\gamma^0 \, e^{-iz(t_2-t_1)}. \tag{3}$$

It obeys the equation

$$[\gamma^\mu(i\partial_\mu - eA_\mu(x_1)) - m] \, S_F(x_1, x_2) = \delta^4(x_1 - x_2) \tag{4}$$

which implies that external field effects are included to all orders. ϕ_n denotes the electron wave function.

The level shift can be expressed as an expectation value of an effective potential U with

$$U(\vec{x}_2) = 4\pi i \alpha \int d(t_2 - t_1) \int d\vec{x}_1 \, D_F(x_2 - x_1) \, Tr[\gamma_0 \, S_F(x_1, x_1)]$$

$$= \frac{i\alpha}{2\pi} \int d\vec{x}_1 \, \frac{1}{|\vec{x}_2 - \vec{x}_1|} \int_{C_F} dz \, Tr \, \mathcal{G}(\vec{x}_1, \vec{x}_1, z). \tag{5}$$

With the vacuum polarization charge density ρ

$$\rho(\vec{x}) = \frac{e}{2\pi i} \int_{C_F} dz \, Tr \, \mathcal{G}(\vec{x}, \vec{x}, z) \tag{6}$$

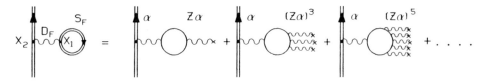

Figure 2. $Z\alpha$-expansion of the vacuum polarization diagram.

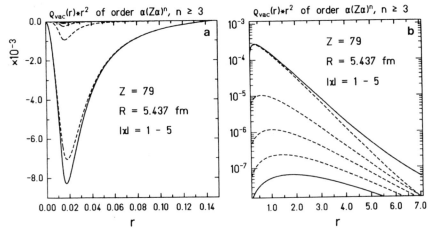

Figure 3. Radial vacuum polarization charge density $\rho_{|\kappa|} \cdot r^2$ of order $\alpha(Z\alpha)^n$ with $n \geq 3$ for the system $Z = 79$ with a nuclear radius $R = 5.437$ fm versus the radial coordinate r in natural units. The various contributions for $|\kappa| = 1, 2, 3, 4$ and 5 are shown separately by the dashed lines. $\rho_{|\kappa|} \cdot r^2$ is given in units of the elementary charge e. The solid line indicates the sum $\rho \cdot r^2$ of the various angular momentum components. a) Linear scale for the range in which the charge density is negative. b) Logarithmic scale to demonstrate the large distance behaviour of $\rho_{|\kappa|} \cdot r^2$. Here the vacuum polarization charge density is positive.

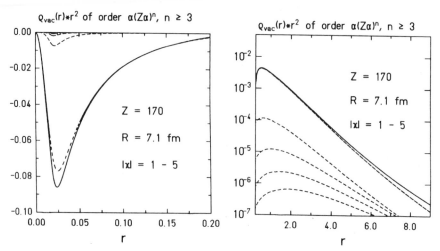

Figure 4. The same as in figure 3 for the almost critical system $Z = 170$ with a nuclear radius $R = 7.1$ fm.

it simply follows

$$U(\vec{x}_2) = -e \int d\vec{x}_1 \frac{\rho(\vec{x}_1)}{|\vec{x}_2 - \vec{x}_1|} \tag{7}$$

The formal expression for ρ still contains the infinite unrenormalized charge. A regularization procedure [5,11] for the total vacuum polarization charge density is to subtract the Uehling contribution which can then be renormalized separately. Expansion of the Green function in eigenfunctions of angular momentum yields for the vacuum polarization charge density of order $\alpha(Z\alpha)^n$ with $n \geq 3$

$$\rho(x) - \rho^{(1)}(x) = \frac{e}{2\pi^2} \int_0^\infty du \left(\sum_{\kappa=\pm 1}^{\pm\infty} |\kappa| \, Re \left\{ \sum_{i=1}^2 \mathcal{G}_\kappa^{ii}(x, x, iu) + \right. \right.$$

$$\left. \left. \int_0^\infty dy \, y^2 \, V(y) \sum_{i,j=1}^2 [\mathcal{F}_\kappa^{ij}(x, y, iu)]^2 \right\} \right) +$$

$$\frac{e}{2\pi} \sum_{\substack{\kappa=\pm 1 \\ -m<E<0}}^{\pm\infty} |\kappa| \left\{ f_1^2(x) + f_2^2(x) \right\}. \tag{8}$$

This equation includes terms from any bound-state pole on the negative real z axis. These terms are picked up as residues in the rotation of the contour of integration. Such terms only appear for superheavy systems where the binding energy of the electron exceeds the electron rest mass. $f_1(x)$ and $f_2(x)$ denote components of the radial Dirac wave function, normalized according to

$$\int_0^\infty dx \, x^2 \, [f_1^2(x) + f_2^2(x)] = 1. \tag{9}$$

$\mathcal{F}_\kappa^{ij}$ are components of the free Dirac Green functions [12,13]. According to Wichmann and Kroll [9] the radial Coulomb Green function components $\mathcal{G}_\kappa^{ij}$ may be represented by solutions of the radial Dirac equation. Expression (8) has been solved numerically.

We have examined the vacuum polarization in the field of a high-Z finite size nucleus. The polarization charge density in coordinate space of order $\alpha(Z\alpha)^n$ with $n \geq 3$ is calculated [11,16]. Energy level shifts of K- and L-shell electrons in hydrogen-like systems are derived. Vacuum polarization corrections to energy eigenvalues of bound leptons have been examined extensively in the past. However, the corresponding influence of vacuum polarization effects on electron levels in atoms has not been completely calculated. One purpose of our work was to carry out a complete calculation of the vacuum polarization of order α in order to provide improved electron binding energy values and to investigate more closely radiative corrections for strongly bound electrons in superheavy systems.

The energy shifts of K- and L-shell electrons in various hydrogen-like systems due to the vacuum polarization of order $\alpha(Z\alpha)^n$ with $n \geq 3$ may be deduced from table 1. The energy correction ΔE usually is expressed via a function ΔF with

$$\Delta E = \frac{\alpha}{\pi} \frac{(Z\alpha)^4}{n^3} \Delta F \, mc^2 \, .$$

n denotes the principal quantum number of the electron state. In table 1 we present the dimensionless quantities ΔF.

Table 1. Higher-order vacuum polarization contributions [11] to the Lamb-shift of K- and L-shell electrons in various hydrogen-like systems. For the nuclear charge distribution we assumed a homogeneously charged spherical shell with a radius R.

$$\Delta F(Z\alpha)$$

System	R[fm]	$1s_{1/2}$	$2s_{1/2}$	$2p_{1/2}$	$2p_{3/2}$
$_{30}$Zn	3.955	0.0020	0.0020	0.0000	0.0000
$_{54}$Xe	4.826	0.0059	0.0064	0.0004	0.0001
$_{82}$Pb	5.500	0.0150	0.0185	0.0035	0.0005
$_{92}$U	5.751	0.0207	0.0272	0.0068	0.0007
$_{100}$Fm	5.886	0.0269	0.0377	0.0118	0.0010
$Z = 170$	7.100	0.518	0.764	3.75	0.017

For fermium we get a noticeable energy shift of about 9 eV for the K_α-line. In conclusion we have computed the vacuum polarization charge density of order $\alpha(Z\alpha)^n$ with $n \geq 3$ for various hydrogen-like systems of the known periodic table of elements. Employing the developped computer code more accurate numbers for the electron Lamb-shift [14,15] in hydrogen-like atoms can be provided.

The computed vacuum polarization charge density for $Z = 79$ (Au) and for the almost critical system $Z = 170$ is depicted in fig. 3 and 4, respectively. The nuclear radius R is indicated. The various contributions for the angular momentum components $|\kappa|$ = 1 - 5 are shown separately. The radial distance r is given in units of the electron Compton wavelength. $\rho_{|\kappa|} \cdot r^2$ is measured in units of the elementary charge e and the inverse Compton wavelength. Part a) shows on a linear scale $\rho_{|\kappa|} \cdot r^2$ in the range where the charge density is negative. The large distance behaviour of $\rho_{|\kappa|} \cdot r^2$ can be taken from figs. 3b and 4b. Please note the logarithmic scale. Here the radial charge density is positive and displays almost an exponential decline in the depicted range. Obviously the ($\kappa = \pm1$) - contribution to the vacuum polarization charge density dominates by about an order of magnitude. A rapid convergence in the κ-summation is indicated. For large distances $(2 \leq r \leq 7)$ $\rho_{|\kappa|} \cdot r^2$ decreases rapidly with different decline constants for the various κ-components. For $Z = 170$ the binding energy of the strongest bound electron state amounts to $E_{1_s} = -1020.895$ keV. The effect of the higher-order vacuum polarization on a K-shell electron in the superheavy system $Z = 170$ results in $\Delta E_{1_s} \simeq 1.46$ keV [11,16], which is completely negligible.

3. VACUUM POLARIZATION CORRECTIONS OF HIGHER ORDER IN MUONIC ATOMS

We also computed the energy shifts in muonic atoms caused by the vacuum polarization of order $\alpha(Z\alpha)^n$ with $n \geq 3$. Nuclear size corrections are taken into account. The calculations are performed for all muonic levels from the $1s_{1/2}$-state up to the $5g_{9/2}$-state in various atoms between $Z = 70$ and $Z = 100$.

The Bohr radius of a bound particle in an atomic orbit is inversely proportional to its rest mass m. Thus to test any deviation from the Coulomb potential at small radial distances it is favorable to measure precisely transition energies of bound muons or pions in heavy atoms. In particular high-lying muonic states, e.g. the $5g_{9/2}$- and the $4f_{7/2}$-state in $_{82}$Pb, are best suited to explore quantum electrodynamical corrections in strong external fields [8]. Despite their small radial expectation values of $< r > \approx 50$ fm these states are hardly influenced by the nuclear extension or by intrinsic degrees of freedom of the nucleus, e.g., nuclear polarization. In addition electron screening corrections play a minor role in these exotic atoms. The vacuum polarization charge density is concentrated close to the nucleus which can be verified by measuring muonic transition energies.

Fig. 5 displays radial probability densities $|\psi r|^2$ in muonic lead. ρ_N indicates the nuclear charge distribution being described by a two-parameter Fermi distribution. The $1s_{1/2}$-muon obviously exhibits a striking overlap with the nuclear interior. The binding energy is extremely sensitive concerning any modification of the nuclear charge distribution. In consequence this state may not be utilized for precision tests of QED. For comparison the dashed line shows $|\psi r|^2$ for a K-shell electron with a radius of about 800 fm. This state has been computed using a Thomas–Fermi potential to account for electron screening. Please note the logarithmic scale for the radial coordinate r. Most important for QED tests are the two muonic states in between, the $5g_{9/2}$- and the $4f_{7/2}$-state. The maximum of their radial probability distribution is located at about 50 fm. The measured transition energy amounts to $\Delta E^{\text{exp}} = 431.353$ keV ± 14 eV. This accuracy allows for high-precision tests of QED in strong fields.

The various QED processes in the interaction of a muon with a nucleus are visualized in fig. 6. The first graph on the right hand side is the ordinary Coulomb interaction. The second graph again represents the Uehling part. The last diagram on the right hand side of the first line as well as the diagrams on the third line are summarized as Källén–Sabry contributions. They are of order $\alpha^2(Z\alpha)$. Their influence on electronic binding energies will be discussed later. The last diagram in fig. 6 of order $\alpha^2(Z\alpha)^2$ represents a Delbrück scattering. For a review of the various contributions we refer to ref. 8. Here we concentrate again on the diagrams in the second line, i.e. on the vacuum polarization of order $\alpha(Z\alpha)^n$ with $n \geq 3$.

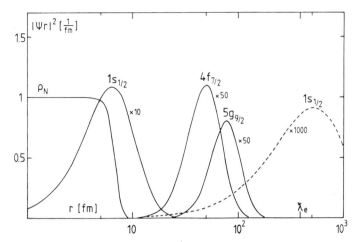

Figure 5. Radial density distributions $|\psi r|^2$ for the muonic $1s_{1/2}-$, $4f_{7/2}-$ and $5g_{9/2}-$state in muonic lead ($Z = 82$) as function of the radial coordinate r in units of fermi. The dashed line plots $|\psi r|^2$ for a K-shell electron. ρ_N indicates the nuclear charge distribution which is described by a two-parameter Fermi distribution.

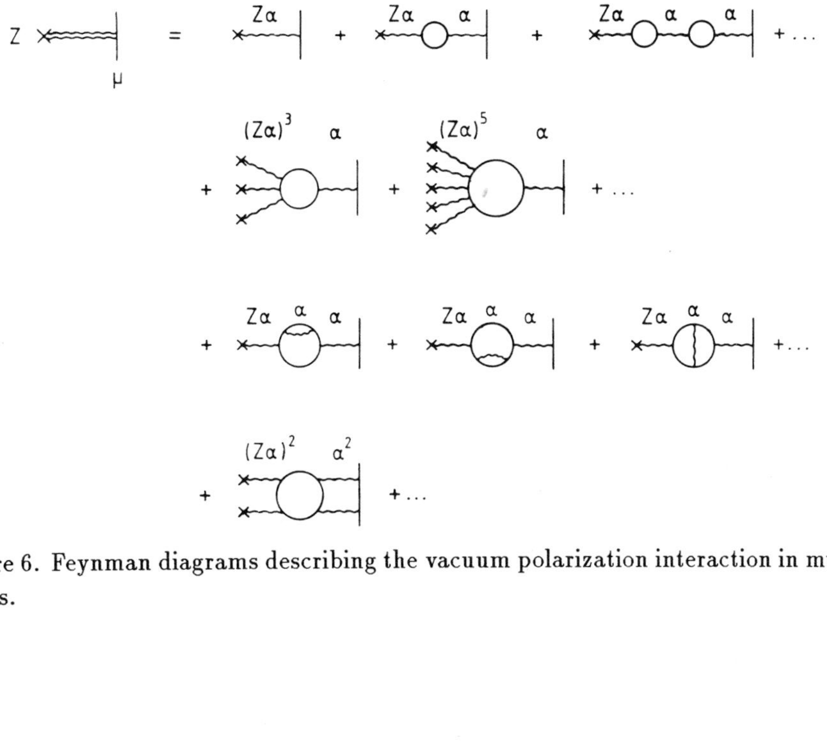

Figure 6. Feynman diagrams describing the vacuum polarization interaction in muonic atoms.

Table 2. Energy corrections in units of eV of muonic bound states caused by the vacuum polarization of order $\alpha(Z\alpha)^n$ with $n \geq 3$. We performed the calculations [17] for all muonic bound states between the $1s_{1/2}$- and the $5g_{9/2}$-state. Muonic atoms ranging from $_{70}$Yb up to $_{100}$Fm are considered.

$$\Delta E \quad (\text{eV})$$

State	$_{70}$Yb	$_{74}$W	$_{79}$Au	$_{82}$Pb	$_{86}$Rn	$_{92}$U	$_{96}$Cm	$_{100}$Fm
$1s_{1/2}$	299.9	354.4	436.0	489.2	562.2	692.8	792.3	900.7
$2s_{1/2}$	130.8	161.9	208.2	239.5	274.1	356.4	420.0	449.0
$2p_{1/2}$	194.4	229.1	296.8	342.3	407.3	521.7	609.3	705.7
$2p_{3/2}$	186.5	230.1	284.0	328.4	392.3	504.3	590.1	684.8
$3s_{1/2}$	65.3	80.4	105.2	122.2	147.3	190.9	227.5	268.8
$3p_{1/2}$	87.1	107.6	141.1	164.1	198.0	256.4	305.1	359.8
$3p_{3/2}$	85.0	106.3	137.7	160.3	193.7	250.8	298.9	352.9
$3d_{3/2}$	96.8	121.9	161.2	188.8	230.4	305.3	365.6	434.3
$3d_{5/2}$	94.0	118.0	155.8	182.2	221.9	293.1	350.6	416.0
$4s_{1/2}$	34.5	43.6	56.9	66.8	81.6	107.1	129.0	154.0
$4p_{1/2}$	43.8	55.7	73.7	85.5	104.7	139.4	165.1	197.2
$4p_{3/2}$	42.9	54.6	72.4	83.9	102.9	137.3	164.6	194.3
$4d_{3/2}$	47.8	61.2	81.9	96.1	118.6	159.7	192.0	230.4
$4d_{5/2}$	46.7	59.7	79.6	93.3	115.1	154.8	185.7	222.8
$4f_{5/2}$	48.1	61.6	82.6	97.5	120.6	163.2	197.5	237.6
$4f_{7/2}$	47.4	60.7	81.2	95.7	118.3	159.7	192.9	231.7
$5s_{1/2}$	19.8	25.0	33.4	39.0	48.1	64.7	77.6	93.3
$5p_{1/2}$	24.5	31.0	41.6	49.0	59.9	80.7	97.5	116.0
$5p_{3/2}$	24.1	30.5	40.9	48.3	59.0	79.7	96.3	114.6
$5d_{3/2}$	26.3	33.7	45.6	54.0	66.8	90.9	109.9	132.7
$5d_{5/2}$	25.8	33.0	44.5	52.7	65.0	88.5	107.4	129.0
$5f_{5/2}$	26.2	33.7	45.7	54.3	67.5	92.3	112.5	135.9
$5f_{7/2}$	25.8	33.2	45.0	53.4	66.3	90.5	110.2	132.9
$5g_{7/2}$	25.5	33.0	44.9	53.4	66.6	91.4	111.6	135.2
$5g_{9/2}$	25.3	32.7	44.4	52.8	65.9	90.2	110.1	133.3

Employing first-order perturbation theory,

$$\Delta E = \int_0^\infty \left(f_1^2(r) + f_2^2(r) \right) U(r) r^2 \, dr , \tag{10}$$

we evaluated the energy correction [17] of muons bound in heavy atoms ranging from $_{70}$Yb up to $_{100}$Fm. In (10) $f_1(r)$ and $f_2(r)$ denote radial components of the bound-state wave function of the muon. Table 2 includes the final energy shifts [17] in units of eV for all muonic levels between the $1s_{1/2}$- and the $5g_{9/2}$-state. The corresponding nuclear radii can be deduced from the table in ref. 14. In muonic lead ($Z = 82$) we obtained as transition energy corrections $\Delta E(5g_{9/2} - 4f_{7/2}) = 42.9$ eV and $\Delta E(5g_{7/2} - 4f_{5/2}) = 44.1$ eV which agree within the quoted uncertainties with the corresponding results published by Borie and Rinker [8] and by Gyulassy [5] . We also investigated modifications of binding energies in pionic atoms caused by the higher-order vacuum polarization potential $U(r)$. Here we have to solve the Klein-Gordon equation incorporating the nuclear Coulomb potential $V(r)$ as well as $U(r)$. For pionic xenon ($Z = 54$) and pionic lead ($Z = 82$) we found ultimately $\Delta E(5g - 4f) = 10.2$ eV and 56.4 eV, respectively.

In conclusion we have presented energy shifts in various exotic atoms caused by higher-order vacuum polarization processes. By comparison with precision experimental data the tabulated numbers may be utilized to test quantum electrodynamics in strong Coulomb fields.

4. THE INFLUENCE OF VACUUM POLARIZATION CORRECTIONS OF ORDER $\alpha(Z\alpha)$ AND $\alpha(Z\alpha)^3$ IN HYDROGEN-LIKE URANIUM

Energy shifts of a bound electron in hydrogen-like uranium caused by vacuum polarization corrections of order $\alpha(Z\alpha)$ and $\alpha(Z\alpha)^3$ are calculated [18]. It is demonstrated that the Wichmann-Kroll correction of order $\alpha(Z\alpha)^3$ dominates for higher electron shells compared with the Uehling contribution.

Usually the Uehling potential provides the dominant vacuum polarization contribution to the Lamb-shift of inner-shell electrons in ordinary atoms as well as in muonic atoms. For large radial distances r from the nuclear charge centre this vacuum polarization potential $V_{11}(r)$ of order $\alpha(Z\alpha)$ displays an exponential decline on a length scale being determined by the electron Compton wavelength ($\lambda_e \approx 386$ fm). Higher-order vacuum polarization corrections were originally discussed by Wichmann and Kroll [9] and later evaluated e.g. in refs. 3 - 8. A striking feature is the asymptotic behaviour of the vacuum polarization charge density of order $\alpha(Z\alpha)^3$, which displays a r^{-7}-dependence at large distances ($r \to \infty$). The corresponding vacuum polarization potential $V_{13}(r)$ declines as [19]

$$V_{13}(r) \stackrel{r \to \infty}{=} \frac{\alpha(Z\alpha)^3}{\pi r} \frac{32}{225} \frac{1}{(2r)^4} . \tag{11}$$

Thus one may obviously expect that the Wichmann-Kroll corrections surpass the Uehling corrections for bound electrons in higher shells. To quantify this insight we computed the corresponding energy shifts for a bound electron in hydrogen-like uranium. The nucleus was assumed to be point-like. The bound-state wave functions have been computed according to Rose [20]. The Uehling potential was evaluated using a representation in terms of modified Bessel functions by Klarsfeld [21]. For the calculation of the vacuum polarization potential of order $\alpha(Z\alpha)^3$ we utilized expressions presented by Blomqvist [22]. Some related technical ingredients are discussed in ref. 14. For radial distances $r > 20\,\lambda_e$ we used the asymptotic form (11).

The computed energy shifts in units of eV are given in table 3. The considered electron levels are signified by the principal quantum number n, the orbital angular momentum quantum number l and by the total angular momentum quantum number j. Already for the 4f-shell the striking long-distance dependence (11) leads to a dominance of the Wichmann-Kroll correction of order $\alpha(Z\alpha)^3$ over the Uehling correction of order $\alpha(Z\alpha)$. However, the tiny absolute value of the energy shifts represents a severe challenge for a possible experimental verification of this exciting QED phenomenon.

Table 3. Energy shifts ΔE_{VP} in units of eV for electron states with quantum numbers n, l and j in hydrogen-like uranium caused by vacuum polarization corrections of order $\alpha(Z\alpha)$ and $\alpha(Z\alpha)^3$.

n	l	j	$\Delta E_{VP}[\alpha(Z\alpha)]$	$\Delta E_{VP}[\alpha(Z\alpha)^3]$
1	0	1/2	-9.800E+01	4.674E+00
2	0	1/2	-1.731E+01	7.748E-01
2	1	1/2	-2.995E+00	1.894E-01
2	1	3/2	-1.265E-01	2.056E-02
3	0	1/2	-5.130E+00	2.272E-01
3	1	1/2	-1.035E+00	6.367E-02
3	1	3/2	-4.831E-02	7.521E-03
3	2	3/2	-1.436E-03	3.748E-04
3	2	5/2	-2.573E-04	1.416E-04
4	0	1/2	-2.117E+00	9.344E-02
4	1	1/2	-4.467E-01	2.726E-02
4	1	3/2	-2.192E-02	3.367E-03
4	2	3/2	-8.775E-04	2.183E-04
4	2	5/2	-1.603E-04	8.118E-05
4	3	5/2	-1.666E-06	3.507E-06
4	3	7/2	-4.336E-07	2.533E-06
5	0	1/2	-1.063E+00	4.683E-02
5	1	1/2	-2.287E-01	1.390E-02
5	1	3/2	-1.155E-02	1.763E-03
5	2	3/2	-5.168E-04	1.262E-04
5	2	5/2	-9.538E-05	4.683E-05
5	3	5/2	-1.450E-06	2.521E-06
5	3	7/2	-3.799E-07	1.753E-06
5	4	7/2	-1.837E-09	1.747E-07
5	4	9/2	-5.696E-10	1.592E-07

5. THE VACUUM POLARIZATION POTENTIAL OF ORDER $\alpha^2(Z\alpha)$

The theoretical values [14,15] for the electron Lamb-shift in hydrogen-like atoms contain uncertainties of various types. The major motivation for our investigations is provided by a possible improvement in the accuracy of these theoretical data. Our investigations dealt with a higher-order vacuum polarization correction which was originally investigated by Källén and Sabry [23] . These authors studied the vacuum polarization process of order $\alpha^2(Z\alpha)$. The corresponding Feynman-diagrams either contain two electron-positron loops or one additional photon line within the ordinary electron-positron loop (cf. fig. 6). The analytical expression for the related vacuum polarization potential was presented by Blomqvist [22] . It yields

$$V_{21}(r) = (Z\alpha) g_2(r). \tag{12}$$

For $r \gg 1$ the potential $g_2(r)$ decreases exponentially [24]. At $r = 20$ a value of 10^{-19} eV is already reached. In first-order perturbation theory the associated energy shifts follow from

$$\Delta E = \int_0^\infty \left(f^2(r) + g^2(r) \right) V_{21}(r) \, r^2 \, dr , \qquad (13)$$

in which $f(r)$ and $g(r)$ denote the small and large component of the relativistic radial wave function, respectively. The energy shift can be represented as

$$\Delta E_{VP} = \frac{\alpha}{\pi} \frac{(Z\alpha)^4}{n^3} \Delta F_{VP} \, mc^2 \qquad (14)$$

The calculated energy shifts of the $1s_{1/2}$-, $2s_{1/2}$-, $2p_{1/2}$- and $2p_{3/2}$-state caused by the vacuum polarization potential of order $\alpha^2(Z\alpha)$ can be deduced from table 1 in ref. 24, in which we tabulate the shift ΔF. As consequence of the attractive interaction the corresponding energy shifts are negative. In particular, we obtain for hydrogen $\Delta F(1s_{1/2}) = -0.00232705$, $\Delta F(2s_{1/2}) = -0.00232713$, $\Delta F(2p_{1/2}) = -3.248 \cdot 10^{-8}$ and $\Delta F(2p_{3/2}) = -9.074 \cdot 10^{-9}$. The vacuum polarization contribution of order $\alpha^2(Z\alpha)$ to the traditional Lamb-shift $E(2s_{1/2}) - E(2p_{1/2})$ in hydrogen amounts to about 236.7 kHz. For the $1s_{1/2}$-state in hydrogen-like uranium we find $\Delta E(1s_{1/2}) \simeq 0.75$ eV, which already represents a sizable binding energy variation. We note, that the energy shift of the $2p_{1/2}$-state is not negligible compared with that of the $2s_{1/2}$-state.

Furthermore we estimated the influence of the vacuum polarization potential of order $\alpha^2(Z\alpha)$ on the energy eigenvalues of the strongest bound electron states in superheavy atoms [24]. To simulate nuclear-size corrections we computed the vacuum polarization potential at about the nuclear radius R and employed this constant value also inside the nucleus. For the nuclear charge distribution we assumed a spherical shell. For the almost critical system Z = 170 with R = 7.1 fm, in which the $1s_{1/2}$-state almost reaches the negative energy continuum, we calculated $\Delta E(1s_{1/2}) = -88.9$ eV. This small value can be completely omitted compared with the huge binding energy of E_{1s}^b = -1020.895 keV .

6. SELF ENERGY

Electronic self energy corrections for high-Z systems were first studied in the pioneering work by Brown and co-workers [25-27]. Their method was further refined and successfully applied in computations of electron energy shifts in high-Z elements by Desiderio and Johnson [28] as well as by Cheng and Johnson [29]. In our calculations we employed these methods, which may be slightly simplified by restriction to K-shell electrons. The energy shift of a $1s_{1/2}$-electron due to the quantum electrodynamical self energy formally is given by

$$\Delta E = 4\pi i \alpha \int d(t_2 - t_1) \int d\vec{x}_2 \int d\vec{x}_1 \, \bar{\phi}_n(x_2) \gamma^\mu S_F(x_1, x_1) \gamma^\nu \phi_n(x_2) D_{\mu\nu}^F(x_2 - x_1) \quad (15)$$

The self-energy correction to be calculated is represented by the Feynman diagram b) in fig. 1. Again the double line indicates the exact electron propagator in the Coulomb field of a nucleus. The next step is to transform propagators and wave functions into momentum space. This admits a decomposition of the self-energy diagram, so enabling infinite mass terms to be identified and removed, leaving the finite observable part of

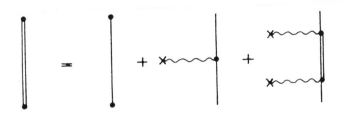

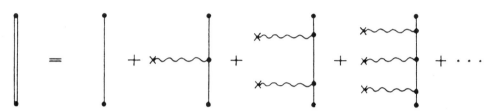

Figure 7. Electron propagator in the external field. Graphical representation of the integral equation for the Coulomb propagator of the electron and the iterated form.

the self energy. We introduce the following Fourier transformations

$$\phi(p) = \int d^4x \, \phi(x) \, e^{ipx} \tag{16}$$

$$A_\mu^{ex}(p) = \int d^4x \, A_\mu^{ex}(x) \, e^{ipx} \tag{17}$$

$$S_F(p_2, p_1) = \int d^4x_2 \int d^4x_1 \, S_F(x_2, x_1) \, e^{-i(p_1 x_1 - p_2 x_2)} \tag{18}$$

The full Feynman propagator in momentum space obeys an integral equation. The result may be represented graphically, where the double line denotes S_F and a single line the free propagator $S_F^{(0)}$. The decomposition of the Feynman propagator may be inserted into the self-energy graph. Calculating the various terms is rather lengthy and not very enlightening. Details of the calculations may be taken from Schlüter [30].

The various existing calculations [12-15,28-32] on the self-energy of K-shell electrons in high-Z systems may be directly compared for mercury ($Z = 80$). The self-energy contribution on the binding energy amounts to about 206 eV. The relative deviation between the different calculations was found to be less than 1%. The obtained energy shifts caused by the self-energy of the strongest bound electron are summarized in fig. 9, where ΔE is plotted versus the nuclear charge number Z. The apparent discrepancy between Mohr's calculation ($\Delta E = 2.586 \pm 0.156$ keV) and our result ($\Delta E = 1.896$ keV) for $Z = 130$ is caused by the neglection of nuclear size effects in ref. [13].

Our most important result was the self-energy shift for 1s-electrons in the superheavy atom with the critical nuclear charge number $Z = 170$. Here the nuclear radius was adjusted so that the K-electron energy eigenvalue differed only by 10^{-3} eV from the borderline of the negative energy continuum. Our numerical calculations [31] for $Z = 170$ yielded $\Delta E_{se} = 10.989$ keV, which still represents only a 1% correction to the total

K-electron binding energy. The sum of all radiative corrections thus almost cancels completely at the continuum boundaries.

We conclude that radiative corrections such as vacuum polarization or self energy may not prevent the K-shell binding energy from exceeding $2mc^2$ in superheavy systems with $Z > Z_{cr} \sim 170$.

7. THE LAMB SHIFT IN HYDROGEN–LIKE ATOMS

With the new GSI – SIS facility it will be posible to produce hydrogen-like high-Z atoms. As a contribution to precision atomic spectroscopy and as a test of quantum electrodynamics in strong fields we evaluated higher–order radiative corrections to the binding energy of inner–shell electrons. In fig. 10 we summarize graphically the contributions to the $1s_{1/2}$ Lamb shift [14]. This figure illustrates the well–known fact that the point–nucleus self energy and the Uehling potential yield the dominant contributions to the Lamb shift for low and intermediate values of Z. The figure also illustrates the fact that nuclear finite size corrections become as important as the self energy toward the end of the periodic table.

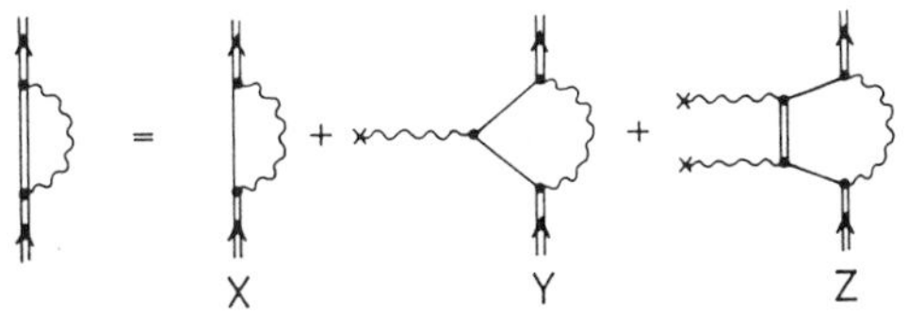

Figure 8. Graphical representation of the self energy in fig. 1b. The upper part of fig. 7 has been inserted. The various terms are denoted by X, Y and Z.

Fig. 11 displays a comparison of our theoretical results for the total 1s Lamb shift in hydrogen–like atoms with available experimental data (cf. e.g. refs. 33-55). The same units as in fig. 10 are employed. The finger points to a very precise measurement [48] of the 1s Lamb shift in a high–Z system. Employing a recoil–ion technique the 1s Lamb shift for hydrogen–like argon ($Z = 18$) could be determined with a relative accuracy of about 1% in fair agreement with the theoretical predictions [14,15].

The ultimate aim of these QED tests for strong Coulomb fields will be a precise determination of the 1s Lamb shift in hydrogen–like uranium. Various considered deviations from ordinary QED corrections e.g. nonlinear extensions of the Dirac equation [56] are expected to be most pronounced in atoms with strong electric fields and high electron densities. However, it was demonstrated [57] that QED tests aiming at utmost

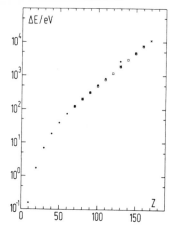

Figure 9. The self-energy shift of K-shell electrons is plotted versus the nuclear charge number Z. ($\bullet \bullet \bullet$) = numerical result of Mohr [13] for 1s-electrons in the Coulomb field of point-like nuclei; ($\square\square\square$) = the computed values of Cheng and Johnson [29]; ($\times \times \times$) = our result [31].

precision are limited by nuclear polarization corrections which amount for the 1s state to about 1 eV in $^{238}_{92}$U compared with a total Lamb shift of about 450 eV.

The precise knowledge of radiative corrections on electron levels is also of definite interest in connection with the prospects of an atomic parity violation experiment in helium–like uranium which ultimately could lead to a more accurate determination of the Weinberg angle. In this system electron wavefunctions display a relatively strong overlap with the nuclear interior which causes a considerable parity violation effect [58] on the almost degenerated electron states $^1S_0$ and $^3P_0$ with opposite parity.

Finally we discuss briefly as a side–remark the influence of the vacuum polarization on nuclear fusion cross sections at astrophysical energies. In heavy–ion scattering the vacuum polarization potential leads to an additional contribution to the Coulomb potential which ultimately results in deviations from ordinary Rutherford or Mott scattering. Subbarrier fusion is extremely sensitive to any correction of the Coulomb potential. Due to the exponential dependence of the tunneling probability on the relative separation between the charge centers tiny changes may cause considerable drifts of the fusion cross section. This may even modify the element synthesis in the universe. As example we show in fig. 12 the ratio $R(E) = \sigma_{\text{stan}}(E)/\sigma_{\text{vac}}(E)$ where σ_{stan} refers to the $^{16}O + ^{16}O$ subbarrier fusion cross section calculated in the standard approach without consideration of vacuum polarization effects. The latter are included in σ_{vac}. The calculations are performed as outlined in ref. 59. For subbarier energies being most relevant for the element synthesis in the universe we obtained modifications of the nuclear fusion cros section in the order of typically 10% - 20%.

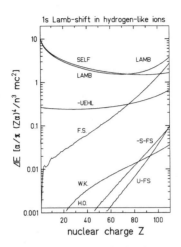

Figure 10. Contributions to the Lamb shift [14] of $1s_{1/2}$ electrons in hydrogen–like atoms versus the nuclear charge number Z. The energy shift ΔE is presented in units of $(\alpha/\pi)\,(Z\alpha)^4/n^3\,mc^2$. LAMB indicates the sum of all contributions. The dominant term (SELF) is provided by the point–nucleus self–energy shift. UEHL denotes the level shift caused by the Uehling potential for point–like nuclei. The energy correction F.S. results from the finite size of the nucleus. The slight irregularities reflect the noncontinuous dependence of the nuclear radius R on the charge number Z. The finite nuclear size correction to the self energy and to the Uehling potential lead to energy shifts S–FS and U–FS, respectively. W.K. denotes the Wichmann–Kroll term and H.O. signifies higher–order corrections incorporating the exchange of two photons. Most of the contributions as well as the total Lamb shift are repulsive. Attractive contributions are indicated by a minus sign.

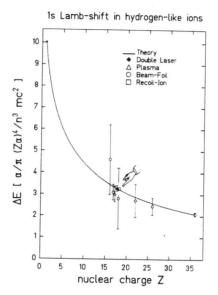

Figure 11. Comparison of theoretical results for the total 1s Lamb shift with available experimental data. The finger points to a precise experimental result of ref. 48.

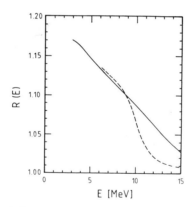

Figure 12. Influence of the vacuum polarization potential on nuclear fusion cross sections for the system $^{16}O + ^{16}O$. Comparison of the ratio of barrier penetrabilities (solid line) with the ratio of cross sections $R(E) = \sigma_{stan}(E)/\sigma_{vac}(E)$ as calculated within the incoming–wave boundary condition model [59].

REFERENCES

1. F.G. Werner, J.A. Wheeler, Phys. Rev. 109, 126 (1958)

2. G. Soff, B. Müller, J. Rafelski, Z. Naturforsch. 29a, 1267 (1974)

3. M. Gyulassy, Phys. Rev. Lett. 33, 921 (1974)

4. M. Gyulassy, Phys. Rev. Lett. 32, 1393 (1974)

5. M. Gyulassy, Nucl. Phys. A244, 497 (1975)

6. G.A. Rinker, L. Wilets, Phys. Rev. Lett. 31, 1559 (1973)

7. L. Wilets, G.A. Rinker, Phys. Rev. Lett. 34, 339 (1975)

8. E. Borie, G.A. Rinker, Rev. Mod. Phys. 54, 67 (1982)

9. E.H. Wichmann, N.M. Kroll, Phys. Rev. 101, 843 (1956)

10. S.S. Schweber, *An Introduction to Relativistic Quantum Field Theory*, (Harper & Row, New York, 1961)

11. G. Soff, P. Mohr, Phys. Rev. A38 (1988) 5066

12. P.J. Mohr, Ann. Phys. (NY) 88, 26 (1974)

13. P.J. Mohr, Ann. Phys. (NY) 88, 52 (1974)

14. W.R. Johnson, G. Soff, Atomic Data and Nuclear Data Tables 33, 405 (1985)

15. P.J. Mohr, Atomic Data and Nuclear Data Tables 29, 453 (1983)

16. G. Soff, Z. Physik D11, 29 (1989)

17. J.M. Schmidt, G. Soff, P.J. Mohr, Phys. Rev. A40, 2176 (1989)

18. G. Soff, P.J. Mohr, Phys. Rev. A40, 2174 (1989)

19. K.-N. Huang, Phys. Rev. A14, 1311 (1976)

20. M.E. Rose, *Relativistic Electron Theory*, (Wiley, New York, 1961)

21. S. Klarsfeld, Phys. Lett. 66B, 86 (1977)

22. J. Blomqvist, Nucl. Phys. B48, 95 (1972)

23. G. Källén, A. Sabry, Mat. Fys. Medd. Dan. Vid. Selsk. 29, 17 (1955)

24. T. Beier, G. Soff, Z. Physik D8, 129 (1988)

25. G.E. Brown, G.W. Schaefer, Proc. Roy. Soc. London Ser. A233, 527 (1956)

26. G.E. Brown, J.S. Langer, G.W. Schaefer, Proc. Roy. Soc. London Ser. A251, 92 (1959)

27. G.E. Brown, D.F. Mayers, Proc. Roy. Soc. London Ser. A251, 105 (1959)

28. A.M. Desiderio, W.R. Johnson, Phys. Rev. A3, 1267 (1971)

29. K.T. Cheng, W.R. Johnson, Phys. Rev. A14, 1943 (1976)

30. P. Schlüter, *Die Diracgleichung in der lokalen Darstellung — Beiträge zur Quantenelektrodynamik starker Felder*, Report GSI-85-15

31. G. Soff, P. Schlüter, B. Müller, W. Greiner, Phys. Rev. Lett. 48, 1465 (1982)

32. P.J. Mohr, Phys. Rev. A26, 2338 (1982)

33. H.W. Kugel, D.E. Murnick, Rep. Prog. Phys. 40, 297 (1977)

34. E. Källne, J. Källne, P. Richard, M. Stöckli, J. Phys. B17, L115 (1984)

35. P. Richard, M. Stöckli, R.D. Deslattes, P. Cowan, R.E. LaVilla, B. Johnson, K. Jones, M. Meron, R. Mann, K. Schartner, Phys. Rev. A29, 2939 (1984)

36. H.F. Beyer, P.H. Mokler, R.D. Deslattes, F. Folkmann, K.-H. Schartner, Z. Physik A318, 249 (1984)

37. S.R. Lundeen, F.M. Pipkin, Phys. Rev. Lett. 46, 232 (1981)

38. H. Gould, D. Greiner, P. Lindstrom, T.J.M. Symons, H. Crawford, Phys. Rev. Lett. 52, 180 (1984)

39. O.R. Wood, C.K.N. Patel, D.E. Murnick, E.T. Nelson, M. Leventhal, H.W. Kugel, Y. Niv, Phys. Rev. Lett. 48, 398 (1982)

40. P. Pellegrin, Y. El Masri, L. Palffy, R. Prieels, Phys. Rev. Lett. 49, 1762 (1982)

41. H.D. Sträter, L. von Gerdtell, A.P. Georgiadis, D. Müller, P. von Brentano, J.C. Sens, A. Pape, Phys. Rev. A29, 1596 (1984)

42. E.S. Marmar, J.E. Rice, E. Källne, J. Källne, R.E. LaVilla, Phys. Rev. A33, 774 (1986)

43. R.D. Deslattes, R. Schuch, E. Justiniano, Phys. Rev. A32, 1911 (1985)

44. P.H. Mokler, Physica Scripta 36, 715 (1987)

45. J.P. Briand, P. Indelicato, M. Tavernier, O. Gorceix, D. Liesen, H.F. Beyer, B. Liu, A. Warczak, J.P. Desclaux, Z. Physik A318, 1 (1984)

46. H. Gould, Nucl. Instr. Meth. B9, 658 (1985)

47. C.T. Munger, H. Gould, Phys. Rev. Lett. 57, 2927 (1986)

48. H.F. Beyer, R.D. Deslattes, F. Folkmann, R.E. LaVilla, J. Phys. B18, 207 (1985)

49. J.D. Silver, A.F. Mc Clelland, J.M. Laming, S.D. Rosner, G.C. Chandler, D.D. Dietrich, P.O. Egan, Phys. Rev. A36, 1515 (1987)

50. D. Müller, J. Gassen, L. Kremer, H.-J. Pross, F. Scheuer, P. von Brentano, A. Pape, J.C. Sens, Europhysics Lett. 5, 503 (1988)

51. J. Gassen, D. Müller, D. Budelski, L. Kremer, H.-J. Pross, F. Scheuer, P. von Brentano, A. Pape, J.C. Sens, Phys. Lett. 147A, 385 (1990)

52. G.W.F. Drake, J. Patel, A. van Wijngaarden, Phys. Rev. Lett. 60, 1002 (1988)

53. C. Zimmermann, R. Kallenbach, T.W. Hänsch, Phys. Rev. Lett. 65, 571 (1990)

54. M.G. Boshier, P.E.G. Baird, C.J. Foot, E.A. Hinds, M.D. Plimmer, D.N. Stacey, J.B. Swan, D.A. Tate, D.M. Warrington, G.K. Woodgate, Phys. Rev. A40, 6169 (1989)

55. J.P. Briand, P. Indelicato, A. Simionovici, V. San Vicente, D. Liesen, D. Dietrich, Europhysics Lett. 9, 225 (1989)

56. D.C. Ionescu, J. Reinhardt, B. Müller, W. Greiner, G. Soff, Phys. Rev. A38, 616 (1988)

57. G. Plunien, B. Müller, W. Greiner, G. Soff, Phys. Rev. A39, 5428 (1989)

58. A. Schäfer, G. Soff, P. Indelicato, B. Müller, W. Greiner, Phys. Rev. A40, 7362 (1989)

59. H.-J. Assenbaum, K. Langanke, G. Soff, Phys. Lett. B208, 346 (1988)

Channeling, Bremsstrahlung and Pair Creation in Single Crystals

Allan H. Sørensen

Institute of Physics, University of Aarhus
DK-8000 Aarhus C, Denmark*   and

Department of Physics, Duke University
Durham, NC 27706, USA

I. Introduction

On the following pages, I shall describe various directional effects associated with the penetration of energetic charged particles through single crystals in directions close to low-index crystallographic directions, that is, close to low-index axes and planes. In particular, I shall focus on the special case of *channeling*: In channeling, the charged particles are so closely aligned with, say, an axis that their motion is completely governed by many correlated collisions with lattice atoms and their flux is prevented from being uniform in the space transverse to the considered crystal direction. In the interaction with a given atomic row or string, a projectile deflects as off a continuum string obtained by smearing the atomic charges uniformly along the string direction ($\hat{z}$). The interaction with the crystal is, in lowest approximation, through a z-independent continuum potential.

Now, why discuss channeling on a school on strong field effects? Well, if you take the viewpoint of an ultrarelativistic channeled electron, say, a few hundred GeV electron, then obviously this particle experiences an immensely strong field as the Lorentz contracted crystal is rushing by. And indeed, if the path of the electron deflected by the atomic string is approximated locally by a segment of a circle and the characteristic frequency for synchrotron radiation is computed, just according to classical electrodynamics, then the corresponding photon energy is easily found to be comparable to the impact energy—or maybe even higher. This is a strong field effect. Similarly, if photons with energies in the hundred GeV region are aimed at an oriented crystal, unusually high rates of conversion into electron-positron pairs are encountered. Or, you may attempt to produce pairs of other light, though more exotic particles as explained in another contribution[1].

With this motivation, I shall in the following Section II go through the basic channeling process. Section III continues by a discussion of channeling radiation and related phenomena. More detailed descriptions as well as detailed reference lists to the literature may be found in the reviews listed as references 2-6 and in the proceedings, ref. 7. The source of those figures originating elsewhere may be found in ref. 6 unless stated explicitly.

---

* Permanent address

Vacuum Structure in Intense Fields, Edited by
H.M. Fried and B. Muller, Plenum Press, New York, 1991

II. Channeling

1. The Continuum Description

Directional effects in aligned single crystals, Fig. 1, appear due to correlation

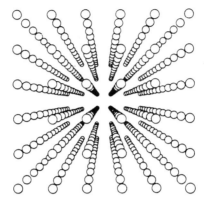

Fig. 1. Axially aligned simple cubic crystal.

between successive soft collisions with target constituents. When a particle moves along, e.g., a low-index axial direction and passes close to one atom experiencing only a slight deflection, it must also pass close to neighboring atoms of the same row.

Let us first consider heavy, positively charged projectiles, Fig. 2a, moving at non-relativistic energies. The angular deflection $\Delta\psi$ associated with the individual binary collisions with target atoms is governed by a screened Coulomb potential, where the screening is due to electrons surrounding the nuclear charge. Each $\Delta\psi$ is small compared to the total deflection, which in itself is small for the major body of the incident beam. The net effect is a gentle steering from many atoms, so that the particle leaves the atomic row or "string" with the same angle as it was approached. (Since the effective action radius of a string is typically smaller than the string interspacing by an order of magnitude, collisions occur in lowest approximation with one string at a time.) With the deflections in the encounters with the individual atoms being very small, the path is well approximated by a straight line during each collision. The corresponding momentum transfer $\Delta\vec{p}$ is then perpendicular to the direction of motion which, in turn, almost coincides with the direction $\hat{z}$ of the string, i.e.,

$$\Delta\vec{p} \simeq -\frac{1}{v}\int_{-\infty}^{\infty} dz\, \vec{\nabla}_{\vec{r}_\perp} V(\vec{r}_\perp, z). \tag{1}$$

Here $\vec{r}_\perp$ denotes the coordinate transverse to the string, and v is the projectile speed. The quantity V denotes the projectile-atom interaction potential (energy). According to Eq. (1), the momentum transfer during the scattering by a single atom may be determined by the integral over the transit time $\Delta t = d/v$ of the force obtained from the potential

$$U(\vec{r}_\perp) = \frac{1}{d}\int_{-\infty}^{\infty} dz\, V(\vec{r}_\perp, z), \tag{2}$$

the distance d being the spacing between atoms on the string. The potential (2) corresponds to that of a charge distribution smeared in the z-direction. Consequently, for motion at small angles to the string, the trajectory of the projectile is governed by the two-dimensional continuum potential obtained by smearing the screened nuclear charges along the z-axis (Figure 2b). The motion through the entire lattice, in turn, is governed by the continuum potential obtained by a superposition of those belonging to single strings.

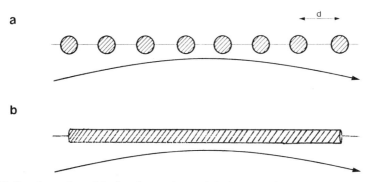

Fig. 2. Deflection of positively charged particle by atomic row upon small-angle impact. (a) Deflection by true row. (b) Same deflection by continuum row.

For energetic ions incident along planar directions in a crystal, a similar continuum approximation applies. In this case the smearing of the atomic charges is over the planes, that is, in two dimensions.

2. Continuum Potentials

For the screened Coulomb potential $V(\vec{r}_\perp, z)$, the so-called "standard" expression has often been used,

$$V(r) = \frac{Z_1 Z_2 e^2}{r} \left\{ 1 - \frac{r}{(r^2 + C^2 a^2)^{1/2}} \right\}, \tag{3}$$

where $Z_1 e$ and $Z_2 e$ denote projectile and target (nuclear) charge, respectively, C^2 is a constant normally set equal to 3 and the screening length a is chosen as $a = 0.89 a_0 (Z_1^{2/3} + Z_2^{2/3})^{-1/2}$, a_0 being the Bohr radius for hydrogen. By insertion of the expression (3) into (2), the standard continuum potential corresponding to a single string is obtained,

$$U(r_\perp) = \frac{Z_1 Z_2 e^2}{d} \ln \left[1 + \left(\frac{Ca}{r_\perp} \right)^2 \right]. \tag{4}$$

The uncertainty in position due to thermal vibrations of target atoms implies a transverse smearing in addition to the longitudinal smearing, Eq. (2). The resulting "thermally averaged" continuum potential is for the axial case given as

$$U_T(\vec{r}_\perp) = \int d^2 \vec{r}'_\perp P(\vec{r}'_\perp) U(\vec{r}_\perp - \vec{r}'_\perp). \tag{5}$$

Within the harmonic approximation for interatomic forces, the probability distribution $P(\vec{r}'_\perp)$ of the target atoms is Gaussian, that is,

$$P(\vec{r}'_\perp) = (\pi \rho^2)^{-1} \exp(-r'^2_\perp / \rho^2), \tag{6}$$

where ρ^2 denotes the two-dimensional mean square thermal displacement from the string. For convenience, we shall immediately drop the subscript T, Eq. (5): All potentials encountered below will be thermally averaged unless otherwise stated.

As a simple approximation to the thermally averaged standard potential for a single continuum string obtained by insertion of the expressions (4) and (6) into Eq. (5), the following analytical form may be used,

$$U(r_\perp) = \frac{Z_1 Z_2 e^2}{d} \ln\left(1 + \frac{C^2 a^2}{r_\perp^2 + \frac{1}{2}\rho^2}\right). \tag{7}$$

Fig. 3a compares this expression to more accurate, frequently used potentials[6]. Note the scale (potential height and width).

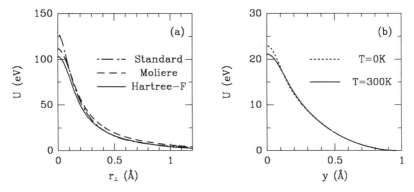

Fig. 3 Continuum potentials ($Z_1 = 1$). (a) Axial potentials for silicon $\langle 111\rangle$ at room temperature. The standard potential (7) is compared to those obtained on the basis of atomic Moliere and Hartree-Fock potentials[6]. (b) Thermally averaged planer potentials (Hartree-Fock) for silicon $\{110\}$. Contributions from nearest neighboring planes are included.

For the planar case the standard continuum potential for a single static plane assumes the form

$$U(y) = 2\pi Z_1 Z_2 e^2 n d_p \left[(y^2 + C^2 a^2)^{1/2} - y\right], \tag{8}$$

where y is the distance from the plane, and d_p the distance between planes. The quantity n is the atomic density. Also for the planes, a thermal averaging should be performed. However, the effect of this smearing is less dramatic than in the axial case since static planar potentials remain finite at the positions of the planes, Fig. 3b (note again the scale).

3. Motion in Continuum Crystal

In the nonrelativistic continuum approximation, there is a complete separation between longitudinal and transverse motion. Since the projectile-string interaction potential (2) is independent of z, the motion parallel to the string is free, i.e., it proceeds with a constant momentum $p_z = \vec{p}\vec{z}$, where $\vec{p}$ denotes the total momentum. As a consequence, the translational energy E_z associated with the longitudinal motion of the projectile of mass M, $E_z = p_z^2/2M$, is conserved. The transverse motion, on the other hand, is governed by U (of the entire lattice). As the total projectile energy E is a constant of motion along with E_z, one is led to the result that the "transverse energy"

$$E_\perp \equiv E - E_z = \frac{p_\perp^2}{2M} + U(\vec{r}_\perp) \tag{9}$$

is a conserved quantity, Fig. 4. Since E in general is very large, compared to U,

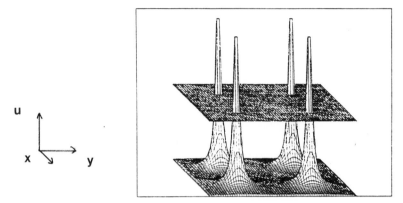

Fig. 4. Four-string continuum potential for Cu $\langle 100 \rangle$ (tops cut). The plane corresponds to given, fixed $E_\perp$; the region of transverse space accessible to particles moving at that $E_\perp$ is defined by the potential not crossing through the plane.

the relation (9) may also be expressed as

$$E_\perp = E\psi^2 + U(\vec{r}_\perp) = \tfrac{1}{2}pv\psi^2 + U(\vec{r}_\perp), \tag{9'}$$

where ψ denotes the local angle to the string, $\psi \ll 1$. According to Eq. (9), $E_\perp$ is composed of the kinetic energy associated with the transverse motion and the potential energy in the interaction with the string. Examples of transverse motion are revealed in Fig. 5.

For particles moving at relativistic speeds, the longitudinal momentum is of course still a constant of motion. As a consequence, the transverse energy, now defined by the relations

$$E_\perp \equiv E - E_z \equiv E - (p_z^2 c^2 + M^2 c^4)^{1/2}, \tag{10}$$

is again conserved. With $p_\perp \ll p_z$ it is easily proven that $E_\perp$ takes the form

$$E_\perp = \frac{p_\perp^2}{2\gamma M} + U(\vec{r}_\perp), \tag{11}$$

where γ is the usual Lorentz factor, $\gamma \equiv (1 - v^2/c^2)^{-1/2}$. Cf. Eq. (9). We may reexpress (11) as

$$E_\perp = \tfrac{1}{2}pv\psi^2 + U(\vec{r}_\perp), \tag{11'}$$

cf. Eq. (9').

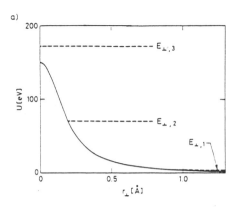
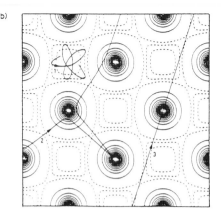

Fig. 5. Transverse motion ($Z_1 = 1$) in $\langle 100 \rangle$ germanium. (a) The potential of an isolated string. (b) Contour plot of the full potential. Examples are sketched in (b) of trajectories corresponding to each of three $E_\perp$-values, (a).

4. Characteristic ("Critical") Angles

One way, although indirect, to observe channeling is to consider the yield of a close-encounter process, that is, a process which requires close contact (on the atomic scale) between projectile and target nuclei (e.g., a nuclear reaction or Rutherford back-scattering). In case $E_\perp$ is somewhat below the potential maximum $U(0)$, the positively charged particle is kept away from the region near, say, the string of high atomic density (Fig. 4-5), whereby the yield of the close-encounter process drops off. On the other hand, if the projectile moves at a transverse energy above barrier, $E_{\perp,3}$ in Fig. 5a, it hits the target atoms essentially as in an amorphous (random) substance. The critical value of $E_\perp$, below which directional effects set in is then $\simeq U(0)$. More precisely, as target nuclei may be encountered off their average position $\vec{r}_\perp = 0$ due to thermal displacements, the slightly lower critical value of $U(\rho)$ could be chosen. Far away from strings, $E_\perp$ is purely kinetic and expressible in the form $E_\perp = \frac{1}{2}pv\psi^2$, where ψ denotes the polar angle of the incident (rather than the local) path relative to the string, cf., Eqs. (9′), (11′). The characteristic angle for directional effects is then $\psi_c \simeq (U(\rho)/\frac{1}{2}pv)^{1/2}$. For any reasonable choice of potential, this result approximates Lindhard's critical angle ψ_1,

$$\psi_c \simeq \psi_1 \equiv \left[\frac{4Z_1 Z_2 e^2}{pvd} \right]^{1/2}. \tag{12}$$

Critical angles ψ_p for planar effects can be obtained in the same way;

$$\psi_p = (4Z_1 Z_2 e^2 n d_p C a/pv)^{1/2}. \tag{13}$$

The planar continuum potential is shallower than that for axes, Fig. 3, resulting in narrower critical angles—typically $\psi_p \sim \psi_1/3$. In the planar case, $E_\perp < U(0)$ corresponds to confinement to one channel.

Figure 6 shows an experimental recording of the backscattering yield for 480 keV protons incident on a tungsten crystal as a function of the angle of incidence to the $\langle 100 \rangle$ axis. The yield is normalized to that measured at incidence far from any

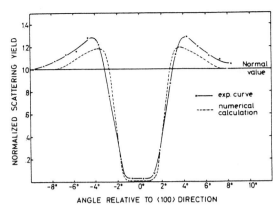

Fig. 6. Backscattering of 480 keV protons in $\langle 100 \rangle$ tungsten[6].

low-index crystallographic directions. A dip appears around zero angle with a half-width close to ψ_1, which amounts to a few degrees in the present case. The maximum extinction approaches two orders of magnitude. For parallel incidence, the only projectiles, which may contribute to close-encounter processes, are those entering the crystal within a distance ρ from the string. The transverse area per string amounts to $(nd)^{-1} \equiv A_0 \equiv \pi r_0^2$. Consequently, the normalized minimum yield for close encounters may be given as

$$X_{\min} = nd\,\pi\rho^2. \tag{14}$$

This rough estimate is in agreement with detailed calculations (see below). The dashed curve in Figure 6 constitutes the detailed theoretical prediction. Clearly the all-over agreement with the empirical data is very satisfactory.

5. Distribution in Transverse Space & Yields

Detailed determination of yield versus angle (cf. Fig. 6) requires knowledge of projectile and target atom distributions in transverse space as well as of the impact-parameter dependence of the considered process. Let us assume the latter to be given. The target-atom distribution is Gaussian, eq. (6), so the quantity to determine is the projectile distribution.

Our strategy in determining the spacial distribution of projectiles is first to determine how they are distributed over transverse energy and then, for given $E_\perp$, determine the spacial distribution. A particle incident at the crystal at an angle ψ_{in} to the considered axial or planar direction enters the crystal with a transverse kinetic energy of $\frac{1}{2}pv\psi_{in}^2$. Entrance at transverse position $\vec{r}_{in}$ further gives it a potential energy $U(\vec{r}_{in})$. Altogether, for this particle the transverse energy (just inside the crystal surface) assumes the value

$$E_\perp = \tfrac{1}{2}pv\psi_{in}^2 + U(\vec{r}_{in}). \tag{15}$$

The distribution of the incident beam over $(\vec{r}_{in}, \psi_{in})$ determines the $E_\perp$-distribution. (The distribution over $\vec{r}_{in}$ is obviously uniform on the atomic scale. Consequently most particles enter the crystal with low potential energy, cf. Fig. 4, whereby the $E_\perp$- distribution peaks at $\frac{1}{2}pv\psi_{in}^2$.)

For an axially channeled beam of particles initially centered around a given transverse momentum, there will be a trend towards statistical equilibrium in transverse phase space. Equal probability in the allowed region of transverse phase space in turn leads to equal probability $P_{E_\perp}(\vec{r}_\perp)$ for finding a particle of given $E_\perp$ anywhere inside the accessible area $A(E_\perp)$ in direct space. Hence

$$P_{E_\perp}(\vec{r}_\perp) = \frac{1}{A(E_\perp)}\Theta(E_\perp - U(\vec{r}_\perp)), \qquad (16)$$

where $\Theta(x)$ is the Heaviside function. The obvious advantage of application of Eq. (16) is that the cumbersome task of calculation of many actual projectile trajectories is avoided.

For the planar case we may immediately write $P_{E_\perp}(y)$ down as

$$P_{E_\perp}(y) \propto \left(\frac{E_\perp}{E_\perp - U(y)}\right)^{1/2}\Theta(E_\perp - U(y)), \qquad (17)$$

since the motion is in one dimension only.

6. Dechanneling

This is not the whole story, though: If you start measuring, e.g., minimum yields, you will quickly find out that, in general, these are higher than theoretical predictions. Why? An important reason is that $E_\perp$- conservation is not exact. On its way through the crystal the projectile has the chance to scatter on point charges (as opposed to the smeared charges in the continuum model)—that is on electrons or (near axes or planes) on thermally displaced nuclei which just by chance happen to be in its way. An alternative way of saying this is that the projectile motion as determined in the continuum model is perturbed by the difference between the true crystal potential (sum of potentials due to point charges) and the continuum potential. In any case, the result of the scattering is that $E_\perp$ on average increases during the passage of the crystal.

To get reliable yields, the z-dependence of the $E_\perp$-distribution need be included in the calculations outlined in the previous section. Doing so, however, it is possible to get good agreement between theory and experiment both in the axial and planar case and for a wide range of crystal thicknesses.

7. Doughnuts

An effect closely related to the trend towards statistical equilibrium discussed in Section 5 is the formation of "doughnuts": When a particle beam enters a crystal at at specific angle close to a string, Fig. 7a, a ring-shaped distribution is observed at the exit side, Fig. 7b, provided the crystal is not too thin. The particles lose their sense of direction in transverse space due to collisions with many different strings (cf. Fig. 5b). Experimental observation of doughnuts is the most direct indication of trend towards equilibrium. Note that the width of the ring, Fig. 7b, is determined partly by dechanneling (non-conservation of $E_\perp$) and by the fact that particles in general enter and exit the crystal at different $\vec{r}_\perp$.

One further very interesting experimental observation is that at GeV energies doughnut formation takes place even far outside the channeling region (out to 10-20 ψ_1). This indicates validity of the continuum model far beyond ψ_1.

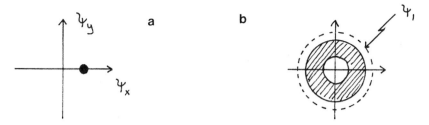

Fig. 7. Incident (a) and exit (b) distributions in angular space.

8. Negative versus Positive Particle Channeling

When negatively charged particles are incident on single crystals, the interaction potential appears equal in shape but of opposite sign as compared to the potential governing the motion of positively charged particles. In the continuum picture, negative particles experience potentials with minima at the center-of-strings or planes. A schematic picture for a planar case is shown in Figure 8.

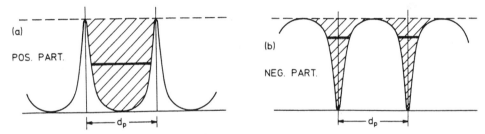

Fig. 8. Planar continuum potentials for particles of (a) positive and (b) negative charge. The channeling region is hatched and typical $E_\perp$-levels for channeled particles are indicated.

Negative, channeled particles are focussed around the nuclei, in contrast to positive particles that are steered away. Hence, in channeling, negative projectiles have increased probability for close encounter processes; the dips in yield observed for close alignment of positive particle beams, Fig. 6, turns into peaks in yield for beams of negatively charged particles. A further consequence of the more frequent contact with target nuclei obviously is a much more severe dechanneling (especially in the axial case) for negative particles than for positive.

The difference in channeling motion for particles of opposite charge is most dramatic in the axial case. Here, positive particles move freely from channel to channel, whereas negative ones are bound to one single string (or, in a few special cases, a pair of strings). Furthermore, the angular momentum for a negatively charged particle channeling around a string is very nearly conserved as the continuum potential it experiences is very nearly rotationally symmetric (unless it is very loosely bound). With angular momentum conservation the distribution (16) cannot a priori be assumed valid.

9. Quantum versus Classical Channeling

So far, we have discussed the channeling phenomena on the basis of a classical description of the particle motion. In order to embark on the question whether such a description is adequate, or whether quantum corrections are essential, consider for the moment a projectile of negative unit charge, $Z_1 = -1$, channeling in an axial direction in a single crystal. Let us assume that it moves at a transverse energy, say, halfway between the potential maximum, which we set to zero, and the potential minimum $-U_0$, i.e. $E_\perp = -U_0/2$. In this case, the transverse excursions from the average string position are less than or about equal to the atomic-screening distance, $r_\perp \lesssim a$. On the other hand, for the transverse momentum, we have $p_\perp^2 \lesssim M\gamma U_0$. Hence,

$$r_\perp p_\perp/\hbar \lesssim (a^2 M\gamma U_0/\hbar^2)^{1/2} = \frac{a}{a_0}\left(\frac{U_0}{e^2/a_0}\right)^{1/2}\left(\gamma\frac{M}{m}\right)^{1/2}, \qquad (18)$$

where m denotes the electron mass and $a_0 = \hbar^2/me^2$ (the Bohr radius). With $a = 0.9a_0 Z_2^{-1/3}$ and $U_0 = Z_2 e^2 \Lambda/d$, where Λ is a numerical factor typically of order 4, we obtain

$$r_\perp p_\perp/\hbar \lesssim \left(Z_2^{1/3}\frac{a_0}{d}\Lambda\right)^{1/2}\left(\gamma\frac{M}{m}\right)^{1/2} \sim \left(\gamma\frac{M}{m}\right)^{1/2} = \left(\frac{E_{\text{tot}}}{mc^2}\right)^{1/2}. \qquad (19)$$

The two expressions to the far right constitute "factor-of-two" estimates. It is now obvious that only for electrons, the product $r_\perp p_\perp$ may approach the quantum limit, that is, only for electrons, a quantal description of the channeling process may be required.

An alternative and slightly more detailed way to judge whether it is essential to treat the channeling process quantum-mechanically is to ask for the number of quantum states corresponding to bound transverse motion, that is, to ask for the number ν of states of transverse energy below the maximum of the continuum potential. The classical description of channeling is approached when the discrete, quantal spectrum of bound states becomes sufficiently dense (high ν) and hence approximates the continuous classical spectrum. An estimate of ν may be obtained as the number of cells of volume h^2, respectively h (Planck's constant), in the classically available phase space specific to a string, respectively a plane, and corresponding to $E_\perp < \max U(r_\perp)$. This procedure leads for negatively charged particles to the (factor-of-two) estimates

$$\text{string:} \quad \nu \sim |Z_1|\frac{\gamma M}{m}$$

$$\text{plane:} \quad \nu \sim \left(|Z_1|\frac{\gamma M}{m}\right)^{\frac{1}{2}}. \qquad (20)$$

For positively charged particles the scaling is the same; the absolute numbers are just higher since positive particles have more space available, cf. Fig. 8.

From the above estimates (20) it is now obvious that the number of bound states of transverse motion is high for heavy projectiles $M \gg m$ at all energies. This justifies the classical treatment of the motion of such particles under channeling conditions. Similarly, a classical treatment of the particle motion is justified for electrons and positrons at GeV energies, $\gamma M/m = \gamma \gg 1$. However, for electrons and positrons channeling at a few MeV, quantum effects may be important.

10. Quantum Channeling

The quantal motion of a relativistic electron is described by the Dirac equation. However, in the analysis of electron channeling, we may apply the simpler Klein-Gordon equation since spin effects turn out to be unimportant. Separate now from the electronic wave function Ψ a rapidly oscillating factor $\exp(ikz)$ corresponding to uniform motion in the channeling direction with energy E,

$$\Psi(\vec{r}_\perp, z) \equiv e^{ikz} w(\vec{r}_\perp, z), \qquad E^2 = (\hbar k c)^2 + m^2 c^4. \tag{21}$$

In the limit of small-angle scattering, the stationary Klein-Gordon equation for Ψ then reduces to a Schrödinger-like equation for the remaining function w,

$$i\hbar \frac{\partial}{\partial t} w(t, \vec{r}_\perp) = H w(t, \vec{r}_\perp), \qquad H(\vec{r}_\perp, t) = -\frac{\hbar^2}{2m\gamma} \Delta_{\vec{r}_\perp} + V(tv, \vec{r}_\perp). \tag{22}$$

In Eq. (22), the time-like coordinate t is defined as $t = z/v$. The final approximation consists in replacing the full crystal potential $V(tv, \vec{r}_\perp)$ by the (thermally averaged) continuum potential $U(\vec{r}_\perp)$. Solutions to Eq. (22) are then of the form

$$w(t, \vec{r}_\perp) = u(\vec{r}_\perp) e^{-iE_\perp t/\hbar} \tag{23}$$

where $u(\vec{r}_\perp)$ is an eigenfunction of H, that is, $u(\vec{r}_\perp)$ is a solution to the "stationary Schrödinger equation"

$$\left[-\frac{\hbar^2}{2m\gamma} \Delta_{\vec{r}_\perp} + U(\vec{r}_\perp) \right] u(\vec{r}_\perp) = E_\perp u(\vec{r}_\perp), \tag{24}$$

corresponding to a well-defined transverse energy $E_\perp$. Clearly Eq. (24) is nothing but the quantized version of Eq. (11).

As an example, Fig. 9a displays the spectrum of bound transverse quantum states obtained by solving the eigenvalue problem (24) for electrons with a kinetic energy of 4 MeV channeled along the $\langle 111 \rangle$ axis of a silicon single crystal at room temperature. In the calculation, the influence of neighboring strings has been neglected. Hence the potential, being that of an isolated string, is rotationally symmetric, $U = U(r_\perp)$, and the eigenfunctions $u(\vec{r}_\perp)$ may be separated in polar coordinates $(r_\perp, \varphi)$ as

$$u(\vec{r}_\perp) = R_{n\ell}(r_\perp) \frac{1}{\sqrt{2\pi}} e^{\pm i\ell\varphi}; \qquad \ell = 0, 1, 2, \ldots, \tag{25}$$

where n is a principal quantum number. We copy the usual notation from atomic spectroscopy by labelling states with $\ell = 0, 1, 2, \ldots$ as $s, p, d, \ldots$ states. Taking into account the degeneracy of non-s states, we find a total of 8 bound states for the present case in very good agreement with the estimate (20). Fig. 9b shows the radial component $R_{n\ell}(r_\perp)$ of the wave function $u(\vec{r}_\perp)$ for the levels of Fig. 9a. In the representation (25), the electron density $|u|^2 = R_{n\ell}^2 / 2\pi$ is independent of angle. In the alternative real wave function representation ($\exp(\pm i\ell\varphi) \to \cos \ell\varphi, \sin \ell\varphi$), the density u^2 shows both radial and angular variations for non$-s$ states, that is, $u^2 = \pi^{-1} R_{n\ell}^2 \cos^2 \ell\varphi$ or $u^2 = \pi^{-1} R_{n\ell}^2 \sin^2 \ell\varphi$.

Clearly, the higher-lying and thereby somewhat delocalized levels in Fig. 9a are influenced by the presence of nearby strings. This is revealed by Fig. 9c which shows the level scheme for 4 MeV electrons incident parallel to the $\langle 100 \rangle$ silicon axis at room temperature. The results of the single-string calculation (left) are compared to results of calculations where neighboring strings are included: a "tight-binding" calculation (middle), copied from usual three-dimensional solid-state physics, including the influence of the four nearest neighboring strings and, a "many-beam" calculation (right) based on expansions on plane waves as in diffraction theory. Apart from a shift in energy corresponding to the difference in the minimum value of the potential (which we could have defined away), we see that the lowest-lying states ($1s$, $2p$) are well determined in the single-string model. Less tightly bound levels, on the other hand, are shifted in position (relative to, e.g., the $1s$ state), and they split up because of loss of complete rotational symmetry. The insert reveals the band structure, that is, the variation with incidence angle of the higher-lying levels.

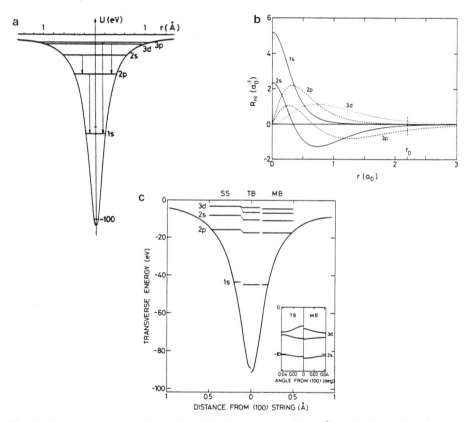

Fig. 9. Quantum channeling; 4 MeV electrons in silicon[6]. (a) Single $\langle 111 \rangle$ string bound states (with radiative dipole transitions indicated). (b) Transverse $\langle 111 \rangle$ wavefunctions. (c) Influence of neighboring strings in $\langle 100 \rangle$ direction.

In Fig. 10 we display another example; the energy bands for 4-MeV electrons penetrating a silicon single crystal close to one of the major planes. The bands have been computed by means of the many-beam technique which is particularly simple and fast for the planar case. Both bound and free levels are shown in the figure.

When electrons are incident on a single crystal in a direction close to that of a major axis or plane, each individual transverse state is populated at a frequency determined by the overlap integral at the target surface of the incident wave with the wave function of the transverse state in question. However, an electron does not necessarily stay in the state it originally chose until it leaves the target on the backside; it may dechannel: During the penetration of the real non-continuum crystal, the electron has a fair chance for suddenly encountering closely a single, thermally displaced target atom. The resulting incoherent, large-angle scattering event disrupts the channeling motion and causes the projectile to dechannel or, in less abrupt cases, to transfer to another channeling state. The lengths traversed during the lifetime for all but the very short-lived ground state ($1s$ is concentrated near atomic positions) are of the order of 1 μm for the cases displayed in Fig. 9.

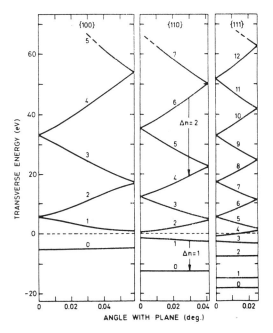

Fig. 10. Transverse energy bands for 4 MeV electrons incident on silicon planes[6]. Typical radiative transitions are identified.

This completes our discussion of channeling.

III. Radiation

As is well known from classical physics, an electrically charged particle emits electromagnetic radiation while accelerated. The instantaneously emitted power P is proportional to the square of the acceleration, that is, $P \propto (F/M)^2$ where F is the force acting at the particle. In the relativistic case for, e.g., transverse acceleration, we find $P \propto (\gamma F/M)^2$. These results immediately imply that only the very lightest particles, electrons and positrons, are of practical interest as radiators.

1. Bremsstrahlung

The radiation emitted by a charged particle penetrating an atomic field is known as bremsstrahlung. At high velocities, the emission process requires a quantum-mechanical description and, in the Born limit $Z\alpha \ll 1$, the cross section for photon emission at energy $\hbar\omega$ upon relativistic electron/positron impact at total energy E reads approximately

$$\frac{d\sigma}{d\hbar\omega} = \frac{16}{3} Z^2 \alpha r_e^2 \frac{1}{\hbar\omega} \left(1 - \frac{\hbar\omega}{E} + \frac{3}{4} \left(\frac{\hbar\omega}{E} \right)^2 \right) \ln(183 Z^{-1/3}). \tag{26}$$

Here $r_e = e^2/mc^2 = \alpha^2 a_0$ denotes the classical electron radius, α is the fine-structure constant, and $Z \equiv Z_2$. The result holds when the atomic (Thomas-Fermi) screening is "complete" which requires $\gamma \gg 1$. Otherwise, the argument of the logarithm is changed. At moderate values of γ, Eq. (26) still holds for soft photons, $\hbar\omega/E \ll 1$. As it stands, the expression (26), which we shall call the Bethe-Heitler cross section, covers the radiation emitted as a result of scattering in the screened field of the atomic nucleus. Furthermore, radiation may be emitted as a result of direct scattering on target electrons. In order of magnitude, this leads to an extra contribution to the cross section of $1/Z$ times the nuclear one, (26).

In addition to the dependence on photon energy, which is essentially as $1/\hbar\omega$ according to Eq. (26), also the dependence of the bremsstrahlung-emission probability on photon-emission angle is of interest: At high energies, photons are emitted in a narrow cone of opening angle $\sim 1/\gamma$ around the initial direction of motion of the projectile.

For a high-energy electron or positron ($\gtrsim 1$ GeV for most materials), the major cause of slowing down is by far emission of bremsstrahlung. It is convenient for later use to introduce the so-called radiation length ℓ_r which is defined as the penetration depth over which such a projectile on an average loses all but a fraction $1/e$ of its initial energy. From Eq. (26) we find that the total stopping cross section $S \equiv \int \hbar\omega \, d\sigma$ is simply proportional to the primary energy, $S = \sigma_0 E$, where σ_0 is energy independent, and the radiation length therefore is simply given as $\ell_r = 1/n\sigma_0$, that is, for a homogeneous (amorphous) medium, we have

$$\ell_r^{-1} = 4Z^2 \alpha n r_e^2 \ln(183 Z^{-1/3}). \qquad (27)$$

Corrections to (27) are of the same relative order as to (26), namely $1/Z$ and $(\alpha Z)^2$. Furthermore, for very light targets, $Z \lesssim 5$, screening should be described more accurately than by the statistical Thomas-Fermi model. Examples of radiation lengths for materials of interest in the present context are 12.2 cm in diamond, 9.36 cm in silicon, 2.30 cm in germanium and 0.35 cm in tungsten.

2. Coherent Bremsstrahlung

When an electron or a positron penetrates a single crystal in a direction close to a major crystallographic axial of planar direction, it has a fair chance to scatter coherently on many target atoms along its way. Channeling, of course, is an example of this. The coherence in scattering is carried on to the radiation emission and, consequently, the emission in a single crystal may greatly exceed the bremsstrahlung emission in a similarly dense but amorphous medium composed of the same type of atoms. We shall study the radiation emission during channeling in the sections following below. In this section, however, let us consider the emission process in the limit where the projectile interaction both with the radiation field and with the crystal atoms may be treated in the Born approximation. According to custom, we call this the case of "coherent bremsstrahlung".

In the perturbation limit, the bremsstrahlung-emission probability is proportional to the square of a second-order matrix element which, in turn, is composed as a sum of terms, each proportional to the product of two first-order matrix elements, one for the interaction with the radiation field and one for the scattering in the atomic field. Since the projectile states are plane waves, we have immediately

$$d\sigma \propto \left| \int V(r) e^{i\vec{q}\vec{r}} d^3\vec{r} \right|^2 \qquad (28)$$

where V is the atomic potential and $\hbar\vec{q}$ is the recoil momentum of the atomic nucleus. In case we consider bremsstrahlung production upon incidence on a total of N atoms, the above potential V needs to be replaced by the total interaction potential

$$V \longrightarrow \sum_n V(\vec{r} - \vec{r}_n). \tag{29}$$

Here $\vec{r}_n$ denotes the position of the n'th atom. In applying the expression (29), we neglect changes appearing in the potential as a result of rearrangements of the outermost parts of the atomic electron cloud which results from mutual atom-atom interactions. This is justified since the target electrons essentially only enter in the screening of the nuclear charge. Substituting the sum (29) for the interaction potential into Eq. (28), we get the following relation between the differential cross sections for photon production at energy $\hbar\omega$ on an isolated atom and on the group of N atoms,

$$\left.\frac{d\sigma}{d\hbar\omega\, d^3\vec{q}}\right|_{N \text{ atoms}} = \left.\frac{d\sigma}{d\hbar\omega\, d^3\vec{q}}\right|_{\text{single atom}} \times \left|\sum_n e^{i\vec{q}\vec{r}_n}\right|^2. \tag{30}$$

For an amorphous medium, the last factor yields just N. On the other hand, for a perfect static single crystal, we find in the limit $N \to \infty$

$$\left|\sum_n e^{i\vec{q}\vec{r}_n}\right|^2 = N\frac{(2\pi)^3}{N_0\Delta}\,|S(\vec{g})|^2 \sum_{\vec{g}} \delta(\vec{q} - \vec{g}). \tag{31}$$

Here $\vec{g}$ denotes a reciprocal lattice vector, $S(\vec{g})$ the structure factor, and N_0 is the number of atoms contained in a unit cell which has the volume Δ. This result is well known from diffraction theory.

The interference structure suggested by Eqs. (30-31) is softened somewhat as a result of thermal vibrations. Furthermore, the interference or coherent radiation spectrum is accompanied by an incoherent contribution (due to thermal diffuse scattering). Except for a slight reduction (by one minus a Debye-Waller factor), this part is identical to Bethe-Heitler bremsstrahlung.

Figure 11 displays coherent bremsstrahlung as computed according to the above description for 10 GeV electrons (or positrons) incident at 1 mrad to the $\langle 110 \rangle$ axis

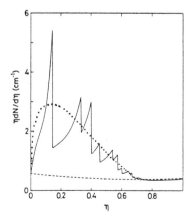

Fig. 11. Coherent bremsstrahlung for 10 GeV $e^{\pm}$ incident on $\langle 110 \rangle$ germanium at 1mrad to the axis.

of a germanium crystal (kept at room temperature). The peak-structured spectrum pertains to incidence parallel to the {110} planes; peaks appear as a result of periodic passing of axes. For incidence far from all planes, the peak structure has washed out (dotted curve) but the enhancement over the Bethe-Heitler rate (dashed line) is still high: The major source of enhancement is coherent scattering on atoms belonging to a given row. Note the incoherent background contribution which is all that is left when $\eta \equiv \hbar\omega/E$ approaches its maximum value of 1.

3. Channeling Radiation; MeV Energies

The perturbation treatment of the projectile-crystal interaction clearly breaks down under channeling conditions. The motion of the channeling particle is governed by the crystal field and cannot be approximated by that of a free particle. For instance, the projectile may oscillate transversely between a set of planes. Hence, under channeling conditions, only the interaction with the radiation field may be treated as a weak perturbation.

At first, let us consider the case of electrons or positrons with energies of a few MeV. As discussed above, a quantal description of the projectile motion through an oriented crystal is then required, especially for electrons. The crystal potential defines the states of motion through eigenfunctions of the transverse, stationary Schrödinger equation (24). The interaction with the radiation field, in turn, allows for transitions between eigenstates through photon emission. Such a situation is familiar from atomic spectroscopy.

The analogy with atomic spectroscopy is particularly close if we transform away the steady longitudinal motion of the projectile, that is, if we take the viewpoint of an observer moving along the channel with the projectile. The corresponding reference frame, moving with the velocity v (or rather, with $v_z \equiv p_z/\gamma m$) with respect ot the laboratory frame, is usually given the somewhat misleading name, the "rest frame". In the rest frame (R), the particle motion is by definition purely transverse, and the energy spectrum is quantized accordingly, cf. Fig. 9a. An electron, say, occupying a higher-lying level may undergo a transition to a lower-lying level by emitting a photon whose energy, when measured in the rest frame, equals the rest-frame energy difference between the electron states of transverse motion (nowhere else to take the energy from).

The rest-frame concept is further useful for establishing the characteristics of channeling radiation. In R the crystal appears Lorentz-contracted along the channeling direction. Consequently, the continuum potential is magnified by a factor of γ, as compared to the laboratory, $U^R = \gamma U$. Also the kinetic-energy term in the transverse-wave equation (24) is magnified by a factory of γ in R as compared to the laboratory frame since in R, the motion is nonrelativistic. As a result, the transverse-energy spectrum is scaled up by γ, $E_\perp^R = \gamma E_\perp$ (while $u(\vec{r}_\perp)$ remains unchanged). The photon energy as measured in R and corresponding to a jump from a state of transverse energy $E_{\perp,i}$ to one of $E_{\perp,f}$ then amounts to

$$\hbar\omega^R = -\Delta E_\perp^R = -(E_{\perp,f}^R - E_{\perp,i}^R) = -\gamma\Delta E_\perp. \tag{32}$$

A Doppler transformation back to the laboratory frame yields the final result

$$\hbar\omega = \frac{-\Delta E_\perp}{1 - \beta\cos\theta_\gamma} \simeq \frac{2\gamma^2(-\Delta E_\perp)}{1 + (\gamma\theta_\gamma)^2}. \tag{33}$$

The expression to the far right in Eq. (33) holds asymptotically for $\beta \equiv v/c \to 1$ but is fairly accurate as soon as E is beyond 2-3 MeV. The quantity θ_γ denotes the emission angle relative to the z-axis.

It is worthwhile noting the factor of γ^2 in Eq. (33); this factor shifts the energy differences $\Delta E_\perp$, which are typically a few tens of eV, into keV photon energies. Obviously, in the laboratory frame the major part of this energy given off by the projectile derives from the longitudinal motion.

Note in passing that also the emission pattern in angular space is conveniently determined by transformation between the two above frames: In R, usual dipole patterns are encountered. However since photon emission at 90° to the z-axis in R corresponds to emission at angle $\sin^{-1}(1/\gamma) \simeq 1/\gamma$ in the laboratory system, channeling radiation is strongly forward directed exactly like normal bremsstrahlung.

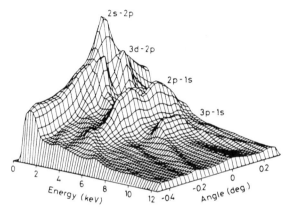

Fig. 12. Experimental recording of axial channeling radiation by 4 MeV electrons incident on $\langle 111 \rangle$ silicon[6].

Fig. 12 shows a collection of channeling-radiation count spectra recorded (this is an experiment!) in the forward direction, $\theta_\gamma = 0$, for 4 MeV electrons incident at various angles to the $\langle 111 \rangle$ axis of a silicon crystal. The quantum jumps are clearly visible above the background of incoherent bremsstrahlung. Numerical solution of the eigenvalue problem (24) yields line energies (33) in exact agreement with experimental observation. Also for yields and linewidths (the latter mainly due to thermal incoherent scattering), nice agreement between theory and experiment has been established.

Having a good theoretical understanding of the experimental results, you may now try to play all sorts of games: if the total beam energy is varied for instance, the line positions will vary—partly because of the factor γ^2 in Eq. (33), and partly because $\Delta E_\perp$ is a function of γ, cf. Eq. (24). The latter fact may be used to tune the beam energy such that a specific state of transverse motion (given symmetry) gets positioned at a desired $E_\perp$ value. Hereby local variations in the target electron density may be scanned by measuring transitions to or from that state. Another possibility is to vary the target temperature; this leads to changes in line positions and linewidths. Furthermore correlations in thermal vibrations may be read out by measuring the variation in linewidth for a planar transition with decreasing angle to a major axis.

If you look back once more to Fig. 12 you will note horseshoe-like structures at higher photon energies. These correspond to free-to-bound transitions, for instance, free-to-3p. Furthermore, straight ridges are seen at low energies for the largest angles. These structures correspond to free-to-free transitions. As such they are obtainable in a perturbation treatment and, indeed, their positions (as well as the yields) are reproduced both by standard coherent bremsstrahlung calculations, cf. previous section, as well as by channeling radiation calculations, cf. Fig. 10.

4. Channeling Radiation; Few GeV

As the primary energy is raised into the GeV region, the number of channeling states becomes large, Section II.9. The level spacing reduces correspondingly. As a result, an essentially continuous transition down through the dense spectrum of states becomes possible. Hence channeling radiation appears as a classical emission process.

In general, classical channeling-radiation spectra are broad and featureless. However, planar positron channeling constitutes a special example: The motion is periodic and, since the interplanar potential is close to harmonic, the oscillation frequency depends only weakly on $E_\perp$. Furthermore, for a purely harmonic potential and non-relativistic rest-frame motion, the radiation goes solely into the first harmonic. Hence we may expect to find cases with the radiation concentrated in a single photon peak. This kind of picture holds for, for example, $\lesssim 1$ GeV positrons incident along the major planes of a silicon crystal.

Anharmonicity of the interplanar potential leads to photon emission at higher harmonics. The effect is magnified when the rest frame motion of the channeled positrons becomes relativistic. The latter case corresponds to $mc^2 \simeq E_\perp^{R\,\max}(\simeq \gamma U^{\max})$ or, equivalently, to $\gamma\psi_p \sim 1$ which may also be expressed as $p_\perp^{\max} \simeq mc$. The quantity $\gamma\psi_p$ scales with $\gamma^{1/2}$ and equals 1 for, for example, 10-GeV positrons incident along the $\{110\}$ planes of a silicon crystal. Relativistic rest-frame motion furthermore tends to give rise to a broading of the observed lines (except around specific γ-values known as γ_{magic} where this effect is balanced by the effect of anharmonicity).

Figure 13 shows channeling radiation from 6.7 GeV positrons incident parallel to the $\{110\}$ plane in a 0.1mm thick silicon crystal kept at room temperature. The

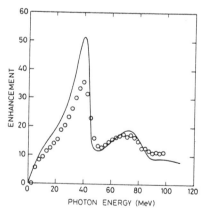

Fig. 13. Planar channeling radiation from 6.7 GeV positrons incident on $\{110\}$ silicon[6].

intensity is normalized to the incoherent bremsstrahlung yield (Bethe-Heitler) which pertains to an amorphous target of the same material. The solid curve is the theoretical prediction according to standard classical electrodynamics formulas. Note the high enhancement. The scaling of the peak position with beam energy is slightly slower than $\gamma^{3/2}$ which applies for non-relativistic rest-frame motion. Photon collimation to angles somewhat less than $1/\gamma$ within the forward direction would sharpen the low-energy side of the photon peak, cf. Eq. (33).

5. Large Angles ($\psi > \psi_{1,p}$) ↔ Coherent Bremsstrahlung

At MeV impact energies, a first Born treatment ("coherent bremsstrahlung") applies all way down to the channeling region. Does the same hold true at GeV energies? An obvious requirement for applicability of the Born approximation is that the radiation is independent of the sign of charge of the radiating particle, cf. Eq. (28). This translates into a requirement of the full particle path being contained inside the radiation cone, that is, it is required that the angular excursions $\Delta\psi$ of the projectile be less than $1/\gamma$. By application of the continuum picture and considering incidence angles ψ far beyond the channeling region ($\psi \gg \psi_{1,p} \gg \Delta\psi$), $\Delta\psi$ corresponding to crossing an axis or a plane is estimated by $\Delta\psi = \gamma^{-1}(U_0/mc^2)\psi^{-1}$ where U_0 is the height (or depth) of the potential. Our requirement for application of the coherent bremsstrahlung scheme then becomes

$$\psi > \max \{\Theta_0; \psi_{1,p}\}, \quad \Theta_0 \equiv \frac{U_0}{mc^2}. \tag{34}$$

We observe that $\Theta_0 = \psi_{1,p}$ for $\gamma \simeq 2mc^2/U_0$ which corresponds to energies of a few GeV in silicon and germanium. At higher energies there will be a gap between the channeling region and the Born region.

A curiosity: If you move somewhat out beyond Θ_0 you may compute radiation spectra according to the first Born approximation. On the other hand, the continuum picture applies far beyond the channeling region, Section II.7, and in this model the particle motion may be considered classical, Section II.9. As long as we stick to soft photon emission, $\hbar\omega \ll E$, the radiation emission then follows classical electrodynamics. Hence, there exists a region outside Θ_0 where both classical and first Born schemes apply! And, indeed, actual computation reveals nice agreement between the two types of calculation as with experiments.

6. Radiation by Multi-GeV Electrons and Positrons

When the energy of the incident electron or positron is raised further into the region of many GeV, the classical description of the radiation process becomes invalid. Clearly, the problem here is not a quantization of the transverse motion of the charged particles since the density of states increases with γ, cf. Section II.9. Instead, the breakdown is associated with the recoil due to the radiation. From the simple classical arguments presented at the beginning of Section III, it is evident that the ratio of the instantaneously emitted power to the primary energy increases with γ. The coherent part of the spectrum extends over an increasing fraction of the available energy $E - mc^2 (\simeq E)$. Correspondingly, a multi-GeV electron (or positron) has a fair chance to emit a photon whose energy amounts to an appreciable fraction of the projectile energy. Such a situation demands a quantum description.

On our way towards getting an understanding of what radiation spectra actually look like at these high energies, we first make a simple (but very useful) observation:

For a channeled particle, the angular excursions made during the penetration of the crystal, are of order the critical channeling angle ψ_1 (ψ_p for planes), which scales as $\gamma^{-1/2}$. On the other hand, the opening angle of the light cone is $\sim 1/\gamma$. Hence at high energies, the "light" emission is as from a car on a hilly road with the beam sweeping up and down as the car gets along; for 150 GeV electrons channeling along the $\langle 110 \rangle$ axis in silicon we have $\gamma^{-1}/\psi_1 \sim 1/12$! So much for the observation; the trick now is that along the classical projectile path coherent photon emission is only possible from points which may be covered by one and the same light cone, that is, from fragments over which the variation of the slope of the tangent is within $\sim 1/\gamma$. With $\gamma\psi_1 \gg 1$, the projection of such fragments on transverse space is small compared with channel widths (or rather, the transverse distances over which the continuum potential varies appreciably) and hence the emission process appears locally as in a constant field. The problem is then reduced to computing the radiation emission in a constant field and subsequently taking the average over the actual field-strength seen by the projectile. The latter part we know how to do (Section II), our problem is the photon emission:

a. Classical synchrotron radiation

Let us start out by reminding ourselves how radiation emission by a highly relativistic electron in a constant electromagnetic fields looks like in the classical limit. Imagine our electron to rotate in a plane perpendicular to a magnetic field of strength B, that is, it rotates at frequency $\Omega_{\mathrm{rev}} = eB/p$ where $p = E/c$ is the particle momentum ($\beta \simeq 1$). With the radius of the circular motion being R, the so-called characteristic frequency ω_c for emission is $\omega_c = 3\gamma^3 c/R$ (with Jackson's convention for the numerical factor[8]) or, by application of the relation for Ω_{rev},

$$\omega_c = 3\gamma^3 eB/p. \tag{35}$$

The frequency distribution of the emitted power takes the form[8]

$$\frac{dP}{d\omega} = \frac{\Omega_{\mathrm{rev}}}{2\pi} 2\sqrt{3} \frac{e^2}{c} \gamma \frac{\omega}{\omega_c} \int_{2\omega/\omega_c}^{\infty} dt\, K_{5/3}(t). \tag{36}$$

Since the modified Bessel function $K_{5/3}$ falls off exponentially for large arguments, ω_c defines effectively the extension of the radiation spectrum.

Although this is all classical, we shall find it useful for later comparisons to convert (36) into a count spectrum ($(\hbar\omega)^{-1}dP/d\hbar\omega$). Also it turns out to be convenient to reexpress the result in terms of some new parameters, namely as

$$\frac{dN}{d\eta} = \frac{\alpha c}{\sqrt{3}\pi\lambda} \frac{1}{\gamma} \left[2K_{2/3}(\xi) - \int_{\xi}^{\infty} K_{1/3}(t)dt \right];$$

$$\eta \equiv \hbar\omega/E, \quad \xi \equiv 2\eta/3\kappa, \quad \kappa \equiv eB\gamma\lambda/mc^2. \tag{36'}$$

The quantity $\lambda = \hbar/mc = \alpha a_0$ is the Compton wavelength of the electron; furthermore we have used the identity of the square-bracket factor in (36') with the integral in (36).

b. Synchrotron radiation & strong fields

When the electron/positron energy becomes sufficiently high or the magnetic field sufficiently strong, photon emission causes a significant recoil. (The pure motion in the field is still classical.) An estimate of when this appears is provided by comparing $\hbar\omega_c$ to the total energy, that is, we expect significant recoils when

$$1 \lesssim \frac{\hbar\omega_c^{\mathrm{classical}}}{E} = 3eB\frac{\lambda}{mc^2}\gamma \equiv 3\kappa, \tag{37}$$

with κ defined in Eq. (36′). Or, in other words, "strong-field effects" appear for $\kappa \gtrsim 1$,

$$\text{strong} \leftrightarrow \kappa \gtrsim 1. \tag{37′}$$

As an example, consider 250 GeV electrons or positrons incident on a germanium crystal along the $\langle 110 \rangle$ direction. Let us estimate κ, replacing B by the electric field strength $\mathcal{E}$. With $\gamma e \mathcal{E} \sim \gamma U_0 / a \simeq 2 \times 10^8 \text{eV}/a_0$ and $\lambda/mc^2 = \alpha a_0 / mc^2 \simeq \frac{3}{2} a_0 / 10^8$ eV we get $\kappa \simeq 3$. Hence in channeling, in the region beyond ~ 100 GeV, we easily encounter cases where the classical radiation spectrum extends beyond the endpoint energy—what then?

The answer is, of course, that you have to do a proper QED calculation. For the case of radiation in a constant magnetic field, such have been made over the last ~ 30 years. In the field of channeling, the so-called semi-classical operator method developed by a group of Novosibirsk physicists has been much in use. The method has the advantage that it may be applied also to cases where the field shows some spatial variation. It makes effective use of the fact that the projectile motion through the crystal (when not perturbed by the radiation field) is classical. However, I do not find the method very instructive for tutorial purposes (and also its derivation remains in parts in the dark for me). Let us stick to simple physical pictures here and by such means try to resolve the question:

c. *Which changes to expect in going from weak to strong fields?*

Below I shall sketch an analysis due to Jens Lindhard[9]. The aim of the analysis is to find some simple substitution rule which, when applied to the expression for the classical emission spectrum, yields the quantum spectrum. The means is a Weizsäcker-Williams type calculation with (i) the classical Thomson cross section for recoilless photon scattering on a resting electron (you move to the instantaneous electron-rest frame) and (ii), the quantum Klein-Nishina cross section. This part is equivalent to Jackson's calculation of bremsstrahlung by the method of virtual quanta[8]. Now, the trick is to bypass the actual calculation of the Weizsäcker-Williams spectrum by equivalating the classical radiation spectrum (known from other sources, that is, from exact calculations as those leading to Eq. (36)) with the result of the Weizsäcker-Williams method with the Thomson cross section. The emerging "translation-rule" is then applied to the outcome of the Weizsäcker-Williams calculation with the Klein-Nishina cross section, and the requested substitution rule is found. It may be noted that since the virtual photon spectrum actually is never calculated in this analysis, the substitution rule is generally valid (that is, not only valid in the case of bremsstrahlung or synchrotron radiation); the only requirement is that somebody should be able to supply you with the classical radiation spectrum for the case of your interest.

Sketch: Let the number of incoming virtual photons in the electron rest system be dN_i (we multiply $(\hbar\omega)^{-1}$ on classical intensities to get number of photons), and let the scattering cross-section be $d\sigma$, then the number of scattered photons dN (also in the rest frame) is proportional to their product, $dN \propto dN_i d\sigma$. Take $g(k', \vec{e}_i') dk'$ to represent the intensity distribution of incoming virtual photons with momentum between k' and $k' + dk'$ and polarization $\vec{e}_i'$ (primed variables are used in the rest frame) and for $d\sigma$ apply the classical Thomson cross section; hence

$$dN_{\text{classical}} = C \frac{dk'}{k'} g(k', \vec{e}_i') \cdot r_e^2 d\Omega' (\vec{e}_i' \cdot \vec{e}_f')^2, \tag{38}$$

where C is a constant, $d\Omega'$ the element of solid angle into which photons are scattered and $\vec{e}'_f$ the direction of polarization after the scattering. Transformation to the laboratory frame yields (after some work)

$$dN_{\text{classical}}(\omega, \vec{n}, \vec{e}_f) = \frac{C}{\gamma}\frac{d\omega}{\omega}g\left(\frac{\omega}{\zeta(\gamma,\theta)}, \vec{e}'_i\right)r_e^2 d\Omega'(\vec{e}'_i \cdot \vec{e}'_f)^2, \tag{39}$$

where $\vec{n} = \vec{k}/k$ and $k'\zeta = k'c[\gamma^2(1 - \cos\theta')^2 + \sin^2\theta']^{1/2} = \omega$, θ' being the photon scattering angle in the electron rest frame.

External sources are assumed to supply us with a full and exact expression for the classical radiation spectrum for the field configuration (or, rather, for the particle path) under consideration, i.e.,

$$dN_{\text{classical}}(\omega, \vec{n}, \vec{e}_f) = F(\omega, \vec{n}, \vec{e}_f)d\omega d\Omega, \tag{40}$$

where we consider F as a known function. Equating this expression with (39), we obtain the requested "translation-rule"

$$\frac{C}{\gamma}\frac{d\omega}{\omega}g\left(\frac{\omega}{\zeta}, \vec{e}'_i\right)r_e^2 d\Omega'(\vec{e}'_i \cdot \vec{e}'_f)^2 = F(\omega, \vec{n}, \vec{e}_f)d\omega d\Omega. \tag{41}$$

Repeat now the calculation leading to (39) but with the classical Thomson cross section replaced by the Klein-Nishina cross section,

$$d\sigma_{KN} = r_e^2 d\Omega'\left[\left(\frac{k'_f}{k'_i}\right)^2 (\vec{e}'_i \cdot \vec{e}'_f)^2 + \frac{k'_f}{4k'_i}\left(1 - \frac{k'_f}{k'_i}\right)^2\right]. \tag{42}$$

Here the first term is a kinematically modified version of the Thomson expression; incident and exit momenta are no longer the same in the rest frame, whereas the second term is due to spin. If we stick to the first term, the laboratory expression to replace (39) becomes

$$dN_{qm}^{(1)}(\omega, \vec{n}, \vec{e}_f) = \frac{C}{\gamma}\frac{d\omega}{\omega}(1 - \eta)g\left(\frac{\omega}{(1 - \eta)\zeta}, \vec{e}'_i\right)r_e^2 d\Omega'(\vec{e}'_f \cdot \vec{e}'_i)^2, \tag{43}$$

where as before $\eta \equiv \hbar\omega/E$. Here, the right-hand side is evidently equal to the left-hand side of (41) if you there make the replacement $\omega \rightarrow \omega/(1-\eta)$ everywhere (except in the $d\omega$-factor). Using the link (41) between g and F, we therefore have

$$dN_{qm}^{(1)}(\omega, \vec{n}, \vec{e}_f) = F_{\text{classical}}\left(\frac{\omega}{1 - \eta}, \vec{n}, \vec{e}_f\right)d\omega d\Omega, \tag{44}$$

where, still, F is assumed to be a known function. This is the central result. We have obtained the quantal radiation spectrum for spin-less particles from the classical spectrum. To get the full spectrum for electrons or positrons, also the second term in (42) has of course to be considered.

d. *Synchrotron case revisited*

Let us now return to the constant field case. In this case our F-function is given by Eq. (36'), and substitution of $\omega/(1 - \eta)$ for ω on the right-hand side only implies a redefinition of the parameter ξ, i.e., application of the result (44) yields

$$\frac{dN^{(1)}}{d\eta} = \frac{\alpha c}{\sqrt{3}\pi\lambda}\frac{1}{\gamma}\left[2K_{2/3}(\xi) - \int_\xi^\infty K_{1/3}(t)dt\right];$$

$$\xi \equiv \xi_{\text{classical}}/(1 - \eta) = 2\eta/3(1 - \eta)\kappa. \tag{45}$$

This expression may now be compared to the spectrum computed, e.g., by the Novosibirsk method[10]

$$\frac{dN}{d\eta} = \frac{\alpha c}{\sqrt{3}\pi\lambda}\frac{1}{\gamma}\left[\left(1 - \eta + \frac{1}{1-\eta}\right)K_{2/3}(\xi) - \int_\xi^\infty K_{1/3}(t)dt\right], \qquad (46)$$

the parameter ξ being defined as in (45). The difference between the two results lies in the square-bracket factors with the Novosibirsk result being in excess by an amount

$$\Delta[\ ] = \frac{\eta^2}{1-\eta}K_{2/3}(\xi). \qquad (47)$$

This extra contribution corresponds to the spin-term in the Klein-Nishina cross section (42) (transformation of (42) to laboratory variables immediately shows a dependence on kinematical factors similar to the front factor in (47)).

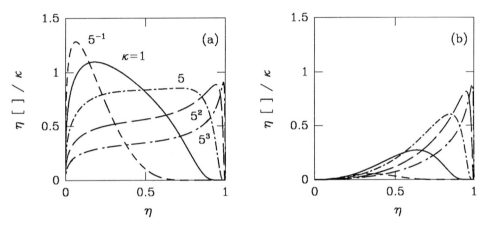

Fig. 14. Strong field synchrotron radiation. (a) Full spectrum, Eq. (46). (b) Spin-$\frac{1}{2}$ contribution, $d(N - N^{(1)})/d\eta$.

Figure 14a displays the radiation spectrum (46) for different values of κ (the coordinate is the square-bracket factor times η/κ; multiplication with η yields power spectrum). For $\kappa \ll 1$, the spectrum is confined to soft photons; for $\kappa \sim 1$, the spectrum extends up to $\hbar\omega \sim E$; for $\kappa \gg 1$ where the classical spectrum has grown far beyond the endpoint energy, the intensity gathers close to the endpoint energy. The asymptotic behavior may be shown to be such that

$$\frac{dE/dt}{E} = \int_0^1 \eta\frac{dN}{d\eta}\,d\eta \propto \kappa^{-1/3} \qquad (48)$$

as $\kappa \to \infty$. In other words, maximum energy loss is expected in the region $\kappa \sim 1$. Fig. 14b shows the spin contribution to dN, Eq. (46), i.e., $dN - dN^{(1)}$. For low κ-values, $\kappa \ll 1$, spin is unimportant as expected (classical). For higher κ-values, $\kappa \gtrsim 1$ the scattering on the electron magnetic moment is essential for the intensity of hard photons.

e. *Radiation by channeled electrons*

We now have all the necessary information to compute radiation spectra for channeled, multi-GeV electrons and positrons. Figure 15 shows power-spectra for

150 GeV electrons channeled along the $\langle 110 \rangle$ axis in a germanium crystal. The ordinate is $\eta dN/d\eta$ (given in cm^{-1}), the abcissa η (integration over η yields the fractional energy-loss rate $E^{-1}dE/dt$). The different curves correspond to different transverse energy values identified through the ratio of the classical turning point (for vanishing angular momentum) to the single string radius r_0 (cf. text before Eq. (14)). For given $E_\perp$, the spacial distribution is assumed uniform inside the accessible area of transverse space, Eq. (16). An electron channeling half way down the potential spends much of its time in regions where the field strength is high; correspondingly, it is seen to radiate much more than a particle just barely bound— an order of magnitude more. On the other hand, even for the latter particle, the radiation intensity is far beyond the (roughly η-independent) Bethe-Heitler level, which in the present case is $\sim \frac{1}{2}$ cm^{-1}! Knowing the radiation spectra for different transverse energies, all that remains in order to compare with experimental results is a final weighting according to the actual $E_\perp$-distribution.

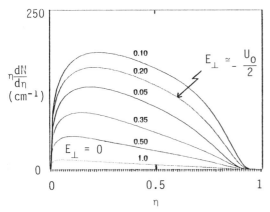
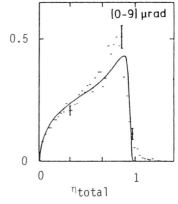

Fig. 15. Radiation by channeled 150 GeV electrons in $\langle 110 \rangle$ germanium. Thin crystal limit.

Fig. 16. Energy radiated by 150 GeV electrons channeled in 0.2mm $\langle 110 \rangle$ germanium[11].

Now, experimentally recorded radiation spectra usually look somewhat differently from a weighted average of the spectra of Fig. 15. An example is displayed in Fig. 16, which shows the energy radiated by 150 GeV electrons incident essentially parallel to the $\langle 110 \rangle$ axis of a 0.2mm thick germanium crystal. The normalization is such that integration over η yields the fractional energy loss, E_{rad}/E_{beam}. In the present case, despite having only 0.2mm medium-Z material, this amounts to as much as $E_{rad}/E_{beam} = 0.3$! (Compare to the Bethe-Heitler yield which predicts only ~ 0.01). This high energy loss, besides being impressive in itself, is the key to the special shape of the spectrum as compared to those of Fig. 15: Channeled electrons simply have a high chance for emitting more than one photon each on the way through the crystal. Since the detector registers two photons emitted by a given electron as one at the sum energy, there will be a pile-up at the high-energy end of the photon spectrum—exactly as seen in the experiment. The curve superimposed on the plot is a theoretical calculation taking multiple-photon emission (with the resulting gradual slowing down of the projectile) into account. Besides being only a first stroke (the transverse energy for a given electron is artificially kept fixed in the passage of the crystal and the statistical distribution (16) is used), this calculation simulates pretty well the experimental findings. (More refined calculations taking into account changes in $E_\perp$ due to radiation emission and scattering have been published by various authors, however, to reproduce accurately the experimental spectra is very difficult due to the high sensitivity to the exact form of the $E_\perp$ distribution, cf. Fig. 15.)

f. *Angular dependence*

Above we found large enhancements over the amorphous rate of the radiation probability for channeled electrons. However, similar enhancements persist far beyond the channeling region—channeling is merely a source of non-uniform particle flux in transverse space. We could pose the question: for which incidence angles to, say, an axis may we expect the constant field approximation to apply? From the discussion at the beginning of this section 6, we would find the answer to be provided by the requirement that the angular excursions $\Delta\psi$ of the charged particle be large compared to $1/\gamma$. But from the discussion in Section III.5 this is immediately seen to correspond to the requirement of the incident angle to the axis be less than Θ_0; so, roughly speaking, we have

$$
\begin{aligned}
\psi > \Theta_0 : \quad & \text{Born approximation} \\
\Theta_0 > \psi > \psi_1 : \quad & \text{constant field approximation (CFA)} \\
\Theta_0 > \psi_1 > \psi : \quad & \text{CFA + channeling flux.}
\end{aligned}
\tag{49}
$$

The angle Θ_0 gives an estimate of how far out from the axis a significant enhancement persists. Typical values for Θ_0 are fractions of a mrad; for Ge $\langle 110 \rangle \sim \frac{1}{2}$ mrad.

7. *Pair Production in Single Crystals at Multi-GeV Energies*

When the probability for hard-photon emission by incoming electrons or positrons is strongly enhanced, you will in general, due to crossing symmetry, also expect a magnification of the inverse process of electron-positron pair creation upon photon impact. Figure 17 reveals an experimental recording of the pair-production yield for 40-150 GeV photons incident on a 0.5 mm thick germanium crystal (cooled to 100K) parallel to the $\langle 110 \rangle$ axis. The yield is normalized to the Bethe-Heitler yield predicted for amorphous germanium. In the 100 GeV-region, the yield raises far beyond the amorphous value.

On Fig. 17, we have furthermore drawn a curve representing the theorist's estimate of the yield. This is obtained in a constant field approximation similar to the one described above for the radiation (with uniform photon flux though), and, in addition, an incoherent background, just taken to equal the Bethe-Heitler yield, is included. Using crossing symmetry, the constant field result (46) for the radiation turns into a pair-production yield of[13]

$$
\frac{dN_{pp}}{d\eta} = \frac{\alpha c}{\sqrt{3}\pi\lambda}\frac{mc^2}{\hbar\omega}\left[\left(\frac{1-\eta}{\eta}+\frac{\eta}{1-\eta}\right)K_{2/3}(\xi)+\int_\xi^\infty K_{1/3}(t)dt\right];
$$

$$
\eta \equiv E_+/\hbar\omega, \quad \xi = \frac{2}{3\eta(1-\eta)\kappa}, \quad \kappa = e\mathcal{E}\frac{\lambda}{mc^2}\frac{\hbar\omega}{mc^2},
\tag{50}
$$

where E_+ is the energy of the outgoing positron (and $\hbar\omega$ the energy of the incoming photon). The Bethe-Heitler yield parallelling (26) is

$$
\frac{dN_{pp}}{d\eta} = \frac{16}{3}Z^2\alpha r_e^2 nc\left(\frac{3}{4}-\eta+\eta^2\right)\ln(183Z^{-1/3}),
\tag{51}
$$

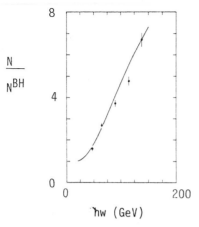

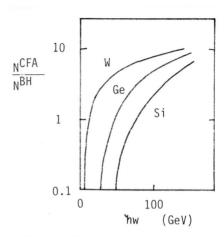

Fig. 17. Electron-positron pair production by 40-150 GeV photons in $\langle 110 \rangle$ germanium[12].

Fig. 18. Pair-production yield as obtained in constant field approximation for various. crystals.

the total "amorphous" pair-production cross section being $7/9n\ell_r$. This construction results in yields in nice agreement with the experimental findings, Fig. 17.

Figure 18 displays further theoretical results for the total pair-production yield as obtained in the constant field approximation, that is, by averaging the expression (50) over the field strengths encountered in the crystal. The results for given crystal pertain to production upon incidence parallel to its strongest axis. With increasing Z, the energy, where the coherent yield starts exceeding the Bethe-Heitler yield, decreases. An estimate[11] of this threshold is obtained by requiring $\kappa_{max} \sim 1$, i.e.,

$$\hbar\omega_{th} \sim \frac{500\text{GeV}}{Z} \frac{\rho d}{a_0^2}, \tag{52}$$

where ρ as before is the thermal vibration amplitude and d the spacing of atoms along the string. On the other hand, at high values of κ, the yield (50) falls off as $\kappa^{-1/3}$, i.e., for any given crystal the asymptotic behavior of the production is

$$N_{pp}^{CFA} \propto (\hbar\omega)^{-1/3} \tag{53}$$

as $\omega \to \infty$; cf. Eq. (48). Hence the production goes through a maximum as ω increases beyond ω_{th}. The maximum enhancement over Bethe-Heitler production may be estimated as[11]

$$\max\left(N_{pp}^{CFA}\Big/N_{pp}^{BH}\right) \sim \frac{10^4}{Z^{4/3}\ln(183Z^{-1/3})}. \tag{54}$$

All in all, the threshold $\hbar\omega_{th}$ is reached earlier with heavier than with light target elements, but the maximum enhancement is less. It should be remembered throughout that while the relative yields displayed on Fig. 18 look pretty much the same for all three targets in the energy region of a few hundred GeV, the absolute values are much different, due to the difference in radiation lengths (cf. end of Section III.1).

Figure 19a displays the variation of the pair-production yield (normalized to Bethe-Heitler) with incidence angle to the $\langle 110 \rangle$ axis of the cooled germanium crystal of Fig. 17. The yields recorded at zero angle are exactly those shown in that figure. The discussion of the angular dependence is parallel to the discussion for the radiation case, Section 6.f and (49) above, except that no variation appears on the scale of the channeling angle since the photon flux is uniform for any angle of incidence. That means, the constant field approximation holds out to $\sim \Theta_0$, and beyond the Born approximation ("coherent pair production") may be used. This is revealed by the theoretical curves shown on Fig. 19a for the first, third and fifth energy intervals: out to $\frac{1}{2}$ mrad ($\sim \Theta_0$), the curves are obtained in the described constant field approach (including background), the variation with angle being due to inclusion of lowest order correction (proportional to angle squared). Beyond 1 mrad standard coherent pair production expressions are applied. The overall agreement is satisfactory, although some overestimation seems to be done in the transition region. Figure 19b furthermore shows the differential pair-production yield as recorded for two different angles of incidence for the group of 100-120 GeV photons from Fig. 19a. Also here the agreement between theory and experiment is satisfactory.

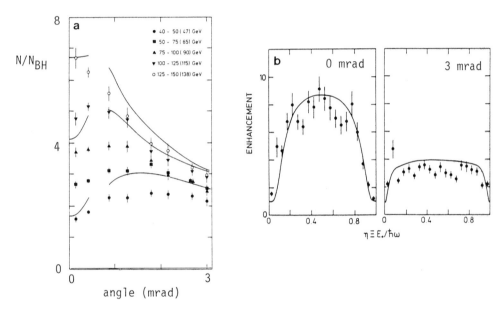

Fig. 19. Pair-production in $\langle 110 \rangle$ germanium[12]. (a) Angular dependence. (b) Differential yield.

IV. Concluding Remarks

Above, we have given a broad introduction to directional effects associated with the penetration of energetic charged particles through aligned single crystals. The discussion covers non-relativistic heavy particle (read non-$e^{\pm}$) channeling as well as the strongly enchanced bremsstrahlung emission by channeled multi-GeV electrons. The emphasis has been on the latter part, the discussion of traditional channeling providing the necessary background.

One field essentially unmentioned is the field of applications. Channeling has very many practical applications, especially in materials science; the interested reader is referred to the book by Feldman, Mayer and Picraux[3]. A more exotic application is GeV beam bending by bent crystals[6,7,14]. Also channeling radiation has potential applications[4-6], as x- and γ-ray sources and for extracting crystal properties (cf. Section III.3). At multi-GeV energies, the highly enhanced shower production along crystal axes may possibly be utilized for construction of compact highly direction-sensitive multi-GeV photon detectors for γ-ray astronomy[6]. The discussion of all of these topics, however, is outside the scope of the present paper, and they have therefore been left out of our discussion.

Acknowledgements

I wish to thank the organizers of the NATO ASI on Vacuum Structure in Intense Fields for inviting me to lecture on such obscure topics as channeling, bremsstrahlung and pair production in single crystals. Also I want to thank Berndt Müller for kindly urging me to create a write-up of my lectures during my stay at Duke University and Julia Clark for the very competent and efficient typing of the manuscript. Jens Lindhard deserves special thanks for many stimulating discussions this spring and for letting me into his ideas of classical versus quantum radiation (ref. 9). Grants from the Danish Natural Science Research Council are acknowledged.

References

1. J. Augustin, S. Graf, W. Greiner and B. Müller, these proceedings.

2. D. S. Gemmel, "Channeling and Related Effects in the Motion of Charged Particles through Crystals", *Rev. Mod. Phys.* **46** (1974) 129.

3. L. C. Feldman, J. W. Mayer, and S. T. Picraux, *Materials Analysis by Ion Channeling*, Academic Press, New York (1982).

4. J. U. Andersen, E. Bonderup and R. H. Pantell, "Channeling Radiation", *Ann. Rev. Nucl. Part. Sci.* **33** (1983) 453.

5. A. H. Sørensen and E. Uggerhøj, "Channeling and Channeling Radiation", *Nature* **325** (1987) 311.

6. A. H. Sørensen and E. Uggerhøj, "Channeling, Radiation and Applications", *Nucl. Sci. Appl.* **3** (1989) 147.

7. R. A. Carrigan, Jr., and J. A. Ellison, eds. *Relativistic Channeling*, NATO ASI Series B 165, Plenum, New York (1987).

8. J. D. Jackson, *Classical Electrodynamics*, Wiley Press, New York, (1975).

9. J. Lindhard, "Quantum Radiation Spectra of Relativistic Particles Derived by Correspondence Principle" (submitted to Phys. Rev.).

10. V. N. Baier, V. M. Katkov and V. M. Strakhovenko, *Phys. Lett.* **A117** (1986) 251.

11. R. Medenwaldt et al., *Phys. Rev. Lett.* **63** (1989) 2827.

12. J. F. Bak et al., *Phys. Lett.* **B202** (1988) 615.

13. V. N. Baier, V. M. Katkov and V. M. Strakhovenko, *Nucl. Sci. Appl.* **3** (1989) 245.

14. S. P. Møller et al., "High Efficiency Bending of 450 GeV Protons Using Channeling", (submitted to Physics Letters).

QED IN STRONG COULOMB FIELDS:
CHARGED VACUUM, ATOMIC CLOCK, AND CORRELATED, NARROW e⁺e⁻-LINES

S. Graf, B. Müller*, E. Stein, J. Reinhardt, G. Soff**, and   W. Greiner

Institut für Theoretische Physik   * Department of Physics
J. W. Goethe Universität   Duke University
Postfach 11 19 32   Durham, NC 27706 USA
D-6000 Frankfurt am Main, Germany

**Gesellschaft für Schwerionenforschung mbH
Postfach 11 05 52
D-6100 Darmstadt, Germany

INTRODUCTION

1. The charged vacuum in supercritical fields

When atomic structure is extrapolated from the known boundary of chemical elements (nuclear charge Z=109) into the region Z=170–190, one finds that the $1s_{1/2}$-state, the atomic K-shell, gains tremendously in binding energy. As shown in Fig. 1 the $1s_{1/2}$-state - and also the next higher state, the $2p_{1/2}$-level - traverses the gap between the positive and negative energy continuum solutions of the Dirac equation, and is predicted to reach a binding energy of $2m_e$=1.022 MeV at the critical nuclear charge Z_c=173±1. The uncertainty derives from our lack of precise knowledge of the extrapolated nuclear charge distribution and from possible radiative corrections of higher order that are not accounted for in the calculations (for a point-like nuclear charge distribution the Dirac Hamiltonian is not self-adjoint for $Z > \alpha^{-1} \approx 137$. The origin of this is the singular behaviour of the vacuum in the point nucleus limit for $Z\alpha > 1$).

What happens at and beyond this critical charge was clarified in the early 1970's by our group at Frankfurt [1, 2] and by another one in Moscow [3]. The transition from a just subcritical 1s-state to the supercritical state is most easily understood in the framework of Fano's theory of configuration interaction and autoionizing states. We start with the reduced Hilbert space of a just critical atom spanned by the 1s-state $|\phi_0\rangle$ and the negative energy continuum of s-wave states $|\phi_E\rangle$, as shown in Fig. 2 (left part):

$$H_C|\phi_0\rangle \approx -m_e|\phi_0\rangle \tag{1}$$

$$H_C|\phi_E\rangle \approx E|\phi_E\rangle \ , \ E < -m_e \tag{2}$$

When a few, Z', protons are added to the nucleus to render the potential supercritical, the 1s-state is drawn into the continuum and only continuum solutions exist:

$$(H_C + Z'U(r))|\Psi_E\rangle = E|\Psi_E\rangle \ , \ E < -m_e \ . \tag{3}$$

The solutions of the supercritical potentials, $|\phi_E\rangle$, are expanded in terms of those of eqs. (1,2):

$$|\Psi_E\rangle = a(E)|\phi_0\rangle + \int_{-\infty}^{-m_e} dE' \ b_{E'}(E)|\phi_{E'}\rangle \ . \tag{4}$$

Vacuum Structure in Intense Fields, Edited by
H.M. Fried and B. Muller, Plenum Press, New York, 1991

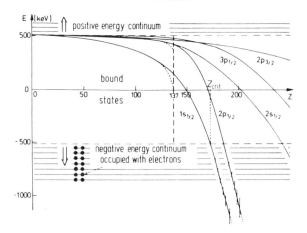

Figure 1. *Atomic binding energies as function of nuclear charge* .

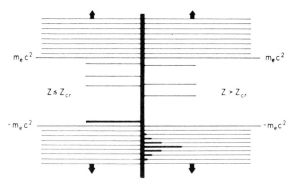

Figure 2. *Transition from Z_c (left) to Z_c+Z' (right).*

Elementary methods for solving integral equations yield the following result for the probability $|a(E)|^2$ of admixture of the critical 1s-state to the supercritical continuum state $|\Psi_E\rangle$:

$$|a(E)|^2 = \frac{1}{2\pi}\Gamma_E \left((E - E_\phi - F(E))^2 + \frac{1}{4}\Gamma_E^2\right)^{-1}, \tag{5}$$

where

$$E_\phi = -m_e + Z'\langle\phi_0|U(r)|\phi_0\rangle \tag{6}$$

$$V_E = Z'\langle\phi_E|U(r)|\phi_0\rangle , \quad \Gamma_E = 2\pi|V_E|^2 \tag{7}$$

$$F(E) = P\int_{-\infty}^{-m_e} dE' \frac{|V_{E'}|^2}{E - E'} . \tag{8}$$

Obviously, the bound 1s-state turns into a resonance in the negative energy continuum located at $E_{res} = E_\phi + F(E_{res})$. The width of the resonance , Γ_E, is of the order 1 keV, corresponding to a lifetime in the range $10^{-18} - 10^{-19}$s. The supercritical situation is shown schematically in Fig. 2 (right part).

The reason why the 1s bound state turns into a resonance is intuitively clear: the vacant K-shell is unstable against pair-decay when the binding energy E_K exceeds twice the electron rest mass. A pair is created, the electron occupying the 1s-state while the positron is emitted freely with kinetic energy $E_p = E_K - 2m_e$. When the K-shell is fully occupied by two electrons the spontaneous decay process is stopped by the action of the Pauli principle. The intuitive picture is easily corroborated by arguments based on second quantized field theory [4, 5].

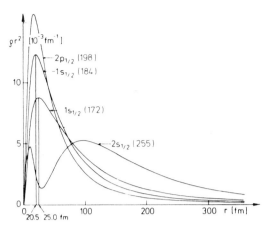

Figure 3. *Real vacuum polarization in comparison with K-shell density at Z_c.*

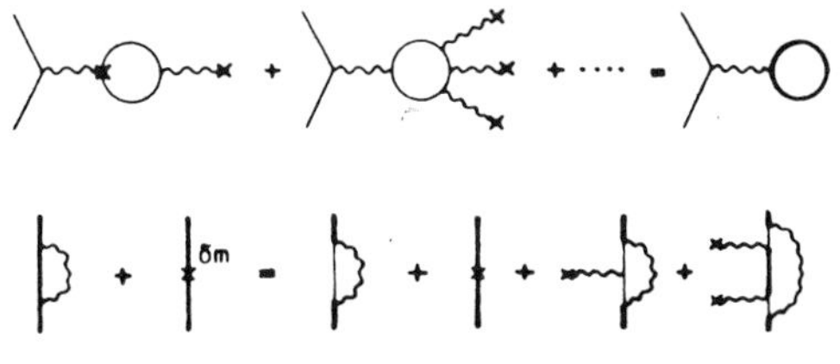

Figure 4. *Vacuum polarization (above) and self-energy corrections (below) to all orders in the nuclear charge.*

As the supercritical K-shell resonance is part of the negative energy continuum, i. e. of the Dirac sea, it is customary to consider it as part of the vacuum. Consequently one speaks of the *charged vacuum* state in supercritical quantum electrodynamics, and the spontaneously occuring process of pair creation involving the vacant 1s-state is known as *decay of the (neutral) vacuum*. That the supercritical vacuum, indeed, contains a nonvanishing charge is illustrated in Fig. 3, showing the vacuum polarisation charge density

$$\rho_{VP}(r) = -|e| \left(\frac{1}{2} \sum_{E<-m_e} \Psi_E^+(r)\Psi_E(r) - \frac{1}{2} \sum_{E>-m_e} \Psi_E^+(r)\Psi_E(r) \right) \tag{9}$$

for the supercritical nuclear charge Z=184. It is very similar to the charge distribution contained in the just subcritical occupied 1s-state at Z=172, which is also shown. Note that the space integral over ρ_{VP} for Z=184 does not vanish, indicating that the vacuum charge is real and not only a displacement charge as for $Z < Z_c$.

So far we have restricted our considerations to the single-particle picture. It is a relevant question whether these conclusions survive when higher order processes of quantum field theory are taken into account. The basic corrections are shown in the Feynman diagrams of Fig 4. It is clear that the usual perturbation expansion in powers of $Z\alpha$ cannot be trusted at the critical Z, hence all orders must be summed by use of exact propagators in the external field (indicated by thick lines). Denoting the exact electron propagator by $G(x,y)$ and the free photon propagator by $D(x,y)$, the corrections to the binding energy of the 1s-state can be written as:

a) vacuum polarisation (Fig. 4a):

$$\Delta E_{1s}^{VP} = -ie^2 \int d^3x \, d^3y \, \bar{\Psi}(x)\gamma^0\Psi(x)D(x-y) \int \frac{d\omega}{2\pi} Tr[\gamma^0 G(y,y';\omega)]_{y\to y'} \tag{10}$$

121

b) self-energy and vertex corrections (Fig. 4b):

$$\Delta E_{1s}^{SE} = +ie^2 \int dt \, d^3x \, d^3y \, \bar{\Psi}(x)\gamma^\mu G(x,y)\gamma_\mu \Psi(y) D(x-y)$$

$$+\delta m \int d^3x \, \bar{\Psi}(x)\Psi(x) . \tag{11}$$

Here $\Psi(x)$ denotes the 1s wavefunction. Eq (10) was evaluated [6, 7] to give a shift:

$$\Delta E_{1s}^{VP}(Z_c) = -10.68\text{keV} \tag{12}$$

at the critical point, increasing the binding energy. Expression (11), which is more difficult to evaluate, was calculated as well [8] to result in a repulsive contribution:

$$\Delta E_{1s}^{SE}(Z_c) = +10.99\text{keV} \tag{13}$$

at the diving point. The almost complete cancellation between the two contributions means that the total shift in the K-shell energy due to field theoretic corrections of α is only $+0.31$ keV, less than 10^{-3} of the total binding energy. It would be highly surprising, if higher orders in α (not $Z\alpha$) would change this picture. It therefore seems clear that the transition to a charged vacuum state must occur at a critical nuclear charge $Z \approx 173$.

It is legitimate to ask the question: How far do we have experimental proof that binding energies comparable to the rest mass of the electron, or twice the rest mass, can actually be achieved? This question actually has two aspects: First, how can the strong binding be set up experimentally and second, how can its presence be observed? The answer to the first question was given around 1970 by the Moscow as well as by the Frankfurt group (see [9] for a historical perspective). In collisions of two very heavy nuclei the electric field of a nucleus with charge $Z_1 + Z_2$ is simulated temporarily. The solution of the Dirac equation with two Coulomb centers showed that the critical binding of $2m_e$ should be reached in, e. g., U+U collisions at a distance $R_c \approx 30$ fm [10]. However, the binding energy varies with time as the nuclei move rapidly on their Rutherford trajectories and it is not so easy to determine the binding energy at a certain internuclear distance experimentally.

An approximate method was, nonetheless, proposed [11], making use of the generalization of Bang and Hansteen's [12] scaling law for direct ionization. In the case of superheavy collision systems, ionization occurs predominantly at the point of closest approach of the nuclei, R_0. One can then show [13, 14, 15], that the ionization probability of the 1s-state on a given scattering trajectory depends - to a good approximation - only on the binding energy at distance R_0:

$$P = D\exp(-\frac{\gamma R_0}{v}E_B^{1s}(R_0)) , \tag{14}$$

where γ is a numerical constant, v the beam velocity, and D is only a function of $Z_1 + Z_2$ which can be obtained by comparison with full-scale numerical calculations. In Fig. 5 we have shown binding energies extracted from measurements of 1s-vacancy production in the Pb+Cm system by Liesen et al. [16], in comparison with results of a two-center Dirac calculation carried out by W. Betz [17]. Although we cannot conclude that critical binding can be achieved, the existence of binding energies in the range between 500 and 800 keV appears to be fairly well established.

For a detailed overview of Vacuum properties in the presence of supercritical fields see [18].

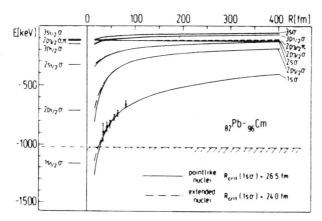

Figure 5. *Binding energy of the $1S_\sigma$-state in Pb+Cm, determined from measured Cm K-hole probabilities with help of the scaling law.*

DYNAMICS OF THE ELECTRON FIELD

2. Supercritical Heavy Ion Collisions

We refer now to the dynamics of the electron field in collision systems where the combined charge of both nuclei is sufficiently large to let the quasimolecular 1s-state enter the Dirac sea at a critical distance R_c. The electronic wavefunction $\Psi_i(\vec{r}, t)$, that satisfies the usual boundary conditions as $t \to -\infty$ and in addition solves the time dependent problem, may be expanded in a basis $\{\psi_j\}$. Specifically we choose this basis to coincide with the set of adiabatic, quasimolecular eigenstates of the two-center Dirac Hamiltonian [19], i.e.

$$\left(E_j(\vec{R}) - \hat{H}_{TC}(\vec{R}, \vec{r}) \right) \psi_j(\vec{R}, \vec{r}) = 0 , \tag{15}$$

$$\Psi_i(\vec{r}, t) = \sum_j a_{ij}(t) \psi_j(\vec{R}(t), \vec{r}) e^{-i\chi_j(t)} , \quad \chi_j(t) = \int_{t_0}^{t} dt \langle \psi_j | \hat{H}_{TC} | \psi_j \rangle . \tag{16}$$

The corresponding time dependent coefficients $a_{ij}(t)$ are determined by solving the coupled channel equations [19]. In order to obtain numerically reliable results for the evolution of the occupation amplitudes in continuum states, describing the resonance of width $\Gamma \approx 1$keV, it would be necessary to include continuum states spaced by much less than 1 keV at several points per 1 fm on the nuclear distance grid. This is far beyond numerical possibilities.

The difficulty can be avoided by employing an improved version of the auto-ionization picture. One artificially constructs a normalizable resonance wavefunction $\tilde{\varphi}_r$ for the supercritical 1s-state, e.g. by cutting off a continuum wavefunction in the centre of the resonance at its first zero (more sophisticated procedures have been devised and are routinely used [19]). In the next step a set of orthogonal states $\tilde{\varphi}_E$ are constructed in the negative energy continuum with the help of a projection technique. Those states are solutions of the projected Dirac equation

$$(H_{TC} - E)\tilde{\varphi}_E = \langle \tilde{\varphi}_r | H_{TC} | \tilde{\varphi}_E \rangle \, \tilde{\varphi}_r \qquad (E < -m_e). \tag{17}$$

Since $\tilde{\varphi}_r$ and $\tilde{\varphi}_E$ do not diagonalize the two-centre Hemiltonian, there exists a non-vanishing static coupling between the truncated 1s-resonance state and the modified negative energy continuum which describes the spontaneous decay of a vacancy in the supercritical 1s-state. The spontaneous decay width is given by the expression

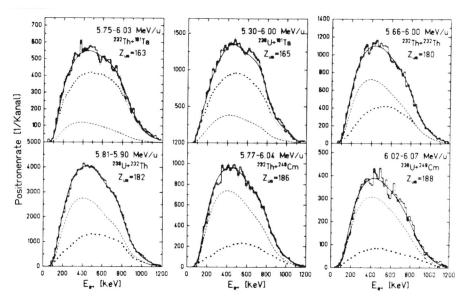

Figure 6. *Total positron spectra for various collision systems. Data: EPOS collabora-tion; dashed lines: QED pair production; dotted lines: nuclear pair conversion; solid lines: sum of both.*

$$\Gamma_E = 2\pi|\tilde{V}_E|^2, \qquad \tilde{V}_E = \langle \tilde{\varphi}_E | H_{TC} | \tilde{\varphi}_r \rangle. \tag{18}$$

Similarly the coupled channel equations for the amplitudes $a_{ik}(t)$ are amended by a "spontaneous" matrix element that does not vanish in the limit where the nuclei do not move:

$$\dot{a}_{ik} = -\sum_{j\neq k} a_{ij}\left(\langle \tilde{\varphi}_k | \frac{\partial}{\partial t} | \tilde{\varphi}_j \rangle + i\langle \tilde{\varphi}_k | H_{TC} | \tilde{\varphi}_j \rangle\right) \exp(i\chi_k - i\chi_j). \tag{19}$$

Careful investigations have shown that the asymptotic amplitudes $a_{ik}(\infty)$ are insensitive to the precise way of constructing the supercritical 1s-state, although the individual matrix elements may differ somewhat for the various procedures. This means that the concept of "spontaneous" pair production has no unique definition in the dy-namical environment of a heavy ion collision, except in the limiting case when the nuclei fuse (at least for some time) into a single compound nucleus. For collisions with-out nuclear contact, when the two nuclei move on hyperbolic Rutherford trajectories, the contribution from the "spontaneous" coupling constitutes only a small fraction, in any event. Accordingly, the calculations do not yield any perceptible change in the predicted positron spectra for such collisions when one goes from subcritical to super-critical systems. The total positron yield at fixed beam energy per nucleon for $Z_u > 137$ is predicted to grow at a very rapid rate that can be parametrized by the effective power law

$$\sigma_p(Z) \propto Z^n, \qquad n \approx 20, \tag{20}$$

showing no discontinuity at the transition to supercriticality. The large value of the exponent demonstrates the entirely nonperturbative nature of pair production in col-lisions of heavy ions, indicating the coherent participation of an average number of 10 virtual photons in the process. The fact that this prediction has been verified in the experiments at GSI is a major confirmation of our ability to accurately treat quantum electrodynamics in strong Coulomb fields by the theoretical methods described above, based on the adiabatic quasimolecular basis and the monopole approximation [19].

124

3. The "Atomic Clock" Phenomenon

When in the course of a heavy ion collision the two nuclei come into contact, a nuclear reaction occurs that lasts a certain time T. The length of this contact or *delay time* depends on the nuclei involved in the reaction and on the beam energy. For light and medium heavy nuclei the nuclear attraction is greater than, or comparable with, the repulsive Coulomb force, thus allowing for rather long reaction times of the order of 10^{-20}s or even more. For very heavy nuclei, in or beyond the Pb region, however, the Coulomb interaction is by far the dominant force between the nuclei, so that delay times are typically much shorter and in the mean probably do not exceed $1 - 2 \times 10^{-21}$s. Nuclear reaction models predict that the delay time increases with the violence of the collision, as measured by inelasticity (negative Q-value), and mass or angular momentum transfer.

A delay in the collision due to a nuclear reaction can lead to observable modifications in atomic excitation processes. The two main observable effects in such collisions are: (a) interference patterns in the spectrum of δ-electrons [20], and (b) a change in the probability for K-vacancy formation [21]. These effects have become known as the *atomic clock* for deep-inelastic nuclear reactions.

The origin of the atomic clock effect is most easily understood in a simple semi-classical model for the nuclear motion, where the nuclear trajectory is described by the classical function $R(t)$ and the only effect of the nuclear reaction is to introduce a time delay T between approach and separation of the nuclei, i.e. $\dot{R}(t) = 0$ for $0 \leq t \leq T$. To retain lucidity of the argument we make use of first-order perturbation theory for the excitation amplitudes a_{ik}:

$$a_{ik}(\infty) = - \int_{-\infty}^{\infty} dt\, \dot{R}(t) \langle \varphi_k | \frac{\partial}{\partial R} | \varphi_i \rangle \exp[i \int_0^t dt' (E_k - E_i)]. \tag{21}$$

The range of the main integral splits into three parts: (a) $t < 0$, (b) $0 \leq t \leq T$, and (c) $t > T$. Because $\dot{R}$ enters as a factor in eq. (21), the median part does not contribute. For the last part one can rewrite $t \rightarrow t + T$ so that the integral runs from 0 to ∞. Because this exit part of the nuclear trajectory is just the time-reverse of the entrance part, i.e. $\dot{R}(-t) = -\dot{R}(t)$, the amplitude from part (c) can be expressed as the complex conjugate of the amplitude $a_{ik}(0)$ from part (a) of the integral, except for a phase factor resulting from the variable substitution in the phase integral in eq. (21). Thus we find the relation:

$$a_{ik}^T(\infty) = a_{ik}(0) - a_{ik}^*(0) \exp[iT(E_k - E_i)], \tag{22}$$

where the energies have to be taken at the distance of nuclear contact. If we write $a_{ik}(0)$ in the symbolic form $a_0 e^{i\alpha}$, the final probability for excitation between states i and k becomes

$$P_{ik}(T) = |a_{ik}^T(\infty)|^2 = 4a_0^2 \sin^2(\frac{1}{2}T\Delta E - \alpha), \tag{23}$$

where $\Delta E = (E_k - E_i)$ is the transition energy. The excitation probability is obviously an oscillating function, either of T for a given transition $i \rightarrow k$, or a function of transition energy ΔE for fixed delay time T.

In reality, of course, things are more complicated. Except in truly elastic collisions the outgoing trajectory is not a precise mirror image of the approaching trajectory. Furthermore, multi-step excitations play an important role in very heavy systems. The total excitation amplitude therefore contains a mixture of contributions from different intermediate states. The simple expression, eq. (22), must then be replaced by the formula

$$a_{ik}^T(\infty) = \sum_j a_{ij}^{in} e^{-iE_j T} a_{jk}^{out}. \tag{24}$$

Finally, the nuclear delay time T is usually not sharply defined, so that an average over a distribution $f(T)$ of delay times has to be taken in eq. (23). The common result of these refinements is a dilution of the interference patterns, i.e. the oscillations become less pronounced. For short delay times and a large uncertainty of T all that remains is a partially destructive interference between the incoming and outgoing branches of the trajectory, observable as a decrease in the K-vacancy yield or a steepening of the slope of the low-energy part of the δ-electron spectrum.

It was demonstrated by O. Graf et al. [22] that collisions of *fully stripped ions* with a small nuclear time delay could serve to trigger supercriticality. As a unique signal for spontaneous pair creation a change in the *angular correlation* between electron and positron is expected that occurs only in supercritical systems: For nuclear delay times of 2×10^{-21}s the normal forward correlation should turn into a backward correlation.

4. Positron Production in Delayed Collisions

In principle, the positron spectra contain the same information about nuclear time delay as the electron spectra. However, because of their low emission probability, positrons are not as useful from a practical point of view, at least in subcritical collision systems. Nevertheless, positron spectra emerging from deep-inelastic heavy ion collisions have been measured [23], and the first experimental observation of the atomic clock phenomenon in heavy ion collisions came in fact from positrons [24]. The yield argument does not necessarily apply to supercritical collision systems for two reasons. Firstly, a reaction-induced nuclear time delay may allow for the detection of spontaneous pair-creation in these systems, as will be discussed below. Secondly, a tiny component of very long reaction times ($T > 10^{20}$s) might become visible in the positron spectrum, because the spontaneous emission mechanism acts as a kind of "magnifying lens" for long delay times [25].

In order to see why this is so, we return to eq. (19) for the amplitudes a_{ik} in a supercritical system, which contained the additional time-independent couplings $\tilde{V}_E$ between the resonant bound state and the (modified) positron continuum states. The presence of this coupling has the effect that the contribution to the integral in eq. (21) from the central time interval $0 \leq t \leq T$ does not vanish any longer. The total amplitude a_E^T for emission of a positron from the supercritical bound state contains therefore an additional term compared with eq. (22):

$$a_E^T(\infty) = a_E(0) - a_E^*(0)\exp[iT(E - E_r)] - \tilde{V}_E \frac{e^{iT(E-E_r)} - 1}{E - E_r}, \qquad (25)$$

where E_r is the energy of the supercritical state when the nuclei are in contact. For $T = 0$ the new term vanishes, but grows rapidly with increasing T. For delay times considerably greater than 10^{-21}s the additional term in eq. (25) begins to dominate over the first two terms , causing the emergence of a peak in the positron spectrum at the energy of the supercritical bound state:

$$|a_E^T(\infty)|^2 = \frac{\Gamma_E}{2\pi}T^2\frac{\sin^2[(E - E_r)T/2]}{[(E - E_r)T/2]^2}, \qquad (T \ll \Gamma_E^{-1}). \qquad (26)$$

The energy distribution has a width $\Gamma(T) = 2\pi/T$ as would be expected on grounds of the uncertainty relation, and the total probability for positron emission grows proportional to T. For extremely long delay times, $T > \Gamma_E^{-1}$, the energy distribution instead of (26) goes over into a Breit-Wigner curve centred at E_r with width equal to the spontaneous decay width Γ_E.

The emergence of a peak in the positron spectrum for strongly delayed supercritical collisions is strikingly demonstrated in Fig. 7. In the subcritical case the delay causes interference patterns like those already known from electron spectra, effectively

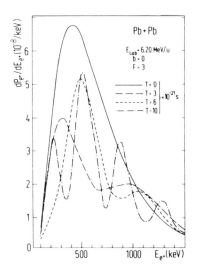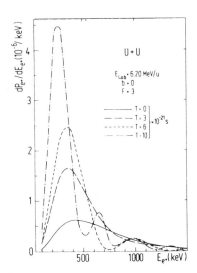

Figure 7. *Effect of a nuclear time delay on the positron spectrum. Left: subcritical system (Pb+Pb). Right: supercritical system (U+U).*

reducing the positron yield. In the supercritical case the positron yield increases dramatically when the delay time exceeds about 3×10^{-21}s. Unfortunately, this is beyond the range accessible for the average time delay in deep-inelastic reactions where not much more than 10^{-21}s has been observed. Still, the situation may not be entirely hopeless, because the intensity of the line structure grows with T and simultaneously becomes more localized at the resonance energy. In principle, even a very small tail of the delay time distribution $f(T)$ could acquire sufficient weight to be visible in the positron spectrum [26].

Such long delay times could only occur if an attractive pocket is present in the internuclear potential for supercritical collision systems, for which no conclusive theoretical or experimental evidence exists at present. But even if a pocket were there, the existence of a sufficiently large tail of long delay times in the distribution $f(T)$ is not ensuréd [27]. Although very narrow positron peaks were obtained for beam energies in a small window around the Coulomb barrier, their intensity was much too low to allow for observation. But again it must be emphasized that these models are too simple to permit definite conclusions for the realistic case.

5. Structures in the Positron Singles Spectrum

When the line structures in the positron spectra were first detected at GSI, they were associated with the spontaneous positron emission line that was predicted by theory for supercritical collision systems with long nuclear time delay. This was quite natural, because that had been the aim and inspiration of the experiments from the beginning. For the first two systems that were investigated, U+Cm and U+U [31, 28], this explanation worked quite nicely; the position of the line agreed rather well with the expected spontaneous emission peak. The measured spectra could be described in the framework of schematic models involving the intermediate formation of a long-lived ($T \approx 5 \times 10^{-20}$s) "giant" di-nuclear system [26, 28, 29].

Of course, the experimentalists were very cautious to make sure that the lines would not be a trivial artefact caused by pair decay of some excited nuclear state. This can be checked experimentally, since a pair-decaying nuclear state can always decay in another way, either by photon emission (if the transition multipolarity is not $L = 0$) or by internal conversion to a K-shell electron. The latter process works for any multipolarity. The branching ratios for the various decays can be calculated essentially model

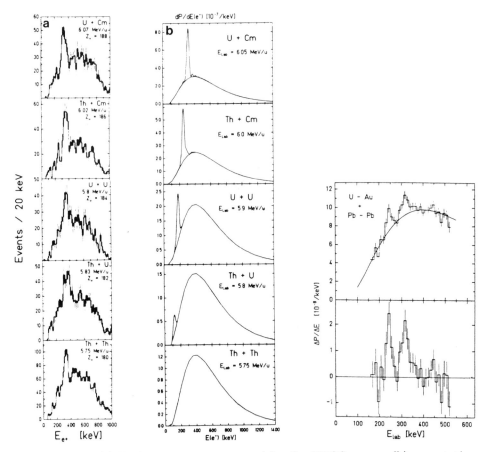

Figure 8. Left: (a) Positron spectra measured by the EPOS group; (b) expectations for spontaneous positron creation. Right: Combined positron spectra from the systems Pb+Pb and U+Au measured by the ORANGE group.

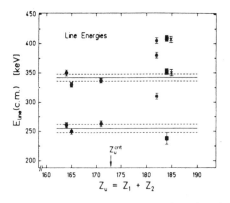

Figure 9. Peak positions measured as function of Z_u by the ORANGE group.

independently, because the nucleus is small compared to the wavelength of the emitted real or virtual photon. The photon and electron spectra were measured simultaneously in the relevant energy range (beyond 1 MeV), but no associated structure was found [31, 28].

The observed line width of 70-80 keV provided a second argument against nuclear pair decay. If the structures were emitted from the scattered nuclei they would have to be Doppler broadened due to the motion of the source. At 45° (lab) scattering angle the broadening would amount to 100 keV, i.e. more than the observed line width even for an intrinsically monochromatic structure. However, the positron spectrum emerging from a normal nuclear pair decay is not monochromatic at all! The energy of the transition is shared between the electron and the positron, and a broad peak develops at the upper end of the positron spectrum only for heavy nuclei due to Coulomb effects. Normal pair conversion could thus be ruled out by line width arguments as well [31].

An intrinsically monochromatic positron line could, in principle, be caused by a process called monoenergetic pair conversion, which can occur if an inner atomic shell is not fully occupied. The electron can then be captured in this bound state, and the positron carries away the full remaining energy. The sharply defined energy is characteristic of a two-body decay $A^* \rightarrow (A+e^-) + e^+$, whereas the normal pair decay into a free electron-positron pair is a three-body decay $A^* \rightarrow A + e^- + e^+$. Although a large number of inner-shell vacancies are created in the heavy ion collision, monoenergetic pair conversion is expected to be strongly suppressed, because the vacancies are filled by transitions from outer shells within about 10^{-17}s. This filling time is at least two orders of magnitude shorter than the lifetime of nuclear excited states. Therefore, a possible origin of the line structures by monoenergetic pair conversion was ruled out, too, on experimental [31] and theoretical grounds [30].

The dependence of the line structures as function of combined nuclear charge $Z_u = Z_1 + Z_2$ afforded a crucial test of the hypothesis that they could be attributed to spontaneous positron production. The line must then occur at the positron kinetic energy corresponding to the energy E_{1s} of the 1s-resonance that is imbedded in the Dirac sea: $E_{peak} = |E_{1s}(Z_u)| - m_e$. The surprising result of such a study by the EPOS collaboration [32] is shown in the left-hand part of Fig. 8: The position of the peak was always in the range 350 ± 30 keV essentially independent of Z_u! For comparison, the expectation for a peak caused by spontaneous pair creation in the strong Coulomb field is also shown in part (b) of the figure. Starting at about 320 keV in the U+Cm system the line should move to lower energies and decrease in intensity, assuming similar nuclear delay times for all systems. Fig. 9 demonstrates that the observed structures fall into three groups, at the positron energies 250, 330, and 400 keV, respectively, with an uncertainty of about 20 keV.

As the details and results of these experiments are covered by separate talks at this conference, we only briefly summarize some of the main results (cf. also earlier conference reports and reviews [33, 34, 35, 36])

- Lines have been observed for a large variety of collision systems, ranging from $Z_u = 163$ (Th+Ta) up to $Z_u = 188$ (U+Cm) and involving nuclei with widely different structure.

- The line positions appear to fall into several groups between 250 and 400 keV; their width is about 70 keV, if all positron emission angles are covered. This value corresponds to the Doppler width of a sharp line emitted by a source moving with center-of-mass velocity.

- A number of lines are common to different collision systems and to both experiments (ORANGE and EPOS).

- Nuclear pair conversion processes ($A^* \rightarrow Ae^-e^+$) appear to be excluded from γ-ray and electron spectra, linewidth, and A-independence.

6. Correlated Electron-Positron Lines

The A- and Z-invariance of the line energies strongly hint at a common source that in itself is not related to the nuclei or the strong electric field, although the strong Coulomb field might play a role in the production of this source. Since no Z-dependence is seen, the most natural candidate for such a source would be some (neutral) object that moves with the velocity of the centre of mass and eventually decays into a positron and a single other particle. (A two-body decay must be invoked to explain the narrow linewidth, as mentioned before.) Could the second decay product simply be a second electron, i.e. could it be that one sees the pair decay of a neutral particle, $X^0 \rightarrow e^+e^-$, with a mass somewhat below 2 MeV [37]?

The search for a correlated electron peak, as performed first by the EPOS collaboration, was successful [38]. Later the existence of this correlated line structure could be confirmed by the ORANGE collaboration as well [39]. We restrict ourselves to a summary of the experimental results that have been accumulated over the last years concerning correlated line structures in electron-positron coincidence spectra:

- Lines at 620 and 810 keV sum energy have been observed by the EPOS and the ORANGE collaboration. A third line at 750 keV was only seen by the EPOS group. The peaks seem to occur at the same positions in various systems, e.g. U+Ta ($Z_u = 165$), U+Th ($Z_u = 182$) and U+U ($Z_u = 184$).

- The width of the sum energy peaks lie in the range 20-40 keV; they are much narrower than the positron singles peaks. The source must move slowly ($\beta_s \leq 0.05$).

- The 810 keV line appears to be caused by back-to-back emission. The 750 keV line in U+Ta could be forward correlated.

- The difference-energy spectra exhibit a broad peak near zero energy, indicating that the lepton pair is not produced inside the strong Coulomb field, or in the vicinity of a third body. Again, the 750 keV line in U+Ta appears to be an exception to this rule.

Finally, it should be mentioned that the line intensity depends very sensitively on the beam energy as depicted in Fig.10

Many features appear to be compatible with the assumption that one observes the pair decay of at least three neutral particle states in the mass range between 1 and 2 MeV. These states must have a lifetime of more than 10^{-19}s (because of the narrow linewidth) and less than about 10^{-9}s (because the vertex of the lepton pair is within 1

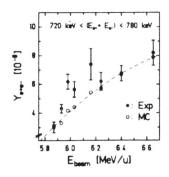

Figure 10. *The intensity of the line as function of the beam energy*

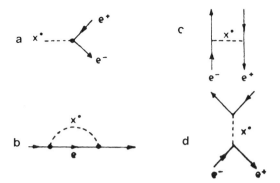

Figure 11. *Feynman diagrams for (a) pair decay of X^0, (b) contribution to the electron anomalous magnetic moment, (c,d) positronium hyperfine splitting.*

cm of the target). Surely, some pieces of data do not really fit into this picture, e.g. the characteristics of the 750 keV line observed in U+Ta. But it is by no means clear at the present time, whether these features provide conclusive evidence against the particle hypothesis, especially if one is not dealing with simple, elementary particle states.

Nevertheless, the very idea that a whole family of neutral particle states in the MeV mass range should have remained undetected through more than 50 years of nuclear physics research is hard to accept for the conservative mind. Most physicists, when first confronted with the GSI data, have therefore tried to explain the data in terms of known nuclear or atomic physics. As mentioned before, nuclear pair decay would be the most natural explanation. However, none of the proposed scenarios has yet stood up against a detailed comparison with the experiments. As long as this is so, nuclear pair decay cannot be considered as a viable alternative to the particle hypothesis. Similar remarks apply to attempts to explain the GSI peaks in terms of atomic physics. None of the ideas that were studied quantitatively have been successful, even if they were based on plausible, but unfounded, ad hoc assumptions.

7. Limits on Light Neutral Bosons from Precision Experiments

Even when the hypothesis of a new neutral particle was first seriously discussed [40, 41], it was recognized that the precision experiments of quantum electrodynamics provide stringent limits on the coupling of such light particles to the electron-positron field and to the electromagnetic field. The strength of this argument lies in the fact that any particle X^0 which decays into an e^+-e^- pair must couple to the electron-positron

Table 1. *Limits on the coupling constant, lifetime, and pair decay width of neutral elementary bosons with mass $M_X = 1.8$ MeV derived from the anomalous magnetic moment of the electron.*

Particle type	Spin J^π	Vertex Γ_i	Max. coupling $\alpha_{Xe} = g_i^2/4\pi$	Min. lifetime τ_X (s)	Max. width Γ_X^{ee} (meV)
Scalar (S)	0^+	1	7×10^{-9}	2×10^{-13}	3.0
Pseudoscalar (P)	0^-	$i\gamma_5$	1×10^{-8}	1×10^{-13}	6.8
Vector (V)	1^-	γ_μ	3×10^{-8}	4×10^{-14}	16
Axial vector (A)	1^+	$\gamma_\mu\gamma_5$	5×10^{-9}	5×10^{-13}	1.4

field. At least in the low-energy limit, the coupling can be expressed by an effective interaction of the form:

$$L_X = g_i(\bar{\psi}\Gamma_i\psi)\phi, \tag{27}$$

where ψ denotes the electron-positron field, ϕ the X^0 field, and Γ_i with $i =$ S,P,V,A stands for the vertex operator associated with the various possible values of spin and parity of the X^0 particle. Given the interaction Lagrangian (27) one can calculate the lifetime of the X^0 particle against pair decay as well as the contributions to QED processes by virtual exchange of an X^0. The most sensitive of these is the anomalous magnetic moment a_e, because of the high experimental accuracy [42]. A reasonable estimate for the 95% confidence limit is $\Delta a_e < 2 \times 10^{-10}$.

As illustrated in Fig. 11, the contribution of a hypothetical X^0 particle to the value of a_e involves two vertices between an electron or positron and an X^0, and is thus proportional to the effective coupling constant $\alpha_{Xe} = g_i^2/4\pi$. The same applies to the decay rate τ_X^{-1}, which involves the square of an amplitude with a single vertex, and to the contribution of an X^0 particle to the hyperfine splitting of the positronium ground states.

The limits derived from these considerations [43] on the X^0-coupling constant and its lifetime are listed in Table 1. Particles with lifetime $\tau_X > 10^{-13}$s cannot be ruled out by this argument. Considering that the experimental conditions only require a lifetime below about 1 ns, there remains an unexplored range of four orders of magnitude in τ_X.

Similar upper limits can be derived for the coupling of an X^0 boson to other known particles [43]. That for the coupling to the muon, $\alpha_{X\mu}$, is about one order of magnitude weaker because of the lower accuracy in the value of a_μ. A limit on the product of the coupling constants to the electron and to nucleons is obtained from the Lamb shift in hydrogen and from the K-shell binding energy in heavy elements, one finds $\alpha_{Xe}\alpha_{XN} < 10^{-14}$. For scalar particles an extremely stringent bound on the coupling to nucleons can be derived from low-energy neutron scattering: $\alpha_{XN} < 10^{-9}$ [44]. Special bounds for vector (gauge) bosons were considered by Zee [45]. Finally, measurements of nuclear Delbrück scattering yield an upper limit on the coupling of a spinless X^0 boson to the electromagnetic field through an effective interaction of the type

$$
\begin{aligned}
L_{X\gamma\gamma} &= g_S(E^2 - H^2)\phi_X \quad &\text{(scalar)} \\
L_{X\gamma\gamma} &= g_P(\vec{E} \cdot \vec{H})\phi_X \quad &\text{(pseudoscalar)}.
\end{aligned}
\tag{28}
$$

The limits are: $g_S < 0.02$ GeV^{-1} and $g_P < 0.5$ GeV^{-1}. They provide lower limits for the lifetime against decay into two photons: $\tau_{\gamma\gamma}(X^0) > 6 \times 10^{-11}$s for a scalar particle and $\tau_{\gamma\gamma}(X^0) > 4 \times 10^{-13}$s for a pseudoscalar particle [46].

8. Inadequacy of Perturbative Production Mechanisms

One consequence of these results was that the particle hypothesis could be rejected off-hand. On the other hand, the condition that the coupling constant between the hypothetical X^0 boson and the particles involved in the heavy ion collision, i.e. electrons and nucleons, must be very small creates severe problems for any attempt to explain the measured production cross section of about 100 μb by a perturbative interaction of the type shown in eq. (27) [40, 41, 43]. Also the cross section for production by the strong electromagnetic fields present in the heavy ion collision falls short by several orders of magnitude, if it is based on the Lagrangian (28) or similar perturbative interactions [47, 48, 49, 50, 51, 52].

A second serious difficulty with the interactions (27) and (28) is that they favour the production of particles with high momenta due to phase space enhancement. For collisions with nuclei moving on Rutherford trajectories the calculated spectra typically are very broad, peaking at velocities $\beta_X > 0.5$. As discussed in the previous chapter, the experiments would require an average particle velocity $\beta_X < 0.05$!

Both these problems could, in principle, be circumvented by the assumption that a very long-lived, excited giant compound nucleus is formed [47, 51], but only at the price of violating other boundary conditions set by the experimental data, e.g. the absence of a much larger peak in the positron spectrum caused by spontaneous pair production [53]. A further counterargument is that the emission of slow particles is tremendously suppressed by phase space factors [52]. One might also consider the possibility that the X^0 particles are somehow slowed down after production, but this cannot be achieved with the interactions discussed above.

Two mechanisms remain, which can conceivably ensure the survival of the particle hypothesis: (1) a form factor that cuts off production at large momenta [54]; and (2) a non-perturbative production mechanism, e.g. production in a bound state around the two nuclei [55, 56, 57]. Both mechanisms require particles with internal structure.

9. Axion Searches

At first the axion [58], i.e. the light pseudoscalar Goldstone boson associated with the breaking of the Peccei-Quinn symmetry required to inforce time-reversal invariance in quantum chromodynamics, seemed like a plausible candidate for the suspected X^0 boson. The interest in an axion was revived when it was realized that there was, indeed, a gap left by previous axion search experiments for a short-lived axion in the mass range around 1 MeV [59]. However, new experimental studies of J/Ψ and Υ decays [60, 61, 62] quickly ruled out the standard axion.

10. Beam Dump Experiments

Beam dump experiments, in particular those with a high-energy electron beam, are an excellent source of rather model-independent bounds on the properties of hypothetical light neutral particles [63]. The initial electron might undergo a bremsstrahlung process. Of course, instead of radiating a photon, the electron can emit some other light neutral particle X^0, if any exists. Except for effects from the particle mass and spin, the expected cross section is given by the cross section for photon radiation, multiplied by the ratio of the coupling constant of the emitted particle to the electron and the electromagnetic coupling constant:

$$\frac{d\sigma_X}{d\Omega dE} = \frac{\alpha_{Xe}}{\alpha}\frac{d\sigma_\gamma}{d\Omega dE}. \tag{29}$$

Table 2. *Excluded ranges of the coupling constant α_{Xe} of a pseudoscalar particle of mass 1.8 MeV derived from beam dump experiments.*

Experiment	Beam	Target	α_{Xe}^{min}	α_{Xe}^{max}
Konaka et al. (KEK) [65]	e^- (2.5 GeV)	W + Fe(2m)	10^{-14}	4×10^{-8}
Davier et al. (Orsay) [66]	e^- (1.5 GeV)	W (10cm)	10^{-11}	10^{-8}
Riordan et al. (SLAC) [67]	e^- (9.0 GeV)	W (10-12cm)	10^{-12}	10^{-7}
Bechis et al. [68]	e^- (45 MeV)		10^{-13}	10^{-10}
Brown et al. (FNAL) [69]	p (800 GeV)	Cu (5.5m)	10^{-10}	10^{-7}

An upper limit for the measured cross section for the X^0-particle cross section hence yields an upper limit for the coupling constant α_{Xe}. A simple formula for the bremsstrahlung cross section has been given by Tsai [64] (see also ref. [70]).

However, for every beam dump experiment there is not only a lower bound for the range of excluded coupling constants but also on upper bound, for the following reason. The lifetime of the hypothetical particle against pair decay is inversely proportional to the α_{Xe}. For sufficiently large values of the coupling constant almost all produced particles therefore decay inside the beam dump, and the e^+e^- pair produced in the decay is absorbed in the target. It is clear that a good value for this other limit requires a short beam dump, whereas high cross-section and low background require a thick target.

Hence, the result of a beam dump experiment is a region of excluded values of α_{Xe}, i.e. the coupling cannot be in the range $\alpha_{Xe}^{min} < \alpha_{Xe} < \alpha_{Xe}^{max}$. In the analysis one assumes that the neutral particles interact so weakly that they pass essentially undisturbed through the target.

The conditions and results of corresponding experiments are listed in Table 2, where the excluded ranges of the coupling constant are given for pseudoscalar particles of mass 1.8 MeV. For a scalar particle the bounds would be similar, but for spin-one particles about one order of magnitude better lower limits would be obtained. Also listed in Table 2 is a proton beam dump experiment performed at Fermilab. Due to the production of secondary electrons and positrons in the target, a limit is obtained also for the coupling to electrons.

Together, the experiments exclude the range of coupling constants α_{Xe} between 10^{-14} and 10^{-7}, corresponding to lifetimes against pair decay in the range 10^{-14}s $< \tau_X < 10^{-7}$s. When combined with the bounds derived from the electron anomalous magnetic moment a_e and by experimental conditions, the beam dump results conclusively rule out any elementary neutral particle as source of the GSI e^+e^- events.

However there is still a "loop-hole" for neutral particles left, as was revealed by an analysis of A. Schäfer, who calculated the bremsstrahlung production cross section for extended particles [63]. He showed that a finite form factor can invalidate the experimental bounds, if the emitted particle has a radius of more than about 100 fm (10^{-11}cm). Basically this comes along with an effective suppression of the e^--X^0-vertex due to the X^0 form factor, when the relevant electronic de Broglie wavelength is small compared to the spatial extension of the X^0.

11. Bhabha Scattering at MeV Energies

All the limits on the possible existence of a light neutral X^0-boson discussed so far were derived assuming that the particle has no internal structure. When one allows

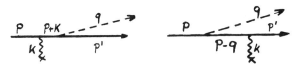

Figure 12. *Feynman diagrams for (a) radiation of a neutral boson by an electron scattering in the Coulomb field of a nucleus.*

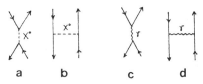

Figure 13. *Feynman diagrams contributing to Bhabha scattering. Only the s-channel diagram (a) is resonant and can compete with the QED process represented by the diagrams (c) and (d).*

for a particle with finite size, they become model dependent, as was mentioned in the previous section. Similar considerations apply to the bounds from the anomalous magnetic moment, Delbrück scattering, positronium hyperfine structure, and so on. The reason for this model dependence lies in the fact that all these processes involve particles off their mass shell, either the electron or the X^0-boson, whereas all particles are on mass shell in the pair decay $X^0 \to e^+e^-$. A form factor, therefore, enters in different ways into these processes.

In order to obtain model-independent bounds it is necessary to consider the process, in which the boson is produced on shell by electrons and positrons that are also on mass shell. This process is called Bhabha scattering on resonance, and represented by the Feynman diagram (a) in Fig. 13. Averaging over electron and positron spin the cross section from this diagram alone is narrowly peaked around the beam energy E_R corresponding in the centre-of-mass system to the rest mass of the X^0-boson:

$$\bar\sigma_X(E) = \frac{\pi\alpha_{Xe}^2 f_{J\pi}(m_X/m_e)}{4(E - E_R)^2 + (m_X\Gamma_X/m_e)^2} \tag{30}$$

where $E_R = (m_X^2/2m_e) - 2m_e$, and $f_{J\pi}(x)$ is a dimensionless function of order unity that depends on spin and parity of the X^0-boson [43]. Right on resonance, i.e. for $E = E_R$, the cross section exhausts the unitarity limit for a single partial wave

$$\bar\sigma_X(E_R) = \frac{\pi\alpha_{Xe}^2 m_e^2 f_{J\pi}}{m_X^2\Gamma_X^2} \approx \frac{\pi}{m_X^2} \cdot \frac{\Gamma_X^{ee}}{\Gamma_X}, \tag{31}$$

unless other final states, such as $\gamma\gamma$, $\gamma\gamma\gamma$ or $\bar\nu\nu$, contribute significantly so that the partial e^+e^- width Γ_X^{ee} is not equal to the total width Γ_X. On the other hand, the QED Bhabha scattering cross section, described by the Feynman diagrams (c) and (d) in Fig. 13, is of the order

$$\bar\sigma_{QED}(E_R) \approx \frac{\alpha^2}{m_X^2} \approx 10^{-4}\bar\sigma_X(E_R). \tag{32}$$

At first glance, therefore, it appears as if the resonance caused by the new particle would give a tremendous signal in Bhabha scattering. In reality, the cross section must

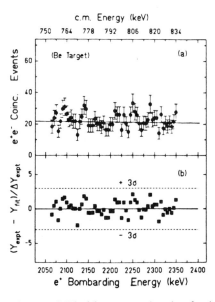

Figure 14. *Energy dependence of Bhabha scattering in the invariant mass region between 1.80 and 1.86 MeV, measured at Grenoble. No structure is seen at an error level of 0.25%.*

be averaged over a finite energy interval ΔE, which depends on the experimental conditions. The QED cross section (32) must then be compared with the energy averaged resonance cross section

$$\frac{1}{\Delta E} \int dE \bar{\sigma}_X(E) = \frac{2\pi^2(2J+1)}{m_X^2 - 4m_e^2} \cdot \frac{\Gamma_X^{ee}}{\Delta E}. \tag{33}$$

In practice, the energy resolution is not determined by the uncertainty in the positron beam energy, but by the Fermi motion of the electrons in the target [71]. In the limiting case of a free electron gas with Fermi momentum k_F, the energy resolution is given by

$$\Delta E = 2k_F\sqrt{2E_R/m_e} = 2k_F\sqrt{\frac{m_X^2}{m_e^2} - 4}. \tag{34}$$

At fixed beam intensity this value can be reduced only at the expense of scattering rate, because the Fermi momentum is related to the electron density n_e in the target, viz. $k_F = (3\pi n_e)^{1/3}$. Using light target materials such as Be, the effective resonance width ΔE is about 30 keV as can be deduced from the Compton profile [72].

The experimental sensitivity can be increased by looking for delayed pair decay of the resonance. This has been done using an active shadow [73] or an energy loss technique [74]. Presently, the highest sensitivity has been achieved by the group working at Grenoble [73]. The upper limit for the e^+e^- width and lifetime of a hypothetical spinless X^0-boson are reported to be:

$$\Gamma_X^{ee} < 8.8 \times 10^{-5} \text{eV}, \qquad \tau_X > 7.5 \times 10^{-12} \text{s}. \tag{35}$$

All structures reported by earlier experiments are clearly excluded by this limit. We can conclude that the Bhabha scattering experiments have now reached a sensitivity that allows to establish limits on new particles which are comparable to those derived

Figure 15. *Diagrams for electromagnetic decay modes of a neutral boson: (a) two-photon decay, (b) three-photon decay, (c) pair decay ($J^\pi = 1^-$), (d) pair decay (other J^π).*

from other QED precision experiments. The crucial advantage of the new limits is that they are measured on mass shell and, therefore, independent of assumptions about the structure of the X^0-boson.

The latest limit (35) is about one order of magnitude away from the bound set by the GSI experiments, i.e. $\tau_X < 10^{-10}$s.

MODELS OF NEW EXTENDED NEUTRAL PARTICLES

12. General Considerations

The postulate of new neutral particles with finite size, or substructure, can simultaneously solve several general difficulties of any explanation of the GSI data in terms of particle decay. These are:

- The fact that several line structures have been seen is naturally explained as the decay of internally excited states of the same particle.

- The small velocity of the pair-decaying source may be explained in two ways: either as a high-momentum cut-off due to the X^0 form factor, if $R_X > 20m_X^{-1} \approx$ 2000 fm; or by production of the X^0-boson in a bound state around both nuclei.

- A composite particle with electrically charged constituents could be efficiently produced by some non-perturbative mechanism that requires the presence of strong Coulomb fields.

- As already argued in the previous section, a general bonus is that all experimental limits are rendered irrelevant for a sufficiently large radius R_X, with the exception of those derived from resonant Bhabha scattering.

Moreover, a general conclusion can be drawn with respect to the competition between two-photon and pair decay. Unless the particle is a bound state of electron-positron pairs, or has a fundamental coupling to the electron field (as the axion would!), the photon decay dominates for all states except those with spin one and negative parity. The argument goes as follows: For electromagnetic decays, two-photon decay and pair decay via a virtual photon (diagrams (a) and (c) in Fig. 15 are both of order α^2, three-photon decay (diagram (b)) is of order α^3 and pair decay via two virtual photons (diagram (d)) is of order α^4. Decay into two photons is possible for all particles except those with spin one, for these three photons are needed in the final state. On the other hand, pair decay via a single virtual photon requires that the particle carries the quantum numbers of the photon, i.e. $J^\pi = 1^-$. Thus pair decay dominates for a vector boson, three-photon decay is the main decay mode for an axial vector boson, and all other particles would predominantly decay into two photons. (This conclusion does not apply if the particle is already composed of one or several electron-positron pairs, which may be emitted without interaction with an intermediate virtual photon.)

Two general routes can be taken by the theorist who wants to construct a model of extended particles in the mass range between 1 and 2 MeV:

- One can speculate that there exists an undiscovered, "hidden" sector of low-energy phenomena within the framework of the standard model of particle physics, i.e. within the $SU(3) \times SU(2) \times U(1)$ gauge theory. This might be a non-perturbative, strongly coupled phase of quantum electrodynamics, low-energy phenomena associated with the Higgs sector of the Glashow-Salam-Weinberg model, or some unknown long-range properties of QCD. It has even been speculated that the standard electromagnetic interaction between charged particles with spin behaves quite differently at short distances than normally assumed in perturbation theory.

- One can invoke new interactions which, for some reason that remains to be explained, do not normally show up in experiments. Examples are many-body forces between electrons and positrons that do not contribute in positronium, or new light fermions that are confined by equally new, medium ranged interaction.

Both roads have been extensively explored during the past three years, overall with little success. One must be aware that any attempt to fit a scheme of new low-energy composite particles into the standard model faces awesome obstacles, viz. the wealth of experimental data and precision measurements accumulated over fifty or more years. For the first class of models, i.e. those models based on some obscure aspect of the standard model itself, the problem is that there is essentially no free parameter. Every conjectured phenomenon can be calculated reliably, at least in principle. Many theorists, who took to this road, have (therefore?) speculated about states which are so complicated in nature that their calculation is, at present, beyond technical means.

Another intensively discussed hypothesis, first put forward by Celenza et al. [75] assumes that QED may possess a second strongly coupled phase resembling in its properties the normal vacuum of QCD and that this new vacuum may be formed in heavy ion collisions. Then the GSI peaks are interpreted as being caused by the decay of 'abnormal QED mesons' [76, 77, 78]. This topic has been covered by several talks at this conference and shall not be reviewed here. Let us only mention that at present no mechanism is known [79, 80] by which the field of the colliding heavy ions can trigger a transition to the new phase (which requires a large value of the fine structure constant α [81]). Furthermore it seems doubtful that the new phase is metastable, i.e. it can exist in the absence of the catalyzing nuclear charges as would be required by the characteristics of a two-body decay.

Although free parameters can be introduced in abundance, the second route is no less treacherous. A new force active in the MeV energy range can potentially show up in every atomic, nuclear, or particle physics experiment. This has led, for instance, to the rejection of speculations about many-body forces between electrons. On the other hand, attempts to fit new interactions with consequences at low energies into the standard model are not entirely hopeless. If some yet unknown interaction at the TeV range could be responsible for the observed mass spectrum of leptons and quarks, as technicolour models suggest, why should it not be possible to construct new low-mass particles in a similar manner?

13. Poly-positronium

In the following we do not attempt to give a complete review of the theoretical attempts to construct models for extended X^0-objects, rather we will concentrate on a few selected models.

There is no mechanism known within the framework of QED for the strong binding required to bring such states far below the threshold of at least $4m_e$. This constitutes the main objection against recent claims that strongly bound states of two electron-positron pairs ("quadronium") hold the key to the solution of the GSI positron puzzle [82]. An equation of Bethe-Salpeter type has been derived for this system [83], but no indication for a strongly bound state has been found. (The $(e^+e^-)^2$ system is known to have a very weakly bound state with binding energy of a few eV [84], which has the structure of an ordinary positronium molecule.)

Strongly bound $(e^+e^-)^n$ states, the so-called "poly-positronium" states, probably would require the assumption that some new, non-QED force exists between electrons and positrons. On this basis, a rather satisfactory phenomenological explanation of the GSI events could be constructed, if the poly-positronium system would have a size of several 100 fm. The states would be expected to be produced in the heavy ion collision by the action of the strong electric fields with a cross section and kinematic charcteristics similar to that of the QED pairs[85].

Is the required new interaction between electrons and positrons compatible with our knowledge of e^+e^- physics? E.g., one might postulate the existence of a short range attractive many-body force that does not act between a single e^+e^--pair, thus avoiding problems in electron-positron scattering at high energy and in the normal positronium system that is well described by QED. The question was systematically studied by Ionescu et al. [86], who considered the limits set by spectroscopic data from heavy atoms on nonlinear interactions of the form

$$L_{int} = \lambda(\bar{\psi}\psi)^n, \tag{36}$$

where n is some integer greater than one. Such forces would contribute measurably to the K-shell binding energy in heavy atoms, if the effective coupling constant λ is too large. The following limits were obtained in this way: $\lambda(n = 2) < 5 \times 10^{-4}$ and $\lambda(n = 3) < 2 \times 10^{-3}$. On the other hand, the values of λ required to support a poly-positronium bound state are at least $8(n = 2)$ or $130(n = 3)$, respectively [86]. Thus, poly-positronium states based on a new e^+e^--interaction of type (36) can be excluded.

For further attempts to relate e^+e^--systems ("Micro-positronium") to the GSI-peaks and the severe problems, see for example [87, 88, 89, 90, 91]. Recently several authers have reported the existence of extremely narrow resonances in the e^+e^--system, using the quasipotential method [92] or the Tamm Dankoff approximation [93]. This stands in contradiction to a general argument by Grabiak, which excludes such resonances in the framework of QED [94].

14. Potential model for the composite X^0-state

We refer now to the "second road" and introduce an X^0-potential-model. This describes the X^0 as a meson-like object built up by a pair of fermions f^+f^-, interacting via yet unknown forces, that are treated in terms of an effective potential $V(r)$. If the rest masses of the constituents account for the major fraction of the total X^0-mass, one may favour a non-relativistic Schrödinger equation in the relative coordinate $\vec{r}$:

$$\left(\frac{\hat{p}^2}{2m_{red}} + V(r) \right) \phi(\vec{r}) = E\phi(\vec{r}) \quad , \tag{37}$$

$m_{red} = \frac{1}{2}m_f$ being the reduced mass of the $f^\pm$. $V(r)$ is the potential between the constitutents to be specified below:

$$V(r) = a_x r \quad , \quad r > r_0 \tag{38}$$

$$V(r) = V_0 = const. \quad , \quad r \leq r_0.$$

Here V_0 is a constant simulating a repulsive interaction of range r_0 between f^+ and f^- ("hard core") the need for which will be explained later. a_x denotes the string tension of the long-range, confining part of the potential which is the analogue of the vacuum pressure B_X in a bag model approach [95]. The parameters m_f and a_x can be used to fit the ground state energy and level spacing of X^0. Invoking the flux tube model it is possible to deduce a correspondence between the bag constant B_X and the string tension a_x which has the form [96]

$$a_x = \sqrt{8\pi B_X \alpha_X} \tag{39}$$

with α_X being the coupling constant of the gauge interaction responsible for confinement. Its precise value is of course unknown, but in analogy with the experience from QCD it may be assumed to be of the order of 1. The values of a_x derived from a bag model calculation fixing B_X and subsequently applying eq.(39) and compared to a calculation within the potential model shows fair agreement. For masses $m_f \approx 800$ keV the resulting string tension is of the order $a_x \approx 0.1$ keV/fm.

PRODUCTION IN HEAVY ION COLLISIONS

15. Production of f^+f^- pairs

If the correlated e^+e^--production in heavy-ion collisions is related to the decay of some neutral particle one has to know the properties of the particle in the presence of the strong electromagnetic fields of target and projectile. As the object here discussed has an electromagnetic substructure and is polarizable its energy is considerably affected by the presence of the strong electric fields of the ions. In order to describe this configuration, we first assume that the particle is centered at the center-of-mass of the two ions. The negative constituent f^- is strongly attracted to the charge center by the Coulomb interaction of the nuclei. On the other hand, the positively charged constituent f^+ is repelled to the outer boundary of the confinement region, enlarging its radius. Neglecting the contribution of the hard core interaction the total energy of the object can be written as

$$E_{X^0}(Z) = E_{f^-}(Z) + E_{f^+}(Z) \tag{40}$$

In order to calculate the energy of the negatively charged constituent E_{f^-} we neglect the influence of the f^+ on the wavefunction of the strongly bound f^- and solve the Dirac equation in the Coulomb field of the two nuclei. Taking the nuclear Coulomb potential $V_C(r)$ in the monopole approximation [97], the Dirac equation

$$(\hat{p} \cdot \gamma - m_f - \gamma^0 V_c(r))\psi_{f-} = 0 \tag{41}$$

was numerically solved. The solution yields an energy $E_{f-}(Z)$ depending on the total charge of the collision system and the internuclear distance R.

In the case of the positively charged constituent, the interaction with the strongly localized f^- has been taken into account by solving the Dirac equation for the f^+ in the Coulomb field of the nuclei including the scalar potential $V(r)$ eq.(38) as an additional central potential. Thus the equation of motion reads

$$(\hat{p} \cdot \gamma - m_f + \tilde{V}(r))\psi_{f+} = 0 \tag{42}$$

with the potential

$$\tilde{V}(r) = \gamma^0 V_c(r) + V(r) \tag{43}$$

By solving eqs.(41,42) for the lowest energy eigenvalues one determines the energy of the X^0-particle in the Coulomb field of the two nuclei. The result in the case of two uranium nuclei can be seen in Fig.17. The total energy of the particle in the strong

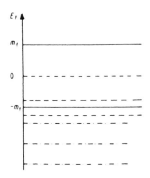

Figure 16. *Schematic energy level diagram of the constituents in the Coulomb field of two heavy ions. The lowermost f^- (dashed dotted lines) is supercritically bound whereas the f^+ (dashed lines) spectrum is discretised due to the confinement.*

electric field is very small, approaching zero energy for small internuclear distances R. If the energy were to decrease below zero, spontaneous supercritical production of the neutral particle could occur.

Even if this does not happen, the small energy gap favours the pair production of bound $f^+ f^-$-pairs in the collision. That point is illustrated in Fig.16 where a single particle level diagram of the $f^\pm$ states is schematically shown. The f^--state of lowest energy is strongly bound with a binding energy which - depending on the total charge of the collision system - may exceed $2m_f$. The f^+ states very much resemble positron states, but their energy levels are discretized due to the confinement in the bag.

As in the case of electrons and positrons the wavefunctions of the constituent particles vary strongly with the nuclear distance due to the rapid change of the Coulomb fields of the nuclei. This yields large dynamical transition matrix elements

$$M_{ij} \sim \langle \varphi_{i,f-} | \partial/\partial t | \varphi_{j,f+} \rangle \tag{44}$$

which mediate the creation of $(f^+ f^-)$-pairs. These pairs are confined due to the potential eq.(38) giving rise to neutral states. For an exact treatment of the electromagnetic production process in principle one has to take into account the confinement interaction (38) between the particles. However, since the extension of the particle is large, $R_X \sim 1000$ fm, an estimate of the total production probability of the $(f^+ f^-)$ states can be obtained by treating the $f^\pm$ as $e^\pm$-like particles with the electron mass replaced by

m_f. The production of f^+f^- pairs can be calculated analogously to dynamical electron-positron production [98].

A numerical calculation in a U-U collision for a constituent mass $m_f = 900$ keV leads to a value of $P_{f^+f^-}$ which is ten times larger than the calculated total e^+-production in the same heavy-ion collision and about three orders of magnitude larger than the corresponding cross section of the observed line structure. The large production probability originates from the fact that in contrast to the case of electrons there are no occupied f^--states in the beginning of the collision and therefore no Pauli suppression for production of f^- particles in bound states occur. These numbers do not include the effect of the hard core repulsion!

Therefore, it can be noted that the large cross section of correlated e^+e^--pairs $\sigma_{e^+e^-} \approx 100\mu$b in heavy-ion collisions can be explained within the model. At this point one should further mention that the similarity of the production process of the bag with dynamical and spontaneous e^+e^- production in the collision suggests that the bag production cross section scales with the total nuclear charge $Z_u = Z_1 + Z_2$ of the collision system like the positron cross section, i.e. roughly

$$\sigma_X \sim (Z_1 + Z_2)^{20} \qquad (45)$$

In contrast to previous reports, it was pointed out on this meeting, that this behaviour is no longer supported by recent experiments.

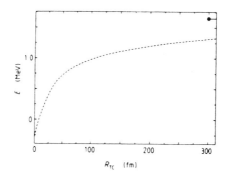

Figure 17. *Energy of the extended particle in the Coulomb field of two Uranium nuclei for m_f=850 keV as function of nuclear separation R. The energy of the particle in the presence of a single U-nucleus is marked to be 1.65 MeV.*

16. Final state effects

When the extended particle has been created as a bound state in the center-of-mass frame of the nuclei one has to consider the break-up of the collision system. Although a detailed study of the dynamics has not yet been performed, one may look at the energy of the particle in the Coulomb field of a single ion in order to get an insight into the strength of binding of the neutral state to target or projectile. One solves eq.(41,42) with the Coulomb potential of a single ion. The energy of the total state, given by eq.(40), is about 1.65 MeV and additionaly depicted in Fig.17. It can be seen that the X^0 in the Coulomb field of a single nucleus still has a binding energy > 100 keV for constituent masses $m_f > 800$ keV.

Combining these results one cannot definitely answer the question what may happen with the produced particle after the collision. Although adiabatically a bag produced in its lowest energy state should be dragged along with a single target or projectile ion, the influence of the dynamics can change that behaviour, especially for states which are not produced in the groundstate but in higher states. The result shows that one needs a dynamical treatment of the f^+f^- states in the collision. Calculations to that point are in progress. It seems to be plausible to expect that a fraction of the produced particles is getting bound by a single ion and another fraction is set free with small velocity with respect to the CM system. That may explain the experimental finding that the difference energy of some of the correlated e^+e^- lines is not centered at $E_{e+} - E_{e-} = 0$ but is shifted to positive values. In some measurements there seem to be indications for two 'peaks' in the difference energy spectrum, one at approximately zero difference energy and one several hundred keV off zero. In addition the decay into e^+e^- pairs from such a bound state would of course not necessarily exhibit a back to back correlation. We will come back to that point in section 20.

17. f^+f^- production in e^+e^- colliders

The next point concerns f^-f^+ production in high-energy electron-positron collisions. As it is known from high-energy experiments, in the collisions jets of hadronic particles may be produced. These jets can be understood in terms of the flux-tube model of QCD.

Since the hypothetical $f^\pm$ constituents should also interact via some confining force one has to consider whether a similar effect producing multiparticle events can occur. Here the analogous effect would be the production of a highly excited (f^+f^-) state which could fragment into a large number of neutral X states subsequently decaying into a shower of e^+e^- pairs. As electron-positron showers of that kind have not been observed in experiments, one has to consider the probability of the break-up of an

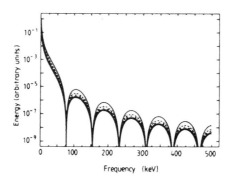

Figure 18. *Energy spectrum (scaled units) of the electromagnetic radiation of a highly excited X^0 with excitation energies 50, 60, ..., 100 GeV, respectively*

excited X^0 particle. The calculation can be performed analogously as it was done in the case of QCD [99]. There the pair creation process in the gluonic fields was treated as a quasiclassical tunneling process, yielding a rate of flux-tube fragmentation per unit volume. For the constituents here discussed the formula reads

$$dP = \frac{a_x^2}{4\pi^3} \sum_{n=1}^{\infty} \frac{1}{n^2} e^{-\pi n m_f^2 / a_x} \tag{46}$$

Taking the values for a_x from table 1, one can compute the fragmentation rate (46) per volume to be negligibly small $(\pi m_f^2 / a_x \approx 250)$.

As the oscillating constituents of a highly excited flux tube are electrically charged the particle may lose its energy by electromagnetic radiation. Treating the state as system of two classical oppositely charged particles interacting via the confinement potential (38) (neglecting the hard core), the power of radiation P is given by the Larmor formula

$$P = -\frac{2}{3} \frac{e^2}{m_f^2} \frac{dp_\mu}{d\tau} \frac{dp^\mu}{d\tau} = -\frac{8}{3} \frac{\alpha}{m_f^2} a_x^2 \tag{47}$$

For a typical value of $a_x = 4.6 \ 10^{-2} \mathrm{keV/fm}$ we get the resulting power $P \approx 10^{13} \mathrm{MeV/s}$. The frequency distribution of the radiation is approximately given by the Fourier transform of the velocity of the oscillating particles:

$$v(t) = \frac{p}{E} = \frac{p_0 \pm a_x t}{\sqrt{(p_0 \pm a_x t)^2 + m_f^2}} \qquad (48)$$

p_0 denotes the initial momentum imparted on the $f^+ f^-$ particles. The $\pm-$sign discriminates the periods of forward and backward motion during the oscillation. In Fig.18 the resulting radiation spectra for different excitation energies are shown. One can see that most of the energy is radiated at low energies < 50 keV where it cannot be easily detected due to the large background of bremsstrahlung radiation in the interaction region.

INFLUENCE OF SHORT RANGE REPULSION

18. Decay of the X^0

The neutral bag state can decay in a similar way as the lowest states of the charmonium system [100]. Since the constituents are electrically charged the object can decay into photons or electrons and positrons depending on the quantum numbers

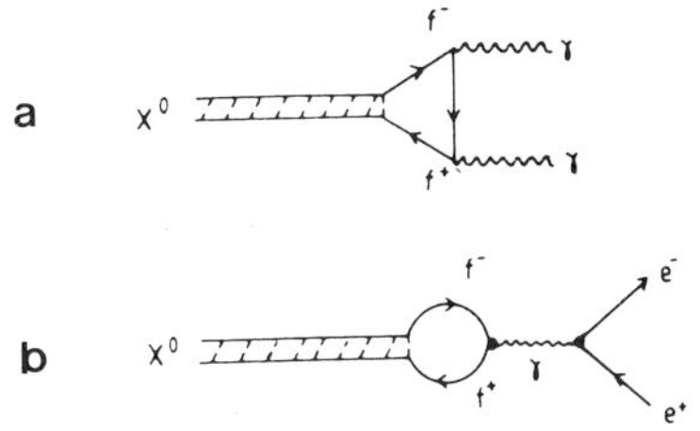

Figure 19. *Feynman diagrams for X^0 decay in a) two photons and b) into a e^+e^- pair.*

m_f	0.85 MeV	0.85 MeV
a_X	9.05×10^{-3} MeV^2	9.05×10^{-3} MeV^2
R_0	4.64 fm	4.64×10^{-3} fm
V_0	340 MeV	234×10^3 MeV
E_{1S}	1.8 MeV	1.8 MeV
E_{2S}	1.88 MeV	1.88 MeV
E_{2P}	1.847 MeV	1.847 MeV
$\langle r \rangle_{1S}$	1560 fm	1560 fm
$\langle r \rangle_{2S}$	2725 fm	2725 fm
$\Gamma_{e^+e^-}$	7.0×10^{-6} eV	2.56×10^{-4} eV
Γ_{rad}	2.4 eV	2.4 eV
$\Gamma_{\gamma\gamma}$	3.1×10^{-5} eV	1.14×10^{-3} eV

Figure 20. *Expectation values and decay width of f^+f^- states for different sets of parameters.*

of the specific state. In the case of an 0^{-+}- ('para'-) state with opposite spins of the constituents the f^+f^- may annihilate into two photons.

The 1^{--}- ('ortho'-) state can annihilate into a virtual photon which subsequently decays into a correlated e^+e^--pair. A calculation of the diagrams in Fig.19 non-relativistic bound states yields the decay widths

$$\Gamma_{0^{-+}\to\gamma\gamma} = \alpha^2 \frac{4\pi}{2\left(\frac{M_X}{2} - m_f - \sqrt{\frac{M_X^2}{4} + m_f^2}\right)^2} \frac{M_X^2}{m_f^2} |\phi_{f\bar{f}}(0)|^2 \qquad (49)$$

$$\Gamma_{1^{--}\to e^+e^-} = \alpha^2 \frac{16\pi}{3}\left(4 + \frac{16}{9}\frac{M_X^2}{m_e^2}\right) \frac{m_e^2}{M_X^5}\left(\frac{M_X^2}{4} - m_e^2\right)^{\frac{1}{2}} |\phi_{f\bar{f}}(0)|^2 \qquad (50)$$

The resulting decay widths and corresponding lifetimes for e^+e^- or two γ-decay are shown in Fig 20 for several parameters of the potential (38). The upper limit for the lifetime of a decaying neutral object set by the heavy-ion collisions is given by the condition that the particle should decay inside of the experimental set-up which yields a value $\tau < 10^{-10}$s which is satisfied by the results shown in Fig. 20.

However, eq. (49) points to a serious problem: Assuming the radial wave functions to be *independent* of the spin configuration, the two-photon decay of the J=0 states is predicted to be faster than the e^+e^- pair decay of the 1^{--}-states! Since in a heavy-ion collision both sets of states should be populated with comparable strength this is in conflict with the non-observation of correlated photon pair in Ref. [103].

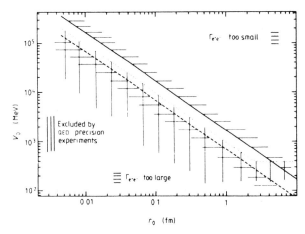

Figure 21. *The range of parameters V_0 and r_0, compatible with QED precision measurements and experimental requirements on the decay width $\Gamma_{e^+e^-}$.*

As already discussed, a lower limit for the lifetime of the particle state can be found from recent Bhabha scattering experiments [73] to give $\tau > 7.5 \cdot 10^{-12}$sec. Comparison with the results from Fig. 20 requires an appropriate enhancement of the hard core. This will further suppress the amplitude for annihilation and therefore enlarge the lifetime. But simultaneously this approach decreases the production crossection in heavy ion collisions and will run into conflict with the line intensities (see section 20).

Another approach to suppress annihilation is an enlargement of the object's spatial extension by lowering the confining strength a_x, since this decreases the amplitude for both particles to be at the same position $\phi_{f\bar{f}}(0)$. For example a tenfold reduction to the value $a_x=5$eV/fm was found to reduce $|\phi_{f\bar{f}}(0)|^2$ by a factor of 10, while R_X increased by about a factor of 3. This factor would serve for consistency with the latest requirements. Such a choice of parameters bears the advantage that the total production cross section in heavy ion collisions remains largly uneffected, since the confinement plays a minor role in this context. The level density of exited states, however, becomes very narrow for small a_x.

19. QED precision tests

Estimating the X^0-contributions to QED precision tests, the anomalous magnetic moment of the muon turned out to impose the most stringent bounds on the model's parameters [101]. The results are governed by the influence of the hard core, that acts like an energy dependent suppression factor at each $f^+ f^-$-vertex.

The range of parameters V_0 and r_0, compatible with the corresponding measurements and experimental requirements on the lifetime $\tau_{e^+ e^-}$ (excluding the latest results [73] from Bhabha scattering, that are concerned in different terms as discussed in the previous section), is depicted in Fig. 21. Our model clearly needs for the hard core.

20. $f^+ f^-$-production in heavy ion collisions

The dynamical calculations described in section 2 treated the fermions as non-interacting. In order to account for short range repulsion we proceed as described in [101] to estimate the total $f^+ f^-$-production. We thus basically solve the coupled channel equations for fermions of mass m_f and refer to the hard core in terms of a Gamov factor, that depends on it's hight V_0 and width r_0. As an example we have chosen the following parameters:

$$V_0 = 234 \text{ GeV} \qquad r_0 = 4.64 \ 10^{-3} \text{fm} \qquad m_f = 850 \text{ keV} \ ,$$

which yields a low-energy Gamow factor $g(2m_0)^2 = 1.7 \cdot 10^{-3}$. The resulting g-2-contributions, as well as the results for Lamb shift in transitions of muonic lead, are compatible with experimental uncertainties [101]. The total rate of dynamical $f^+ f^-$-pair production now yields a fraction of about 1.7% of the dynamically induced $e^+ e^-$ pairs, which compares reasonably well with the experimental value.

This is where increasing the hard core influences the f-pair production in heavy ion collisions. In contrast, enlarging the spatial extension of the object, by lowering the strength of the confining potential leads to smaller level spacing which justifies the continuum approximation even better. To some extent we are therefore in the position to decouple the requirements originating from Bhabha-scattering (small decay width) and from heavy-ion experiments (large production rate).

After the collision, a fraction of the created composite particles may remain bound to one of the separated nuclei. The X^0 ground state and possibly various excited bound states will be populated according to the collision dynamics. Annihilation of a bound X^0 could in principle give rise to $e^+ e^-$ coincident lines at discrete energies below M_{X^0}. Our calculations (Fig.17) predict that the energy shift is of the order of -150 keV for the ground state (f^- in a 1s orbit around the nucleus) and perhaps -50 keV for a possible excited state (f^- in a 2s orbit). Precise numbers for the mass shifts require three-body calculations. Whether the excited states show up in the $e^+ e^-$-sum energy spectrum depends on the competition between annihilation and radiative de-excitation. Since the bound X^0 decay proceeds in the vicinity of the nucleus, its two-body characteristics (angular correlation, $e^+ e^-$ energy difference) will be disturbed. Precisely this has been observed in recent experiments [102] for all lines except the highest one (809 keV). However, first calculations [104], referring to a X^0 bound by an Uranium nucleus in its ground state, indicate that the angular correlation of $e^+ e^-$ pairs, emerging from the X^0 decay, is still preferrably back to back. The ("two-body-") peak at 180^0 is broadenend by roughly $\pm 25^0$, while the difference-energy lies approximately in the range of $\Delta E = |E_{e^+} - E_{e^-}| \lesssim 100 \text{keV}$ (both in the C.M.-frame of the binding nucleus). Even though there is a three body decay, this angular correlation would still give rise to narrow lines in the sum energy spectra.

The role of pair decay from a bound X^0-state is further called into question by the competing one-photon annihilation process which in this case is kinematically allowed. For the ground state the width $\Gamma_{X^0 \to \gamma}$ is about one order of magnitude larger than $\Gamma_{X^0 \to e^+ e^-}$. However, the value of $\Gamma_{X^0 \to \gamma}$ depends on the Fourier transform of the X^0-wavefunction, since this enters to weight the available frequencies for a photon on the mass shell. The suggested spatial enlargement of the X^0 will therefore cause a smaller decay width $\Gamma_{X^0 \to \gamma}$. Nevertheless the bound-state-decay hypothesis appears to be unlikely.

21. High-energy e^+e^--collisions

Next, we consider the influence of intermediate X^0-production in high-energy e^+e^--collision ($E_{Beam} >$ several GeV) experiments on the value for the ratio R of hadronic production cross section and muon pair creation

$$R = \frac{\sigma_{e^+e^- \to \text{hadrons}}}{\sigma_{e^+e^- \to \mu^+\mu^-}} = 3 \sum_f Q_f^2 \ , \tag{51}$$

with Q_f being the quark charge with flavour f. The sum includes all flavours whose pair creation threshold lies below the available beam energy.

However as discussed before, the preferred values of the hard core are $V_0 > 100$ GeV. Then, for energies below $E \leq V_0$, the barrier generates a suppression factor in the f^+f^- creation compared to the production of non-interacting particles such, that at presently accessible energies the modifications arising from f^+f^--pair production should be very small compared to the experimental uncertainties ($\Delta R \approx 10\%$). The same argument also holds in other high-energy experiments as long as the energy of possible f^+f^--pair creation lies below the barrier V_0.

CONCLUSIONS

We discussed the properties of the QED vacuum state in the presence of strong Coulomb fields, i. e. the spantaneous transition from the neutral- into the charged vacuum, when the external fields aquire supercriticality. In order to test theory, we explored the dynamics of the electron field in heavy ion collisions and related the results to those from corresponding experiments. In terms of this approach it was not possible to explain the observed sharp e^+e^--coincident lines. For this reason we discussed a model of an extended neutral particle X^0 with a mass of about ~ 1.8 MeV whose decay into an electron-positron pair may serve as an explanation of this lines as measured at GSI. The extended object is assumed to be a bound state consisting of new electrically charged constituents (f^+,f^-) interacting via a confining hypothetical force which was implemented by adopting a potential model exhibiting a linearly rising potential. The most appropriate rest masses of the constituents $f^\pm$ turn out to be relatively large, i.e. $m_f > 800$ keV. For a given value of m_f the coefficient a_x of the confining potential and the corresponding radii R_X of the bag can be determined favouring large bags with $R_X > 1000$ fm. The resulting level structure of free and bound X^0-states is comparable to the spacing of the measured coincidence lines. The lifetime of the particle lies within the experimental limits.

Due to its large extension the neutral bag should not show up in beam-dump experiments but will get stuck in the dump until it decays. Since one assumed a confining interaction between the constituents the theoretical possibility of fragmentation of a highly excited X-bag into many X subsequently producing an electron-positron shower was discussed. It could be shown that a process of that kind is suppressed for high rest masses of the constituent particles. However, the state may be deexcited through electromagnetic radiation of low-energy photons within a time of 10^{-8} s and afterwards decays electromagnetically .

Our assumption that the composite object contains electrically charged constituents, has several essential consequences:

i) The virtual contributions of the X^0 to QED high-precision experiments like the leptonic anomalous magnetic moment are potentially large. *Only* if one assumes an interaction potential between the constituents which exhibits, in addition to long range confinement, a repulsive hard core the contributions remain small enough. Comparison with experimental results requires that the hard core be to extremely high and localized, with preferred values $V_0 > 100$ GeV, and $r_0 < 10^{-2}$ fm.

ii) Due to electric polarization the energy of the particle may decrease from 1.8 MeV near to zero energy during a heavy-ion collision. That implies large production cross sections of the X^0 in a collision of large-Z nuclei. The most striking success of our model is that it *predicts* a cross section for dynamical production of f^+f^--pairs in heavy ion collisions, that compares reasonably well with the experimental requirements. When target and projectile separate, the bag may stay in the c.m. system of the nuclei or can be dragged along by a single nucleus. In the latter case the e^+e^--decay of the particle in the Coulomb field of one nucleus (which is in competition with the one photon decay) should result in a) higher energies for the positron compared to the correlated electron; b) an e^+e^- angular correlation which is not necessarily back to back; and c) a strong dependence of the line occurrence on the detailed collision dynamics, which will influence the population of excited bound states.

One may speculate that the short-range repulsion originates somehow from phenomena at the scale of electroweak symmetry breaking. It should be noted, however, that we have not yet been able to derive a working model for the hard core from an elementary interaction. Another possibility to explain the suppression of the relative f^+f^- wavefunction at zero separation might be a strong coupling to an electrically neutral sector at $r < r_0$, which could be described by an imaginary potential. Such neutral sectors would naturally appear in the context of a gauge theory of the f^+f^--interaction.

One may summarize that the model is adequate to explain quite a number of the puzzling experimental features of the positron lines found at GSI. If these are in fact caused by neutral particle decays, as many but not all experimental data indicate, one may hope to retain here a working model which could explain the experimental results, without running into contradiction with other well established data.

References

[1] B. Müller, H. Peitz, J. Rafelski, W. Greiner, Phys. Rev. Lett. 28: 1235 (1972)

[2] B. Müller, J. Rafelski, W. Greiner, Z. Phys. 257: 62, 183 (1972)

[3] Ya. Zel'dovich, V. S. Popov, Sov. Phys. Usp. 14: 673 (1972)

[4] J. Rafelski, B. Müller, W. Greiner, Nucl. Phys. **B68**: 585 (1974)

[5] L. P. Fulcher, A. Klein, Ann. Phys. (N. Y.) 84: 335 (1974)

[6] M. Gyulassy, Nucl. Phys. **A244**: 497 (1975)

[7] G. A. Rinker, L. Wilets, Phys. Rev. **A12**: 748 (1975)

[8] G. Soff, P. Schlüter, B. Müller, W. Greiner, Phys. Rev. Lett. 48: 1465 (1982)

[9] W. Greiner, Opening remarks in: Quantum Electrodynamics of Strong Fields, NATO-ASI at Lahnstein, ed. W. Greiner, Plenum, N.Y. (1983)

[10] B. Müller, J. Rafelski, W. Greiner, Phys. Lett. **47B**: 5 (1973)

[11] G. Soff, B. Müller, W. Greiner, Phys. Rev. Lett. 40: 540 (1978)

[12] J. Bang, J. H. Hansteen, K. Dan. Vidensk. Selsk. Mat. Fys. Medd. 31: No 13 (1959)

[13] B. Müller, G. Soff, W. Greiner, V. Ceausescu Z. Phys. **A285**: 27 (1978)

[14] B. Müller, J. Reinhardt, W. Greiner, G. Soff Z. Phys. **A311**: 151 (1983)

[15] F. Bosch Z. Phys. **A296**: 11 (1980)

[16] D. Liesen et al., Phys. Rev. Lett. **44**: 983 (1980)

[17] W. Betz, PhD-thesis (unpublished), Frankfurt (1980)

[18] W. Greiner, B. Müller, and J. Rafelski, "Quantum Electrodynamics of Strong Fields", Springer, Berlin-Heidelberg (1985).

[19] J. Reinhardt, B. Müller, and W. Greiner, Phys. Rev. A 24:103 (1981).

[20] G. Soff, J. Reinhardt, B. Müller, and W. Greiner, Phys. Rev. Lett. 43:1981 (1979).

[21] R. Anholt, Phys. Lett. B 88:262 (1979).

[22] O. Graf, J. Reinhardt, B. Müller, W. Greiner, and G. Soff, Phys. Rev. Lett. 61: 2831 (1988)

[23] R. Krieg, E. Bozek, U. Gollerthan, E. Kankeleit, G. Klotz-Engmann, M. Krämer, U. Meyer, H. Oeschler, and P. Senger, Phys. Rev. C 34:562 (1986).

[24] H. Backe, P. Senger, W. Bonin, E. Kankeleit, M. Krämer, R. Krieg, V. Metag, N. Trautmann, and J.B. Wilhelmy, Phys. Rev. Lett. 50:1838 (1983).

[25] J. Rafelski, B. Müller, and W. Greiner, Z. Phys. A 285:49 (1978).

[26] J. Reinhardt, U. Müller, B. Müller, and W. Greiner, Z. Phys. A 303:173 (1981).

[27] U. Heinz, U. Müller, J. Reinhardt, B. Müller, and W. Greiner, J. Phys. G 11:L169 (1985).

[28] M. Clemente, E. Berdermann, P. Kienle, H. Tsertos, W. Wagner, C. Kozhuharov, F. Bosch, and W. Koenig, Phys. Lett. B 137:41 (1984).

[29] U. Müller, G. Soff, T. de Reus, J. Reinhardt, B. Müller, and W. Greiner, Z. Phys. A 313:263 (1983).

[30] P. Schlüter, T. de Reus, J. Reinhardt, B. Müller, and G. Soff, Z. Phys. A 314:297 (1983).

[31] J. Schweppe, A. Gruppe, K. Bethge, H. Bokemeyer, T. Cowan, H. Folger, J.S. Greenberg, H. Grein, S. Ito, R. Schule, D. Schwalm, K.E. Stiebing, N. Trautmann, P. Vincent, and M. Waldschmidt, Phys. Rev. Lett. 51:2261 (1983).

[32] T. Cowan, H. Backe, M. Begemann, K. Bethge, H. Bokemeyer, H. Folger, J.S. Greenberg, H. Grein, A. Gruppe, Y. Kido, M. Klüver, D. Schwalm, J. Schweppe, K.E. Stiebing, N. Trautmann, and P. Vincent, Phys. Rev. Lett. 54:1761 (1985).

[33] "Quantum Electrodynamics of Strong Fields", W. Greiner, ed., NATO Advanced Study Institute Series B, vol. 80, Plenum, New York (1981).

[34] "Physics of Strong Fields", W. Greiner, ed., NATO Advanced Study Institute Series B, vol. 153, Plenum, New York (1986).

[35] H. Backe and B. Müller, Positron production in heavy ion collisions, in: "Atomic Inner-Shell Physics", p. 627, B.Crasemann, ed., Plenum Press, New York (1985).

[36] J.S. Greenberg and P. Vincent, Heavy ion atomic physics - experimental, in: "Treatise on Heavy Ion Science", vol. 5, p. 141, D.A. Bromley, ed., Plenum Press, New York (1985).

[37] A. Schäfer, J. Reinhardt, B. Müller, W. Greiner, G. Soff, J. Phys. G: Nucl. Phys. 11 (1985) L69-L74

[38] T. Cowan, H. Backe, K. Bethge, H. Bokemeyer, H. Folger, J.S. Greenberg, K. Sakaguchi, D. Schwalm, J. Schweppe, K.E. Stiebing, P. Vincent, Phys. Rev. Lett. 56:444 (1986).

[39] E. Berdermann, F. Bosch, P. Kienle, W. Koenig, C. Kozhuharov, H. Tsertos, S. Schuhbeck, S. Huchler, J. Kemmer, and A. Schröter, Monoenergetic (e^+e^-) pairs from heavy-ion collisions, preprint GSI-88-35 (1988).

[40] A. Schäfer, J. Reinhardt, B. Müller, W. Greiner, and G. Soff, J. Phys. G 11:L69 (1985).

[41] A.B. Balantekin, C. Bottcher, M. Strayer, and S.J. Lee, Phys. Rev. Lett. 55:461 (1985).

[42] R.S. Van Dyck, P.B. Schwinberg, and H.G. Dehmelt, The electron and positron geonium experiments, in: "Atomic Physics 9", p. 53, ed. R.S. Van Dyck and E.N. Fortson, World Scientific, Singapore (1984).

[43] J. Reinhardt, A. Schäfer, B. Müller, and W. Greiner, Phys. Rev. C 33:194 (1986).

[44] R. Barbieri and T.E.O. Ericson, Phys. Lett. B 57:270 (1975); U.E. Schröder, Mod. Phys. Lett. A 1:157 (1986).

[45] A. Zee, Phys. Lett. B 172:377 (1986).

[46] A. Schäfer, J. Reinhardt, W. Greiner, and B. Müller, Mod. Phys. Lett. A 1:1 (1986).

[47] A. Chodos and L.C.R. Wijewardhana, Phys. Rev. Lett. 56:302 (1986).

[48] B. Müller and J. Rafelski, Phys. Rev. D 34:2896 (1986).

[49] Y. Yamaguchi and H. Sato, Phys. Rev. C 35:2156 (1987).

[50] A. Schäfer, B. Müller, and J. Reinhardt, Mod. Phys. Lett. A 2:159 (1987).

[51] D. Carrier, A. Chodos, and L.C.R. Wijewardhana, Phys. Rev. D 34:1332 (1986).

[52] A. Schäfer, J. Reinhardt, B. Müller, and W. Greiner, Z. Phys. A 324:243 (1986).

[53] B. Müller and J. Reinhardt, Phys. Rev. Lett. 56:2108 (1986).

[54] S. Barshay, Mod. Phys. Lett. A 1:653 (1986).

[55] B. Müller, Positron production in heavy ion collisions - A puzzle for physicists, in: "Intersections between Particle and Nuclear Physics", D.F. Geesaman, ed., AIP Conf. Proceed. 150:827 (1986)

[56] S. Brodsky and M. Karliner, private communication.

[57] L.M. Krauss and F. Wilczek, Phys. Lett. B 173:189 (1986).

[58] R.D. Peccei and H.R. Quinn, Phys. Rev. D 16:1791 (1977); S. Weinberg, Phys. Rev. Lett. 40:223 (1978); F. Wilczek, Phys. Rev. Lett. 40:279 (1978).

[59] N.C. Mukhopadhyay and A. Zehnder, Phys. Rev. Lett. 56:206 (1986).

[60] G. Mageras, P. Franzini, P.M. Tuts, S. Youssef, T. Zhao, J. Lee-Franzini, and R.D. Schamberger, Phys. Rev. Lett. 56:2672 (1986).

[61] T. Bowcock et al. (CLEO collaboration), Phys. Rev. Lett. 56:2676 (1986).

[62] H. Albrecht et al. (ARGUS collaboration), Phys. Lett. B 179:403 (1986).

[63] A. Schäfer, Phys. Lett. **211B**, 207 (1988)

[64] Y.S. Tsai, Phys. Rev. D 34:1326 (1986).

[65] A. Konaka, et. al Phys. Rev. Lett. 57:659 (1986).

[66] M. Davier, J. Jeanjean, and H. Nguyen Ngoc, Phys. Lett. B 180:295 (1986).

[67] E.M. Riordan, et. al Phys. Rev. Lett. 59:755 (1987).

[68] D.J. Bechis, T.W. Dombeck, R.W. Ellsworth, E.V. Sager, P.H. Steinberg, L.J. Tieg, J.K. Joh, and R.L. Weitz, Phys. Rev. Lett. 42:1511 (1979).

[69] C.N. Brown, et. al Phys. Rev. Lett. 57:2101 (1986).

[70] H.A. Olsen, Phys. Rev. D 36:959 (1987).

[71] J. Reinhardt, A. Scherdin, B. Müller, and W. Greiner, Z. Phys. A 327:367 (1987).

[72] A. Scherdin, J. Reinhardt, W. Greiner, B. Müller, Low energy e^+e^- scattering, UFTP preprint 248 (1990), to be published in Rep. Prog. Phys.

[73] S. M. Judge, B. Krusche, K. Schreckenbach, H. Tsertos, P. Kienle, Phys. Rev. Lett. **65**, 972 (1990)

[74] E. Widmann, W. Bauer, S. Connell, K. Maier, J. Major, A. Seeger, H. Stoll, F. Bosch, preprint

[75] L.S. Celenza, V.K. Mishra, C.M. Shakin, and K.F. Liu, Phys. Rev. Lett. 57:55 (1986).

[76] D.G. Caldi and A. Chodos, Phys. Rev. D 36:2876 (1987).

[77] Y.J. Ng and Y. Kikuchi, Phys. Rev. D 36:2880 (1987).

[78] C.W. Wong, Phys. Rev. D 37:3206 (1988).

[79] E. Dagotto and H.W. Wyld, Phys. Lett. B 205:73 (1988).

[80] R.D. Peccei, J. Solà, and C. Wetterich, Phys. Rev. D 37:2492 (1988).

[81] J. Kogut, E. Dagotto, and A. Kocić, Phys. Rev. Lett. 60:772 (1988).

[82] J.J. Griffin, J. Phys. Soc. Jpn. **58**, 427 (1989)

[83] S.K. Kim, B. Müller, and W. Greiner, Mod. Phys. Lett. in print (1988).

[84] E.A. Hylleraas and A. Ore, Phys. Rev. 71:493 (1947); Y.K. Ho, Phys. Rev. A 33:3584 (1986).

[85] J.M. Bang, J.M. Hansteen, and L. Kocbach, J. Phys. G 13:L281 (1987).

[86] D.C. Ionescu, J. Reinhardt, B. Müller, W. Greiner, and G. Soff, J. Phys. G 14:L143 (1988).

[87] A.O. Barut and J. Kraus, Phys. Lett. B 59:175 (1975); A.O. Barut, Nonperturbative treatment of magnetic interactions at short distances and a simple magnetic model of matter, in: Ref. [33], p. 755.

[88] C.Y. Wong and R.L. Becker, Phys. Lett. B 182:251 (1986).

[89] K. Geiger, J. Reinhardt, B. Müller, and W. Greiner, Z. Phys. A 329:77 (1988).

[90] H.W. Crater, C.Y. Wong, R.L. Becker, and P. VanAlstine, Non-perturbative covariant treatment of the e^+e^- system using two-body Dirac equation from constraint dynamics, preprint, Oak Ridge (1987).

[91] C.Y. Wong, On the possibility of QED (e^+e^-) resonances at 1.6-1.8 MeV, in: "Windsurfing the Fermi Sea", Vol.2, T.T.S. Kuo and J. Speth, Eds., Elsevier, Amsterdam (1987), p. 296.

[92] B.A. Arbuzov, E.E. Boos, V.I Savarin, S.A. Shicharin, Phys. Lett. B240, 477 (1990)

[93] J.R. Spence, J. P. Vary, to be published

[94] M. Grabiak, B. Müller, and W. Greiner, Ann. Phys. in print.

[95] S. Schramm, B. Müller, J. Reinhardt, and W. Greiner, Mod. Phys. Lett. A 3:783 (1988).

[96] K. Johnson, C.B. Thorn, Phys. Rev. **D13**, 1934 (1976)

[97] G. Soff, W. Greiner, W. Betz, B. Müller, Phys. Rev. **A20**, 169 (1979)

[98] J. Reinhardt, B. Müller, W. Greiner, Phys. Rev. **A 24**,103 (1981)

[99] N.K. Glendenning, T. Matsui, Phys. Rev. **D28**, 2890 (1983)

[100] V.A. Novikov, L.B. Okun, M.A. Shifman, A.I. Vainshtein, M.B. Voloshin, V.I. Zahkarov, Phys. Rep. **41**, 1 (1978)

[101] S. Graf, S. Schramm, J. Reinhardt, B. Müller, W. Greiner, J. Phys. G **15**, 1467 (1989)

[102] H. Bokemeyer, P. Salabura, D. Schwalm, K. E. Stiebing preprint GSI-89-49 (1989)

[103] K. Danzmann, W.E. Meyerhof, E.C. Montenegro, X.-Y. Xu, E. Dillard, H.P. Hülskotter, F.S. Stephens, R.M.Diamond, M.A. DelePLanque, A.O. Macchiavelli, J. Schweppe, R.J. McDonald, B.S. Rude, and J.D. Molitoris, Phys. Rev. Lett. 59:1885 (1987).

[104] E. Stein, Diploma thesis

Production of hypothetical X^0-mesons in high-energy channeling reactions

JÜRGEN AUGUSTIN[1,2], STEFAN GRAF[2], WALTER GREINER[2]
AND BERNDT MÜLLER[1]

[1] Physics Department, Duke University, Durham, NC 27706, USA
[2] Institut für Theoretische Physik, Joh. Wolfg. Goethe-Universität
Postfach 111932, 6000 Frankfurt am Main 11, Germany

Introduction

The narrow, correlated e^+e^- line structures observed in heavy ion collisions near the Coulomb barrier still long for an convincing explanation [1]. A possible interpretation of the experiments is the pair-decay of a yet unknown neutral particle X^0 with a mass of about 1.8 MeV. However, is now clear that the X^0 *cannot* be an elementary particle, but an extended object with internal structure has not been ruled out [2]. Recently a model for a mesonlike X^0 particle consisting of two elementary charged fermions f^+ and f^- has been proposed [3]. This model describes many features of the experimental data, including a quantitative estimate of the measured cross sections. Therefore it would be usefull if the exsistence of these exotic particles could be confirmed by different experiments.

Since the creation of the f^+f^--pairs is due to the strong electromagnetic fields accompanying heavy ion collisions, it is interesting to investigate other experimental situations where strong electromagnetic fields are important. A promising candidate for X^0-production is the channeling of ultrarelativistic particles through oriented crystals [4]. Here we estimate rates for f^+f^--pair creation by channeled photons and electrons.

The model

Before we go into the details of our calculation we review briefly the basic features of the X^0-model. The f^+f^--pair is bound by a weak, but confining potential with a short range repulsion. This hard core is an essential aspect of the model, because it prevents contradictions between the exsistence of new elementary fermions and QED precission tests. However, for our purposes the hard core can be described by a phenomenological Gamov factor, which takes the suppression of the relative f^+f^--wave function at the origin into account.

For the description of the channeling process we confine ourselves to axial channeling with perfect alignment. Since the pair creation is most likely taking place near a crystal axis, where the electric fields are strongest, the single string approximation can be applied [5]. In this model the crystal is represented by a single row of atoms,

Vacuum Structure in Intense Fields, Edited by
H.M. Fried and B. Muller, Plenum Press, New York, 1991

whose electrostatic potentials are averaged along the crystal axis (from now on called the z-axis). Therefore the $f^+ f^-$ particles are moving freely in the z-direction, while their transverse motion is determined by an effective Schrödinger equation, which follows from the underlying Dirac equation in the ultrarelativistic limit [5]:

$$\left[-\frac{\hbar^2}{2\gamma m} \left(\frac{\partial^2}{\partial x^2} + \frac{\partial^2}{\partial y^2} \right) - eV(\rho) \right] \chi = \epsilon \chi \ , \tag{1}$$

ϵ and $\gamma = p_z/mc$ denote the transverse energy eigenvalue and the Lorentz factor of the channeled particle, respectively. $V(\rho)$ has been calculated out of the standart or Lindhard potential [6] (for more details about channeling see also the article of A. Sørensen in this volume). Also thermal vibrations of the lattice atoms have been taken into account by substituting $\sqrt{\rho + u^2}$ for ρ, u^2 beeing the two-dimensional mean square displacement of the crystal atoms.

How can the pair creation by an photon be calculated in such an external field? Since we have assumed that the photon enters the crystal without transvers momentum, the f^- must be created in a transvers bound state. Then the f^+ is forced by the confinement potential and the crystal fields to follow the f^- and to stay in the space between the rows of atoms. Hence the size of the X^0 inside the crystal is of the order of 1Å.

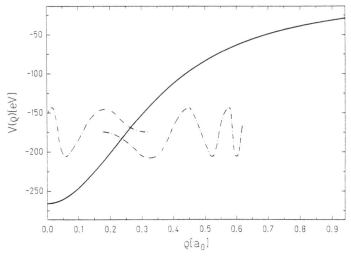

Figure 1. Transverse potential for the $\langle 110 \rangle$ axis of germanium. The $f^\pm$ wave functions are shown only schematically.

Figure 1 depicts a typical transverse potential and also shows $f^+ f^-$ wave functions. It is clear that only the region where the wave functions overlap enters the calculation of the pair creation amplitude. In addition the potential may be assumed to have constant slope in the region of overlap. Therefore the $f^+ f^-$ pair creation can be treated (exept for the Gamov factor) like the $e^+ e^-$ pair creation in a constant electric field. The total rate for X^0 creation by a photon of energy $\hbar\omega$ is then given by formula (32d) of [5]:

$$W(\hbar\omega) = \frac{\alpha}{3\pi\sqrt{3}} \frac{(m_f c^2)^2}{\hbar^2 \omega} \int_0^1 dy \, \frac{9 - y^2}{1 - y^2} \, K_{\frac{2}{3}} \left(\frac{8 E_0 m_f c^2}{3\hbar\omega E(\rho)(1 - y^2)} \right) \ . \tag{2}$$

Here m_f and $E_0 = m_f^2 c^3 / e\hbar$ denote the $f^{\pm}$ mass and the critical field, respectively, while $E(\rho)$ is the electric field of the atomic string at a distance ρ. Since the X^0 can be created in various transverse states the expression (2) must be averaged over ρ. The expression (2) can also be applied to the pair creation by a channeled electron, if the contributions of the virtual photons are neglected. In other words, this second order process is split up into two consecutive first order processes. Denoting the spectral distribution of the intermediate photons by $n(\omega)$, the total pair creation rate by an electron of energy ε then reads:

$$F(\varepsilon) = \int_{x_{min}}^{1} dx\, n(x)\, \overline{W}(x) \ , \quad x = \hbar\omega/\varepsilon \ . \tag{3}$$

The rate $\overline{W}$ is obtained from (2) by averaging over ρ. The simplest possible choice for $n(\omega)$ is the Weizsäcker-Williams spectrum, which one gets by replacing the electromagnetic field of the incoming electron by an equivalent cloud of photons:

$$n(\omega) = \frac{\alpha}{\pi\omega} \left(\frac{\hbar\omega}{\varepsilon}\right)^2 \left[K_0\left(\frac{\hbar\omega}{\varepsilon}\right) K_2\left(\frac{\hbar\omega}{\varepsilon}\right) - K_1^2\left(\frac{\hbar\omega}{\varepsilon}\right)\right] \ . \tag{4}$$

However, this method neglects the fact that in some materials channeled electrons can emit most of their energy into one single photon. We also calculated $F(\varepsilon)$ applying more realistic photon spectra, which have been computed recently by Kononets et al. [7].

Results

The $f^+ f^-$-pair creation rates turn out to depend strongly on the parameters of the crystal field. They become larger with increasing nuclear charge Z and decreasing lattice constant d. The results shown in figure 2 are calculated for the $\langle 110 \rangle$ axis of germanium at $293° K$ (the values for d and u are taken from [6]). F and $\tilde{F}$ denote the $f^+ f^-$ creation rate by an electron based on the Weizsäcker-Williams and Kononets' photon spectrum, respectively. The corresponding rates for the $\langle 110 \rangle$ axis of silicon are smaller by a factor of about 100.

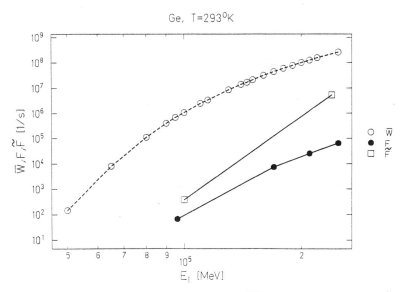

Figure 2. Rates for $f^+ f^-$ creation by a photon ($\overline{W}$) and an electron ($F, \tilde{F}$). The calculation is done for the $\langle 110 \rangle$ axis of germanium.

In order to obtain total pair creation probabilities one has to multiply $\overline{W}$ and F by the above mentioned Gamov factor G and the time τ needed by the incoming particle to pass through the crystal, i.e. the crystal length divided by c. Assuming for example a crystal length of 1 mm, $G = 10^{-3}$ and 240 GeV energy of the incoming particle, we get $G\tau\overline{W} = 10^{-6}$, $G\tau\tilde{F} = 10^{-8}$ and $G\tau F = 10^{-10}$. Together with a particle flux of $10^7 e^-/s$, which is available at the SPS at CERN, this yields considerable numbers of X^0 produced in a high energy channeling experiment.

But with which signal could an X^0 be identified experimentally? As mentioned earlier the f^+f^- pair channels through the crystal with a separation of about 1Å, which corresponds to an highly excited X^0 state of mass $M_X \approx 10$ MeV. If the pair carries a total kinetic energy of 100 GeV, such an X^0 would leave the crystal with a Lorentz factor $\gamma_X \approx 10^4$. It would then reach its ground state after about $5 \times 10^{-13} s$, radiating off 10^3 photons with a typical energy of 10 keV in the X^0 rest frame [3]. This corresponds to 100 MeV photons in the laboratory frame, which could be directly observable. In addition the lifetime of the X^0 of $10^{-11} - 10^{-10} s$ yields a decay length of about 30-300 m. Therefore e^+e^- pairs created by X^0 decay should be distinguishable from those, which are produced inside the crystal.

Finally it is worth mentioning that the lifetime of the X^0 is roughly the same as the lifetime of parapositronium, which is $0.8 \times 10^{-10} s$ [8]. Since the probability for $e^{\pm}$-creation is substantially higher than for $f^{\pm}$-creation, it is not clear that positronium formation can be neglected.

References

[1] T. Cowan et al. (EPOS collab.), *Phys. Rev. Lett.* **54**, 1761 (1985);
    T. Cowan et al. (EPOS collab.), *Phys. Rev. Lett.* **56**, 444 (1986);
    H. Tsertos et al. (ORANGE collab.), *Z. Phys.* **A328**, 499 (1987).

[2] A. Schäfer, *J. Phys.* **G15**, 373 (1989).

[3] S. Graf, S. Schramm, J. Reinhardt, B. Müller and W.Greiner, *J. Phys.* **G15**, 1467 (1989);
    S. Schramm, B. Müller, J. Reinhardt and W. Greiner, *Mod. Phys. Lett.* **A3**, 783 (1988);
    See also the lectures by W. Greiner at this school.

[4] J. Augustin, S. Graf, B. Müller and W. Greiner, *Z. Phys.* **A336**, 353 (1990).

[5] J. C. Kimball and N. Cue, *Phys. Rep.* **125**, 69 (1985).

[6] D. S. Gemmell, *Rev. Mod. Phys.* **46**, 129 (1974).

[7] Yu. V. Kononets and V. A. Riabov, *Nucl. Instr. Meth.* **B48**, 269 and 274 (1990).

[8] A. I. Achieser and W. B. Berestezki, Quantenelektrodynamik, Frankfurt: Harri Deutsch 1962.

PAIR CONVERSION IN SUB- AND SUPERCRITICAL POTENTIALS

Ch. Hofmann[1], J. Reinhardt[1], G. Soff[2], W. Greiner[1], and P. Schlüter[3]

[1]Institut für Theoretische Physik, Universität Frankfurt
Postfach 111 932, D-6000 Frankfurt a.M. 11, Germany

[2]Gesellschaft für Schwerionenforschung (GSI), Planckstraße 1
Postfach 110 552, D-6100 Darmstadt, Germany

[3]Siemens AG, Zentralabteilung Forschung und Entwicklung,
Postfach 830 953, D-8000 München 83, Germany

1. INTRODUCTION

We study the angular correlation of electrons and positrons which are emitted in internal pair conversion (IPC). This process is not only of interest in nuclear physics where our work supplements the early calculations of Rose [1] who employed the plane wave Born approximation. Since in different experiments at the Gesellschaft für Schwerionenforschung (GSI) two collaborations (ORANGE, EPOS) [2] have measured narrow peak structures in electron and positron spectra, the various pair creating processes have to be considered in more detail. The spectra are recorded in heavy-ion collisions with energies near the Coulomb barrier. The origin of these narrow lines is presently not understood although various explanations have been proposed [3].

Internal pair conversion of the colliding nuclei as the source of the puzzling e^+-peaks has been excluded. In the accompanying δ-electron spectrum and γ-ray spectrum one should detect additional structures which are not observed. The triangle shaped electron and positron spectra do not fit to the narrow peaks in the electron and positron single spectra. These items were discussed extensively by P. Schlüter et al.[4]. And, finally, according to the calculations of Rose [1] electron and positron should be emitted preferentially with an opening angle centered around zero degree, whereas in the GSI-experiments an opening angle of 180° seem to be correlated with the peak structures.

But internal pair conversion is besides the dynamically produced electrons and positrons the main process which contributes to the spectra [5]. The ORANGE- and EPOS-detectors enable the measurement of the spectra in different opening angle bins [2]. It is important to determine theoretically the angular distribution of electrons and positrons taking into account the Coulomb distortion. For the dynamically produced electrons and positrons this was accomplished by O. Graf et al. [6]. For IPC the calculations with special regard to supercritical systems will be presented here [7].

2. THEORETICAL BACKGROUND

In heavy-ion collisions close to the Coulomb barrier the nucleus can be Coulomb-excited. An excited state then can decay electromagnetically by emission of a photon. But it is also possible that an electron in a bound state of the atomic shell, e.g. the K-shell, is emitted into the continuum. This process is denoted as internal conversion (K-shell conversion) and is described theoretically as a scattering of the bound state electron into the continuum by a virtual photon which is emitted by the nucleus. If the energy of the nuclear transition is greater than twice the electron rest mass pair creation becomes possible. In the language of Dirac's hole picture an electron of the negative energy continuum is scattered into the positive energy continuum leaving behind a hole, which is a positron.

Experimentally the angular correlation is determined by measuring the number of electron-positron pairs emitted in IPC per angle interval $d\cos\theta$ and positron energy interval dE normalized to the number of photons that are emitted by the same nuclear transition. This quantity can only be defined in electric (E) and magnetic (M) IPC of nuclear transitions with angular momentum $L > 0$. For electric monopole ($E0$) conversion no photons are emitted. Here one normalizes to the number of electrons which are emitted by K-shell conversion of the same transition. The double differential conversion coefficient is defined by

$$\frac{d\beta}{dE\,d\cos\theta} = \frac{dP_{e^+e^-}/dE\,d\cos\theta}{P_\gamma} \tag{1}$$

for EL- and ML-conversion with $L > 0$, and for $E0$-conversion

$$\frac{d\eta}{dE\,d\cos\theta} = \frac{dP_{e^+e^-}/dE\,d\cos\theta}{P_{e^-}}. \tag{2}$$

P_γ is the γ-emission probability, P_{e^-} is the probability of emitting a bound state electron. $P_{e^+e^-}$ denotes the pair conversion probability

$$P_{e^+e^-} \propto \sum_{initial\ states} \sum_{final\ states} |U_{if}|^2\, \delta(\omega - E - E') \tag{3}$$

with the retarded transition amplitude [8]

$$U_{if} = -\alpha \int_{-\infty}^{+\infty} d\tau_n \int_{-\infty}^{+\infty} d\tau_e \left[\rho_n(\vec{r}_n)\,\rho_e(\vec{r}_e) + \vec{j}_n(\vec{r}_n) \cdot \vec{j}_e(\vec{r}_e) \right] \frac{\exp\{i|\vec{r}_n - \vec{r}_e|\omega\}}{|\vec{r}_n - \vec{r}_e|}. \tag{4}$$

The first term in the square brackets indicates the Coulomb interaction of the nuclear transition charge density ρ_n with the electron transition charge density ρ_e. The second term describes the coupling of the nuclear and the electron transition current densities $\vec{j}_n$ and $\vec{j}_e$. For the electron transition charge and current densities we write the explicit forms $\rho_e = \Psi_f^\dagger \Psi_i$, $\vec{j}_e = \Psi_f^\dagger \vec{\alpha} \Psi_i$. The factor $\exp\{i|\vec{r}_n - \vec{r}_e|\omega\}/|\vec{r}_n - \vec{r}_e|$ is the photon propagator in coordinate space.

We treat the electromagnetic interaction of the electrons with the radiation field of the nuclear transition in lowest order perturbation theory. If we neglect in our treatment the external potential V we reproduce the results obtained by Rose within Born approximation. But the coupling constant $Z\alpha$ can be of the order of one in the case of heavy ions or even greater than one if we assume that the colliding heavy ions form a combined system. Since we want to describe those cases we fully take into account the external potential using electron and positron wave functions Ψ_f, Ψ_i, which are distorted by the Coulomb potential of the nucleus.

The Dirac equation is solved with the following boundary conditions. Electron and positron are detected at a large distance from the target. The detector is assumed to measure direction and energy of the leptons. Electron and positron should therefore be described by momentum eigenfunctions, i.e. asymptotically by plane waves. Regarding the deformation of the wave function by the Coulomb potential this leads us to the Coulomb-distorted plane waves [8]

$$\Psi_{\vec{p},\lambda}^{(\pm)} = \sum_{\kappa,\mu} a_{\kappa\mu}^{(\pm)}(\Omega,\lambda)\, \chi_{W\kappa\mu} \tag{5}$$

for an electron of energy W, momentum $\vec{p}$ with direction Ω and polarization λ. The wave functions here are given in terms of their partial wave decomposition into the spherical Dirac spinors $\chi_{W\kappa\mu}$. The expansion coefficients read

$$a_{\kappa\mu}^{(\pm)}(\Omega,\lambda) = (Wp)^{-1/2} i^l e^{\pm i\delta_\kappa} \sum_m (l\,1/2\,j|m\,\lambda\,\mu)\, Y_{lm}^*(\Omega). \tag{6}$$

We choose the $\Psi^{(-)}$-waves for electron and positron which corresponds to a minus-sign of the phase shift δ_κ [9]. It can be shown that these waves describe outgoing particles which behave asymptotically for $t \to \infty$ and $r \to \infty$ as Coulomb distorted plane waves.

We insert these wave functions into the matrix element and thus get a decomposition of the transition amplitude

$$U_{if} = \sum_{\kappa,\kappa'} \sum_{\mu,\mu'} a'^{*}_{\kappa\mu}\, a_{\kappa\mu}\, U_{if}^{sph} \tag{7}$$

where U_{if}^{sph} is the matrix element evaluated with the spherical Coulomb waves. With the use of the multipole decomposition of the photon propagator this yields finally the following form of the transition amplitude

$$U_{if}^{sph} = \sum_{L>0,M} \{U_{if}^{(e)}(L,M) + U_{if}^{(m)}(L,M)\} + U_{if}^{E0} \tag{8}$$

where the matrix elements $U_{if}^{(e)}$ and $U_{if}^{(m)}$ describe the electric and the magnetic pair conversion, respectively, with an angular momentum $L > 0$ of the nuclear transition. U_{if}^{E0} describes the electric monopole pair conversion.

We obtain $dP_{e^+e^-}/dE\,d\cos\theta$ if we do not perform the energy and angular integration [7]. Dividing by the γ-emission probability or, in the E0-case, the probability of emitting a bound state electron this yields finally the following expressions

$$\frac{d\beta}{dE\,d\cos\theta}(\tau,L) = \frac{1}{2}\frac{d\beta}{dE}\left(1 + \sum_I \frac{a_I}{a_0} P_I(\cos\theta)\right) \tag{9}$$

with $\tau = E, M$ for electric and magnetic pair conversion. The $P_I(\cos\theta)$ are the Legendre-Polynomials. For the E0-conversion it simply results

$$\frac{d\eta}{dE\,d\cos\theta} = \frac{1}{2}\frac{d\eta}{dE}(1 + \epsilon\cos\theta) \tag{10}$$

The angular correlation functions separate into the parts $d\eta/dE$ and $d\beta/dE$ which describe the spectra, and the parts that determine the anisotropy of the emitted pairs

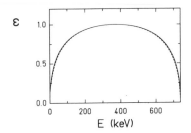

Figure 1. Anisotropy factor ϵ versus the kinetic positron energy for E0-conversion in $^{90}_{40}$Zr and the transition energy $\omega = 1.76$ MeV.

represented by the anisotropy coefficients a_I/a_0 and ϵ. The differential pair conversion coefficients with respect to the positron energy have been calculated by Schlüter [10] and Soff [11]. The anisotropy coefficients are given by lenthy expressions involving multiple sums over angular momenta [7]. Since we employ wave functions distorted by the Coulomb-potential of the nucleus these coefficients depend on the nuclear transition energy ω, the positron energy E, and – in contrast to the Born approximation – also on the nuclear charge number Z.

3. RESULTS

We discuss first the angular correlation for the electric monopole conversion. In this case we have to consider only one anisotropy factor ϵ The results can be easily transferred to the conversion incorporating higher multipoles.

Fig. 1 shows the dependence of the anisotropy factor ϵ on the kinetic positron energy E for the E0-transition $\omega = 1.760$ MeV in $^{90}_{40}$Zr. Also the Born approximation result (dashed line) is plotted. Only a small deviation is visible. ϵ remains positive over the entire positron energy range. This corresponds to the fact that electrons and positrons are preferentially emitted in the same direction. The anisotropy of the pairs has its maximum for a symmetric splitting of the nuclear transition energy onto electron and positron; $\epsilon \approx 1$. If either the electron or the positron gets the maximum kinetic energy the angular distribution is isotropic, $\epsilon \approx 0$. If we go to nuclei with higher charge Z or lower transition energy ω the deviations from the results of our calculation to those obtained within Born approximation become stronger. For example we plot in Fig. 2 the anisotropy factor ϵ versus the kinetic energy for a uranium nucleus for an assumed nuclear transition energy $\omega = 1.3$ MeV. Again the dashed curve represents the result obtained in Born approximation.

But let us now study the case of nuclear charges around $Z_{cr} = 173$. These nuclear charges are not present in nature but may be produced transiently in collisions of highly charged nuclei. Let us assume that such a hypothetical combined nuclear system lives long enough to allow the observation of pair conversion. Fig. 3 shows ϵ versus E for one subcritical nuclear charge $Z = 170$ and three supercritical nuclear charges $Z = 174, 182, 188$. One recognizes that the maximum of the anisotropy factor ϵ for

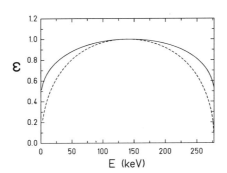

Figure 2. Anisotropy factor ϵ for E0-conversion in $^{238}_{92}$U assuming a nuclear transition energy $\omega = 1.3$ MeV.

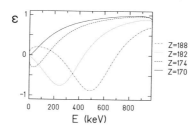

Figure 3. ϵ plotted for E0-conversion assuming a transition energy $\omega = 2$ MeV and several critical and supercritical nuclear charge numbers Z.

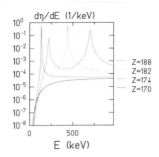

Figure 4. Differential pair conversion coefficient versus the kinetic positron energy plotted for the same transition energy and the same nuclear charge numbers Z.

$Z = 170$ is shifted to higher positron energies and that ϵ for $Z > 173$ no longer is positive definite. This implies that for these nuclear charges the angular distribution is centered around $180°$, i.e. electron and positron can be emitted back-to-back as in the case of a particle decay.

If we compare Fig. 3 with Fig. 4 where the differential conversion coefficient $d\eta/dE$ is plotted versus the kinetic positron energy we see that the zeros of the function $\epsilon(E)$ correspond exactly to the positions of the peaks in the function $d\eta/dE$. These peaks arise from the fact that for nuclear charges $Z \geq 173$ the atomic states become imbedded into the negative energy continuum. The first state, which "dives" into the Dirac sea for $Z = 173$ is the $1s$-state forming the s-resonance, for higher charge Z the $2p$-state begins to "dive" forming the p-resonance, and so on [12]. The $E0$-conversion is a penetration effect and the pair creation is determined by the overlap of the electron wave functions with the nuclear interior. For a resonant state the electron density inside the nucleus and thus the pair conversion probability is strongly enhanced. ϵ is proportional to the inverse of $d\eta/dE$ and is therefore suppressed at the resonance energies. Since the wave function of a resonant state undergoes a rapid jump in the phase shift by π and $\epsilon \propto \cos\delta$, δ the phase shift, the anisotropy factor changes its sign at the resonance energies.

A similar behaviour occurs in the case of electric and magnetic IPC of higher multipole transitions. Here the overlap of the electron wave functions with the electromagnetic field of the photon has to be calculated. Since this field is centered around the nucleus this overlap is enhanced if the wave function of the negative energy electron becomes a resonant state wave function.

4. CONCLUSIONS

We have considered the angular correlation of electrons and positrons in internal pair conversion taking into account the distortion of the leptons' wave functions by the Coulomb potential of the nucleus. As expected we obtain large deviations compared with the results calculated within Born approximation if we consider high nuclear charges or low nuclear transition energies. For IPC corresponding to nuclear transitions of electric or magnetic type with angular momentum $L > 0$ our calculations yield an increase of produced pairs at opening angles around $\theta = 180°$ and a decrease at $\theta = 0°$ compared to the Born approximation.

The behaviour in the case of electric monopole IPC is different. The anisotropy of the pairs here is enhanced compared to Born approximation. $E0$-conversion and EL-, ML-conversion ($L > 0$) give similar results for the supercritical region of nuclear charges, $Z > 172$. In this region electron and positron can be emitted back-to-back.

One might be tempted to take the IPC in supercritical fields as explanation for the narrow line structures measured at the GSI. But there are several severe arguments against this interpretation. Firstly, if the peaks in the positron spectrum stem from IPC in supercritical collision systems then the electrons and positrons are emitted isotropically, not back-to-back. Furthermore, if the ions would stick together a time long enough ($> 10^{-15}$ sec) that IPC becomes measurable than the positron spectrum should be dominated by the peak of the spontaneous positron production [12].

Besides these two points one should see in the spectra a dependence of the peaks on the nuclear charge, since the position of the resonances depend strongly on the nuclear charge. However, the opposite is observed; the peaks measured at the GSI remain for all nuclear charges even for subcritical ones at the same positions.

REFERENCES

1. M. E. Rose, Phys. Rev. 76, 678 (1949)

2. W. Koenig, E. Berdermann, F. Bosch, S. Huchler, P. Kienle, C. Kozhuharov, A. Schröter, S. Schuhbeck, H. Tsertos, Phys. Lett. B218, 12 (1989)
   H. Bokemeyer, P. Salabura, D. Schwalm, K. E. Stiebing, in O. Fackler, J. Tran Thanh Van eds., *Tests of Fundamental Laws in Physics*, Series: Moriond Workshops, (Editions Frontiers, Gif-sur-Yvette, 1989), p. 77

3. B. Müller, in R. Marrus ed., *Atomic Physics of Highly Ionized Atoms*, (Plenum Press, New York, 1989), p. 39
   A. Schäfer, J. Phys. G15, 373 (1989)

4. P. Schlüter, U. Müller, G. Soff, Th. de Reus, J. Reinhardt, W. Greiner, Z. Phys. A323, 139 (1986)

5. M. Krämer, B. Blank, E. Bozek, E. Ditzel, E. Kankeleit, G. Klotz-Engmann, C. Müntz, H. Oeschler, M. Rhein, P. Senger, Phys. Rev. C40, 1662 (1989)

6. O. Graf, J. Reinhardt, B. Müller, W. Greiner, G. Soff, Phys. Rev. Lett. 61, 2831 (1988)

7. C. Hofmann, J. Reinhardt, W. Greiner, P. Schlüter, G. Soff, Preprint GSI-90-31, Phys. Rev. C, in print

8. M. E. Rose, *Relativistic Electron Theory*, (Wiley, New York, 1961)

9. H. A. Bethe, L. C. Maximon, Phys. Rev. 93, 768 (1954)

10. P. Schlüter, G. Soff, W. Greiner, Phys. Rep. 75, 327 (1981)

11. G. Soff, P. Schlüter, W. Greiner, Z. Phys. A303, 189 (1981)

12. B. Müller, J. Rafelski, W. Greiner, Z. Phys. 257, 62 (1972)

PRODUCTION OF EXOTIC PARTICLES IN ULTRARELATIVISTIC HEAVY-ION COLLISIONS

M. Greiner[1,2], M. Vidović[3], J. Rau[3], C. Hofmann[3], and G. Soff[1]

[1]Gesellschaft für Schwerionenforschung (GSI), Planckstraße 1
Postfach 110 552, D-6100 Darmstadt, Germany

[2]Institut für Theoretische Physik, Justus Liebig Universität
Heinrich-Buff-Ring 16, D-6300 Gießen, Germany

[3]Institut für Theoretische Physik, Johann Wolfgang Goethe-Universität
Postfach 111 932, D-6000 Frankfurt am Main, Germany

1. INTRODUCTION

Future pp-supercolliders in principle will also provide the opportunity to accelerate heavy ions. Ion energies up to 3.5 TeV/nucleon will be attainable at the Large Hadron Collider (LHC) at CERN. At the Superconducting Supercollider (SSC) in Texas heavy-ion energies up to 8 TeV/nucleon would be possible. The strong electromagnetic fields prevailing in these ultrarelativistic heavy-ion collisions can give rise to the production of exotic particles. The considered particles are for example heavy leptons, mesons intermediate vector bosons and even Higgs bosons or supersymmetric particles. They are effectively created via elementary two-photon processes. A simple estimate for the available photon energy yields

$$\hbar\omega_0 = \frac{\hbar c \gamma}{R} \simeq 200 \text{ GeV} \tag{1}$$

assuming a Lorentz contraction factor $\gamma \simeq 8000$ at the SSC and $R \simeq 7$ fm for a lead or uranium nucleus. Thus, centre of momentum energies of a few hundred GeV can be reached in collisions of two virtual photons contained in the Coulomb fields carried along by the colliding nuclei, which represents a necessary prerequisite to produce some of these exotic particles. In particular we evaluate in the following total cross sections [1,2] for the formation of Higgs bosons and supersymmetric particles via the method of equivalent photons [3-7].

With respect to Higgs bosons we also discuss the impact parameter dependence of the production probability. We want to avoid central collisions, which would give rise to direct hadronic reactions.

Vacuum Structure in Intense Fields, Edited by
H.M. Fried and B. Muller, Plenum Press, New York, 1991

2. THE METHOD OF EQUIVALENT PHOTONS

For $\gamma \gg 1$ the electromagnetic field of a nucleus becomes almost transverse. The electric field strength is perpendicular to the magnetic field strength and both are perpendicular to the momentum vector $\vec{p}$ of the moving source. This situation displays some similarity with a real photon. Thus we can regard the colliding nuclei as efficient carriers of real photons. Employing the associated method of equivalent photons [3-7], also known as Weizsäcker-Williams method, one can easily estimate the production cross section of elementary particles in ultrarelativistic heavy-ion collisions. The total cross section σ_{AA}^f to produce the final state f can be expressed as $(\hbar = c = 1)$

$$
\begin{aligned}
\sigma_{AA}^f &= \int d\omega_1 \, d\omega_2 \, n_A(\omega_1) \, n_A(\omega_2) \, \sigma_{\gamma\gamma}^f(\omega_1, \omega_2) \\
&= \int d\omega_1 \, d\omega_2 \, n_A(\omega_1) \, n_A(\omega_2) \int ds \, \delta(4\omega_1\omega_2 - s) \sigma_{\gamma\gamma}^f(s) .
\end{aligned}
\tag{2}
$$

$\sigma_{\gamma\gamma}^f$ indicates the elementary two-photon cross section to produce the final state f at centre of momentum energy $\sqrt{s}$. Since the photons are massless, one finds $s = 4\omega_1\omega_2$, which is guaranteed by the δ-function in (2). $n_A(\omega_i)$ represents the number of equivalent photons of the nucleus A with charge Z; it is given by [8]

$$
n_A(\omega) = \frac{2 Z_A^2 \alpha}{\pi \, \omega} \int_{\omega/\gamma}^{\infty} dk \, \frac{k^2 - (\omega/\gamma)^2}{k^3} \, F_A(k)^2 = \frac{2 Z_A^2 \alpha}{\pi \, \omega} f(\omega/\omega_0)
\tag{3}
$$

with $\omega_0 = \gamma/R$ from (1). $F_A(k)$ denotes the elastic nuclear charge form factor. If the nuclear charge distribution is described by a homogeneously charged sphere with radius R it results

$$
\begin{aligned}
f(x) &= \int_x^{\infty} dz \, \frac{9(z^2 - x^2)}{z^5} \, j_1^2(z) \\
&= \frac{3}{16 \, x^6} + \frac{3}{8 \, x^4} - \left(\frac{3}{16 \, x^6} + \frac{7}{40 \, x^2} + \frac{1}{20} \right) \cos(2x) \\
&\quad - \left(\frac{3}{8 \, x^5} + \frac{1}{10 \, x^3} - \frac{9}{20 \, x} - \frac{x}{10} \right) \sin(2x) - \left(1 + \frac{x^2}{5} \right) \mathrm{Ci}\,(2x)
\end{aligned}
\tag{4}
$$

with the abbreviation $x = \omega/\omega_0$, $j_1(z)$ being a spherical Besselfunction and $\mathrm{Ci}(2x)$ being the cosine-integral function

$$
\mathrm{Ci}\,(2x) = - \int_{2x}^{\infty} d\tau \, \frac{\cos \tau}{\tau} .
\tag{5}
$$

Because of the factor Z^2 in the equivalent photon spectrum (3), the cross section (2) rises with Z^4. Therefore it becomes apparent that, in spite of the suppression due to the nuclear form factor, heavy ions are much more effective for the production of exotic particles than protons or electrons.

3. LEPTON PAIR AND BOSON PAIR PRODUCTION

The two-photon cross section for the production of a lepton pair reads [6]

$$
\begin{aligned}
\sigma_{\gamma\gamma}^{l^+ l^-}(s) = \frac{4\pi \alpha^2}{s} &\left[\left(1 + \frac{1}{y} - \frac{1}{2y^2} \right) 2 \ln \left(\sqrt{y} + \sqrt{y-1} \right) \right. \\
&\left. - \sqrt{1 - \frac{1}{y}} \left(1 + \frac{1}{y} \right) \right] \Theta(y - 1)
\end{aligned}
\tag{6}
$$

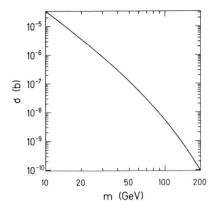

Figure 1. Total cross section σ for lepton pair production in U + U collisions with E_{ion} = 8 TeV/nucleon. σ is plotted versus the rest mass m of a hypothetical new lepton.

with the abbreviation

$$y = \frac{s}{4m^2}, \tag{7}$$

where m denotes the particle mass. The minimum value of $\sqrt{s}$ is $(\sqrt{s})_{min} = 2m$. Similary one finds for the two-photon production cross section of a pair b^+b^- of scalar bosons

$$\sigma_{\gamma\gamma}^{b^+b^-}(s) = \frac{2\pi\alpha^2}{s}\left[\sqrt{1 - \frac{1}{y}}\left(1 + \frac{1}{y}\right) - \right.$$
$$\left. - \frac{1}{y}\left(2 - \frac{1}{y}\right)\ln\left(\sqrt{y} + \sqrt{y - 1}\right)\right]\Theta(y - 1). \tag{8}$$

Inserting (6) and (8) into (2), it is now easy to evaluate total cross sections for pair creation in relativistic collisions of heavy nuclei [see also ref. 7]. Let us first consider the lepton pair production at the planned RHIC collider with 100 (150) GeV/u U-beams. For a μ-pair we find $\sigma_{UU}^{\mu^+\mu^-} = 364\,(538)$ mb and correspondingly for a τ-pair $\sigma_{UU}^{\tau^+\tau^-} = 5.3\,(20.2)\,\mu$b.

To take into account the possible existence of still heavier members in the lepton hierarchy we computed for $E_{ion} = 8\,$TeV/u (SSC) the production cross section of lepton pairs as function of the lepton mass m. In figure 1 we depict σ for U + U collisions as function of m. For $m < 200$ GeV the cross sections are obviously large enough to be measurable. – In the computation of the production cross section for proton pairs $(p\bar{p})$ and for pion pairs $(\pi^+\pi^-)$ we neglected form factors of the generated particles. Consequently we treated for example p and $\bar{p}$ as point-like Dirac fermions. For the mentioned RHIC energies the cross sections result to be $\sigma_{UU}^{\bar{p},p} = 140\,(352)\,\mu$b and $\sigma_{UU}^{\pi^+\pi^-} = 25.7\,(38.7)$ mb, respectively.

4. PRODUCTION OF INTERMEDIATE VECTOR BOSONS

Now we turn the discussion to the formation of W-bosons in the electromagnetic field of colliding nuclei. The total cross section for W-boson production follows from the

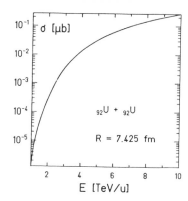

Figure 2. Total cross section σ for W^+W^- pair production in U + U collisions as function of the ion energy in units of TeV/nucleon.

elementary two-photon process [9]

$$\sigma_{\gamma\gamma}^{W^+W^-}(s) = \frac{4\pi\alpha^2}{s}\sqrt{1-\frac{1}{y}}\left\{4y\left(2+\frac{3}{8y}+\frac{3}{8y^2}\right)\right.$$
$$\left. + \frac{3/(2y)-3/(4y^2)}{\sqrt{1-1/y}}\ln\left[\frac{1-\sqrt{1-1/y}}{1+\sqrt{1-1/y}}\right]\right\} . \qquad (9)$$

The results which are displayed in figure 2 are obtained simply by plugging the elementary expression (9) into (2). For U + U collisions at 8 TeV/nucleon we computed $\sigma \simeq 0.1~\mu b$. We can compare this cross section with corresponding results for W^+W^--pair production in proton-proton collisions at supercollider energies ($E = 40$ TeV at the SSC). According to Eichten et al [10] one gets $\sigma_{W^+W^-} \simeq 0.1$ nb, which is approximately three orders of magnitude smaller. This striking difference basically is caused by the factor $Z_1^2 Z_2^2 \sim 10^8$ entering into the formula for the electromagnetic production with heavy ions.

5. PRODUCTION OF SUPERSYMMETRIC PARTICLES

Supersymmetry [11-13] is one of the most promising attempts to unify all fundamental forces in nature. It involves a symmetry between bosons and fermions which allows for a unified description of matter, built of fermions, and interactions, carried by bosons. In order to unify the strong, weak and electromagnetic interaction with gravity one has to connect internal symmetries with space-time symmetry in a non-trivial way. Due to the theorem of Coleman and Mandula [14] this is impossible in the framework of symmetries described by standard Lie groups. Supersymmetry is a way of obtaining a non-trivial unification of space-time and internal symmetries [15]. Another important feature is that supersymmetric theories are free from infinties at least on the one-loop level; as boson and fermion loops contribute to radiative corrections with opposite signs many divergencies cancel pairwise. – An immediate consequence of supersymmetry is the existence of supersymmetric partners to all known elementary particles, the so-called superpartners (sparticles), which have exactly the same properties except that their spin

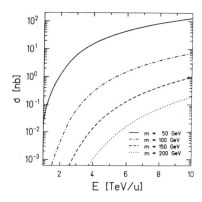

Figure 3. Production cross section of winos $\tilde{W}^{\pm}$ as function of the ion energy E for Pb + Pb collisions. For the rest mass of the produced winos we assumed values between 50 GeV and 200 GeV.

differs by $\frac{1}{2}$. Evidently since no light charged supersymmetric particles have been observed, supersymmetry must be broken.

Formally one distinguishes the sparticles from ordinary particles by introducing the so-called R-parity which is +1 for particles and -1 for sparticles. In most supersymmetric theories this quantum number is multiplicatively conserved. Immediate consequences are: (a) sparticles have to be produced in pairs, (b) heavy sparticles decay into lighter sparticles and (c) the lightest supersymmetric particle is stable.

As we consider production of sparticles in heavy-ion collisions, i.e. by virtue of electromagnetic interaction, it is obvious to focus on charged sparticles. In particular, we compute the production cross section for sleptons $\tilde{L}$ and winos $\tilde{W}$, the superpartners of the leptons and the charged intermediate vector bosons. Since global supersymmetry and conventional internal symmetries decouple, internal quantum numbers like charge, weak isospin etc. remain unaffected by supersymmetry transformations. Therefore the interaction of charged sparticles with the electromagnetic field can be described by minimal coupling as usual. Especially sleptons behave like ordinary charged scalars and winos like ordinary charged Dirac fermions in the electromagnetic field. So in order to evaluate total production cross sections in ultrarelativistic heavy-ion collisions one simply has to insert (6) and (8) into (2), m now being a parameter for the unknown slepton and wino mass, respectively.

Lower mass limits are provided by existing high-energy accelerators where sparticles have not yet been observed. So far upper limits are not available, but if supersymmetry is to be relevant to the physics of the electroweak scale (mass ratios), then the supersymmetric partners of the known fundamental fields must have masses that are no more than a few times the scale $(\sqrt{2}G_F)^{-1/2} = 247$ GeV of the electroweak symmetry breaking. Hence these masses are supposed to lie in the energy region which can be explored by the planned supercolliders .

With (2), (6) and (8) we computed the cross sections as displayed in figures 3 and 4. For rest masses around $m = 100$ GeV we obtained for $E_{ion} = 8$ TeV/u $\quad \sigma_{UU}^{\tilde{W}^+\tilde{W}^-} \approx 10\,\mathrm{nb}$ and $\sigma_{UU}^{\tilde{l}^+\tilde{l}^-} \approx 1\,\mathrm{nb}$. The corresponding results [16] for proton-proton collisions at $E = 40$ TeV are $\sigma_{pp}^{\tilde{W}^+\tilde{W}^-} \lesssim 0.1\,\mathrm{nb}$ and $\sigma_{pp}^{\tilde{l}^+\tilde{l}^-} \lesssim 0.01\,\mathrm{nb}$.

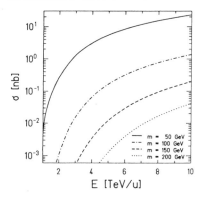

Figure 4. The same as in figure 3 for the production of sleptons $\tilde{l}^{\pm}$.

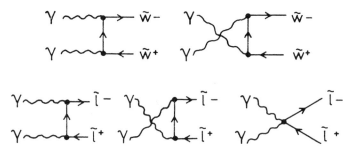

Figure 5. Feynman diagrams for the two-photon production of winos $\tilde{W}$ and sleptons $\tilde{l}$.

Figure 5 shows the lowest-order two photon processes for the formation of sleptons $\tilde{l}$ and winos $\tilde{W}$, while possible decay channels are depicted in figure 6 assuming that either photinos or Higgsinos (left) or sneutrinos (right) are the lightest supersymmetric particle (LSP). As a common signature, one always finds a high-energetic lepton together with some missing energy and momentum, because the neutrino and the LSP cannot be detected.

6. PRODUCTION OF NEUTRAL PARTICLES (HIGGS BOSONS)

In the case that the final state is a single neutral particle X^0, the cross section for the process $\gamma\gamma \to X^0$ is narrowly peaked around the particle mass M_{X^0}. It can then be well approximated by [17]

$$\sigma_{\gamma\gamma}^{X^0}(s) = \frac{8\pi^2}{M_{X^0}} \Gamma_{X^0 \to \gamma\gamma} \delta(s - M_{X^0}^2), \tag{10}$$

where $\Gamma_{X^0 \to \gamma\gamma}$ denotes the partial two-photon decay width of the particle produced with rest mass M_{X^0}. For the total production cross section in heavy-ion collisions it results

$$\sigma_{AA}^{X^0} = F(M_{X^0})/f_{X^0}^2 \tag{11}$$

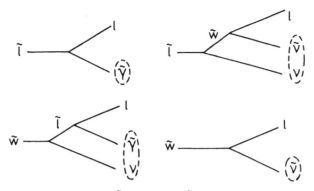

Figure 6. Decay modes of sleptons $\tilde{l}$ and winos $\tilde{W}$. Left: The photino is stable. Right: The sneutrino is stable.

with the definition $(x_0 = M_{X^0}/(2\omega_0))$

$$F(M) = \frac{Z^4 \alpha^4}{\pi^3} \int_{x_0}^{\infty} \frac{dx}{x} f(x) f(x_0^2/x) \tag{12}$$

and the decay constant

$$f_{X^0}^2 = \frac{\alpha^2 M_{X^0}^3}{64 \pi^3 \Gamma_{X^0 \to \gamma\gamma}} . \tag{13}$$

We found $\sigma_{UU}^{\pi^0} = 66$ mb (90mb), $\sigma_{UU}^{\eta_c} = 1.16$ mb (1.95 mb), $\sigma_{UU}^{\eta_b} = 1$ μb (2 μb) at E_{ion} = 3.5 TeV/u (8 TeV/u), respectively.
In the evaluation of the production cross section of Higgs bosons we employed

$$\Gamma_{H \to \gamma\gamma} = \frac{\alpha^2}{8\sqrt{2}\,\pi^3} G_F m_H^3 |I|^2 \tag{14}$$

with the Fermi coupling constant $G_F/(\hbar c)^3 = 1.16637 \cdot 10^{-5}$ GeV^{-2} and $|I|$ being of order unity [18].

Figure 7 shows the computed cross section σ_H for Higgs-boson production in U + U collisions as function of E_{ion}. Depending on the unknown Higgs-boson mass m_H, which has to be larger than 41.6 GeV [19], typical values in the order of 1 nb are obtained (cf. also ref. 20). Again we can compare this prediction with calculated values for proton - proton collisions at $E = 40$ TeV. Here the gluon-gluon fusion leads typically to [10] $\sigma_{pp}^H \simeq 0.01$ nb.

7. IMPACT-PARAMETER DEPENDENCE OF THE HIGGS BOSON PRODUCTION PROBABILITY

Looking again closer on the production of Higgs bosons or supersymmetric particles, one might question whether the main contribution to the electomagnetic cross section really results from impact parameter being larger than twice the nuclear radius [21]. Central collisions should be omitted because of nuclear reactions taking place, which would mask a clean experimental signal. The Equivalent Photon Method which has been employed

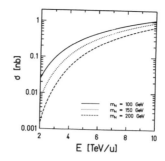

Figure 7. Production cross section for Higgs bosons in U + U collisions as function of E_{ion}. For the Higgs boson mass m_H we assumed 100 GeV (solid line), 150 GeV (dotted line) and 200 GeV (dashed line), respectively.

to calculate the production cross sections only yields total cross sections, because the wave front of the equivalent photons is by definition extended to infinity in transversal directions. In order to overcome this short-coming, we develop an impact-parameter dependend Equivalent Photon Method for the Higgs production.

The electromagnetic potential A^μ of an extended charge A moving with constant velocity on a straight line is proportional to its four-velocity u^μ :

$$A_A^\mu(k_A) = Z_A\, e^{ik_A^\nu \cdot b_\nu}\, r_A(k_A)\, u_A^\mu \qquad (15)$$

with

$$r(k) = 8\pi^2 e\, \delta(k^\nu u_\nu)\, \frac{F(-k^\mu k_\mu)}{k^\sigma k_\sigma}\ . \qquad (16)$$

b is the impact parameter and $F(-k^2)$ the elastic nuclear charge formfactor. Both electromagnetic potentials of the colliding heavy ions enter linear in the corresponding lowest-order S-matrix element. In order to make an identification with the real $\gamma\gamma \to H$ vertex, one has to transform the velocity u^μ into a polarisation vector ϵ^μ. Because of gauge invariance this is achieved by adding a quantity proportional to the photon momentum to the velocity,

$$u^\mu \quad \to \quad u^\mu + \alpha k^\mu = h(k)\,\epsilon^\mu \qquad (17)$$

with

$$h(k) = \frac{\gamma |\vec{k}_\perp|}{\omega} \qquad (18)$$

as $v \to c$. Using then explicetly the real $\gamma\gamma \to H$ vertex

$$\epsilon_1^\mu\, \Gamma_{\mu\nu}\, \epsilon_2^\nu = \frac{g_H}{2}\left[(k_1^\sigma k_{2\sigma})(\epsilon_1^\tau \epsilon_{2\tau}) - (k_1^\sigma \epsilon_{2\sigma})(k_2^\tau \epsilon_{1\tau}) \right]\ , \qquad (19)$$

neglecting transversal components of the photon momenta and introducing a polarisation averaged $\gamma\gamma \to H$ S-matrix element

$$\overline{S}_{\gamma\gamma}^{H} = (2\pi)^2 \delta^4(p_H - k_1 - k_2) \frac{1}{\sqrt{2}} \frac{g_H}{2} (k_1 \cdot k_2) , \tag{20}$$

we derive the following expression for the differential cross section with respect to the impact parameter (cf. also ref. 22)

$$\frac{d^2\sigma}{db^2} = 2 \left(\frac{Z_1 Z_2 e^2}{\pi} \right)^2 \int \frac{d\omega_1}{\omega_1} \int \frac{d\omega_2}{\omega_2} \int \frac{d^2q}{(2\pi)^2} e^{iq \cdot b}$$
$$\cdot \sum_{ij} \eta^{ij}(\omega_1, \vec{q}) \, \eta^{ij}(\omega_2, -\vec{q}) \cdot \sigma_{\gamma\gamma}^{H}(\omega_1, \omega_2) \tag{21}$$

with

$$\eta^{ij}(\omega, \vec{q}) = \int \frac{d^2 k_\perp}{(2\pi)^2} \frac{F(-k^2)}{k^2} \frac{F(-(k-q)^2)}{(k-q)^2} k_\perp^i (k_\perp - q)^j . \tag{22}$$

Integrating the differential cross section (21) over all impact parameters, yields the Weizsäcker-Williams cross section (2) with (3) and (10).

Figure 8 shows the differential cross section for the Higgs boson production in a Pb + Pb collision at SSC energies, where the cross section $\sigma_{\gamma\gamma}^{H}$ of eq. (10) and (14) has been used (solid curve). The rather curious form of this curve, especially the dip at $b \simeq R$ traces back to the fact, that the electric fields of the two colliding nuclei have to be parallel in order to produce the scalar Higgs boson [23]. – The dotted curve, also shown in figure 8 differs from the solid curve by the substitution $k_\perp^i (k_\perp - q)^j \rightarrow \frac{1}{2} |\vec{k}_\perp| |\vec{k}_\perp - \vec{q}|$ in expression (22). It originates from a non-justified application of the real $\sigma_{\gamma\gamma}^{H}$ cross section, where polarisation averaging has not been considered.

Introducing a sharp cutoff impact parameter, $b_c = 2R$, and thus discarding central collisions, the reduced cross section

$$\tilde{\sigma} = \int d^2 b \frac{d^2\sigma}{db^2} \theta(b - 2R) \tag{23}$$

is only by a factor of 2.1 (2.6) for Pb + Pb at LHC energies and 1.5 (1.7) at SSC energies for a Higgs mass of 100 (150) GeV smaller than the corresponding equivalent photon cross sections. These results agree quite well with those obtained by other groups [22,24-26]. Thus, for an expected LHC and SSC luminosity of 10^{28} cm^{-2} sec^{-1} and a running time of 10^7 sec/year ($\cong 1/3$ year) still about 42 (6) produced Higgs bosons with an assumed mass of 100 GeV are to be expected for SSC (LHC).

8. CONCLUSIONS

We have demonstrated that in peripheral heavy-ion collisions with $E_{ion} \simeq 10$ TeV per nucleon photon energies up to $\hbar\omega \simeq 200$ GeV will be available. Using the Equivalent Photon Method we calculated total cross section for the two-photon production of exotic particles such as Higgs bosons and supersymmetric particles; these cross sections are of the order $\lesssim 1$ nb. The exclusion of central collisions has no dramatic consequences for these processes. The production cross sections utilizing the strong electromagnetic field of heavy ions are generally larger than corresponding results for hadronic reactions in central proton-proton collisions. Also the double-pomeron exchange for the Higgs boson in peripheral nuclear collisions is negligible with respect to the analogous two-photon process [27].

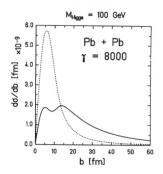

Figure 8. The differential cross section for Higgs-boson production (solid curve). The dotted curve differs from the solid curve by the substitution $k_{\perp}^{i}\,(k_{\perp}-q)^{j}\;\to\;\frac{1}{2}|\vec{k}_{\perp}|\,|\vec{k}_{\perp}-\vec{q}|$ in expression (22).

The disadvantage of a possible lower luminosity of a heavy-ion beam might be compensated by the lack of overwhelming hadronic showers as in proton-proton collisions. In heavy-ion collisions one can possibly trigger to peripheral events without direct hadronic reactions.

Some care has to be taken for the Higgs bosons. After its production it mainly decays into a $b\bar{b}$ quark pair. Therfore, as a serious backround, $b\bar{b}$ productions via two-photon fusion comes up. A careful discussion [28] devoted to this problem shows that setting sharp windows on the $b\bar{b}$ invariant mass the signal $\gamma\gamma \to H \to b\bar{b}$ can be deduced (cf. also ref. 29).

In future investigations one should elaborate in more detail new methods to determine the impact parameter b in ultrarelativistic heavy-ion collisions. Coulomb dissociation of nuclei in ultrarelativistic heavy-ion collisions may serve as a measure of b. Similarly the intensity of nucleus-nucleus bremsstrahlung as well as the number of generated e^{+}, e^{-}-pairs can be used as indicator for b. In fixed target experiments the recoil of the target nucleus may be utilized to determine b.

The two-photon Feynman diagrams with virtual photons ($q^{2} \neq 0$) have to be evaluated. In addition higher-order processes such as radiative corrections of many-particle final states should be taken into account.

Finally it might be worth-while to study explicitly central collisions in which we have to deal with hundreds of individual parton collisions.

REFERENCES

1. M. Grabiak, B. Müller, W. Greiner, G. Soff, P. Koch, J. Phys. G15, L25 (1989)

2. J. Rau, B. Müller, W. Greiner, G. Soff, J. Phys. G16, 211 (1990)

3. E. Fermi, Z. Physik 29, 315 (1924)

4. E.J. Williams, Proc. Roy. Soc. A139, 163 (1933)

5. C. Weizsäcker, Z. Physik 88, 612 (1934)

6. V.M. Budnev, I.F. Ginzburg, G.V. Meledin, V.G. Serbo, Phys. Rep. 15, 181 (1975)

7. C.A. Bertulani and G. Baur, Phys. Rep. 161, 299 (1988)

8. G. Soff, J. Rau, M. Grabiak, B. Müller and W. Greiner, in "The Nuclear Equation of State", Part B, eds.: W. Greiner and H. Stöcker, p.579, (Plenum Press, New York, 1989)

9. M. Katuya, Phys. Lett. 124B, 421 (1983)

10. E. Eichten, I. Hinchliffe, K. Lane and C. Quigg, Rev. Mod. Phys. 56, 579 (1984)

11. M.F. Sohnius, Phys. Rep. 128 (1985) 39

12. H.J.W. Müller-Kirsten and A. Wiedemann, Supersymmetry, (World Scientific, Singapore, 1987)

13. J. Rau, Supersymmetrie, GSI Report 89-20 (1989)

14. S. Coleman and J. Mandula, Phys. Rev. 159, 1251 (1967)

15. R. Haag, J.T. Lopuszansk and M.F. Sohnius, Nucl. Phys. B88, 257 (1975)

16. S. Dawson, E. Eichten, C. Quigg, Phys. Rev. D31, 1581 (1985)

17. S.J. Brodsky, T. Kinoshita and H. Terazawa, Phys. Lett. 25, 972 (1970); Phys. Rev. D4, 1532 (1971)

18. R. Bates and J.N. Ng, Phys. Rev. D33, 657 (1986)

19. D. Decamp et al. (ALEPH Collaboration), Phys. Lett. 246B, 306 (1990)

20. E. Papageorgiu, Phys. Rev. D40, 92 (1989)

21. R.N. Cahn and J.D. Jackson, preprint LBL-28592, Berkeley (1990)

22. B. Müller and A.J. Schramm, preprint DUK-TH-90-7, Durham (1990)

23. see for example: T.D. Lee, Particle Physics and Introduction to Field Theory, (Harwood Academic Publishers, Chur, 1981)

24. E. Papageorgiu, preprint RAL-90-037, Oxon (1990)

25. G. Baur and L.G. Ferreira Filho, preprint, Jülich (1990), subm. to Nucl. Phys.

26. J.S. Wu, C. Bottcher, M.R. Strayer and A.K. Kerman, preprint ORNL/CCJP/90/02, Oak Ridge (1990), subm. to Particle World

27. B. Müller and A.J. Schramm, preprint DUK-TH-90-8, Durham (1990)

28. M. Drees, J. Ellis, D. Zeppenfeld, Phys. Lett. 223B, 454 (1989)

29. G. Altarelli and M. Traseira, Phys. Lett. B245, 658 (1990)

APPLICATION OF THE PROPER-TIME METHOD

TO NON-PERTURBATIVE EFFECTS IN QED

Alan Chodos

Center for Theoretical Physics
Yale University
217 Prospect Street
New Haven, CT 06511

Prefatory Note: This article is based on lectures given at the NATO Advanced Study Institute on Vacuum Structure in Intense Fields. Originally, the lectures also included a discussion of the production of scalar and pseudoscalar particles by a background field. Since this is very nearly textbook material, I have decided in this written version to concentrate on the topic of Schwinger's proper time representation for the electron's Green function and its possible application to the understanding of the GSI experiments. I have, however, attempted to retain the pedagogical flavor of the lectures as they were originally given.

I. INTRODUCTION

One thing we have learned at this meeting is that, although the existence of several narrow lines in the GSI data appears to be firmly established,[1] a more detailed look[2] at the experimental results does not show any clear pattern leading to an unambiguous phenomenological description of what is going on. For example, not so long ago it seemed that the trend of the data was that the e^+e^- pairs were emitted back-to-back and with equal energy; more recent analysis shows, however, that this is definitely not true of all the peaks.

Notwithstanding (or perhaps because of) this kind of uncertainty, it appears to be possible to account for the data by postulating that, in the heavy-ion collisions, an object is made that is extended[3] (with a radius of 1000 fm or so) and fairly fragile, and that likes to decay into e^+e^- pairs. Such an object would have a spectrum of excited states that might explain the multiple peaks, and, because it is so extended, it could naturally evade the constraints imposed on the existence of a point particle by sensitive tests of QED (such as g-2 measurements)[4] and by the beam-dump type of searches[5] that have been carried out.

One is not, at this stage, forced to adopt such a model; other proposals remain viable[6]. However, if one pursues this idea as a likely candidate for explaining the data, the next question to be faced is what the nature of this extended object is. At one end of the spectrum of possibilities is the model proposed by the Frankfurt group, which boldly goes beyond QED and postulates the existence of new quark-like object, f, of mass about 800 keV/c^2, with the following properties: (a) it is integrally charged, and therefore subject to ordinary electromagnetic forces; (b) it is confined; (c) the force between f and $\bar{f}$ involves an additional strong short-range repulsion of unexplained origin.

This model is described in Greiner's lectures in this volume, and elsewhere in the literature,[7] so I shall not dwell on it further. It remains to be seen whether Nature has indeed been sly enough to tuck away a whole new family of composite particles in this mass range.

In contrast to the Frankfurt model, other authors have, perhaps more conservatively, sought to find an explanation within QED itself. Some years ago it was suggested that at short distances there might be strong magnetic forces[8] that would dominate over the Coulomb term and that could give rise to new quasi-bound states. More recently, a model has been proposed[9] in which the extended object is to be thought of as "quadronium", that is, a state of $e^+e^-e^+e^-$, which is strongly bound by many-body forces (long ago it was shown by Wheeler[10] that $e^+e^-e^+e^-$ is bound by ordinary Coulombic forces, but such a state would have a mass of very nearly $4m_e$ and hence would have nothing to do with the observed GSI spectrum).

My own motivation stems from another idea, that the effect of the heavy ions is to produce strong and rapidly varying electromagnetic fields, which in turn induce a phase transition[11,12] in the QED vacuum over some region of space. The extended object is then to be thought of as a blob of new phase, in which there can be states analogous to positronium in ordinary QED, although with a rather different spectrum. After the heavy ions have scattered, the blob of new phase becomes a region of "false vacuum" which can decay by tunneling into the true perturbative vacuum of ordinary QED. When this happens, the positronium-like state appears as an ordinary e^+e^- pair.

The crux of any of these suggestions is to be able to demonstrate that the postulated phenomenon does, in fact, occur. In the case of strong magnetic forces, not only is there no evidence to support them, there is a virial theorem[13] for QED that apparently rules them out. In the case of quadronium, there is no evidence for the mysterious many-body forces that would be needed to make the model work, but the problem of actually solving for the ground state of two electrons and two positrons is sufficiently difficult that the best one can say is that the problem is in limbo.

The case of the QED phase transition is not a priori any easier, but here at least we have persuasive theoretical evidence, both analytic[14] and numerical,[15] that QED does undergo a phase transition as a function of the coupling constant α. When $\alpha > 1$, QED enters a new, non-perturbative phase in which chiral symmetry is spontaneously broken and which is probably confining. Tuning the coupling constant, which is theoretically easy but experimentally impossible, is of course not the same as adjusting the background field, which is more or less the reverse. Nevertheless, one is encouraged to examine what is known about QED in an arbitrarily strong background field.

It is here that Schwinger's proper-time formalism[16] becomes of interest. Using it, Schwinger derived an expression for the electron's Green function, that sums up the contribution to all orders of a constant background field. He went on to derive therefrom the effective Lagrangian of QED as a function of the background field.

The good part about this is that these expressions are non-perturbative in the sense that no expansion need be made in the strength of the field. The bad part is that the analysis is necessarily restricted to constant fields, and it is quite possible that the rather rapid variation of the field present in a typical GSI scattering event may be an important feature that should not be neglected. Unfortunately, attempts to amend Schwinger's formulae[17] to take the variation of the field into account have not led to very tractable results.

This article is organized as follows: first we review Schwinger's original work, commenting on some of the properties of his Lagrangian. We then extract therefrom an effective charge,[18] and extend Schwinger's work to include the effects of a chemical potential,[19] which, we argue, may be a relevant parameter in the GSI experiments.

II. THE SCHWINGER LAGRANGIAN

Following Schwinger (Phys. Rev. 82, 664 (1951)) we study the Green function $G_A(x,y)$ of an electron propagating in a constant background electromagnetic field; i.e. G_A satisfies

$$(i\not{\partial} - e\not{A} - m)G_A = 1 \tag{1}$$

where $\partial_\mu A_\nu - \partial_\nu A_\mu = F_{\mu\nu}$ and $F_{\mu\nu}$ is a constant.

Having computed G_A, Schwinger went on to deduce the Lagrangian $\mathcal{L}$ by integrating the equation

$$\delta\Gamma = ie \, \text{Tr} \, \delta\not{A} \, G_A \tag{2}$$

where $\Gamma = \int d^4 x \, \mathcal{L}$ and "Tr" stands for a diagonal summation on both Dirac and spacetime indices:

$$\text{Tr} \, O_{\alpha\beta}(x,y) = \sum_{\alpha,\beta} \int dx \, dy \, \delta_{\alpha\beta} \, \delta(x-y) \, O_{\alpha\beta}(x,y) \ . \tag{3}$$

Note that the solution for Γ is formally

$$\Gamma = -i \, \text{Tr} \ln \mathcal{M} \tag{4}$$

where $\mathcal{M} = i\not{\partial} - e\not{A} - m$. This follows immediately because $\delta\Gamma = -i \, \text{Tr} \, \mathcal{M}^{-1} \, \delta\mathcal{M}$, and $\delta\mathcal{M} = -e \, \delta\not{A}$ while $\mathcal{M}^{-1} = G_A$.

From this formula we can also see that Γ is the same as what, in more modern terms, would be called the one-loop effective action for QED. Recall that to define the effective action[20] one starts with the path integral formula for the logarithm $W[J]$ of the partition function:

$$e^{iW[J]} = \int DA_\mu D\Psi \overline{D\Psi} \, e^{iS_{QED} + i(J,A)} \ . \tag{5}$$

Here S_{QED} is the usual classical action for QED (we suppress reference to gauge fixing and ghost terms, which should be included as well), $J_\mu(x)$ is an external classical source, and (J,A) is shorthand for $\int d^4x \, J_\mu(x)A^\mu(x)$.

In the path integral, we expand about the classical solution given by

$$\Psi(x) = \overline{\Psi}(x) = 0 \tag{6a}$$

and

$$A_\mu(x) = A_\mu^{cl}(x) \tag{6b}$$

where

$$\partial_\mu F^{\mu\nu(cl)}(x) = -J^\nu(x).$$

i.e. we write $A_\mu(x) = A_\mu^{cl}(x) + \grave{A}_\mu(x)$ and treat $\grave{A}_\mu(x)$ as the new integration variable.

Because A_μ^{cl} is a classical solution, there will be no terms linear in $\grave{A}_\mu$. To one loop, we discard all terms higher than quadratic in $\grave{A}_\mu$. There is then a Gaussian path integral to do, which yields

$$W[J] = -\frac{1}{4} F_{\mu\nu}^{(cl)} F^{\mu\nu(cl)} + (J, A^{cl}) - i\, Tr\, \ln\, (i\slashed{\partial} - e\slashed{A}^{cl} - m).\tag{7}$$

Then we define $\overline{A}_\mu = \frac{\delta W}{\delta J^\mu}$ and $\Gamma = W - (J, \overline{A})$. Now $\overline{A}_\mu = A_\mu^{cl} + \Delta A_\mu$, but it is easily seen that ΔA_μ plays no role to one loop. Therefore

$$\Gamma(\overline{A}) = -\frac{1}{4} \overline{F}_{\mu\nu} \overline{F}^{\mu\nu} - i\, Tr\, \ln\, (i\slashed{\partial} - e\slashed{\overline{A}} - m)\ .\tag{8}$$

The first term is the Maxwell (0-loop or classical) contribution, and the second term is the Schwinger (1-loop) correction.

One corollary of this observation is that it provides us with a graphical expansion for the Schwinger Lagrangian. It is well-known[21] that the effective action is the sum of all one-particle irreducible graphs. Thus the Schwinger Lagrangian can be obtained by performing the sum

$$\Gamma^{(1)}(\overline{A}) = \sum_{n=0}^{\infty} \qquad\qquad\qquad\qquad n \text{ vertices}\tag{9}$$

where each propagator is the free Dirac propagator and each vertex represents the insertion of $ie\slashed{\overline{A}}$.

Schwinger did not evaluate Γ either by doing the path integral or by directly summing an infinite set of graphs. Rather, he used the proper-time technique, which I shall now sketch.

Let
$$\pi_\mu = i\,\partial_\mu - eA_\mu.\tag{10}$$

Then
$$G = (\pi_\mu \gamma^\mu - m)^{-1}$$

$$= (\pi_\mu \gamma^\mu + m)\, [(\pi\cdot\gamma)^2 - m^2]^{-1}\ .\tag{11}$$

In these equations, the inverse operators are defined via the $i\epsilon$ prescription:

$$G = (\slashed{\pi} + m)\, \frac{1}{(\pi)^2 - m^2 + i\epsilon}\ .\tag{12}$$

Now

$$\frac{1}{A + i\epsilon} = -\,i \int_0^\infty ds\, e^{i(A + i\epsilon)s}\ .\tag{13}$$

Note that the sign in the exponent is dictated by the $i\epsilon$ prescription. Thus

$$G = (\slashed{\pi} + m)(-i) \int_0^\infty ds\, e^{-ism^2}\, e^{-iHs}\tag{14}$$

where $H = -(\gamma\cdot\pi)^2$, and

$$G(x,y) = -i<x|(\not{\pi}+m)\int_0^\infty ds\, e^{-ism^2} e^{-iHs}|y> = (-i)(i\not{\partial}-e\not{A}+m)\int_0^\infty ds\, e^{-ism^2}<x|e^{-iHs}|y> . \quad (15)$$

The essence of the problem is the evaluation of

$$U_A(x,y) = <x|e^{-iHs}|y> . \quad (16)$$

This can be regarded as a quantum mechanical problem in which s is the evolution variable, called proper time, $x_\mu(s)$ are the coordinates and $p_\mu(s)$ the associated momenta:

$$[x_\mu(s), p^\nu(s)] = -i\, \delta_\mu{}^\nu . \quad (17)$$

One wants to compute the transition amplitude $<x(s)|y(0)>$ governed by the Hamiltonian H. One way to do this is to write the matrix element as a path integral[22]

$$U(x,y;s) \equiv <x(s)|y(0)> = \int Dx\, Pe^{i\int_0^s ds'L(\dot{x},x)} \quad (18)$$

where $L(x,\dot{x}) = \frac{1}{4}\dot{x}_\mu\dot{x}^\mu + eA_\mu\dot{x}^\mu + \frac{e}{2}\sigma\cdot F$, the boundary condition is $x`^\mu(s) = x^\mu$; $x`^\mu(0) = y^\mu$ and the symbol P denotes path ordering, which is required because L carries Dirac indices. This approach is useful to do numerical work (which is necessary if one takes $F_{\mu\nu}$ to be a function of x), but following Schwinger we shall proceed using operator techniques.

Before doing so, however, note the following:

(a) if we transform to Euclidean space, $x^0 \to -ix^4$ then all the γ-matrices can be chosen Hermitean and the Hamiltonian becomes $(\gamma\cdot\pi)^2$, which is non-negative. This Hamiltonian is supersymmetric. To see this, define the charges $Q_\pm = \frac{1}{2}(1\pm\gamma^5)\gamma\cdot\pi$ and observe that

$$Q_\pm^2 = 0 \quad (19a)$$

$$\{Q_+, Q_-\} = H . \quad (19b)$$

This is the algebra of quantum-mechanical supersymmetry.[23] As is well-known, supersymmetry is spontaneously broken if and only if the lowest eigenvalue E_0 of H is greater than zero. Now the lowest eigenvalue can be extracted by computing

$$U(x,y; s = -i\tau) = <x|e^{-H\tau}|y> \quad (20)$$

in the limit $\tau \to \infty$. In this limit

$$U \sim e^{-E_0\tau} \quad (21)$$

so the signal for supersymmetry breaking is to see whether U decays exponentially.

(b) To study spontaneous chiral symmetry breaking in QED, one wants to evaluate

$$<0|\overline{\Psi}\Psi(x)|0>$$

in the limit that m → 0. Since

$$G_A(x,y)_{\alpha\beta} = <0|T\Psi_\alpha(x)\overline{\Psi}_\beta(y)|0> \qquad (22)$$

we have $<0|\overline{\Psi}\Psi(x)|0> = -\text{tr } G_A(x,x)$ where the trace is on the Dirac indices only. From equation (15) we have, noting that the trace of an odd number of γ matrices vanishes,

$$<\overline{\Psi}\Psi(x)> = m \int_0^\infty d\tau \, e^{-m^2\tau} \text{ tr } <x|e^{-H\tau}|x> \qquad (23)$$

where once again we have let $s = -i\tau$. In order for this to tend to a finite, non-vanishing limit as m → 0, it is easy to see[24] that we require $\text{tr}<x|e^{-H\tau}|x> \underset{\tau\to\infty}{\longrightarrow} c/\tau^{1/2}$. If the falloff is more rapid, then $<\overline{\Psi}\Psi>\to 0$ and chiral symmetry is unbroken; if less, then $<\overline{\Psi}\Psi>\to\infty$.

Returning now to the problem of actually computing U, we write the Heisenberg equations of motion

$$\frac{dx^\mu}{ds} = i[H,x^\mu] = 2\pi^\mu \qquad (24)$$

$$\frac{d\pi^\mu}{ds} = i[H,\pi^\mu] = 2e\pi^\nu F_\nu{}^\mu - ie \frac{\partial}{\partial x^\nu} F^{\nu\mu} - \frac{i}{2} e \, \sigma_{\lambda\nu} \frac{\partial F^{\lambda\nu}}{\partial x_\mu} \qquad . \qquad (25)$$

When $F_{\mu\nu} = $ const., the last two terms in the equation for $\frac{d\pi^\mu}{ds}$ vanish, and one can then solve these equations explicitly. To obtain the matrix element, one integrates the equation

$$i \frac{\partial}{\partial s} <x(s)|y(0)> = <x(s)|H|y(0)> \qquad (26)$$

with suitable boundary conditions, and obtains

$$<x(s)|y(0)> = \frac{-i}{(4\pi)^2} e^{ie\int_y^x dx'^\mu A_\mu(x')} \frac{1}{s^2} e^{-L(s)} \exp[\frac{i}{4}(x-y)^\mu[eF \coth eFs]_{\mu\nu}(x-y)^\nu] \exp[\frac{i}{2} e s \sigma_{\mu\nu}F^{\mu\nu}] \quad (27)$$

where $L(s) = \frac{1}{2} \text{ tr ln } (\frac{\sinh eFs}{eFs})$ (trace on the indices of $F_{\mu\nu}$). The Schwinger Lagrangian, which up to an additive constant is given by

$$\mathcal{L} = \frac{i}{2} \int_0^\infty \frac{ds}{s} \, e^{-im^2s} \text{ tr } <x|e^{-iHs}|x> \quad , \qquad (28)$$

then becomes

$$\mathcal{L} = \frac{1}{32\pi^2} \int_0^\infty \frac{ds}{s^3} \, e^{-im^2s} \{e^{-L(s)} \text{ tr } e^{\frac{i e \sigma \cdot Fs}{2}} - 1\} \qquad (29)$$

where we have chosen the constant to make $\mathcal{L}$ vanish in the absence of a background field. Unfortunately, this expression is still logarithmically divergent, signalling the need for renormalization. After performing the trace in L(s) and the trace over $e^{\frac{i}{2} e \sigma \cdot Fs}$ (note that the former is a trace over   the vector indices   carried by

F, whereas the latter is over the spinor indices carried by σ) one has

$$L = \frac{1}{8\pi^2} \int_0^\infty \frac{d\tau}{\tau^3} \, e^{-m^2\tau} \, [e_0^2\tau^2 K_+ K_- \coth(e_0\tau K_-)\cot(e_0\tau K_+) - 1] \tag{30}$$

where we have once again rotated the contour to run along $\tau = is$, and where

$$K_\pm = [(\mathcal{F}^2 + \mathcal{G}^2)^{1/2} \pm \mathcal{F}]^{1/2} \tag{31}$$

and $\mathcal{F} = \frac{1}{2}(\vec{E}^2 - \vec{B}^2)$, $\mathcal{G} = \vec{E}\cdot\vec{B}$. In anticipation of the renormalization, we have also denoted the charge by e_0.

The full Lagrangian, including the Maxwell term, is then

$$L = \mathcal{F}[1 + \frac{e_0^2}{12\pi^2} \int_0^\infty \frac{d\tau}{\tau} \, e^{-m^2\tau}] - \frac{1}{8\pi^2} \int_0^\infty \frac{d\tau}{\tau^3} \, e^{-m^2\tau} \, [e_0^2\tau^2 K_+ K_- \coth(e_0\tau K_-)\cot(e_0\tau K_+) - 1 + \frac{2}{3} e_0^2\tau^2 \mathcal{F}]. \tag{32}$$

Note that what we have done is to add and subtract the term linear in $\mathcal{F}$ in the Schwinger Lagrangian. The coefficient of $\mathcal{F}$ in the first term is then logarithmically divergent, but the rest of the expression is now finite. The whole Lagrangian can be rendered finite by a simultaneous renormalization of the potential and the charge:

$$A_\mu^{ren} = A_\mu(1 + e_0^2 C)^{1/2} \tag{33}$$

$$e^2 = \frac{e_0^2}{1 + e_0^2 C} \tag{34}$$

where $C = \int_0^\infty \frac{d\tau}{\tau} \, e^{-m^2\tau}$.

We note in passing that, in Euclidean space, the trace of the matrix element when x=y is given by

$$\mathcal{M}(\tau) = tr <x|e^{-H\tau}|x> = \frac{|\mathcal{G}| e^2}{4\pi^2} \coth \tau\widehat{K}_+ \coth \tau\widehat{K}_- \tag{35}$$

where $\widehat{K}_\pm = e[\widehat{\mathcal{F}} \pm (\widehat{\mathcal{F}}^2 - \mathcal{G}^2)^{1/2}]^{1/2}$ and $\widehat{\mathcal{F}} = \frac{1}{2}(\vec{E}^2 + \vec{B}^2)$, $\mathcal{G} = \vec{E} \cdot \vec{B}$. Thus as $\tau \to \infty$

$$\mathcal{M}(\tau) \to \frac{|\mathcal{G}| e^2}{4\pi^2} \tag{36}$$

so there is no evidence for spontaneous supersymmetry breaking. Likewise, if $\mathcal{G} = 0$, we observe that

$$\mathcal{M}(\tau) = \frac{e^2}{4\pi^2\tau} (2\widehat{\mathcal{F}})^{1/2} \coth[(2\widehat{\mathcal{F}})^{1/2} e\tau] \tag{37}$$

so that no chiral symmetry breaking occurs. If $\mathcal{G} \neq 0$, one finds instead $<\overline{\Psi}\Psi> \to \infty$ as $m \to 0$. This can be understood, however, not as evidence for spontaneous chiral symmetry breaking, but rather as an explicit breaking due to the anomaly, since

$$\partial_\mu j^{\mu 5} = \frac{e^2}{2\pi^2} \mathcal{G} \ . \tag{38}$$

Finally let us derive from Schwinger's Lagrangian the amplitude for pair production in a constant electric field. For this, we return to the Minkowski-space expressions, and, taking $\vec{B} = 0$, we find $K_- = 0$ and $K_+ = |\vec{E}| = E$. Recall that the partition function $e^{iW[J]}$ has the physical interpretation as the vacuum-to-vacuum transition amplitude $<0_+|0_->_J$ in the presence of the source J. Therefore

$$|<0_+|0_->_J|^2 = e^{-2ImW} = e^{-2Im\Gamma} \tag{39}$$

since $\Gamma = W + (A,J)$.

Since $2Im\Gamma = \int d^4x \, (2Im\mathcal{L})$ we see that $2Im\mathcal{L}$ has the interpretation of giving the probability of particle production per unit volume per unit time. Now the proper time expression for $\mathcal{L}$ is of the form

$$\mathcal{L} = \int_0^\infty ds \, e^{-im^2s} f(s) \tag{40}$$

where the asymptotic behavior of the integrand is dictated by the e^{-im^2s} factor. Thus to shift the contour one must close in the lower half-plane as shown in the figure.

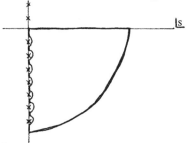

The poles in the figure come from the factor coth eEs, which has poles at eEs = -i nπ, n = 0, 1, 2, ... Since the contour has been rotated down in the clockwise direction, it is clear that the contour must run to the right of the poles on the negative imaginary s axis. There is actually no pole for n = 0, because it is cancelled by the other terms in the integrand.

Defining the real variable $\tau = is$ we see that the Lagrangian takes the form

$$\mathcal{L} = -\frac{1}{8\pi^2} \int_0^\infty \frac{d\tau}{\tau^3} e^{-m^2\tau} [e\tau E \cot e\tau E + ...] \tag{41}$$

where the ... denotes other terms that are non-singular but that cancel the pole at $\tau = 0$, and the contour runs slightly above the poles at $e\tau E = n\pi$. Let $x = e\tau E$. Then the relevant part of the Lagrangian is

$$\mathcal{L}_{rel.} = -\frac{e^2E^2}{8\pi^2} \int_0^\infty \frac{dx}{x^2} e^{-\frac{m^2x}{eE}} \cot x. \tag{42}$$

The contribution from the poles at $x = n\pi$ is

$$\mathcal{L}_{pole} = -\frac{e^2E^2}{8\pi^2} \sum_{n=1}^\infty \frac{1}{n^2\pi^2} e^{-m^2n\pi/eE} \int_{C_n} \frac{dy}{y} \tag{43}$$

where C_n is a small half-circle centered at $n\pi$ and circulating clockwise from $y = n\pi - \epsilon$ to $n\pi + \epsilon$.

Therefore $\displaystyle\int_{C_n} \frac{dy}{y} = -i\pi$ and, with $\alpha = \dfrac{e^2}{4\pi}$ we obtain

$$2\mathrm{Im}\,L = \frac{\alpha E^2}{\pi^2} \sum_{n=1}^{\infty} \frac{1}{n^2}\, e^{-m^2 n\pi/eE} \tag{44}$$

where the nth term may be thought of as the contribution from the production of n pairs. It is interesting to see that the pair production amplitude is non-vanishing for any E > 0, however small. This is of course because we have taken $\vec{E}$ to be a constant throughout space, so there is an infinite amount of energy available in even a very weak field. On the other hand, this formula clearly shows that the rate is greatly suppressed unless $eE \geq m^2$.

Various generalizations of Schwinger's pair-production formula have appeared over the years. One interesting one is discussed in the contribution to this volume by Martin.

III. EFFECTIVE CHARGE

As mentioned in the introduction, it is known that QED undergoes a phase transition if the charge becomes sufficiently large. In this section, we use the Schwinger Lagrangian to define an effective, field-dependent charge.[18] We shall discover that this charge indeed grows with the magnitude of the field, but at a rate that is too small to induce of itself the strong-coupling phase transition in the GSI experiments.

The key to our definition is to observe that in the final, renormalized expression for the Lagrangian,

$$L = \mathcal{F} - \frac{1}{8\pi^2} \int_0^{\infty} \frac{d\tau}{\tau^3}\, e^{-m^2\tau} [e^2\tau^2 K_+ K_- \coth(e\tau K_-)\cot(e\tau K_+) - 1 + \frac{2}{3} e^2\tau^2 \mathcal{F}] \tag{45}$$

one can rescale one more time,

$$A_\mu \to \frac{1}{e}\, A_\mu \tag{46}$$

the effect of which is to isolate all the charge dependence in the Maxwell term. Calling the rescaled Lagrangian L', we have

$$L' = \frac{1}{e^2}\, \mathcal{F} + L_1\,(\mathcal{F},\mathcal{G})\ . \tag{47}$$

We note further that, as a consequence of the renormalization described earlier, L_1 contains no additional piece linear in $\mathcal{F}$. (By parity invariance, there can also not be any terms linear in $\mathcal{G}$). Thus $\dfrac{1}{e^2}$ appears as the coefficient of the term linear in $\mathcal{F}$, and can be written as

$$\frac{1}{e^2} = \frac{\partial L'}{\partial \mathcal{F}}\bigg|_{\mathcal{F}=\mathcal{G}=0}\ . \tag{48}$$

This lends itself to a straightforward generalization:

$$\frac{1}{e^2(\mathcal{F}_0,\mathcal{G}_0)} = \frac{\partial L'}{\partial \mathcal{F}}\bigg|_{\mathcal{F}=\mathcal{F}_0;\ \mathcal{G}=\mathcal{G}_0}\ . \tag{49}$$

As a consequence one has (dropping the subscripts on $\mathcal{F}$ and $\mathcal{G}$)

$$\frac{e^2(\mathcal{F},\mathcal{G})}{4\pi} \equiv \alpha(\mathcal{F},\mathcal{G}) = \frac{\alpha}{1 - \alpha X} \tag{50}$$

where

$$X = -4\pi \frac{\partial \mathcal{L}_1}{\partial \mathcal{F}} \quad .$$

$$= \frac{1}{2\pi} \int_0^\infty \frac{d\tau}{\tau} e^{-m^2\tau} \left[\frac{\tau K_+ K_-}{K_+^2 + K_-^2} \left\{ \frac{K_- \cot(\tau K_+)}{\sinh^2(\tau K_-)} - \frac{K_+ \coth(\tau K_-)}{\sin 2(\tau K_+)} \right\} + \frac{2}{3} \right] \quad . \tag{51}$$

Here we want the effective charge to be real, so we define the integral as a principal value.

Note that if we take $\mathcal{L}$ seriously as an effective Lagrangian for QED, suitable, for example, for generating equations of motion (to do this we must forget that the derivation of $\mathcal{L}$ was limited to constant fields), then we have

$$\mathcal{L} = \mathcal{F} + \mathcal{L}_1 \,(e^2\mathcal{F}, e^2\mathcal{G}) \quad . \tag{52}$$

Denote $\dfrac{\partial \mathcal{L}_1}{\partial(e^2\mathcal{F})} = \phi; \dfrac{\partial \mathcal{L}_1}{\partial(e^2\mathcal{G})} = \gamma.$ (The X defined above is just $-4\pi\phi$). Then

$$D_i = \frac{\partial \mathcal{L}}{\partial E_i} = E_i(1 + e^2\phi) + B_i(e^2\gamma) \tag{53}$$

and

$$H_i = -\frac{\partial \mathcal{L}}{\partial B_i} = B_i(1 + e^2\phi) - E_i(e^2\gamma) \quad . \tag{54}$$

Thus the system obeys the usual set of Maxwell's equations written in terms of $\vec{E}, \vec{B}, \vec{D}$ and $\vec{H}$. The non-linearities of the medium are characterized by the functions ϕ and γ.

The function $\alpha(\mathcal{F},\mathcal{G})$ may be calculated numerically, and one finds in general that it grows as a function of the magnitudes of $\vec{E}$ and $\vec{B}$. However, the growth is only logarithmic,[25] so that there is no chance that the charge will be large enough at the field strengths attained in the GSI experiments to trigger a phase transition. The origin of this logarithmic growth can be seen in $\mathcal{L}$ itself. If for simplicity we take $\vec{E} = 0$ we have $K_+ = 0$, $K_- = \sqrt{2}\,|\vec{B}|$, and

$$\mathcal{L} = -\frac{1}{2} B^2 - \frac{1}{8\pi^2} \int_0^\infty \frac{d\tau}{\tau^3} e^{-m^2\tau} \left[(e\tau K_-)\coth(e\tau K_-) - 1 - \frac{1}{3} e^2\tau^2 B^2 \right] \quad . \tag{55}$$

In the integral, let $e\tau B = x$, so that

$$\mathcal{L} = -\frac{1}{2} B^2 - \frac{e^2 B^2}{8\pi^2} \int_0^\infty \frac{dx}{x^3} e^{-\frac{m^2 x}{eB}} \left[\sqrt{2}\,x\coth(\sqrt{2}\,x) - 1 - \frac{1}{3} x^2 \right] \quad . \tag{56}$$

For $eB/m^2 \gg 1$, the last term in the integrand dominates and one finds that

$$\mathcal{L}(B) \sim -\frac{1}{2} [1 + O(\alpha)]B^2 + \frac{\alpha}{6\pi} B^2 \ln \frac{eB}{m^2} \quad . \tag{57}$$

Thus ultimately the second term will become larger than the first, but only for magnetic fields for which

$$\frac{eB}{m^2} \approx e^{\frac{3\pi}{\alpha}} \quad . \tag{58}$$

Similar results hold for the electric field case. It is interesting to note that the term giving rise to the $B^2 \ln B$ behavior was introduced in the process of renormalization; however, the $\ln B$ itself comes from the large τ behavior of the integrand and hence is an infrared effect rather than an ultraviolet one.

IV. THE CHEMICAL POTENTIAL

The role of the chemical potential μ is to describe a constant background density of ρ of particles. If N is the operator corresponding to this density:

$$\rho = <N> \tag{59}$$

then, given a specific value fo ρ, one replaces the hamiltonian H by H - μN, and adjusts μ to insure that eq. (59) is satisfied.[26]

The reason for our interest in the chemical potential is that the heavy-ion collisions studied at GSI necessarily take place in an environment populated by a large number of electrons. The target ion is in fact an atom, with 90 electrons or so. The incident ion is only partially ionized, with perhaps 50 or 60 electrons. As the collision evolves, some of these electrons may be compressed into a rather small volume, leading to effectively large values of μ. Of course, just as a constant background field does not really describe the time-dependent fields present in the GSI experiments, so, too, the chemical potential, which is designed to describe an equilibrium density, is not adequate to account for the charge density that changes in a complicated way as the collision proceeds. Nevertheless, for simplicity, in this section we shall discuss the case of QED in the presence of a chemical potential and a constant background field.[19]

As we shall see below, when there is only the chemical potential and no background field, the relationship between density and μ is

$$\rho = \frac{(\text{sgn}\mu)}{3\pi^2} \; \theta(\mu^2 - m^2)(\mu^2 - m^2)^{3/2} \tag{60}$$

In order for μ to be relevant to the GSI experiments, it should be on the order of 1 MeV, since that is the mass scale of the observed phenomena. Let us assume that in the course of the collision, some number N of the available electrons become compressed into a sphere of radius R, so that $\rho = \frac{3N}{4\pi R^3}$. Then we find, with μ = 1 MeV, that

$$N^{1/3} \cong 225 \; R \tag{61}$$

where R is measured in Ångstroms. Now the available N is presumably not more than (and probably significantly less than) 100, because only a fraction of the electrons will be involved in the compression. We conclude that if R is in the neighborhood of 10^{-2}Å = 10^3f, then it is possible to achieve $|\mu| \geq 1$ MeV. It is interesting to recall that 10^3f is a typical size for the extended object that is supposed to give rise to the GSI events.

In QED, the replacement H $\rightarrow$ H - μN where N is the fermionic charge $\int d^3\Psi^\dagger\Psi$ leads to the modified Green function equation

$$[i\not{\partial} - e\not{A} - m - \mu\gamma^0] \; G = 1 \quad . \tag{62}$$

It appears that, formally, μ corresponds to $A^0 \to A^0 + \frac{1}{e}\mu$, which is nothing but a gauge transformation $A_\mu \to A_\mu + \frac{1}{e}\partial_\mu\Lambda$ with $\Lambda = \mu t$. However, this ignores the boundary conditions associated with solving eq. (62). It is incorrect simply to use the standard $i\epsilon$ prescription, because that corresponds to having the levels with $E < 0$ filled and those with $E > 0$ empty, whereas μ has the significance of filling all the levels with $E < \mu$. The correct prescription is encoded in the propagator[26]

$$G(x - y',\mu) = \int \frac{d^4p}{(2\pi)^4} \frac{(\tilde{p} + m)\, e^{-ip\cdot(x-y)}}{(\tilde{p}_0 + i\epsilon\ \mathrm{sgn} p_0)^2 - \vec{p}^2 - m^2} \tag{63}$$

where $\tilde{p} = (p_0 - \mu, \vec{p})$. What we wish to compute then is the same set of graphs depicted in eq. (9), but now with the free Dirac propagator replaced by the propagator of eq. (63).

In the case that there is no background field at all, the evaluation of the Lagrangian can be done straightforwardly and leads to the expression

$$\mathcal{L}(\mu) = \frac{1}{3\pi^2}\ \theta(\mu^2 - m^2) \int_m^{|\mu|} d\mu`\ \theta(\mu`^2 - m^2)\ (\mu`^2 - m^2)^{3/2}\ . \tag{64}$$

Since $\rho = \dfrac{\partial\mathcal{L}}{\partial\mu}$ one obtains eq. (60).

When a magnetic field is present,[19] one can write the Green function as

$$G = (\gamma\cdot\tilde{\pi} - m) \frac{1}{(\gamma\cdot\tilde{\pi})^2 - m^2 + i\,\epsilon\ (p_0 - \mu)\ \mathrm{sgn}\ p_0} \tag{65}$$

where $\tilde{\pi}_\lambda = \pi_\lambda - \mu\delta_{\lambda 0}$. The fact that

$$[p_0, \tilde{\pi}_\lambda] = 0 \tag{66}$$

allows us to write

$$G = (\gamma\cdot\tilde{\pi} - m) \left[\frac{1}{(\gamma\cdot\tilde{\pi})^2 - m^2 + i\,\epsilon}\ \theta[(p_0 - \mu)\ \mathrm{sgn}\ p_0] \right.$$
$$\left. + \frac{1}{(\gamma\cdot\tilde{\pi})^2 - m^2 - i\,\epsilon}\ \theta[(\mu - p_0)\ \mathrm{sgn}\ p_0] \right] \tag{67}$$

and then make use of the proper time formalism:

$$G = (\gamma\cdot\tilde{\pi} - m) \left[-i\int_0^\infty ds\ e^{is[(\gamma\cdot\tilde{\pi})^2 - m^2]}\ \theta[(p_0 - \mu)\ \mathrm{sgn}\ p_0] \right.$$
$$\left. + i\int_0^\infty ds\ e^{-is[(\gamma\cdot\tilde{\pi})^2 - m^2]}\ \theta[(\mu - p_0)\ \mathrm{sgn}\ p_0] \right]\ . \tag{68}$$

From this follows the Lagrange density

$$L = \frac{i}{2} \int_0^\infty \frac{ds}{s} \ \text{tr} \ \Big\{ <x|e^{is[(\gamma \cdot \widetilde{\pi})^2 - m^2]} \ \theta((p_0 - \mu) \ \text{sgn} \ p_0)|x>$$

$$+ \ <x|e^{-is[(\gamma \cdot \widetilde{\pi})^2 - m^2]} \ \theta(\mu - p_0) \ \text{sgn} \ p_0|x> \Big\} \quad . \tag{69}$$

To evaluate this, one

(i) inserts a complete set of $(p_0, \vec{x})$ eigenstates; (ii) does a part of the p_0 integral by writing

$$\theta[(p_0 - \mu) \ \text{sgn} \ p_0] = 1 - \theta[(\mu - p_0) \ \text{sgn} \ p_0]$$

and

$$\int_0^\infty dp_0 \ e^{-is(p_0 - \mu)^2} = \left(\frac{\pi}{s}\right)^{1/2} \ e^{i\pi/4} \ ; \tag{70}$$

(iii) evaluates the matrix elements using

$$\text{tr} \left[e^{i\sigma \cdot Fs/2} \right] <x|e^{is\vec{\pi}^2}|x> = \frac{4sB \ \text{cotsB}}{(4\pi s)^{3/2}} \ e^{-3\pi i/4} \quad , \tag{71}$$

and similarly for the term with $e^{-is\vec{\pi}^2}$. The result of this is then

$$L(B,\mu) = \frac{1}{8\pi^2} \int_0^\infty \frac{ds}{s^3} \ e^{-ism^2} \ [sB \ \text{cotsB} - 1]$$

$$+ \ 2 \ \text{Re} \ \Big\{ \frac{e^{3\pi i/4}}{8\pi^{5/2}} \int_0^{|\mu|} dp_0 \int_0^\infty \frac{ds}{s^{5/2}} \ e^{is(p_0^2 - m^2)} \ [sB \ \text{cotsB} - 1] \Big\}$$

$$+ \ L(\mu) \quad . \tag{72}$$

The first term is Schwinger's Lagrangian for the case of a magnetic field, and the second is the correction due to the simultaneous presence of B and the chemical potential. The third term is just $L(\mu)$, to which L reduces when B = 0. Note that the last two terms are finite as they stand; the only infinity resides in the first term, and is handled by renormalization exactly as described earlier. This is to be expected, because the addition of a chemical potential is, as we have seen, fundamentally a change in the boundary condition imposed on the propagator, and should in no way disturb the short-distance properties of the theory.

By rotating the contour in the second term of eq. (72), one can render the real part explicit, and show that this term can be written as

$$\frac{\theta(\mu^2 - m^2)}{4\pi^{5/2}} \int_m^{|\mu|} dp_0 \int_0^\infty \frac{d\tau}{\tau^{5/2}} \ e^{-(p_0^2 - m^2)\tau} \ [\tau B \ \text{coth}\tau B - 1]$$

which demonstrates that there is no contribution if $\mu^2 < m^2$. This is as expected, because for $\mu^2 < m^2$ one is filling non-existent levels in the gap - m < E < m.

If one takes the propagator $G(x,y;B,\mu)$ in the presence of a magnetic field and chemical potential, one can Fourier transform in the variable x-y and obtain thereby a function $\widetilde{G}(p_0,\vec{p})$. One finds that there are poles in p_0^2 at the locations

$$E_n^2(p_z) = m^2 + p_z^2 + 2nB, \qquad n = 0, 1, 2, \dots \qquad (73)$$

where the magnetic field has been oriented in the z direction. These correspond to the well-known Landau levels[27] occupied by a particle in a constant magnetic field. The chemical potential does not affect the position of these levels.

In order to confront the GSI data, it is necessary to be able also to include an electric field.[28] The present formalism, however, runs into trouble when this is done. The reason for this is simple to see. The propagator can still be written exactly as in eq. (65) if we use the $A_0 = 0$ gauge. However, we now have

$$[p_0, \widetilde{\pi}_\lambda] \neq 0 \qquad (74)$$

because A_λ contains dependence on x^0. Therefore it is incorrect to write eq. (67) (or any trivial modification of it, such as a different ordering of the factors). What one can do, however, is to employ a trick due to Feynman.[29] Consider the expression

$$\frac{1}{A + i\epsilon\, B}$$

where $[A,B] \neq 0$. Feynman tells us to endow each of the operators A and B with dummy index λ running from 0 to 1, and to replace A and B in the expression with $\mathcal{A} = \int_0^1 A(\lambda)d\lambda$ and $\mathcal{B} = \int_0^1 B(\lambda)d\lambda$ respectively. Whenever they act, $A(\lambda) = A$ and $B(\lambda) = B$; the role of λ is to prescribe the ordering of the various factors of A and B. One has the equality

$$\frac{1}{A + i\epsilon\, B} = \Lambda \left[\frac{1}{\mathcal{A} + i\epsilon\, \mathcal{B}} \right] \qquad (75)$$

where Λ denotes λ-ordering: all oeprators are to be arranged from left to right in order of decreasing λ. The advantage of this is that under the Λ symbol, factors may be manipulated as if they commute. One therefore has

$$\frac{1}{A + i\epsilon\, B} = \Lambda \left[\frac{1}{\mathcal{A} + i\epsilon}\, \theta(\mathcal{B}) + \frac{1}{\mathcal{A} - i\epsilon}\, \theta(-\mathcal{B}) \right]$$

$$= \left(\frac{-1}{2\pi} \right) \int_0^\infty ds \int_{-\infty}^\infty \frac{dq}{q - i\eta}\, \Lambda \left[e^{+is(\mathcal{A} + q\mathcal{B})} - e^{is(\mathcal{A} + q\mathcal{B})} \right] \quad . \qquad (76)$$

Here the proper time expression has been used for $\dfrac{1}{\mathcal{A} \pm i\epsilon}$ and the θ-functions have been replaced by their Fourier representations.

But now the λ-ordering is irrelevant, because the only operator that occurs is $\int_0^1 d\lambda\, [A(\lambda) + qB(\lambda)]$. As long as we insure that A + qB acts as a unit, there is no need for an ordering prescription. We have, then, finally,

$$\frac{1}{A + i\epsilon B} = \frac{-1}{2\pi} \int_0^\infty ds \int_{-\infty}^\infty \frac{dq}{q - i\eta} \left[e^{is(A + qB)} - e^{-is(A + qB)} \right] . \tag{77}$$

We have resolved the ordering problem at the cost of adding an extra integral. For our problem, $A = (\gamma \cdot \tilde{\pi})^2 - m^2$ and $B = (p_0 - \mu)$ sgn p_0. In order to bring the proper-time method to bear on this problem, it is convenient to replace sgn p_0 by p_0 itself. This should make no difference, because p_0 and sgn p_0 have the same sign and hence lead to the same $i\epsilon$ prescription for the Green function, eq. (65). Then

$$A + qB = -\tilde{\pi}^2 + qp_0(p_0 - \mu) - m^2 + \frac{1}{2} \sigma \cdot F$$

$$\equiv H - m^2 + \frac{1}{2} \sigma \cdot F , \tag{78}$$

and H, being quadratic in the momenta, is a Hamiltonian whose dynamics we can solve in the same manner as described earlier. The details of this procedure are much too lengthy to be set down here; I shall, instead, record the result for the Fourier transform of the Green function with respect to (x-y). One has

$$G(p) = G_+(p) + G_-(p) \tag{79}$$

where

$$G_+(p) = -\frac{-1}{2\pi} \int_0^\infty ds \, e^{-is(m^2 + p_x^2 + p_y^2)} \exp[\tfrac{i}{2} \sigma \cdot Fs]$$

$$\otimes \int_{-\infty}^\infty \frac{dq}{q - i\eta} \frac{1}{\cosh(Es\sqrt{1 + q})} \exp\left\{ \frac{-isq^2\mu^2}{4(1 + q)} - \frac{i \, (p_z - \frac{E\alpha}{2})^2}{E(\sqrt{1 + q} \coth(Es\sqrt{1 + q})} \right.$$

$$\left. + \frac{i \, (p_0 + \frac{\mu}{2} (\frac{2 + q}{1 + q}))^2 \sqrt{1 + q}}{E \coth(Es \sqrt{1 + q})} \right\}$$

$$\otimes \left[p_x \gamma_1 + p_y \gamma_2 + (p_z - \frac{E\alpha}{2})\gamma_3 - (p_0 + \mu)\gamma^0 - m \right.$$

$$\left. + \frac{(p_0 + \frac{\mu}{2} (\frac{2 + q}{1 + q})) \sqrt{1 + q} \, \gamma_3}{\coth(Es \sqrt{1 + q})} - \frac{(p_z - \frac{E\alpha}{2})\gamma^0}{\sqrt{1 + q} \coth(Es \sqrt{1 + q})} \right] \tag{80}$$

and where $G_-(p)$ is obtained by letting $s \to -s$ everywhere (and also $ds \to -ds$). The parameter α denotes $(x_0 + y_0)$; the original $G(x,y)$ was not function only of x-y. We have taken $\vec{E} = E\hat{z}$.

The following remarks are in order: (a) despite appearances, there is no singularity in the integrand at $q = -1$; one can check this by combining the various singular terms; (b) if one sets $\mu = 0$, G_- vanishes and G_+ reproduces Schwinger's result; (c) if one sets $E = 0$, $G_+ + G_-$ reduces to the usual expression, eq. (63), for the propagator in the presence of a chemical potential; (d) if neither E nor μ vanish, however, the integrand possesses an infinite number of essential singularities at the points q_n for which

$$Es\sqrt{1 + q_n} = i(n + \frac{1}{2})\pi, \qquad n = 0, 1, 2, ... \tag{81}$$

(one can, by convention, take the imaginary part of $\sqrt{1 + q} > 0$ for $q < -1$, since there is no cut).

A careful analysis[28] shows that upon closing the contour in the q plane appropriately at infinity, one must go around these singularities in a manner such that they are included within the contour of integration. Although the integral is probably well-defined, its evaluation is extremely difficult and we know of no technique that will allow us to proceed analytically. It has been suggested[30] that these difficulties may be associated with the fact that the simultaneous presence of a homogeneous electric field and a chemical potential does not constitute a properly posed physical problem. Nevertheless, one can reasonably argue that, at least approximately, these ingredients do represent the environment in which the GSI heavy-ion collisions are taking place.

Finally, we remark that we have also succeeded in carrying through the proper-time method in the case of general constant $F_{\mu\nu}$, although of course the expressions so obtained are, if anything, more complicated than the pure electric field case and hence equally unrewarding of further analytical effort.

In these lectures we have explored the extent to which Schwinger's proper time techniques may be applied to the analysis of physical situations that are, at least approximately, relevant to understanding the GSI experiments. We have discovered an effective charge that grows with increasing field strength, although much too slowly to explain the GSI data.

We have also investigated the role of the chemical potential, and have found that although the pure magnetic field case is quite tractable, when an electric field is added the straightforward application of the proper time method leads to expressions that are very difficult to analyze. Thus we are as yet unable to say whether these non-perturbative expressions contain within them a possible explanation for the dynamics that gives rise to the e^+e^- peaks.

ACKNOWLEDGEMENTS

I wish to thank Kenneth Everding, whose thesis contains many of the calculations underlying the material in these lectures. I am grateful to Daniel Caldi and David Owen, who have collaborated on and stimulated much of the research that I have talked about. I thank Janos Polonyi for first arousing my interest in the chemical potential, and Jack Greenberg for his continual willingness to engage in discussions about the experimental situation. In addition I wish to thank Denis Carrier, Charles Sommerfield, Saeed Vafaeisefat and L.C.R. Wijewardhana for their contributions to the work that was described in these lectures.

Finally, a great deal of thanks is due to Herb Fried and Berndt Mueller, who organized a most stimulating meeting in a most enjoyable location.

REFERENCES

1. P. Salabura, et al., Phys. Lett. B245, 153 (1990); W. Koenig, et al., Phys. Lett. B218, 12 (1989) and references therein.

2. See ref. 1 and the contribution of Bokemeyer and Koenig to this volume.

3. A Schaefer, J. Phys. G 15, 373 (1989).

4. A. Schaefer, et al., J. Phys. G 11, L69 (1985).

5. E.M. Riordan, et al., Phys. Rev. Lett. 59, 755 (1987); A. Konaka, et al., Phys. Rev. Lett. 57, 659 (1986).

6. For example, see: Y. Hirata and H. Minakata, Phys. Rev. D34, 2493 (1986); ibid. D35, 2619 (1987); A. Iwazaki and S. Kumano, Phys. Lett. B212, 99 (1988); J.M. Cornwall and G. Tiktopoulos, Phys. Rev. D39, 334 (1989); H.M. Fried and H.-T. Cho, Phys. Rev. D41, 1489 (1990). This is but a small sample; many other interesting models have been proposed.

7. S. Schramm, B. Mueller, J. Reinhardt and W. Greiner, Mod. Phys. Lett. $\underline{A3}$, 783 (1988).

8. A.O. Barut in "Physics of String Fields", Walter Greiner, ed., (1987); C.Y. Wong and R.L. Becker, Phys. Lett. $\underline{B182}$, 251 (1986).

9. J.J. Griffin, Proc. 5th Int. Conf. on Nucl. React. Mech., Varenna, E. Gadioli, ed., p. 669 (1988); J.J. Griffin, J. Phys. Soc. Jpn. $\underline{58}$, Suppl. p. 427 (1989).

10. J.A. Wheeler, Proc. N.Y. Acad. Sci. $\underline{48}$, 219 (1946).

11. D.G. Caldi and A. Chodos, Phys. Rev. $\underline{D36}$, 2876 (1987); Y.J. Ng and Y. Kikuchi, Phys. Rev. $\underline{D36}$, 2880 (1987).

12. L.S. Celenza, V.K. Mishra, C.M. Shakin, and K.F. Liu, Phys. Rev. Lett. $\underline{57}$, 55 (1986).

13. M. Grabiak, B. Mueller and W. Greiner, Ann. Phys. $\underline{185}$, 284 (1988).

14. A. Guth, Phys. Rev. $\underline{D21}$, 2291 (1980); J. Froehlich and T. Spencer, Commun. Math. Phys. $\underline{83}$, 411 (1982).

15. C.B. Lang, Nucl. Phys. $\underline{B280}$, 225 (1987); E. Dagotto and J. Kogut, ibid. $\underline{B295}$, 123 (1988); Phys. Rev. Lett. $\underline{59}$, 617 (1987); J. Kogut, E. Dagotto and A. Kocic, ibid. $\underline{60}$, 772 (1988). See also Dagotto's contribution to this volume.

16. J. Schwinger, Phys. Rev. $\underline{82}$, 664 (1951). Se also W. Heisenberg and H. Euler, Z. Physik $\underline{98}$, 714 (1936).

17. L.H. Chan, Phys. Rev. Lett. $\underline{54}$, 1222 (1985); ibid. $\underline{57}$, 1199 (1986); Phys. Rev. $\underline{D38}$, 3739 (1988); H.W. Lee, P.Y. Pac and H.K. Shin, Phys. Rev. $\underline{D40}$, 4202 (1989).

18. A. Chodos, D.A. Owen and C.M. Sommerfield, Phys. Lett. $\underline{B212}$, 491 (1988).

19. A. Chodos, K. Everding and D.A. Owen, to appear in Physical Review D.

20. R. Jackiw, Phys. Rev. $\underline{D9}$, 1686 (1974) and references therein.

21. See S. Coleman, "Aspects of Symmetry", Cambridge University Press, 1985, sec. 3.3.

22. R.P. Feynman, Phys. Rev. $\underline{84}$, 108 (1951).

23. E. Witten, Nucl. Phys. $\underline{B188}$, 513 (1981).

24. J. Kogut and L. Susskind, Phys. Rev. $\underline{D11}$, 395 (1975).

25. R.D. Peccei, J. Solà and C. Wetterich, Phys. Rev. $\underline{D37}$, 2492 (1988).

26. See, for example, E.V. Shuryak, Phys. Rep. $\underline{61}$, 73 (1980).

27. See C. Itzykson and J.-B. Zuber, "Quantum Field Theory", (McGraw-Hill, 1980), pp. 67-68.

28. K. Everding, Yale Ph.D. thesis, 1990 (unpublished); A. Chodos, K. Everding and D.A. Owen, manuscript in preparation.

29. R.P. Feynman, ref. 22.

30. Berndt Mueller, summary talk at this meeting.

A NEW PHASE OF QED IN STRONG COUPLING:

A GUIDE FOR THE PERPLEXED

Elbio Dagotto

Institute for Theoretical Physics
University of California at Santa Barbara
Santa Barbara, CA 93106, USA

ABSTRACT

Recently, Quantum Electrodynamics (QED) in strong coupling has attracted much attention. By numerical and analytical techniques it has been found that a phase transition takes place separating the "normal" (weak coupling) phase of QED from a region where chiral symmetry is spontaneously broken in the massless limit (strong coupling). This critical point (second order phase transition) opens the possibility of a nontrivial continuum limit for QED. In these lectures I will briefly review lattice gauge theories results in strongly coupled electrodynamics. The motivation is presented, lattice techniques and results are described and the physical interpretation of the numerical data is discussed. Many other topics like QED in three dimensions, QED in strong fields, different nonperturbative techniques, etc. are also briefly reviewed. Most of this work has been done in collaboration with A. Kocić, J. Kogut and S. Hands.

I. INTRODUCTION

In these lectures I will present a very brief but hopefully self-consistent description of the present status of numerical studies of QED (Quantum Electrodynamics) in strong coupling. As an introduction, different nonperturbative methods are briefly described and a short review of lattice gauge theories is made. The bulk of the lectures corresponds to the application of the lattice formalism to strongly coupled QED. I will describe the motivation for this study, the techniques used, old and new numerical results and their physical interpretation. The main purpose of these talks is to give to the students an overview of this topic trying to avoid too many technical details (that may scare them out of the field!). Quite possibly experts in strongly coupled gauge theories will not find much new information in these lectures that is not already published in papers or preprints. However, due to the informality of the present review perhaps they may find useful some of my comments that otherwise would not appear in print in formal journals like Nuclear Physics or Physical Review. I also would like to warn the reader that the references presented in this review are certainly incomplete and he/she should follow the chain of references to get a complete picture of what has been done thus far in this quickly growing area of research.

Vacuum Structure in Intense Fields, Edited by
H.M. Fried and B. Muller, Plenum Press, New York, 1991

I.a Nonperturbative Methods in Quantum Field Theories. A Quick Overview

We are all familiar with the approach to Quantum Mechanics using the path integral method of Feynman[1]. In this formalism we are typically interested in calculating the mean value of some operator, lets denote it by $\hat{M}$, which is given as a ratio of functional integrals

$$< \hat{M} > = \frac{\int \mathcal{D}\phi M(\{\phi\}) e^{iS(\{\phi\})}}{\int \mathcal{D}\phi e^{iS}} \tag{1}$$

where $\mathcal{D}\phi$ denotes the "sum over all configurations" of some bosonic or fermionic field ϕ and S is the action of the theory under study which typically depends on some coupling constants, masses or other parameters in addition to the field ϕ.

In perturbation theory the method for calculating $< \hat{M} >$ is fairly well defined. The action is divided into a "free" piece (which contains the noninteracting part obtained by sending to zero all the couplings) and the potential term. The noninteracting sector corresponds (typically) to gaussian integrals which can be exactly computed. The exponential of the potential term is expanded in powers of the coupling constant. In this way the calculation of, for example, electron-electron scattering is approximated to lowest orders by a few Feynman diagrams. In QED we know that this approach works very well and we always present QED (treated in perturbation theory) to new students as a great achievement of theoretical physics (which certainly it is!). For example, the calculation of the anomalous magnetic moments of leptons is in remarkable agreement with experiments up to many digits. Actually one of the leaders of these calculations, Dr Kinoshita (private communication), told me recently that perhaps he will not continue the calculation of higher order corrections because experimental techniques can not yet reach the necessary accuracy required to test the theory.

However, and regretfully, perturbation theory is not the end of the story in Quantum Field Theories (QFT). Let me present a short list of theories that can not be attacked by the procedure described above:

i) QCD is a very typical example. There we have phenomena like confinement which is certainly absent when QCD is treated in perturbation theory. The linear term in the potential between two heavy quarks (string tension) has nonperturbative origin. Other phenomena associated with the theory of strong interactions like the hadronic spectrum, the breaking of chiral symmetry and the deconfining phase transition at finite temperature are all difficult to reach by a truncated series expansion in powers of the coupling. A very simple way to visualize these difficulties is just to remember that renormalization group calculations predict that any physical observable of QCD behaves as a function of the coupling constant (g) like e^{-1/g^2}. This functional form has an essential singularity at $g = 0$ and it can not be expanded in powers of g;

ii) You will learn in these lectures that QED in the strong coupling region (rather than in weak coupling) also has interesting nonperturbative physics resembling QCD in many cases;

iii) There are many other examples. To name a few: triviality of $\lambda\phi^4$ theories; superconductivity in a model described by the BCS theory like the negative U Hubbard model (where singularities of the type $e^{-1/U}$ are also present); the study of critical phenomena where singularities of the power-law $(T - T_c)^\alpha$ or essential singularity type $e^{-1/(T-T_c)}$ exist; compactifications of bizarre manifolds in superstrings theories; and the list goes on and on.

Faced with a nonperturbative problem, what can we do besides panicking? Many wise people just ignore them constraining their analysis to the formulation or reformulation of the problem at the Lagrangian level, the careful study of symmetries and other important but not too hard to obtain details, write a paper and move to the next fancy topic. But if you really want to understand a problem you have to make the effort of doing an actual calculation. Below I will present a short list of some popular nonperturbative approaches. There are many variations and modifications on these methods and I do not claim that the list is complete in any sense.

1) The simplest approach is to try to push perturbation theory as much as we can. In many cases in a perturbative calculation, even when many terms of the series expansion are known, spurious singularities (in the complex plane of the coupling constant we are expanding in) spoil the convergence of the series. It is desirable to make an extrapolation (analytic continuation) away from that region of parameter space. A method to do that is to use the *Padé* (or Borel) methods of extrapolation where typically the truncated series available are written as a ratio of polynomials with coefficients fixed by requiring consistency when this ratio itself is expanded. Then, the truncated series and the Padé approximant now coincide in "weak" coupling with the advantage that the Padé approximant contains poles in the strong coupling region which may correspond to physically interesting cases (phase transitions for example). In general, this approach can be used without spending too much time but the results are not under much control. As examples let me just mention two: i) when the trick is applied to the Ising model in 2 or 3 dimensions (the expansion is in $1/T$, T being the temperature rather than the coupling) the technique provides good results. Universality between the square and triangular ferromagnetic lattices has been found in this way; ii) lattice gauge theories in strong coupling are a counterexample. Here the expansion is in the inverse of the coupling constant and due to the complexity of the theory not many terms are available. The extrapolation using Padé approximants never worked well (were unstable) and after Monte Carlo methods were available for lattice gauge theories the technique of series expansions supplemented by Padé approximants was basically abandoned.

2) Still trying to make use of perturbation theory as much as we can, let us try to sum at least a subset of the infinite number of Feynman graphs typically contributing to some quantity at all orders. If this subset is itself composed of an infinite set of graphs then we can perhaps get nonperturbative results out of it and hope that the neglected diagrams will not change the physical results drastically (only in rare cases it is possible to make quantitative statements about this last assumption). Typical examples of this type of approach are the "ladder" and "RPA" approximations. In the ladder summation approach to QED (or similar theories where matter interact through the interchange of gauge particles) we sum the non-crossing diagrams

while in the RPA approach (also called one loop renormalization group calculation) the set of diagrams we sum contain pairs electron - positron

Typically the diagrams we can sum correspond to geometric series. There are ways to make this approach more systematic by using the *Schwinger-Dyson* equations.

3) It is very common to attack many body problems by *variational* techniques. In this case a wave function is presented as a candidate for the ground state

and physical properties are extracted from it. In condensed matter problems this technique is widely used for the analysis of magnetism and superconductivity. The crystal lattice provides a natural regularization for this wave function. In QFT we can use also variational wave functions in the context of lattice gauge theories but in general in the continuum it is difficult even to write a simple Ansatz.

It is obvious that the success of a variational wave function resides in how "close" it is from the actual ground state of the system. Special care must be taken about what criterion is used to judge whether a given wave function is a good approximation or not to a problem. A very common criterion is related with the energy i.e. the lower the energy the better. Regretfully, this is not enough. There are many examples of wave functions that are very close in energy to the ground state of a given system (known for example by numerical calculations) but that differ drastically from it in the correlation functions. The reason is that the main contribution to the energy of a system comes from short distance effects rather than long distance physics. However, most of the important physical properties of a model reside in that long distance behavior. A very famous example these days is the RVB Ansatz of Anderson for the ground state of the two dimensional Heisenberg model. Its energy is very close to that found numerically by Monte Carlo simulations but it has only short range correlations (at least in its minimal version as a sum of dimers of length one) rather than infinite range correlations as in the Néel state which is the actual ground state of the model. Then, care must be taken in judging the quality of a given Ansatz based only on the energy.

4) *Mean field* techniques are also very good methods to attack a problem[2]. In some sense they can be thought of as variational techniques. The idea is to approximate the many body problem by that of a single particle moving in an external background that mimics the effects of the rest of the degrees of freedom (d.o.f.). This background is obtained by requiring self-consistency. For example, in a system of ferromagnetic spins all but one spin is frozen to a mean value. Then, by requiring that the magnetization of the unfrozen spin is equal to that of the rest we arrive to a self-consistent equation (usually highly nonlinear). It is assumed that fluctuations around the mean-field solution do not destabilize it (this assumption is not checked in most of the calculations). The mean-field method is very popular in condensed matter. For example, it is widely believed that many difficult problems (like high Tc superconductors) will have a clear description once the correct degrees of freedom are considered in a mean-field approximation.

5) The *renormalization group* technique is much more powerful than the methods described thus far[3]. There are many ways to present it and do actual calculations with this method. The basic idea is to try to integrate out the high frequency degrees of freedom of a problem and construct an effective model describing the physics of the infrared modes which are the important ones at large distances. This technique has produced remarkable results for some models (e.g. Kondo model) and it helped in understanding critical phenomena very much. However, it is a tool that can not be immediately applied to any problem at hand. The integration of high frequency modes is in general difficult.

6) Finally we arrive to what is considered these days to be the most powerful approach to many body problems: *computer simulations*. The basic idea is that we have to define the problem at hand in terms of a finite but large number of degrees of freedom. By systematically increasing this number it is hoped that we will arrive to a good answer for the actual problem where, of course, the number of d.o.f. is infinite.

In condensed matter there is a very natural way to proceed. In a crystal we have a regular lattice with spins or fields located at their sites or links. The bulk limit is recovered by analyzing lattices of different volumes. However, note that with present day computers this limit is not always simple to obtain or smooth. For example, a problem of classical spins (say the XY model) in two dimensions can be typically studied on lattices of 256 X 256 sites while fully quantum mechanical problems like the two dimensional Hubbard model which has fermionic variables can be analyzed only with 16 X 16 sites lattices with present day computers[4]. Then, extreme care must be taken in the extrapolations to the bulk limit since these lattices may not be large enough. In condensed matter, another very popular numerical technique is the *Lanczos* approach[5] where the Hamiltonian matrix for a finite lattice (as large as possible) is diagonalized exactly. Knowing the whole spectrum of a problem allows calculations at finite temperature and real time response functions.

What happens in particle physics? To reduce the problem of a quantum field theory on a continuous manifold to a finite number of d.o.f. we follow K. Wilson[6] and discretize Euclidean space-time into a finite lattice (usually regular). This method has proven to be very useful in the analysis of many problems and is the subject of the next section.

II. LATTICE GAUGE THEORIES: A BRIEF REVIEW

II.a Path Integral in Quantum Mechanics

In Quantum Mechanics (QM) we are interested in calculating, for example, the probability amplitude for a particle that starts at time t_a in coordinate x_a to be at time t_b at position x_b. The particle is subject to some given potential. In the Path Integral approach this amplitude arises from calculations of quantities like

$$Z = \sum_{\text{paths from (a) to (b)}} e^{\frac{i}{\hbar} \int_{t_a}^{t_b} dt L}, \tag{2}$$

where the sum include all paths joining points a and b and the phase factor weights differently each one of these trajectories (L is the Lagrangian of the particle, for example, $L = \frac{1}{2}(\dot{x}^2 - \omega^2 x^2)$ in the case of the harmonic oscillator).

In practice to do a sum over "all paths" is not clearly defined. Feynman told us to do the following. Discretize the time interval into N_t pieces of length ϵ such that $t_b - t_a = \epsilon N_t$. Then, the sum over paths is simply reduced to an integral over the coordinate x for each time slice. We have to write the Action also in terms of a discrete time. For example, the time derivative appearing in the kinetic energy should be written as a finite difference. Finally, and for practical reasons let me just make a rotation to Euclidean space by replacing $t \to i\tau$. This replacement does not change the *static* properties of the Schrödinger equation although the time evolution is very different. The advantage of this rotation is that in practice dealing with exponentials rather than phase factors in the path integral is much easier. Then, after some work Z is written in our example as,

$$Z = \left(\prod_{i=1}^{N_t} \int_{-\infty}^{+\infty} dx_i \right) e^{-\frac{\epsilon}{2\hbar} \sum_{i=1}^{N_t} \{ (\frac{x_{i+1} - x_i}{\epsilon})^2 + \omega^2 x_i^2 \}}, \tag{3}$$

where x_i is the x coordinate at time i. In front of the integral the limits $\epsilon \to 0$, $N_t \to \infty$ such that $\epsilon N_t = t_b - t_a$ are understood. Then, the problem of quantum mechanics has been reduced in this way to the calculation of a (still difficult) multiple integral. Now we finally know what we are doing since we have a regularized theory with a finite number of degrees of freedom i.e. in the case presented above as many as N_t. It can be shown in many examples that starting from eq.(3) in the $N_t \to \infty$ limit we can exactly recover results of QM obtained in the more standard Hamiltonian formalism (like for example the spectrum of the harmonic oscillator).

II.b Generalization to Quantum Field Theories

Our main interest in particle physics is the analysis of QFT's. In this case we always have to deal with infinite number of particles since if enough energy is available pairs will be created with a finite probability out of the vacuum. Then, the formalism described above for QM needs to be generalized to many degrees of freedom. As a typical and simple example of a many body problem let us consider a set of particles interacting through springs as shown in the figure. For simplicity, the particles are assumed to be able to move only perpendicularly to the array of springs in the z direction (i.e. in the $x - y$ plane we have a rigid lattice)

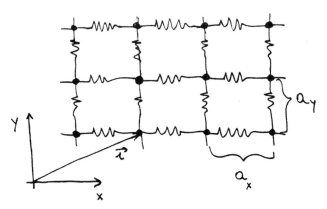

Figure 1) An example of a many body problem.

$\phi_{\vec{i}}$ is the z coordinate of the particle $\vec{i}$. Let us assume the Lagrangian contains a kinetic energy term and a harmonic potential term. Following the recipe of Feynman outlined above, time is made discrete and thus the kinetic energy can be written as a finite difference in time, one per particle. The potential energy contains a sum over nearest neighbors pairs of particles. Formally the path integral is

$$Z = \sum_{\text{sum over configurations of } \{\phi\}} e^{-\frac{1}{\hbar}S_{\text{Euclidean}}}. \tag{4}$$

After the discretization of the time coordinate and the lattice action, the final result corresponds to

$$Z = \left(\prod_i^{N} \int_{-\infty}^{+\infty} d\phi_i \right) e^{-\frac{1}{\hbar}S_{\text{Euclidean}}}, \tag{5}$$

where $i = (\vec{i}, \tau)$ contains the site and time indices and

$$S_{\text{Euclidean}} = \sum_{i=1}^{N} \left[\frac{(\phi_i - \phi_{i+a_0})^2}{2m} + V(\{\phi_i\}) \right], \tag{6}$$

where V is the potential term (equal time) and a_l is the lattice spacing in the direction l ($l = \hat{x}, \hat{y}$), a_0 is the lattice spacing in time, N_s is the number of particles in the array of springs or equivalently the number of sites in it and $N = N_s.N_t$ The limits $a_0 \to 0$, $a_l \to 0$, $N_s \to \infty$, $N_t \to \infty$ are understood in the measure if a continuum limit is taken both in time *and* space. Usually eq.(5) is abbreviated in papers and text books using a "continuum" notation both in space and time as

$$Z = \int \mathcal{D}\phi e^{-\int d\tau d^d x \mathcal{L}(\{\phi(x,\tau)\})}, \tag{7}$$

where d is the spatial dimension of the problem ($d = 2$ in the example above). A very important point to remark is that the path integral on the lattice in Euclidean time resembles very much a problem of Statistical Mechanics where the role of the Hamiltonian is played by the Action, temperature (T) is replaced by h and thus $e^{-S/h}$ by $e^{-\mathcal{H}/T}$. This analogy allows the use of techniques of condensed matter in particle physics and has proven to be quite useful.[7]

II.c Continuum Limit

Recovering the continuum limit of a lattice discretized field theory is not an easy task. One would naively assume that it is just a matter of taking the lattice spacing $a \to 0$ but in the most general case we would obtain either zeros or infinities as an answer to physical observables by doing so. The correct way to approach the continuum limit can be understood by the following example. Suppose that in QCD we want to study the interaction of two heavy quarks, one located at the origin and the other at distance L. In the path integral approach space-time is discretized on a fine grid of lattice spacing a. Suppose that between the two quarks there are N points. Then, we would like to send a to zero, N to ∞ such that the product is kept constant equal to L. But at the level of the Statistical Mechanics problem where the lattice spacing is just an arbitrary constant (taken traditionally equal to 1) this means that we want two particles located at distance $N \to \infty$ to "see" each other (interact). This is very difficult to achieve from local vertices as those contained in the Lagrangian *unless* the correlation length of the problem is as large as N itself. In the language of Statistical Mechanics an infinite correlation length exists only for particular values of couplings and masses i.e. when we have a *second order phase transition*!

Then, the strategy of a lattice gauge theory approach to a given problem is the following:

i) Select the bare Lagrangian you want to study,

ii) Discretize it (regulate the path integral) i.e. put the theory on a lattice,

iii) Find the phase diagram of the lattice model and its second order phase transitions,

iv) Approach some of these second order phase transitions to get the continuum limit. At this step it is important to remark that you may have many options. For example, let us imagine a theory with two coupling constants g_1 and g_2 and let us assume that after a careful work we found that the phase diagram contains four phases e.g. one superconducting, another confining and breaking chiral symmetry, a third one that confines but does not break chiral symmetry and a fourth one where no symmetry is broken and particles behave like free, as shown in fig. 2. The three arrows in the figure denote three possible continuum limits (if the lines of phase transitions are assumed to be of second order). The properties of the continuum theory we would get depend on from what phase we start the approach to the second order transition. It is very important to stress once more that a Lagrangian is not enough to define a QFT. We need in addition the point in parameter space where the continuum limit is taken and the phase from where we approach it. Below we will see that QED has a second order phase transition at some coupling constant of order 1 and at that point two different types of continuum limits can be taken, one where the theory breaks chiral symmetry and the other where it does not. Finally results should be independent of "small" details like the shape of the lattice used to write the path integral. For example, if instead of a regular hypercubic lattice a distorted random lattice is used, physical results should be the same. This is the concept of *universality* which is crucial in all this approach i.e. details of the regularization do not matter in the continuum limit. In the next section we will try to apply this technique to the specific nonperturbative study of QED.

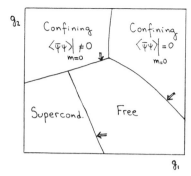

Figure 2) A possible phase diagram of an idealized theory showing four different phases. g_1 and g_2 are coupling constants. The arrows denote possible ways of approaching a continuum limit.

III. QED IN STRONG COUPLING

III.a Why Studying QED in Strong Coupling is Important?

QED in strong coupling is a subject of much interest in particle physics these days. Our motivation when we started the analysis of this theory on a lattice was to try to answer the following question: is there a $non - asymptotically$ free as well as $non - trivial$ field theory in four space-time dimensions? Growing evidence suggests that the simplest possibility given by $\lambda\phi^4$ is trivial i.e. the renormalized charge goes to zero when the cutoff is removed (although this topic is still debated). What happens in QED? This theory has a similar perturbative behavior as $\lambda\phi^4$. At the level of the standard one loop resumming of bubble diagrams (known as RPA approximation in condensed matter), the renormalized (e_R^2) and bare (e_0^2) charges are connected by

$$e_R^2 = \frac{e_0^2}{1 + \frac{e_0^2}{3\pi} ln(\frac{\Lambda^2}{m^2})}, \tag{8}$$

where Λ is a cutoff introduced in the theory and m is an infrared scale. When $\Lambda \to \infty$ the renormalized charge vanishes (keeping the bare charge constant) and thus the theory may be trivial. The screening due to the fermionic loops is so large that a test charge introduced in the "vacuum" of QED is "swallowed" by it. In other words, the Dirac sea rearranges itself such that the charge observed at large distances is zero. Of course, we do not know whether higher order terms may modify this structure. It may also occur that the β function of the theory is of the form i.e. a new zero of the β function may exist away from weak coupling (let me remind the reader that the β function is proportional to the derivative of the renormalized charge with respect to the cutoff. When the β function has a zero it means that the renormalized charge is cutoff independent. The results are regularization independent only at zeros of the β function. Then, as emphasize in the previous section a QFT is defined by a Lagrangian *and* a zero of the β function. A given theory may have many of these zeros or places where a continuum limit can be taken). For couplings larger than that corresponding to the new zero the theory may behave like QCD. At least in investigations of compact pure gauge QED on a lattice evidence was found that in strong coupling the theory was confining and thus a phase transition should exist between strong and weak coupling. Regretfully, including fermions in this compact action the transition was found to be of first order[8] and thus a new formulation had to be developed (as described below) in order to take the continuum limit.

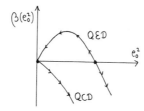

Figure 3) Speculative form of the β function of QED in 3+1 dimensions. Also shown is the β function of QCD in the same dimension.

Independently, using the Schwinger-Dyson equations in the quenched ladder approximation it has been shown[9,10] that the anomalous dimensions of scalar and pseudoscalar fermion composite operators become *large* and *negative* as the bare coupling is increased. This opens the possibility of finding renormalizable four-Fermi interactions in the continuum limit of QED at its critical point where chiral symmetry is spontaneously broken in the chiral limit. This feature may have interesting applications for technicolor models[11,12]. It also illustrates very nicely the fact that labelling a theory as "nonrenormalizable" based on perturbation theory (analyzing the behavior of integrals in momentum space) may be very misleading. A theory can be nonrenormalizable in weak coupling but renormalizable in strong coupling.

Let me also clarify that in principle the problem of the "triviality" of QED is phenomenologically *irrelevant*. We strongly believe that QED is not realized in Nature as an isolated field theory but only as part of a grand unified theory (still to be found). Then, whether we can completely remove the cutoff in QED or not is only a conceptual academic problem (unless the speculations about technicolor models mentioned below are correct) since there will always be a finite cutoff related with the next scale of physics implicit in the problem. The effects of the loop corrections to the charge become important only at enormous energies. Another point that deserves special clarification is the following: I have been asked many times whether our results below imply that the old ideas of Baker et al. about *finite* electrodynamics were finally correct. In my opinion, this is not the case. As you will see below we still have infinities in the theory even in strong coupling and only through a process of dimensional transmutation a finite (and arbitrary) scale appears in the theory. Only if an experiment is conducted in the strong coupling region of QED we can fix that scale and make predictions. Otherwise, only ratios of observables can be calculated. Then, the new developments in the area of QED do *not* suggest that, for example, the famous renormalized charge near 1/137 can be calculated from first principles. We still need *one* experiment to fix the scale as it happens in QCD with Λ_{QCD}.

III.b Lattice Formulation of QED. Some Details

As a first step towards a nonperturbative (mostly numerical) study of QED we need to write the theory on a lattice. The pure gauge part of the action of QED is defined on a lattice as follows[13]:

$$-\frac{1}{4}\sum_{\mu,\nu}\int d^4x\, F^2_{\mu\nu}(x) = -\frac{1}{2e^2}\sum_{\text{plaquettes}} A^2_{\mu\nu}(x) \tag{9}$$

where by "plaquette" we understand the elementary squares of a hypercubic lattice (each one defined by a site index x and two space-time direction indices μ, ν). e is the coupling constant. On the lattice it is common to use the notation $\beta = 1/e^2$. The "plaquette" variable $A_{\mu\nu}(x)$ is defined as the circulation of the gauge field around the plaquette i.e.

$$A_{\mu\nu}(x) = A_\mu(x) + A_\nu(x + \hat{\mu}) - A_\mu(x + \hat{\nu}) - A_\nu(x), \tag{10}$$

$A_\mu(x)$ is the gauge field (which since it is a vector field it is defined on the links of the lattice needing site and direction indices). In eq.(9) a rescaling of the gauge field $A \to A/(ea)$ has been done (a is the lattice spacing). $\hat{\mu}, \hat{\nu}$ are unit vectors in the directions of the lattice axis.

A very important property of this lattice formulation is that it keeps gauge invariance explicitly for any a i.e. the lattice regularization does not break this local symmetry which is considered to be a very important symmetry to preserve in the process of renormalization (dimensional regularization shares this property with the lattice but it is defined only in perturbation theory). Of course, the lattice regularization breaks explicitly other global symmetries like Lorentz invariance that we hope is recovered when the lattice spacing is sent to zero in a controlled way as discussed below.

We can also write in a discrete form the interaction term between fermions and photons

$$\int d^4x \sum_\mu \bar{\psi}(x)\gamma^\mu(\partial_\mu - ieA_\mu(x))\psi(x) \to \frac{1}{2}\sum_{x,\mu}(\bar{\psi}(x)e^{iA_\mu(x)}\gamma^\mu\psi(x + \hat{\mu}) - c.c.). \tag{11}$$

Then, the final form for QED on a lattice after absorbing the lattice spacing everywhere is

$$Z = \left(\prod_{x,\mu}\int_{-\infty}^{+\infty} dA_\mu(x)\right)\left(\prod_x \int d\bar{\psi}(x)d\psi(x)\right) e^{-S_{\text{lattice}}}, \tag{12}$$

where S_{lattice} is the sum of eq.(9) and (11) with the addition of a mass term for fermions. $\bar{\psi}$ and ψ are Grasmann anticommuting variables defined on the sites of the lattice. The number of sites of the space-time lattice is denoted by N. Eq.(12) is just a complicated but well-defined multiple integral.

After writing the path integral in a controlled way we have to select the best technique for its study. Over the years various people have tried with all the methods I described in section I applied to QCD. The best has been the computer simulations since in most of the others it is very difficult to know how close you are from the continuum limit. To perform a numerical study of lattice QED the first step is to integrate out the fermions. The action is already quadratic in the Grassmann field and thus we immediately replace the fermionic integral by

$$Z = \int \mathcal{D}A e^{-\frac{1}{2}\sum_p A_p^2} \det D(\{A\}),\tag{13}$$

where $A_p = A_{\mu\nu}(x)$ and the operator D is defined by

$$D_{x,y}(\{A\}) = m\delta_{x,y} + \frac{1}{2}\sum_{\mu}\eta_{\mu}(x)[e^{ieA_{\mu}(x)}\delta_{y,x+\hat{\mu}} - e^{-ieA_{\mu}(y)}\delta_{y,x-\hat{\mu}}],\tag{14}$$

where m is the bare mass of the fermions. It is very important to remark that in lattice gauge theories the presence of fermions introduces the problem of *doubling* of species i.e. although we discretize the action corresponding to one fermionic degree of freedom, when the continuum limit is taken we obtain more particles. Intuitively the reason is that the relation energy-momentum on a lattice of finite size and with local interactions is periodic and continuous between momenta $k = 0$ and $k = 2\pi$. Then, if the energy vanishes at some momentum k_1, by continuity it has to vanish elsewhere at some other momentum k_2. It can be shown that these zeros of the dispersion relation are very important for identifying how many particles we have in the continuum limit. One possible way to reduce this problem corresponds to the use of *staggered* fermions where basically the spin d.o.f. is diagonalized and the spin index is later drop. In eq.(14) this is what we have done i.e. starting with eq.(11) we integrated out the fermions and diagonalize the spin index i.e. we wrote the fermionic operator into four identical blocks and consider only one block simply by dropping the other three. The lattice action obtained in this way has only 4 species in the continuum if we start with one while without doing this trick it would have been 16 (in four dimensions). The factor $\eta_{\mu}(x) = (-1)^{x_0 + \dots + x_{\mu-1}}$ is the only remnant of the gamma matrices of the original formulation[14].

At this point we have reduced the problem to that of a pure gauge model having a very complicated long-range interaction hidden in the determinant. To study this model numerically it would be difficult to use a standard Monte Carlo technique since it would require many times the (costly) evaluation of determinants. The best approach is to use stochastic differential equations like the Langevin method or some of its variations. For simplicity I will describe the Langevin approach. In this case a new "time" variable (t) is introduced in the problem and the gauge fields depend on this new parameter through the following differential equation

$$\frac{dA_{\mu}(x,t)}{dt} = -\frac{\partial S}{\partial A_{\mu}(x,t)} + \xi_{\mu}(x,t),\tag{15}$$

where by x we mean the complete four vector that includes Euclidean time. $\xi_{\mu}(x,t)$ is a random variable with the property $< \xi^2 >= 1$ and $< \xi >= 0$ i.e. it is a gaussian random number.

Like for the path integral, the coupled set of differential equations (eq.(15)) also need to be written in a discrete form for the computer simulation. A new "lattice" spacing $\delta\tau$ in the time direction needs to be introduced. Care must be taken with this parameter since if it is too large it introduces systematic errors in the final results. Using the Fokker-Planck formalism it can be shown that the system of equations (15) in the large t limit have the property that the probability P of a given configuration is,

$$P(\{A_\mu(x)\}) \underset{t\to\infty}{\sim} e^{-S_{\text{lattice}}(\{A_\mu\})}. \tag{16}$$

Where is the bottleneck of the simulation? The problem lies in the fermionic determinant. When the derivative of the action with respect to the gauge field is taken we need to calculate the inverse of the fermionic operator D. This calculation is typically done using the conjugate gradient technique which is very efficient but anyway it consumes an appreciable amount of computer time in the simulation specially for small fermionic masses. In particular for $m = 0$ the operator D has a zero eigenvalue and thus the chiral limit can only be obtained by working with small masses and extrapolating to zero mass.

Summarizing, what is being done in the computer is the process of iterating the coupled Langevin equations (as many as sites of your lattice) monitoring different observables like the energy or condensates until they reach equilibrium. In that regime, information is accumulated for, typically, many thousands of time steps $\delta\tau$ and final results for mean values of operators are obtained. In practice, instead of the Langevin equation we use the "Hybrid" algorithm which is more efficient[8].

III.c Compact Formulation

In eq.(9) we have the so-called *noncompact* way of defining pure gauge QED on a lattice which seems very natural. However, there are other ways. In general a lattice action only has the constraint of reproducing the correct continuum limit when the lattice spacing a goes to zero, thus the higher order terms in a are in principle arbitrary. Many actions on the lattice can have the same $a \to 0$ limit and differ in those higher order terms. Due to this ambiguity there is another very popular formulation of QED on the lattice which is the *compact* one. In this case the pure gauge action is

$$S_{\text{lattice}} = -\frac{2}{e^2} \sum_{\mu>\nu} [1 - cos(A_{\mu\nu}(x))]. \tag{17}$$

It can be easily shown that this action has the same continuum limit as the noncompact version. Actually, expanding the cosine up to second order we recover exactly the noncompact action. The higher order terms contain higher powers of a. The gauge fields in this case can be constrained to the interval $(0, 2\pi)$ and this is the origin of the name compact. This action is gauge invariant and it can be generalized easily to nonabelian theories. The noncompact action does *not* have this property i.e. we do not know a noncompact nonabelian lattice action that preserves exactly gauge invariance for a finite lattice spacing. The compact action is highly nontrivial even in the pure gauge limit where it can be shown that in strong coupling is *confining*. It has a first order phase transition at a finite coupling separating it from the standard weak coupling phase. This occurs with and without fermions and thus this formulation is not useful for our purpose of obtaining a continuum limit.

III.d How Things Can Go Wrong Even if You are Careful

There are many ways in which a lattice simulation can have problems (discarding trivial ones like having a bug in the program). Let me mention here two very general ones and then in the rest of the lectures you will find problems specifically related with QED. Suppose that after spending some time you have written the lattice version of your favorite QFT and following the steps described above you have integrated out the fermions. What happens if the fermionic determinant is not positive definite? Then, the proof of converge of the Fokker-Planck equation does not work (it requires an explicitly positive probability) and also other techniques like

Monte Carlo simulations can not be applied. This problem is not merely academic. For example, QCD with a finite density of quarks (nonzero chemical potential) has a complex determinant spoiling the possibility of a good numerical study. Other theories with problems are for example QFT's with Chern-Simons terms whose Euclidean version is complex. Models in external electric fields usually suffer from the same trouble and many very important problems in condensed matter have their bottleneck in the nonpositivity of the fermionic determinant. For example, the Hubbard model which some people believe can explain high Tc superconductivity can not be studied in the physically relevant region of finite doping of holes and low temperatures due to the so called "sign" problem of the fermionic determinant.

A second infamous trouble in lattice gauge theories is the presence of first order phase transitions. Suppose that after writing the discrete version of your favorite QFT and doing the (very time consuming) numerical simulation you find that the phase transition you are interested in to recover the continuum limit is of *first* order i.e. discontinuous. In such a case the correlation length in units of the lattice spacing is finite and thus when $a \to 0$ the correlation in physical units vanishes! We do not know of any way to extract an interacting continuum theory out of a first order transition. As mentioned before, this problem occurs in QED in the compact version.

III.e Continuous Transition with the Noncompact Formulation

When the formulation of QED on the lattice is changed from the compact to the noncompact action, better results are found. Let me remind the reader what we are looking for in strongly coupled QED. Our original purpose was to find some new zero of the β function in order to obtain a continuum limit which may be nontrivial. A zero of this function is usually associated with a phase transition and each phase transition has an order parameter. Since we know from previous experience that QCD and QED are similar in strong coupling and QCD with light fermions breaks chiral symmetry then it is natural to look for a similar behavior in QED. Besides, the Schwinger-Dyson equations approach predict this type of phenomenon i.e. the chiral condensate is nonzero at zero fermionic bare mass in strong coupling and then it vanishes at a finite coupling. Can we observe similar results in lattice noncompact QED?

In fig.4 I show the condensate $< \bar{\psi}\psi >$ as a function of β (now β is the inverse of the coupling constant) using a lattice of $8^3 \times 16$ sites and extrapolating results linearly to zero mass using $m = 0.04$ and 0.02. This particular result has been obtained in the *quenched* approximation i.e. neglecting fermionic loops. This is achieved by simply forgetting the fermionic determinant in eq.(13) in the generation of new configurations. In general eq.(13) can be analytically continued to any number of fermions by the replacement $det D \to det^{N_f/4} D$ where N_f is now a continuous variable denoting the number of "flavors" or leptons of the theory and the factor 4 comes from the doubling problem.

The result shown in fig.4 suggests that QED has the phase diagram shown in fig.5 at $m = 0$ as a function of the coupling constant since the chiral condensate is nonzero in strong coupling but zero in weak coupling with a critical coupling $\beta_c = 1/e_c^2 \sim 0.36$.

This is exactly the phase transition we have been looking for. Note the very important detail that the condensate vanishes with continuity i.e. there are no indications of a first order phase transition and thus in principle we can obtain a continuous limit. As described below it is important to know the functional form of the condensate near the critical point. This old result shown in fig.4 suggested an essential singularity behavior but new and better results presented in section IV suggest a power-law behavior with a critical coupling smaller than 0.36. The qualitative picture Fig. 5 is unchanged and has been confirmed by other groups.

There is a crucial difference between the situation where the charge of the

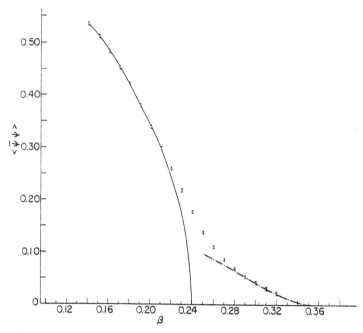

Figure 4) Chiral condensate as a function of β. The solid line is a mean field fit and the dashed line is a fit assuming $\langle \bar{\psi}\psi \rangle \sim e^{-\frac{1}{\sqrt{\beta_c - \beta}}}$.

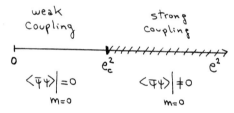

Figure 5) Schematic phase diagram of QED as a function of the coupling constant at zero fermionic bare mass.

elementary excitations of a problem is increased with respect to the case where an external charge is increased. In the former, pair creation may occur trying to quickly screen the external source. In the latter, however, each new pair that is created is itself critical and thus pair creation does not stop (chain reaction) forming a condensate until the density is so high that the background of pairs itself produces screening of charge. Test particles moving in this background acquire an effective mass proportional to the condensate.

In short, the results obtained with the noncompact formulation are very encouraging since we have observed a *continuous* chiral transition similar to that predicted by the Schwinger-Dyson approach.

III.f Technicolor

One of the possible applications of the new phase of QED is in technicolor models. Let me remind the reader that technicolor theories are defined by the standard model without Higgs fields and a higher energy sector containing fermions which are condensed like in QCD. This condensate plays the role of a Higgs field at low energies and it produces masses for the W and Z particles. In the early 80's it was found that these ideas had the problem that the predictions for neutral flavor changing currents (n.f.c.c.) were too large compared with experiments and the method was basically abandoned. However, recently it has been claimed[11,12] that if the "technigroup" is U(1) defined in strong coupling then the changed in the anomalous dimension of 4-fermi operators near that transition alleviate the problem of n.f.c.c. (a somewhat related approach is "walking" technicolor). Then, applications of QED in strong coupling may arise if we do not try to identify the U(1) charge with the e.m. charge (i.e. with the theory of electrons and photons) but if it is taken as a new abelian subgroup of a grand unified theory. This is a very promising idea.

IV. NEW NUMERICAL RESULTS

IV.a Old Problems and Some Solutions

In this section I will discussed the current status of numerical studies of Quantum Electrodynamics mainly in the quenched approximation. Some of these results are still not published or in preparation. However, the picture emerging out of them is very robust and it deserves a detailed discussion. Our "old" results on QED in strong coupling (section III.e) were produced under the assumption that the chiral condensate behaves *linearly* as a function of the electronic bare mass. This assumption was motivated by the predictions of the Schwinger-Dyson approach in the absence of four-Fermi interactions[9]. Based on this assumption our preliminary results shown in the previous section suggested that the scaling law of the condensate (extrapolated to zero mass) as a function of the bare coupling follows an exponential behavior (essential singularity) also in agreement with the results of Miransky et al. However, it was later shown by different groups[15,16,17] including ourselves[18] that $< \bar{\psi}\psi >$ as a function of the bare mass m is not linear. Fig.6 is a typical example. Linear or quadratic extrapolations produce considerable differences in the chiral limit if masses around 0.02 (lattice units) or larger are used. Then, this seemingly harmless detail is in fact an important complication for the analysis of chiral symmetries in QED. Let me remind the reader that we can not calculate the condensate at zero mass since the fermionic operator has a zero mode in this case. Also note that the smaller the mass the longer it takes for the conjugate gradient technique to produce the chiral condensate and thus we can not consider masses arbitrarily small.

To avoid this problem there are two ways to proceed: i) find a "theory" that you trust and extract from it the functional form for $< \bar{\psi}\psi >= f(m)$ to make the extrapolation; and/or ii) reduce considerably the values of masses being studied in order to minimize the ambiguities in the $m \to 0$ limit result. The second approach is, of course, unbiased and this is what we basically did[19]. For that purpose we developed a new algorithm that works very efficiently for the quenched case and noncompact action[18,20]. The idea is that in the quenched case the action is quadratic in the gauge field (gaussian integral). Then, it is enough to generate a set of random gaussian numbers (one per momenta and direction), multiply them by the square root of an appropriate photonic propagator (to avoid a trivial zero mode due to gauge invariance it is necessary to fix a gauge. In our case we considered the Feynman gauge. Of course, gauge invariant quantities are independent of this gauge fixing) and then transform the gauge configuration to coordinate space (we use a Fast Fourier Transform subroutine). If the correlation between random numbers is neglected, then the configurations produced in this way are all *independent* and critical slowing down is avoided. This is a considerable advantage over the previous hybrid algorithm simulations for the special case of $N_f = 0$.

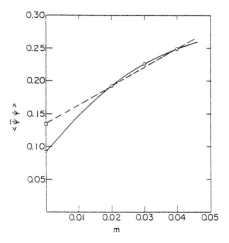

Figure 6) Chiral condensate as a function of the bare mass for $\beta = 0.22$, two flavors and a 10^4 lattice. The dashed line is a linear extrapolation while the continuous line corresponds to a quadratic extrapolation.

What type of results do we expect to find? One of the simplest scaling laws we can test corresponds to the behavior of the chiral condensate as a function of the bare mass *at* the critical point. In general for a second order phase transition we expect a power-law behavior of the type $< \bar{\psi}\psi > (\beta = \beta_c) = Am^{1/\delta}$ where δ is the associated critical exponent. If $\delta = 3$ then we say that the theory has "mean-field" behavior since this is the value that a mean-field approximation (or a gaussian theory) predicts for this model. We associate mean-field behavior with "triviality" (although if a theory has mean-field behavior, only by studying the logarithmic corrections around it we can decide on the issue of triviality. However, such a calculation would be too difficult numerically). If, on the other hand, convincing evidence is presented that δ is different from 3 then the theory would be a serious candidate for nontrivial behavior in the continuum limit. The main trouble in finding this critical exponent lies in the accurate determination of the critical coupling of the model. This is the main source of uncertainty in the evaluation of δ.

Another scaling law that can be studied numerically is the behavior of the chiral condensate as a function of the coupling constant in the zero mass limit. If there is a power-law behavior the associated exponent is usually called β (which should not be confused with the inverse of the coupling constant on the lattice). If this exponent is 0.5 then the theory is mean-field. Another possible scenario would occur if we find a different functional dependence of the condensate with the coupling e.g. an essential singularity like in an XY model transition in two dimensions.

In fig.7 we observe $< \bar{\psi}\psi >$ as a function of the bare mass m at $\beta = 0.250$ (β is the inverse of the coupling in the lattice language). Due to the use of the new algorithm results have been extracted[19] for masses as small as $m = 0.0007$ in lattice units. This simulation has been done using the *noncompact* formulation and staggered fermions. Results for two different lattice sizes show small finite size effects. A distinction between the results for 16^4 and 24^4 lattices can be made only for masses smaller than 0.0025.

Fig.7 shows clearly that the condensate at zero mass for this value of β is nonzero and thus the critical beta (β_c) is larger than 0.250. This result is important since in a recent paper it has been conjectured[17] that the critical point of the theory is precisely in the vicinity of 0.250. Measuring the exponent δ at this point it was

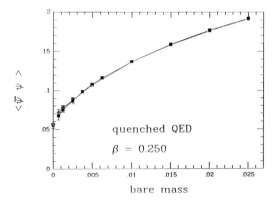

Figure 7) Chiral condensate as a function of the bare mass at $\beta = 0.250$ for quenched QED using lattices with 16^4 and 24^4 sites (full and open squares, respectively). Also shown (circle) is the Lanczos result of ref. 21.

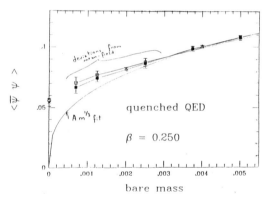

Figure 8) Chiral condensate as a function of the bare mass at $\beta = 0.250$ for quenched QED using lattices with 16^4 and 24^4 sites (full and open squares, respectively). Also shown (circle) is a Lanczos result[21]. The triangles correspond to results of the DESY group[17]. The thin solid line is the best mean-field fit using masses larger than 0.005.

found that in some region of masses the theory follows a mean-field behavior and thus the possibility of triviality of QED would exists. Fig.7 shows that such a conclusion needs some revision. Consider for example fig.8 where an expansion of the previous figure is presented.

We have found that if masses larger than 0.005 are used then it is possible to fit the data very well with a power law having $\delta = 3$ at $\beta = 0.250$. However, including the new "generation" of small masses there are clear deviations from mean-field. Then, to observe the actual scaling law of the theory it is necessary to study very small masses and be very close to the critical point. This is a dangerous potential problem for studies in the unquenched case where we expect the scaling window to be even smaller due to the screening effect of the fermionic loops.

Repeating a similar calculation as that presented for $\beta = 0.250$ but for many other values of the coupling β, we can get a good idea of the behavior of the condensate in the chiral limit (zero mass). The result is shown in fig.9.

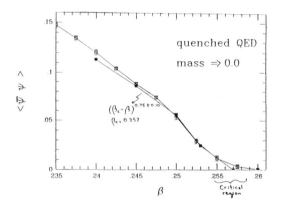

Figure 9) Chiral condensate extrapolated to zero mass (full squares) as a function of the coupling β. Also shown are Lanczos results[21] (open squares).

The calculations for $\beta = 0.253$ and 0.255 are still in progress so those results are tentative. A smooth extrapolation of the results obtained for $\beta \leq 0.250$ predicts that the critical coupling is in the vicinity of $\beta = 0.256$ or 0.257. Assuming that $\beta_{critical} = 0.257$ the zero mass condensate can be fitted with a power law having a critical exponent 0.75 ± 0.10. Having found approximately where the critical coupling is, then it is easy to get the other exponent δ. The result is shown in fig.10.

The best fit of the data produces $\delta \approx 2.1$ while a fit with $\delta = 3$ is clearly deviated from the (numerically found) chiral condensate. The result $\delta = 2.1$ is very stable and can be obtained using the very small region of masses or even masses larger than 0.005.

In fig.11 we show the optimal value of the power law fit $< \bar{\psi}\psi >= Am^{1/\delta}$ where now δ is *not* the critical exponent 2.1 but another constant in the fit which together with A has to be found by minimization of a χ-square calculation.

This result was obtained by using masses larger or equal than 0.005. We observe that if the critical coupling is found to be 0.250 as claimed by the DESY group then the theory is mean-field like (δ close to 3) so their scenario is self-consistent[17]. However, if evidence is found that the critical coupling is larger, then the critical exponent is non mean-field. The main source of error for δ is precisely the error in the determination of the critical coupling. Our estimation is that our numerical results predict $\delta = 2.1 \pm 0.1$.

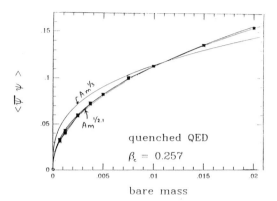

Figure 10) Chiral condensate as a function of the bare mass at $\beta = 0.257$ (which we found is the critical coupling). The data corresponds to lattice of 16^4 sites (full squares) and 24^4 sites (open squares). Also shown is an attempt to fit the data with a mean-field law.

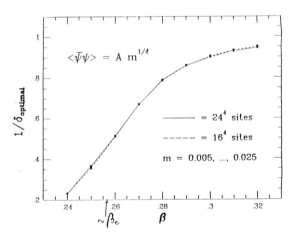

Figure 11) Optimal value of $1/\delta$ for different couplings in the fit $< \bar{\psi}\psi >= Am^{1/\delta}$ (both A and δ are free parameters in the fit).

IV.b Physical Interpretation and Future Work

Why the lattice simulation does not reproduce the quenched ladder predictions for pure QED in 4 dimensions? A possible explanation of our results is the following: suppose that the lattice version of QED contains (hidden somewhere) an attractive four-Fermi interaction[21,22]. Its origin can be dynamical or it may come from the (usually) discarded higher order "irrelevant" operators originated by the discretization of the lattice action. One way to visualize how this effect can occur is by noticing that deep in strong coupling i.e. at $\beta = 0$ the gauge fields can be integrated exactly and the resulting pure fermionic interaction corresponds to a vector square attractive four-Fermi term that may strongly influence the results close to the critical point. Another way to visualize the possible appearance of contact interactions in the action of QED is the way in which we prepare the configurations in the new algorithm described above. The photon propagator we used is a lattice version where roughly p^2 is replaced by $2(1 - cos(p))$ (in a one dimensional analog of the problem). The lattice propagator is *flat* at $p \sim \pi$ and that corresponds to a contact term. Then, it would not be surprising if lattice QED corresponds in the continuum to a combination of QED plus a four-Fermi interaction term. Usually in QCD we do not care about these type of terms but in QED we are precisely trying to study if they become relevant! Then, most of the standard philosophy of lattice gauge calculations has to be carefully reviewed for the strongly coupled QED analysis.

How large is this lattice generated contact term? To answer this question let us analyze the predictions of the Schwinger-Dyson approach[23] when an explicit four-Fermi interaction is included in the continuum formulation. We found that $\delta = 2.1$ corresponds to a point approximately in between the NJL (Nambu-Jona-Lasinio) limit and the results of Miransky et al.[9] and Bardeen, Leung and Love[10]. The actual prediction for δ coming out of the Schwinger-Dyson approach is

$$\delta = \frac{2 + \sqrt{1 - \frac{\alpha}{\alpha_c}}}{2 - \sqrt{1 - \frac{\alpha}{\alpha_c}}} \tag{18}$$

where α is the e.m. coupling and α_c its critical value[21,22]. For $\alpha = 0$ i.e. at the NJL point we get $\delta = 3$, a mean-field theory. For $\alpha = \alpha_c$ we recover the result $\delta = 1$ of Miransky et al. and Bardeen et al. The lattice simulation result corresponds to an e.m. coupling roughly $\alpha = \frac{1}{2}\alpha_c$. All the other critical exponents can be obtained as a function of α. In particular it was found that they satisfy hyperscaling relations[21]. For example the exponents δ and β are related by $\beta = 1/(\delta - 1)$. Finding (numerically) evidence for the existence of these relations is an indirect way to check the Schwinger-Dyson equations predictions.

Many other interesting points have been recently discussed in the literature[21,22]. In particular, accepting the idea that there is an important (hidden) four-Fermi term in the lattice action then the fact that the scaling window is so small can be easily understood[21]. It is also particularly interesting to discuss what occurs when the four-Fermi interaction at $\alpha = 0$ is fined tuned near its critical coupling such that the quadratic divergences are eliminated. In such a case only logarithmic divergences remain. When QED is turned on the logarithms are traded by a finite number. In this way pions are no longer pointlike particles and they become interacting[21].

If this scenario is correct then the next steps to follow are clear. We have to modify the lattice action in order to eliminate the contribution of the lattice generated four-Fermi term. One very natural way to do this is by explicitly introducing a *repulsive* four-Fermi interaction. In fig.12 we show what we believe is the correspondence that exists between the lattice results and the Schwinger-Dyson approach.

In the lattice simulation without an explicit contact term we are obtaining results compatible with a point in the middle of the phase diagram on the right-hand side of the figure due to the hidden four-Fermi lattice interaction (point B). In the absence of gauge fields we expect to obtain a "mean-field" theory (NJL point) in both cases (point A) Then, clearly for $G_0 < 0$ (G_0 being the coefficient of the four-Fermi term which if negative is repulsive) the exponent δ will reduce its value until eventually $\delta = 1$ is found (point C). This scenario seems correct but technically it is difficult to introduce a repulsive contact term because in principle the positivity of the fermionic determinants may be spoiled. Work is in progress along these lines. A similar line of research regarding modifications of the photon propagator in the Feynman gauge that we used in our program is under consideration. Also it should be remarked that if using different lattice regularizations we find similar results i.e. δ systematically close to 2.1 for example, then the possibility of an isolated fixed point (rather than a line of fixed points) somewhere in the middle of the phase diagram (G_0, α) should not be excluded.

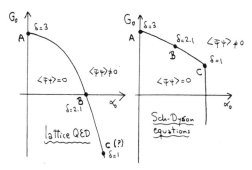

Figure 12) Correspondence between lattice simulation results and predictions of the Schwinger-Dyson approach. α_0 is the electromagnetic coupling and G_0 the four-Fermi interaction coupling.

IV.c Unquenched Simulations

After the results presented for the quenched case, it is clear that a simulation including dynamical fermions would be very difficult specially regarding the search for a nontrivial scaling region. Any simulation with dynamical fermions that did not reach the same masses and lattice sizes we studied in quenched QED is in doubt. Then, in this section I will mainly report on results for the phase diagram of QED changing the number of fermions N_f. Even without analyzing in detail the scaling region, we nevertheless found interesting results. In fig.13 the phase diagram of noncompact QED in the plane (N_f, β) is shown[24].

The critical coupling necessary to break chiral symmetry increases (β decreases) with N_f which is a manifestation of screening of the charge due to fermionic loops. For small N_f we have the region where we hope to observe deviations from mean-field as in the quenched case. Our previous studies for $N_f = 8$ have shown that the chiral condensate follows very well the mean-field curve almost up to the critical point[18]. Even if there is a tiny region of non mean-field behavior near the critical point, in practice it would be very difficult to observe it and thus we label $N_f = 8$ as a "mean-field" case. For $N_f = 4$ our early studies showed that like for $N_f = 8$, there were small chances of observing deviations from mean-field. Then, we would not be surprised if for $N_f = 4$ present day computer simulations show mean-field behavior. Currently, we are concentrating on the case of $N_f = 2$ to try to analyze the continuum limit.

A very interesting detail of the phase diagram fig.13 is the existence of a tricritical point T. For larger N_f the chiral transition is of first order and actually when $N_f = 30$ it just disappears i.e. even with an infinite bare charge there is no chiral condensate. This result can be a lattice artifact or it may survive the continuum limit thus implying that a theory of massless leptons has an upper limit of how many species can exist if a phase that breaks chiral symmetry is required.

Let me finish this section with a *wild* speculation. Looking at fig.13 and forgetting about the intrinsic meaning of N_f as an integer labeling the number of flavors, it looks as a traditional phase diagram obtained for example in a condensed matter problem having two coupling constants. Typically the RG trajectories in those phase diagrams are highly nontrivial and strongly influenced by the presence of a tricritical point. The question is if in our case where N_f is not a coupling we can anyway think of RG trajectories for this problem. In other words, would it be possible that N_f is also renormalized like masses or couplings? Imagine for example that dynamically some flavors acquire a very large mass and others remain massless. The low energy effective theory may then have less particles than the original one and thus the physics is governed by the results at lower N_f. Then, let me just challenge the reader to think about the following: can N_f be *renormalized* in a quantum field theory?

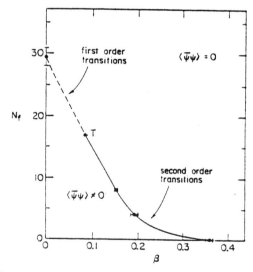

Figure 13) Phase diagram of QED in 3+1 dimensions in the plane (N_f, β). T is a tricritical point.

IV.d QED in 2+1 Dimensions

Lately our group and others have been studying also QED in 2+1 dimensions[25]. The motivation for such an analysis is three-fold: i) first it is believed that QED in 2+1 dimensions has many similarities with QCD in 3+1 dimensions specially in the sense that chiral symmetry is broken; ii) some recent results in the context of the Schwinger-Dyson approach have challenged the previous folklore of this model[26]. It was believed for some time that for all values of N_f chiral symmetry was broken spontaneously. However, a reexamination of the results have opened the possibility for the existence of a critical N_f beyond which chiral symmetry is restored. This number is predicted to be near 3.2. Can we observe such a behavior on a lattice simulation?; iii) recent results in the context of high Tc superconductivity have shown a close relation between a very popular model in this area, the Heisenberg model, with lattice gauge theories in 2+1 dimensions. More specifically it has been found that the Heisenberg model can be rewritten exactly as a SU(2) lattice gauge theory in 2+1 dimension defined in strong coupling ($\beta = 0$) and with dynamical staggered fermions of zero mass[27]. The analog of the Néel phase of antiferromagnets is the chirally broken phase of lattice gauge theories. We believe (and have checked recently) that there are no major differences between QED and $SU(2)$ in 3 dimensions and thus a study of the abelian case would be relevant also for the nonabelian theory.

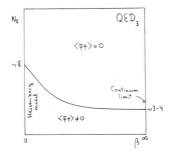

Figure 14) Schematic phase diagram of QED in 2+1 dimensions in the plane (N_f, β).

Our results have already appeared in the literature so let me just schematically present our most important conclusions. They are summarized in fig.14 where, like for the case of 3+1 dimensional QED, we present the phase diagram in the plane (N_f, β).

For a small number of fermions we find that chiral symmetry is spontaneously broken. The simulation has many of the problems of its counterpart in 3+1 dimensions but we have the advantage that we know where the continuum limit is and what are the scaling laws. All this information comes from dimensional analysis since the coupling constant has units in 2+1 dimensions. Then, it can easily be deduced that on the lattice the continuum limit is recovered at $\beta = \infty$. Our numerical results suggest that the critical number of flavors at $\beta = \infty$ beyond which there is no symmetry breaking is close to the analytic prediction. On the other side, at exactly $\beta = 0$ the critical number of fermions is near 8 (the transition still being of second order). The strong coupling limit is the region of interest for the Heisenberg model. The fact that chiral symmetry is broken here indicates that the ground state of the Heisenberg model is Néel like in agreement with Monte Carlo and Lanczos calculations directly performed on the spin model.

Why there is a critical number of flavors? Let me introduce the following rough but hopefully correct idea. In the large N_f limit when $N_f e_0^2$ is kept constant, the photon propagator is modified by bubble diagrams. Simply due to dimensional analysis the first correction to the photon propagator modifies it from $1/k^2$ to $1/(k^2 + cN_f e_0^2 k)$ where c is a number. Remember that e_0^2 has units of k. Then, truncating the photon propagator at this point implies that that the behavior at large distances of the effective potential between two heavy test charges changes from $ln(r)$ without fermions to $1/r$ at large distances when fermions are included. The problem resembles now 3+1 dimensional QED where we know a critical coupling exists. In the language of the 2+1 dimensional problem the role of the charge is played by $1/N_f$ and thus there is a critical N_f. Of course, this is an speculative idea that needs more thought but it seems to capture correctly some of the physics of the problem.

IV.e QED in a Strong External Coulomb Field

Not much is known about QFT's in external arbitrary fields. In particular we would like to analyze the reaction of the new phase of QED to external sources. This study is mainly motivated by the experimental observation of electron and positrons correlated narrow peaks in heavy-ion collisions[28] at GSI. It has been conjectured[29] that these unexpected spectrum comes from a phase transition in QED from the weak coupling phase to the new phase. This transition is supposed to be triggered by the presence of the two heavy ions. Then, it is in principle natural to mimic their effect by the study of QED in the presence of a *static* Coulomb center with an arbitrary (large) charge. The simulation of a given theory in an external electric field is difficult mainly because the positivity of the fermionic determinants is lost in this case. However, there are other ways to attack the problem i.e. some reliable mean-field methods, simulations in the quenched approximation (where the determinant is not used in the generation of configurations) and others. More specifically in QED it can be shown that an electric field in Euclidean time in the path integral formulation plays a role very similar to that of a chemical potential. Actually an electric field is exactly analogous to a coordinate dependent chemical potential. This is reasonable since we know an electric field creates pairs and a chemical potential does the same work. Of course, the electric field produces this effect conserving charge while a uniform chemical potential does not. This is not a contradiction because due to the periodicity on the lattice the electric field at the boundary is of opposite sign as in the bulk and there is where creation of antiparticles takes place. For more details the reader should consult the original reference on this problem[30]. Making a long story short, let me just present what we think is the phase diagram of QED in a strong electric field (see fig.15).

In this figure we plot β (inverse of coupling constant) in the horizontal axis and Z (the potential is assumed to be Z/r) in the vertical axis. The condensates and densities are all measured in the vicinity of the charge. It consists of three parts: i) Phase I where chiral symmetry is broken (new phase of QED) and the fermionic density n_f (not to be confused with the number of flavors) is zero; ii) phase II where chiral symmetry is respected and the density is zero (standard phase of QED) and iii) phase III where chiral symmetry is not broken but there is a finite density of particles due to pair creation. The position of the transition between phases II and III depends on the mass of the particles. In fig.15 it is assumed to be finite i.e. a finite electric field is necessary to begin pair creation. The important point of fig.15 is that the phase that breaks chiral symmetry is *suppressed* by the external field since it tries to break the electron-positron pairs of the condensate. Then, the explanation of the narrow peaks at GSI will require more work since the realistic region for the heavy ion collisions i.e. small coupling and large fields does not break chiral symmetry. In this workshop we have listened to talks by various people proposing different external gauge configurations that may produce the opposite effect as the Coulomb field. Then, a *very* interesting question with many important implications is the following: is there an external field (at this point no matter how bizarre and unrealistic) that *reduces* the value of the critical coupling necessary to break chiral symmetry in QED? Work is in progress along these lines.

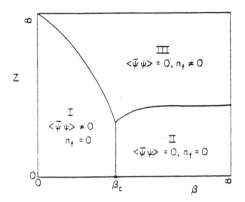

Figure 15) Schematic phase diagram of QED in an external Coulomb field. The meaning of the different phases is explained in the text.

V. SUMMARY

The main conclusions of these lectures are the following:

- *Quenched* QED in 4 dimensions: i) there is a phase transition separating the weak coupling regime of QED from a phase where chiral symmetry is broken. The transition is continuous and these results have been confirmed by many groups; ii) At the critical coupling the chiral condensate as a function of the electron bare mass follows a power law with a critical exponent $\delta = 2.1 \pm 0.1$ i.e. we observe deviations from mean-field behavior. No indications of an essential singularity have been observed. The scaling region is very narrow and large lattices, small masses and couplings very close to the critical point are needed to observe asymptotic scaling.

- *Unquenched* QED in 4 dimensions: its N_f dependence is very interesting presenting a tricritical point. For $N_f = 8$ and 4 it would be very difficult to find results different from mean-field.

- QED in 3 dimensions: numerical results suggest the existence of a critical number of flavors around 3 or 4 beyond which chiral symmetry is not broken.

- QED in strong fields: the simplest case of electrodynamics in a strong Coulomb field does not explain the new results of the heavy ion collisions. But it is not clear if *any* external field will suppress the new phase of QED.

We have many topics in preparation for the near future. In particular we would like to push ideas similar to those presented for fermionic QED for the case of *scalar* QED. A recent computer simulation of *noncompact* scalar QED has shown[31] the existence of a tricritical point in the Higgs-Coulomb phase transition (no confining phase in this case). Currently, we are investigating critical exponents in this model searching for indications of nontriviality.

I firmly believe that the analysis of QFT in unusual "environments" like strong couplings or fields is a very important topic of research. Only now with the development of lattice gauge theories techniques and Schwinger-Dyson equations methods we are able to do some rough (but under control) calculations and many interesting results are emerging. It would be very important to find applications (experimental or theoretical) for the interesting new phase of QED.

ACKNOWLEDGMENT

I thank my many collaborators and colleagues for discussions and comments about the physics of quantum field theories. This project was supported in part by the NSF grants PHY89-04035 and supplemented by funds from NASA.

REFERENCES

1. R. Feynman and A. Hibbs, "Quantum Mechanics and Path Integrals", Mc Graw-Hill (1965).

2. For a review of this technique in particle physics see J. M. Drouffe and C. Itzykson, Phys. Rep. **38**:133 (1978).

3. K. Wilson and J. Kogut, Phys. Rep. **12**:75 (1974).

4. A. Moreo et al., Phys. Rev. **B 41**:2313 (1990).

5. E. Dagotto, NSF-ITP-90-70, invited review for Int. Journal of Mod. Physics B.

6. K. Wilson, Phys. Rev. **D 14**:2455 (1974).

7. J. Kogut, Rev. Mod. Phys. **55**:775 (1983).

8. E. Dagotto and J. Kogut, Nucl. Phys. **B 295**:123 (1988) and references therein.

9. P. I. Fomin, V. P. Gusynin, V. A. Miransky and Yu. Sitenko, Riv. Nuovo Cimento **6**:1 (1983); V. A. Miransky, Nuovo Cimento **90A**:149 (1985).

10. C. N. Leung, S. T. Love and W. A. Bardeen, Nucl. Phys. **B273**:649 (1986); C. N. Leung, S. T. Love and W. A. Bardeen, Nucl. Phys. **B323**:493 (1989).

11. R. Holdom, Phys. Rev. Lett. **62**:997 (1989).

12. K. Yamawaki, M. Bando and K. Matumoto, Phys. Rev. Lett. **56**:1335 (1986); M. Bando, T. Morozumi, H. So and K. Yamawaki, Phys. Rev. Lett. **59**:389 (1987).

13. E. Dagotto and J. Kogut, Phys. Rev. Lett. **60**:772 (1988); Phys. Rev. Lett. **61**:2416 (1988); Nucl. Phys. **B317**:253 (1989); Nucl. Phys. **B317**:271 (1989).

14. N. Kawamoto and J. Smit, Nucl. Phys. **B192**:100 (1981).

15. A. M. Horowitz, Nucl. Phys. B (Proc. Suppl.) **9**:403 (1989).

16. S. P. Booth, R. D. Kenway and B. J. Pendleton, Phys. Lett. **B228**:115 (1989).

17. M. Gockeler, R. Horsley, E. Laermann, P. Rakow, G. Schierholz, R.J Sommer and U. J. Wiese, Nucl. Phys. **B334**:527 (1990).

18. E. Dagotto, A. Kocić and J. B. Kogut, Nucl. Phys. **B331**:500 (1990).

19. E. Dagotto, S. Hands, A. Kocić and J. B. Kogut, in preparation.

20. S. Hands, J. B. Kogut and E. Dagotto, Nucl. Phys. **B333**:551 (1990).

21. A. Kocić, S. Hands, J. B. Kogut and E. Dagotto, NSF-ITP-90-62, to appear in Nucl. Phys. **B**.

22. W. A. Bardeen, S. Love and V. A. Miransky, NSF-ITP-90-63.

23. K. I. Kondo, H. Mino and K. Yamawaki, Phys. Rev. **D** **39**:2430 (1989).

24. E. Dagotto, A. Kocić and J. Kogut, Phys. Lett. **B** **232**:235 (1989).

25. E. Dagotto, A. Kocić and J. Kogut, Phys. Rev. Lett. **62**:1083 (1989).

26. T. Appelquist, D. Nash and L. Wijewardhana, Phys. Rev. Lett. **60**:2575 (1988) and references therein.

27. E. Dagotto, E. Fradkin and A. Moreo, Phys. Rev. **B38**:2926 (1988) and references therein.

28. T. Cowan et al., Phys. Rev. Lett. **56**:444 (1986).

29. D. Caldi and A. Chodos, Phys. Rev. **D** **36**:2876 (1987); J. Ng and Y. Kikuchi, Phys. Rev. **D** **36**:2880 (1987).

30. E. Dagotto and H. W. Wyld, Phys. Lett. **B** **205**:73 (1988).

31. M. Baig, E. Dagotto, J. Kogut and A. Moreo, Phys. Lett. **B242**:444 (1990).

CONTINUUM LIMIT OF QUENCHED QED. CRITICAL EXPONENTS AND

ANOMALOUS DIMENSIONS

Aleksandar Kocić

Department of Physics
University of Arizona
Tucson, AZ 85721, U.S.A.

1. INTRODUCTION

Recently, there has been an increased interest in strongly coupled QED. This was motivated by the discovery of an ultraviolet stable fixed point at strong couplings [1]. If this fixed point would turn out to be non - gaussian, then QED would be the first nontrivial nonasymptotically free theory in four dimensions. The importance of such a result would be twofold. First, the old question of the existence of QED would be settled. Of course, this would be the case provided that the low energy limit of the theory actually describes photons and electrons; apriori, there is no reason to assume this. Second, and we feel a more important issue, is its paradigmatic value. Within that context QED, or better the $U(1)$ gauge theory, would be a paradigm for other nonasymptoticaly free theories. Of special interest would be nonabelian gauge theories with many flavors so that at weak couplings beta - function is positive (or vanishing). These theories are at present considered as viable candidates for technicolor unification schemes [2].

Let us see what we should expect from a theory that is strongly coupled at short distances. We know that in asymptotically free theories pointlike structures are simple objects because they are free. At large distances, they have complicated form factors and are strongly interacting etc. On the other hand, lack of asymptotic freedom at high energies leads to some quite unexpected consequences. Within that context the most fundamental problem is to understand how truly pointlike structures can interact strongly. If this is indeed the case, then the short distance regime of the theory is responsible for the existence of the fixed point whose character is decisive for the emergence of an interacting low energy limit. For this to happen, the high frequency modes must play the crucial role at low energies, contrary to the expectations based on the perturbative decoupling theorems [3]. Such a scenario is possible if the theory has the anomalous dimensions.

Before geting to the more technical part, we review some concepts that we will be using along. Those are the beta - function, continuum limit, anomalous dimensions and decoupling theorems.

Vacuum Structure in Intense Fields, Edited by
H.M. Fried and B. Muller, Plenum Press, New York, 1991

In quantum field theory, the field is a fluctuating variable. Because of the causality it exhibits independent fluctuations at adjacent space - like separations and appears rough at short distances. Therefore, at short distances it is impossible to describe the theory in terms of the smooth configurations. If we are to keep the traditional formulation with minimum modifications, it is necessary to cut off the high frequency fluctuations and to commit ourselves only to a region below the cutoff (Λ). The knowledge of the physics beyond the cutoff is buried in the Λ - dependence of the coupling constant. A change in Λ can be compensated by an appropriate adjustment of the coupling constant so that the low energy physics is kept unchanged. Theories in which such a manipulation is possible are called renormalizable. An important quantity that contains information about the cutoff dependence of the coupling constant is the beta - function. It is defined as,

$$\beta(g) = \Lambda \frac{dg(\Lambda)}{d\Lambda}. \tag{1.1}$$

If we solve this equation for g and call $g_B = g(\Lambda)$ and $g_R = g(\mu)$, where μ is some infra red scale, then we obtain,

$$\ln\left(\frac{\Lambda}{\mu}\right) = \int_{g_R}^{g_B} \frac{dg}{\beta(g)}. \tag{1.2}$$

Next, we want to take the continuum limit, either traditionally $\Lambda \to \infty$ or (better) $\mu/\Lambda \to 0$. Since we are sending the left hand side to infinity, the beta - function must have a zero (fixed point) for this procedure to make sense. (Of course, this zero must be such that $1/\beta$ has a nonintegrable singularity.) In that case the integrand in eq.(1.2) is singular and both sides can diverge simultaneously. We should mention that the beta - function always has a zero at the origin. If this is the only zero, then we should distinguish between the two possibilities: increasing and decreasing tendency. Assume the following form: $\beta(g) = \frac{1}{2}bg^3$ and solve equation (1.1). The result is,

$$b\ln\left(\frac{\Lambda}{\mu}\right) = \frac{1}{g_R^2} - \frac{1}{g_B^2} \tag{1.3}$$

The two cases give qualitatively different continuum limits:

1) $b > 0$ leads to $g_R \to 0$ for any g_B. The origin is the ultra violet unstable fixed point. This is the case of nonasymptoticaly free theories without an ultra violet stable fixed point. This scenario implies a trivial continuum limit.

2) $b < 0$ leads to $g_B \to 0$ and arbitrary $g_R \neq 0$. The origin is an ultra violet stable fixed point. This is the statement of asymptotic freedom.

Needless to say the existence of a nonperturbative fixed point in the first case could very well change the nature of the continuum limit.

Anomalous dimensions and the decoupling theorems

Consider a relativistic particle in a Coulomb center. A stable bound state is formed due to the balance of the attraction and the zero - point repulsion. At short distances both of these contributions scale the same way (for a bound state of size R, the energy is $E \approx 2p - \alpha/r \approx (1 - \alpha)/R$). If α is increased, the wave function is enhanced near the origin. This is recorded in its anomalous scaling:

at short distances the s - wave behaves as $\psi(r) \sim 1/r^{1-\sqrt{1-\alpha^2}}$. The quantity $\eta = 1 - \sqrt{1-\alpha^2}$ is the anomalous dimension. The probability of finding the particle near the origin $P(r) \sim r^2|\psi(r)|^2$ increases with coupling. The canonical phase space factor r^2 is modified by the anomalous dimension giving $P(r) \sim r^{2(1-\eta)}$. Clearly, when $\eta = 1$, the probability of particle sitting at the center is unsuppressed i.e. $P(r) = \mathcal{O}(1)$ as $r \to 0$. In this way the short distance physics becomes visible to the particle. Now, it is not difficult to see what will happen when we add to the Coulomb potential a short - range scalar interaction $V_{SR} = ge^{-Mr}/r$ with screening mass M being much larger than any scale in the problem. The corrections to the energy levels due to such a short range interactions can be calculated, for example, in perturbation theory. The leading term scales as $< \psi|V_{SR}|\psi > \sim 1/M^{2(1-\eta)}$. Clearly, at weak couplings this contribution will be suppressed as $1/M^2$ (modulo logarithmic corrections). This is an example of a standard (perurbative) decoupling theorem. However, as the coupling increases and the anomalous dimension approaches its maximum value $\eta = 1$, the suppression factor disappears at the critical couplling $\alpha = 1$, and the short range modification of the Coulomb potential gives unsuppressed correction to the bound state energy [4]. We note that the main reason for nondecoupling of the heavy mass is the fact that the vector interaction is sufficiently strong at short distances.

How do these concepts generalize to field theory? The heavy modes in field theory are typically represented by the higher dimensional operators i.e nonrenormalizable interactions. When studying a theory in four dimensions we truncate our lagrangian with operators of dimension four. This is because in perturbation theory the contribution of the higher dimensional operators is suppressed as some power of the lattice cutoff. However, as we will see, this will not be the case away from perturbation theory: certain higher dimensional operator will acquire anomalous dimension and will become renormalizable. Their precise structure will be constrained by the global symmetries and conservation laws of the theory. Let us illustrate these points by an example from the scalar field theory. The relevant operator with highest canonical dimension in this case is ϕ^4. Since this is a composite operator, we take $\phi^2(x)\phi^2(y)$ assuming the limit $x \to y$ and do the operator product expansion,

$$\phi^2(x)\phi^2(y) \approx \frac{C_0}{(x-y)^{2d_{\phi^2}}} + \frac{C_2\phi_R^2}{(x-y)^{d_{\phi^2}}} + \frac{C_4\phi_R^4}{(x-y)^{2d_{\phi^2}-d_{\phi^4}}} + \frac{C_6\phi_R^6}{(x-y)^{2d_{\phi^2}-d_{\phi^6}}} + ..., \quad (1.4)$$

where the notation is standard; d_{ϕ^n} denote the scaling dimensions of the corresponding composite operators and ϕ_R^n are renormalized operators. We note that if $d_{\phi^4} > 2d_{\phi^2}$ the series truncates with the quadratic term and renormalized theory is noninteracting. Conversely, an interacting local limit exists if $d_{\phi^4} \leq 2d_{\phi^2}$. If we define the anomalous dimension as the deviation from the canonical one e.g. $d_{\phi^4} = 4 + \eta_{\phi^4}$, and take, for simplicity, $d_{\phi^2} = 2$, then the condition for nontriviality becomes $\eta_{\phi^4} \leq 0$. Analogous criteria can be established for the anomalous dimensions of the higher order terms in the expansion in eq.(1.4). If these criteria are met, it would mean that the renormalized quartic coupling is determined not only by the ϕ^4 interactions, but, through the operator mixing, by the induced ϕ^6 etc. vertices. If this is the case, then the action should contain all the operators which can become relevant [5].

In massless quenched QED short distance fluctuations induce a chirally symmetric four - fermi interaction. The scaling dimension of the contact interaction is $d_{(\bar{\psi}\psi)^2} = 6 - \eta_{(\bar{\psi}\psi)^2} = 4 + 2\sqrt{1-\alpha/\alpha_c}$ [6]. To study the phase diagram in this case we have to start with the gauged four - fermi lagrangian $L = L_{QED} + G_0[(\bar{\psi}\psi)^2 + (\bar{\psi}i\gamma_5\psi)^2]$. One particular reason for this choice is the possibility that lattice simulations of QED incorporate these higher point interactions which are amplified by the strong gauge coupling and give a significant contribution to the low energy physics. We should therefore examine the importance of the induced contact interactions. The question remains how to obtain the anomalous dimensions in a universal way. We know from the statistical mechanics and renormalization group considerations that critical exponents can be expressed in terms of the anomalous dimensions. This dependence is general and does not depend on the system in question. Thus, in what follows, we will concentrate on the methods how to extract the critical exponents both from the lattice and continuum theory. Once we calculate the critical exponents, we will obtain the equation of state.

Recall that the equation of state records the response of the system's order parameter M to an external symmetry-breaking field h;

$$h = M^{\delta} \mathcal{F}(\Delta T M^{-1/\beta}), \qquad (1.5)$$

where $\Delta T = (T - T_c)$ is the temperature measured relative to the critical point $T = T_c$. In our case $M = <\bar{\psi}\psi>, h = m$ and $T - T_c = \alpha - \alpha_c$. At the critical point there is a singular response, $M(T = T_c) \sim h^{1/\delta}$. If ΔT is not precisely zero but $\Delta T M^{-1/\beta}$ is sufficiently small, then (1.1) can be expanded:

$$h = \mathcal{F}(0)M^{\delta} + \mathcal{F}'(0)\Delta T M^{\delta-1/\beta} + \cdots \qquad (1.6)$$

Note that when there is no external field, equation (1.2) predicts the familiar spontaneous magnetization curve $M \sim (T - T_c)^{\beta}$ and the susceptibility $\partial M / \partial h|_{h=0} \sim (T - T_c)^{-\gamma}$.

As an illustration consider a simple model of the effective potential for the order parameter M near the critical point and in a small external field,

$$V(M) = -hM + \frac{1}{2}a(T)M^2 + \frac{g}{\delta+1}M^{\delta+1} + \cdots \qquad (1.7)$$

The equilibrium state occurs at the extremum of the effective potential, $V'(M) = 0$. This condition gives

$$h = a(T)M + gM^{\delta} + \cdots \qquad (1.8)$$

If we define the critical temperature as the point that separates two phases, this translates into the condition for the saddle point of the effective potential, $V''(M = 0)|_{T=T_c} = 0$. In that case the value of T_c is determined from the equation $a(T_c) = 0$. Away from criticality, the response to the external field is always linear irrespective of the nature of the phase transition. The *subleading* term $M^{\delta+1}$ in the effective potential contains the information about the fluctuations in the critical region. Thus, the precise value of the critical exponent δ is sensitive to the character of the critical point. It is given by $\delta = \left(\frac{d\log h(M)}{d\log M}\right)_{T=T_c}$. The requirement that $T = T_c$ ensures that the leading term is removed from the expression for $h(M)$ (1.6), and the fluctuating part is isolated. Near a second order phase transition other quantities also scale as power laws. The remaining critical exponents are defined as follows:

$$M \sim (T_c - T)^{\beta}, \quad \xi \sim (T_c - T)^{-\nu}, \quad \chi \sim (T_c - T)^{-\gamma}, \qquad (1.9)$$

where M is the order parameter, ξ correlation length and $\chi = \frac{\partial M}{\partial h}|_{h=0}$ susceptibility. Dimensional analysis based on the divergent correlation length ξ in the critical region leads to the hyperscaling relations,

$$2\beta + \gamma = d\nu, \;\; 2\beta\delta - \gamma = d\nu, \;\; \beta = \frac{1}{2}\nu(d - 2 + \eta). \qquad (1.10)$$

where d is the dimensionality of the system and η measures the deviation from canonical scaling in the correlation function near the critical point; e.g. in a scalar theory $< \phi(x)\phi(0) > \sim 1/|x|^{d-2-\eta}$, and η is *twice* the anomalous dimension of the field ϕ. In four dimensions in the absence of fluctuations we have mean field scaling behavior with critical exponents $\beta = 1/2$, $\nu = 1/2$, $\gamma = 1$ and $\delta = 3$. These values are compatible with the hyperscaling relations (1.10) when $\eta = 0$. The effective potential (1.3) with $\delta = 3$ becomes of Landau - Ginzburg type with $a(T) \sim (T - T_c)$. All these results are standard in the study of critical phenomena [7].

Can our analytic work and the lattice simulations be directly compared? It is interesting to speculate on a correspondence which we can support with numerical results. Although the bare lattice action is pure QED, the high frequency fast modes in the system may be expected to generate a more complicated effective action controlling the physics of the system's longer wavelengths. It is natural to conjecture that a four - fermi term is induced in the lattice simulation analogous to that found in the continuum approach. From the point of view of the continuum theory, by changing the coupling α in lattice simulations, we may be moving along some path in the (α, G_0) - plane and cross the critical line at a point that does not necessarily correspond to the "collapse" fixed point of pure QED. It is easy to understand why this is happenenig: the lattice photon propagator has the following form,

$$D^{(lattice)}(k) = \frac{1}{1 - cos(ka)}. \qquad (1.11)$$

At low momenta ($ka \ll 1$) we recover the standard continuum limit behavior $D(k) \sim 1/k^2$. However, near the first Brillouin's zone ($ka \approx \pi$) we have $D(k) \sim const$. So, at high momenta it looks like we have a contact interaction.

The extraction of the critical scaling behavior of quenched QED by numerical simulation is a nontrivial issue, as witnessed by the fact that our group [8] and the DESY/Jülich group [9] have published different interpretations based upon seemingly very similar data. Our fits favor values of $\delta = 2.2(1)$, and $\beta = 0.8(1)$, both of which are distinct from their mean field values, which were used as the basis for data fitting in reference [9]. Thus the numerical results do appear to support some kind of interacting continuum limit.

2. CALCULATION OF THE CRITICAL EXPONENTS FROM THE GAP EQUATION

Analytical studies of chiral symmetry breaking in QED using the Schwinger - Dyson equation in the ladder approximation [1,6] have been the main motivation behind recent attempts to study this problem using lattice techniques [8-11]. The most important results of the analytical approach are: the interpretation of the critical coupling as an ultraviolet fixed point; and the observation that the anomalous dimensions of fermion composites grow as the coupling increases and become large and negative near the fixed point. An immediate consequence of the latter result is that certain higher dimensional fermionic operators become relevant and enter the renormalized action through operator mixing. If we imagine doing momentum shell renormalization, then by integrating out the fast modes, we generate an effective action that, in addition to the interactions of the original type, contains higher dimensional terms. In vectorlike theories, typical higher dimensional interactions are represented by four - fermi operators. In perturbation theory, standard decoupling theorems [3] imply that the fast modes decouple. In that case we can neglect the contact terms in the effective action since they have no influence on low energy physics. However, this may not be correct beyond perturbation theory. If the composite operators acquire anomalous dimensions, as is the case in strongly coupled QED, then canonical power counting does not apply and decoupling is governed by the total scaling dimensions.

First, we summarize some basic facts about the Schwinger-Dyson approach to this problem. In Landau gauge, the fermion self-energy satisfies,

$$\Sigma(p^2) = m - G_0 < \bar{\psi}\psi > +3e^2 \int_q \frac{1}{(p-q)^2} \frac{\Sigma(q^2)}{q^2 + \Sigma^2(q^2)} \tag{2.1}$$

where we have neglected vertex corrections (this is the difference between quenched and ladder approximations). After performing the angular integration, and changing variables to $x = p^2$, (2.1) becomes a boundary value problem,

$$\frac{d}{dx}(x^2 \Sigma'(x)) + \frac{\alpha}{4\alpha_c} \frac{x}{x + \Sigma^2(x)} \Sigma(x) = 0, \tag{2.2a}$$

$$(1+g)\Lambda\Sigma'(\Lambda^2) + \Sigma(\Lambda^2) = m, \tag{2.2b}$$

where we have used the following notation: $\alpha_c = \pi/3$; $g = (G_0\Lambda^2/\pi^2)\alpha_c/\alpha$; and Λ is the ultra-violet cutoff, together with the observation that $< \bar{\psi}\psi > = \Lambda^4 \Sigma'(\Lambda^2)/3\pi\alpha$. Note that the bare mass m and four - fermi coupling g enter the gap equation through the boundary condition only. Comparing the equations (1.2) and (2.2b) we see that *the boundary condition (2.2b) is in fact the equation of state.* Consequently, critical exponents are determined from the asymptotic form of the fermion self - energy only. Let us solve the system (2.2) for $\alpha \leq \alpha_c$. For large values of momentum the nonlinearity present in the gap equation can be neglected by replacing $x + \Sigma^2(x) \rightarrow x$. The ultra-violet asymptotics of the self-energy is then determined by a linear differential equation which can easily be solved leading to $\Sigma(x) \sim \frac{A\mu^2}{\sqrt{x}} \sinh(\frac{1}{2}\sqrt{1 - \alpha/\alpha_c} \ln(x/\mu^2))$, where A is a numerical constant and μ some infra-red scale. We will use the abbreviation $\theta = \sqrt{1 - \alpha/\alpha_c} \ln(\Lambda/\mu)$ to simplify the equations that follow. Then, the boundary condition gives

$$m = \frac{A\mu^2}{2\Lambda}[(1-g)\sinh\theta + (1+g)\sqrt{1 - \alpha/\alpha_c}\cosh\theta], \tag{2.3}$$

and the order parameter is given by:

$$3\pi\alpha\frac{< \bar{\psi}\psi >}{\Lambda^2} = \frac{A\mu^2}{2\Lambda}[-\sinh\theta + \sqrt{1 - \alpha/\alpha_c}\cosh\theta]. \tag{2.4}$$

We use equation (2.4) to express the boundary condition (2.3) in terms of $< \bar{\psi}\psi >$, rather then μ. Let us rewrite the boundary condition as,

$$m = C[1 - g + (1+g)\sqrt{1 - \alpha/\alpha_c}]\mu^{2 - \sqrt{1 - \alpha/\alpha_c}} + D\mu^{2 + \sqrt{1 - \alpha/\alpha_c}}, \tag{2.5}$$

where C and D are some constants. The critical line $1 - g + (1+g)\sqrt{1 - \alpha/\alpha_c} = 0$ (or equivalently $G \equiv g\frac{\alpha}{\alpha_c} = (1 + \sqrt{1 - \alpha/\alpha_c})^2)$ [12] is shown in fig.1. Now, we examine the behavior of physical

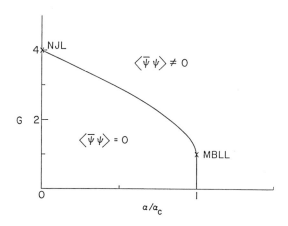

FIGURE 1. Phase diagram of QED extended by a chirally invariant four - fermi interaction.

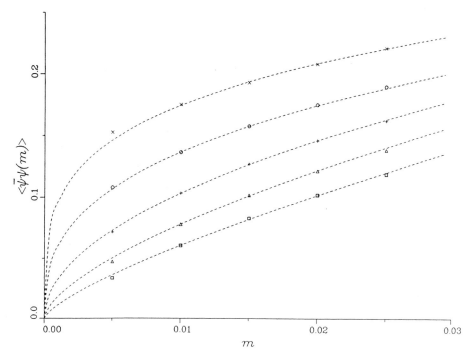

FIGURE 2. Plot of $< \bar{\psi}\psi(m) >$ against m, obtained via conjugate - gradiient studies on a 16^4 lattice, for the β values 0.24 (crosses), 0.25 (circles), 0.26 (pluses), 0,27 (triangles), 0.28 (squares). The dashed lines are the power - law fits.

quantities in the critical region. We introduce the notation $1 - g + (1 + g)\sqrt{1 - \alpha/\alpha_c} = (T - T_c)/T_c$. Then, in the chiral limit, the equation of state gives, $1/\mu \sim (T_c - T)^{\frac{-1}{2\sqrt{1-\alpha/\alpha_c}}}$, and we identify the critical exponent ν:

$$\nu = \frac{1}{2\sqrt{1 - \alpha/\alpha_c}}. \tag{2.6}$$

Using the relation $< \bar{\psi}\psi > \sim \mu^{2 - \sqrt{1-\alpha/\alpha_c}}$, we obtain the equation of state,

$$m = a(T - T_c) < \bar{\psi}\psi > + b < \bar{\psi}\psi >^{\frac{2 + \sqrt{1-\alpha/\alpha_c}}{2 - \sqrt{1-\alpha/\alpha_c}}} + \cdots \tag{2.7}$$

Equations (2.7) and (1.2) have the same form after appropriate correspondences are made. According to the discussion in the introduction, the exponent δ and β can be just read off:

$$\delta = \frac{2 + \sqrt{1 - \alpha/\alpha_c}}{2 - \sqrt{1 - \alpha/\alpha_c}}, \quad \beta = \frac{2 - \sqrt{1 - \alpha/\alpha_c}}{2\sqrt{1 - \alpha/\alpha_c}}. \tag{2.8}$$

Finally, we evaluate the susceptibility and the exponent γ. In analogy to a magnetic system, the susceptibility is defined as $\chi = \left(\frac{\partial < \bar{\psi}\psi >}{\partial m}\right)_{m=0}$. From eq.(2.8) we get then $\chi \sim (T_c - T)^{-1}$, implying $\gamma = 1$.

From the expressions for the critical exponents it is easy to check that they satisfy the hyperscaling relations. Eliminating the coupling constant dependence in eqs.(2.6,7,8), we obtain:

$$\beta = \frac{1}{\delta - 1}, \quad \nu = \frac{1}{4}\frac{\delta + 1}{\delta - 1}. \tag{2.9}$$

These are the first two hyperscaling relations of equation (1.4) with $\gamma = 1$. In general $\nu = \gamma(\delta + 1)/4(\delta - 1)$, $\beta = \gamma/(\delta - 1)$. It is possible to express the critical exponents in terms of the anomalous dimension of $\bar{\psi}\psi$. If we use $\eta/2 = \eta_{\bar{\psi}\psi} = 1 - \sqrt{1 - \alpha/\alpha_c}$, then

$$\beta = \frac{1}{2}\frac{1 + \eta_{\bar{\psi}\psi}}{1 - \eta_{\bar{\psi}\psi}}, \quad \delta = \frac{3 - \eta_{\bar{\psi}\psi}}{1 + \eta_{\bar{\psi}\psi}}, \quad \nu = \frac{1}{2}\frac{1}{1 - \eta_{\bar{\psi}\psi}}. \tag{2.10}$$

The result $\gamma = 1$ remains unchanged. We see from (2.10) that the values of the exponents vary from their mean field values at $\alpha = 0$ when there are no fluctuations ($\eta_{\bar{\psi}\psi} = 0$, so $\bar{\psi}\psi$ has its canonical scaling dimension), to the essential singularity behavior in the fluctuation dominated region at $\alpha = \alpha_c$ where $\eta_{\bar{\psi}\psi} = 1$ and the scaling dimension of $\bar{\psi}\psi$ is 2.

A comment about the $\gamma = 1$ result is in order. There is a connection between the exponents ν and γ and the ladder approximation. To display this fact, we examine the relation between γ, ν and the anomalous dimensions; $\gamma = \nu(2 - \eta)$, and $1/\nu = 2 - \eta_{(\bar{\psi}\psi)^2}$. When these two equations are combined, they lead to $\gamma = \frac{2 - \eta}{2 - \eta_{(\bar{\psi}\psi)^2}}$. The fact that $\gamma = 1$ implies that $\eta_{(\bar{\psi}\psi)^2} = \eta = 2\eta_{\bar{\psi}\psi}$, which is a characteristic of the factorization scheme employed in the ladder approximation. Since we are using this factorization property to make conclusions about the renormalizability of the four - fermi interactions, this property of the ladder approximation needs to be examined more thoroughly. A measurement of the exponent γ in lattice simulations would therefore be important in establishing the validity of the ladder approximation.

3. LATTICE SIMULATIONS: CRITICAL COUPLING AND SCALING

First we describe our fits for the exponent δ. Exactly at criticality ($\beta = \beta_c$), we expect power-law scaling of the form $< \bar{\psi}\psi(m) > \sim m^{1/\delta}$ and $\rho(\lambda) \sim \lambda^{1/\delta'}$, where as described below, $\rho(\lambda)$ is the spectral density function of the gauge-covariant Dirac operator. In the phase with broken chiral symmetry, however, both $< \bar{\psi}\psi(m) >$ and $\rho(\lambda)$ should have discontinuities at the origin, and hence a power-law fit is inappropriate away from criticality. Therefore in principle we can find β_c simply by finding the best fit to a power-law form. Of course, since we only have simulation data for certain rational values of β, in practice we can strictly only set bounds on β_c. We will see, however, that the power-law scaling hypothesis enables further progress. Our fits fall naturally into two subsections:

(i) Fits of the form $< \bar{\psi}\psi(m) >= Am^{1/\delta}$

The chiral condensate as a function of bare electron mass m can be found by inverting the propagator matrix:

$$< \bar{\psi}\psi(m) >= \frac{1}{V}\mathrm{tr}(\rlap{/}{D}[U] + m)^{-1}, \tag{3.1}$$

where V is the volume of spacetime, $U_{x,\mu} \equiv \exp(ieA_{x,\mu})$, and the real link variables $\{A\}$ are generated as described in [13]. Using a least-squares fitting routine we can obtain the "optimal" values of A and δ. Figure 2 gives gives the plot of $< \bar{\psi}\psi >$ vs. m for several values of β in the neighbourhood of the critical coupling. Since, in the critical region the order parameter is not a simple power, but a polynomial $m = a(\beta - \beta_c) < \bar{\psi}\psi > +b < \bar{\psi}\psi >^\delta$. So, the first term in this equation is the error that we commit with the ansatz $m \sim < \bar{\psi}\psi >^\delta$. Being proportional to $\beta - \beta_c$ this error changes its sign as we increase the coupling from subcritical to supercritical. Therefore, in order to determine the value of the critical coupling, we look for the change of systematics at low values of m. This tendency is apparent in figure 2: if we monitor the order parameter at the smallest value of mass $m = 0.005$, the we see that the lower three points lie below wheras the upper two lie above the fit. From here we conclude that critical coupling lies between 0.25 and 0.26.

If the critical β_c is near 0.25 then the best exponent is $1/\delta_{opt} \simeq 0.36$ which is very close to the mean field result $1/3$ and consistent with the analysis of [9]. However, if $\beta_c = 0.26$ then the best exponent is $1/\delta_{opt} \simeq 0.51$ in agreement with the previously reported results [8], where it was pointed out that for $\beta > 0.26$ the $m = 0.005$ point lies below the fit, whereas for $\beta < 0.26$ this point lies above the fit (this was taken as evidence that the best value of $\beta_c \simeq 0.26$: here we are content to note that it is the origin of the systematic effect of the previous paragraph). Thus, the position of the critical coupling seems to be of paramount importance in determining whether lattice QED is described by mean field exponents or not.

(ii) Fits of the form $\rho(\lambda) = B\lambda^{1/\delta'}$

Using the basis of eigenmodes $\mid n >$ of the Dirac operator $\rlap{/}{D}$ having eigenvalue $i\lambda_n$, $< \bar{\psi}\psi >$ has the following form:

$$< \bar{\psi}\psi(m) >= \frac{m}{V}\int_{-\infty}^{+\infty} d\lambda \frac{\rho(\lambda)}{\lambda^2 + m^2}, \tag{3.2}$$

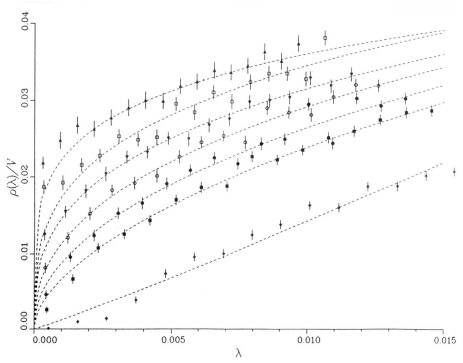

FIGURE 3. Plot of $\rho(\lambda)/V$ versus λ obtained by binning eigenvalue data from a 16^4 lattice. The β values shown are 0.2475 (up triangles), 0.2500 (squares), 0.2525 (down triangles), 0.2550 (circles), 0.2575 (filled circles), 0.2600 (filled squares) and 0.2700 (diamonds). The dashed lines are the power - law fits.

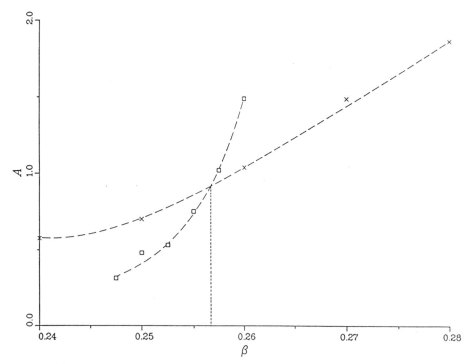

FIGURE 4. Plot of the power - law amplitude A versus β, as estimated directly from conjugate - gradient data (crosses), and as infered from the eigenvalue data (aquares), showing the prefered value of β_c. The dashed lines are merely to guide the eye.

where $\rho(\lambda)$ is the spectral density function. In the limit $m \to 0$, (3.2) yields the formula

$$< \bar{\psi}\psi(0) >= \frac{\pi}{V}\rho(0),$$ (3.3)

implying that $\rho(0) > 0$ signals chiral symmetry breaking. What can be said at criticality? Let us postulate a power-law scaling $\rho(\lambda)/V = B\lambda^{1/\delta'}$. Equation (3.2) then gives

$$< \bar{\psi}\psi(m) >= Bm \int_{-\infty}^{+\infty} d\lambda \frac{\lambda^{1/\delta'}}{\lambda^2 + m^2} = \frac{B\pi}{\cos(\pi/2\delta')} m^{1/\delta'}.$$ (3.4)

We thus have $\delta = \delta'$ and a relation between the amplitudes $A = B\pi/\cos(\pi/2\delta')$.

The same trends are observed as in the $< \bar{\psi}\psi(m) >$ fits; namely the fitted $1/\delta'$ increase with β. However, the actual values of the $1/\delta'$ disagree with those for $1/\delta$, being systematically smaller for $\beta \leq 0.25$, and larger for $\beta \geq 0.26$. Once again, we prefer to look for systematic effects in order to locate the best fit, rather than using χ^2 which we take merely to exclude $\beta_c = 0.27$. The plot of $\rho(\lambda)$ vs. λ for several value of the coupling constant is given in figure 3.

There is a consistency check between fits *(i)* and *(ii)* arising from equation (3.4). We can compare the fitted amplitude A from fit *(i)* with its value predicted from B and δ' from fit *(ii)*. This comparison is shown in figure 4, where it is apparent that the fitted A varies less sharply with β than the "theoretical" A. A linear interpolation shows that the two are in agreement for $\beta \simeq 0.2566$, which happily lies within the bounds we have already obtained. We estimate the error from the width of the range of inclusion of β_c in fit *(ii)*. We estimate $1/\delta$ at criticality by interpolation to be 0.452 *(i)* or 0.455 *(ii)*, with a statistical error of approximately 0.02, which we feel is probably comparable with any remaining systematic error.

Thus, by a variety of means, plus the assumption of power-law scaling at criticality, we have obtained the bounds $0.2550 < \beta_c < 0.2575$, with a "best" value $\beta_c = 0.257(1)$. The corresponding value of the critical exponent $\delta = 2.2(1)$. In deriving these values we have not only fitted for the exponent itself, but also used the fitted values of the amplitudes in a novel way to enhance our accuracy.

Next, inspired by the analytic predictions relating the various exponents, we return to the estimates of $< \bar{\psi}\psi >$ in the chiral limit in an attempt to extract the exponent β. Recall from reference [8] that near the transition, the enhanced curvature of $\rho(\lambda)$ for small λ makes an unambiguous extraction of $< \bar{\psi}\psi >$ via equation (3.3) impracticable in the absence of any theoretical input. For $\beta \geq 0.25$, we resorted to two different extrapolation techniques and took the difference between them as a measure of systematic error. Therefore power-law fits of the form

$$< \bar{\psi}\psi >= C(\beta_c - \beta)^\beta$$ (3.5)

(take care not to confuse the two different βs!) should not include any data for $\beta > 0.25$. In figure 5 we plot $< \bar{\psi}\psi >$ values extracted from eigenvalue data from a 16^4 lattice using an extrapolation method which takes no account of any possible finite volume effects (Cf. "method *(i)*" of figure 8 in reference [8]), together with an unconstrained fit of the form (3.5) to the seven points in the range $\beta \in [0.235, 0.25]$. The resulting $\beta_c = 0.2555(19)$, the exponent $\beta = 0.77(12)$, and the $\chi^2 = 0.4$. The fit is stable if we either exclude the upper or lower points of the range, or fix β_c to the value obtained

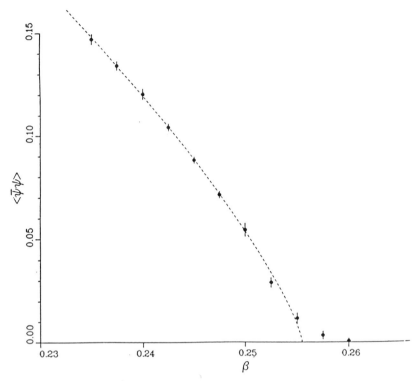

FIGURE 5. Plot of $< \bar{\psi}\psi >$ in the chiral limit versus β using the eigenvalue data fits. The dashed line is the power - law fit described in the text.

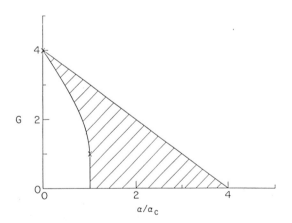

FIGURE 6. The phase diagram of figure 1, this time also showing the mean field prediction for the critical line (4.2). The shaded area gives rough indication of the critical region of the theory.

in our previous fits, with the fitted exponent varying between 0.74 and 0.88, so that the statistical errors accommodate any possible systematic ones. The value of β_c is also entirely consistent with our previous estimate. We therefore can quote a value for the exponent β of 0.8(1), which is quite distinct from the mean field value of $1/2$. The two exponents we have fitted so far give excellent agreement with the relation (2.9) between β and δ: setting the latter equal to 2.2 yields $\beta = 0.833$, well within our margin of error. Note that these estimates are completely independent of each other, and that the relative accuracy of the estimate for β is much the worse of the two.

To obtain the susceptibility exponent, we note that it can be expressed in terms of the spectral function,

$$\chi = \frac{1}{V} \int_{-\infty}^{+\infty} d\lambda \rho(\lambda) \frac{\lambda^2 - m^2}{(\lambda^2 + m^2)^2} . \tag{3.6}$$

Unlike for the order parameter, the integrand in eq.(3.6) does not simplify in the chiral limit and there is no way to extract the exponent γ directly short of having data for $\rho(\lambda)$ over the entire spectrum, which would be prohibitively expensive. Instead, we prefer to test the validity of the equation of state (1.8), which we rewrite:

$$m = a(\beta_c - \beta) < \bar{\psi}\psi > + b < \bar{\psi}\psi >^\delta . \tag{3.7}$$

As explained in the previous section, this form actually presupposes $\gamma = 1$, so that any evidence that (3.7) is a universal scaling relation will also support this value of γ.

The parameters used were $b = 1.184$, $\delta = 2.19$, and $\beta_c = 0.257$, all in close agreement with the values derived above. To conclude this section, we have succeeded in extracting estimates for the critical exponents δ and β from numerical simulations, and found that they are distinct from their values in mean field theory. We have also found evidence for the applicability of the equation of state derived in the previous section in a rather narrow scaling window around the transition.

4. SCALING WINDOW, ANOMALOUS DIMENSIONS AND NONTRIVIALITY OF THE CONTINUUM LIMIT

In this section we briefly discuss the expected width of the scaling region in quenched QED by a comparison of the results of section 2 with the behavior predicted by a mean field treatment. Only at the NJL point discussed in the introduction do the two coincide. We go on to explain how it may be that even tiny deviations from the pure NJL point, parameterized by a QED action with $\alpha \ll 1$, can result in very different behavior, and the possibility of a non-trivial continuum limit.

Away from the critical line at strong couplings the fermion self-energy is approximately constant. For constant Σ, the gap equation in its integral form (2.1) reads:

$$\Sigma = \frac{\alpha}{4\alpha_c}(1+g)\frac{1}{\Lambda^2} \int_0^{\Lambda^2} dy \frac{y\Sigma}{y + \Sigma^2} . \tag{4.1a}$$

It can be solved easily, giving

$$\Sigma = \frac{\alpha}{4\alpha_c}(1+g)\Sigma\left(1 - \frac{\Sigma^2}{\Lambda^2}\ln(\Lambda^2/\Sigma^2)\right). \tag{4.1b}$$

There is a non-trivial solution for Σ when the effective coupling exceeds some critical value. The line that separates the regions with trivial and non-trivial solutions is given by

$$G = 4 - \frac{\alpha}{\alpha_c} \tag{4.2}$$

For comparison we rewrite the critical line: $G = (1 + \sqrt{1 - \alpha/\alpha_c})^2$ When the gauge coupling vanishes, both curves start at the same point $G \to 4$ (Recall that we defined $g = G(\alpha_c/\alpha) = (G_0\Lambda^2/\pi^2)(\alpha_c/\alpha)$). This is the pure four - fermi theory that exhibits mean field behavior: $\beta = 1/2$, $\nu = 1/2$, $\delta = 3$, $\gamma = 1$. This limit is characterized by the absence of fluctuations and vanishing anomalous dimensions ($\eta = 0$). As α increases, the two curves approach the horizontal axis at different rates and the region between them widens (see figure 6). When $G = 1$, the width of the critical region is $2\alpha_c$, and for $G = 0$, it is $3\alpha_c$.

The narrow scaling region observed in the lattice simulations might be an indication that, by adjusting the gauge coupling only, the induced four - fermi interaction is such that we do indeed cross the critical line at some point on the critical line given by (2.6). The point at which this happens is the one at which the critical exponents match. If we take the numerical estimates for δ and β of section 3 seriously, then this point corresponds to a value of $\alpha/\alpha_c \simeq 0.44$.

It is well known that the pure four - fermi theory has a trivial continuum limit in four dimensions. One would expect that triviality occurs because the interaction is represented by operators of dimension six that decouple at low energies. Typically, suppression factors for these operators would scale as $1/\Lambda^2$, since $1/\Lambda$ is the range of the interaction. However, one can fine tune the coupling constant so that this large screening is removed. In the case of a continuous symmetry, this is guaranteed by the Goldstone theorem. What remains are logarithmic divergences, and they ultimately make the theory trivial. The modes other than the Goldstone bosons (and their chiral partners) are pushed up in energy with masses that scale like the cutoff. Their decoupling is canonical.

To illustrate these points, consider the $\alpha = 0$ limit of the theory. For $G \approx G_c = 4$ equation (4.1) leads to

$$\Sigma \sim \Lambda(G - G_c)^{1/2}. \tag{4.3}$$

Since for $G > G_c$ the theory breaks chiral symmetry spontaneously, the emerging Goldstone bosons couple to the fermions with strength given by the Goldberger - Treiman relation:

$$g_{\pi e^+ e^-} = \frac{\Sigma}{f_\pi}, \tag{4.4}$$

where f_π is the pion decay constant. To estimate its value, we shall use the Pagels - Stokar formula [14]. In general, if the fermion self-energy is allowed to vary with momentum, as will happen when we turn on α, the value of f_π is given by

$$f_\pi^2 \sim \int_0^{\Lambda^2} dx \, x \Sigma(x) \frac{\Sigma(x) - \frac{1}{2}x\Sigma'(x)}{\left(x + \Sigma^2(x)\right)^2} \tag{4.5}$$

For constant Σ the integral in (4.5) has a logarithmic divergence in the ultraviolet, leading to $f_\pi^2 \sim \Sigma^2 \ln(\Lambda/\Sigma)$. Consequently, the pion decouples in the continuum, ie: $g_{\pi e^+ e^-}^2 \sim \frac{1}{\ln(\Lambda/\Sigma)}$, when $(\Lambda \to \infty)$. If, in addition, we recall that the pion radius is $r_\pi \sim 1/f_\pi$, the resulting continuum theory is that of free pointlike mesons.

What happens when $\alpha > 0$? Let us fix G and increase α, and move horizontally on the (α, G) - plane in figure 6. As long as we are in the strong coupling region, an increase of α amounts to an increase in the strength of attraction, thus resulting in a larger dynamical mass, f_π, etc. If we are to keep the low energy physics fixed, i.e. keep the composite Goldstone bosons massless, with a short-ranged interaction, then we must have just enough attraction to compensate the zero - point motion. In this case we follow the mean field curve (4.2a). If, on the other hand, the interaction is operative over a finite range, then massless composites can be formed with less attraction than before because there is less zero - point energy for it to counter. Thus, the true critical curve (4.2b) lies below the mean field one. In this case, the effective strength of the attraction is traded for its finite range. As a consequence, bound states will have finite radii, forcing f_π to be smaller than in the pure four - fermi case.

Therefore, we anticipate that even the tiniest α will change the mean field scenario qualitatively. Technically, this occurs because a gauge interaction softens the ultra-violet asymptotics of the fermion self-energy. According to equation (2.3) the leading term in the expansion of $\Sigma(p)$ at large momenta is

$$\Sigma(p) \sim \mu \left(\frac{\mu}{p}\right)^{1-\sqrt{1-\alpha/\alpha_c}} \tag{4.6}$$

With this ultra-violet behavior for the $\Sigma(p)$, the integral in (4.5) leads to a convergent expression for f_π. We see that in this case the Goldstone bosons do not decouple in the continuum and the theory has a nontrivial scattering matrix etc. At first glance, this is a somewhat surprising result since the anomalous dimension of the four - fermi interaction $(\eta_{(\bar\psi\psi)^2} = 2 - 2\sqrt{1-\alpha/\alpha_c})$ is not large if α is small. These operators remain irrelevant so long as $\alpha < \alpha_c$, and as such should not affect the low energy physics. However, a cooperation between the two interactions occurs in this case. QED at weak couplings cannot break chiral symmetry by itself, but can modify short distance physics in a non-trivial way. On the other hand, the four - fermi interaction can induce spontaneous breaking of chiral symmetry, but can not survive the continuum limit alone. By fine tuning the two couplings α and G in such a way that they obey (2.6), a non-trivial continuum theory emerges.

5. CONCLUSIONS

The present results indicate that the problem of the existence of QED is more difficult than we initially anticipated, the origin of the main difficulty being that the nonasymptotically free nature of the theory modifies short distance physics in a nontrivial way. The most significant consequence of this are the large anomalous dimensions of higher dimensional operators and, therefore, the amplification of short distance physics. This means that we should interpret the lattice results with a grain of salt. It is easy to understand this warning remark if we recall that the lattice action is in fact the continuum action plus irrelevant operators. If some of these irrelevant operators become renormalizable by virtue of large anomalous dimensions, then the emerging critical theory might have little to do with the original one. This being the case with QED, it is not very likely that, starting with the simple lattice action at the fixed point, we will recover the theory of electrons and photons at low energies. In fact the renormalization group trajectories i.e. the lines of fixed physics will look different depending on what we keep fixed at low energies.

To make the matter even more complicated, we mention that even in the noncompact formulation the effective lattice monopoles appear because the fermions couple only to the compact structures of gauge fields. Their effect is amplified at low energies and they can drive the phase transition before the strong vector forces become operative. This is not happening in the quenched theory: the monopole transition occurs at somewhat stronger coupling than chiral symmetry breaking [15]. However, in full theory, effect of monopoles increases with N_f, the number of light dynamical fermions. The phase transition is completely driven by lattice monopoles and we never get to see the true strong coupling regime of QED [16]. Before any attempt is made to tie the lattice results with real electrons and photons, we need to understand which lattice artifacts will be amplified and which won't. Different lattice regularized actions give different critical behavior and, therefore, different low energy limits as demonstrated in reference [15].

In this study we have seen that once pure QED is extended by inclusion of a chirally-symmetric four - fermi interaction, then the essential singularity behavior originally expected exactly at the MBLL point may be modified to a more conventional power-law form along a whole line of fixed points in the (α, G) plane. We have argued that it is natural for the lattice-regulated form of QED to incorporate effective four - fermi interactions. For example, at strong coupling ($\beta = 0$), the link variables of U(1) lattice gauge theory can be integrated over exactly leading to an explicit attractive four -fermi term in the fermion action. Contrary to perturbative intuition, the new terms survive renormalization group flow and have important effects in the vicinity of the transition. We have derived the critical exponents along the critical line as functions of α/α_c: a particularly satisfying feature is that they satsify hyperscaling, which one would hope for if there is indeed an underlying interacting field theory.

How does this result affect the numerical search for a non-trivial fixed point in lattice QED? Now that power-law scaling is a possibility, the crucial issue is to find deviations from mean field behavior in the critical exponents. In section 3 we analyzed our numerical data in terms of power-law behavior, and indeed found deviations from mean field which we feel are robust. Moreover the observed values of the exponents δ and β are consistent with the relation (2.9) derived in section 2. Thus, there seems to be reasonable evidence that the chiral transition in lattice QED is actually described by a point on the critical line in the *interior* of the phase diagram figure 1. Of course, we cannot as yet eliminate other possibilities, but we find the one presented here more plausible than the supposition of pure mean field theory [9]. In fact, recent small mass simulations [17] indicate unambiguously that the asumption of reference [9] that $\beta_c = 0.250$ is the critical point for the quenched theory is unjustified and, consequently, mean - field exponents are definitely ruled out.

As emphasized in section 4, the region of scaling is very small near the NJL point, and this may be precisely why numerical studies have proved difficult. We are also aware of the apparent arbitrariness of the value 2.2 of the critical exponent δ. If there is really a line of fixed points in quenched QED then other lattice regularizations or any other "small" change in the problem may lead to a different exponent. To make further progress an improved lattice formulation may be desirable. One might think of constructing a version which better fits the continuum formulation by suppressing lattice artifacts. A step in this direction has already been made [15]; in this study a "linearized" version of the fermi – gauge action was implemented. The results found are that the chiral transition persists, occuring at a stronger value of the bare fine structure constant, and that the exponent $\delta \simeq 2.5$, closer to the mean field value. We do not find it easy to account for this in the present framework.

A very interesting possibility would be that the present results $\delta \simeq 2.2$, $\beta \simeq 0.8$ are the values of the exponents at a new fixed point of the model not predicted by the ladder approach. Further work will be needed to clarify this issue.

ACKNOWLEDGEMENTS

This work was done in collaboration with Elbio Dagotto, Simon Hands and John Kogut. The reviewed results were reported in a preprint *Equation of State and Critical Exponents in Qenched Strongly Coupled QED* , AZPH-TH/90-27. The author was supported by DOE contract DE-FG02-85ER40213. The computer simulation data reviewed here were obtained at the Pittsburgh Supercomputer Center, the National Center for Supercomputing Applications and the National Magnetic Fusion Energy Computer Center.

REFERENCES

[1] P.I. Fomin, V.P. Gusynin, V.A. Miransky and Yu.A. Sitenko, Riv. Nuovo Cimento **6** (1983) 1; V. A. Miransky, Nuovo Cimento **90A** (1985) 149.

[2] See for example *Proceedings of the 1988 Inyternational Workshop on New Trends in Strongly Coupled Gauge Theories, Nagoya*, eds. M. Bando, T. Muta and K. Yamawaki (World Scientific Pub. Co., Singapore, 1989).

[3] T. Appelquist and J. Carrazone, Phys. Rev. **D11** (1975) 2856.

[4] E. Dagotto, A. Kocić and J.B. Kogut, Phys. Rev. Lett. **62** (1988) 1001.

[5] E. Dagotto, A. Kocić and J.B. Kogut, Phys. Lett. **237B** (1990) 268.

[6] C.N. Leung, S.T. Love and W.A. Bardeen, Nucl. Phys. **B273** (1986) 649; Nucl. Phys. **B323** (1989) 493.

[7] D.J. Amit, *Field Theory, the Renormalization Group, and Critical Phenomena*, (McGraw-Hill, 1978).

[8] S.J. Hands, J.B. Kogut and E. Dagotto, Nucl. Phys. **B333** (1990) 551.

[9] M. Göckeler, R. Horsley, E. Laermann, P. Rakow, G. Schierholz, R. Sommer and U.-J. Wiese, Nucl. Phys. **B334** (1990) 527.

[10] S.P. Booth, R.D. Kenway and B.J. Pendleton, Phys. Lett. **228B** (1989) 115.

[11] E. Dagotto, A. Kocić and J.B. Kogut, Phys. Rev. Lett. **61** (1988) 2416; Nucl. Phys. **B317** (1989) 253.

[12] K.-I. Kondo, H. Mino and K. Yamawaki, Phys. Rev. **D39** (1989) 2430.

[13] E. Dagotto, A. Kocić and J.B. Kogut, Nucl. Phys. **B331** (1990) 500.

[14] H. Pagels and S. Stokar, Phys. Rev. **D20** (1979) 2947.

[15] S.J. Hands, J.B. Kogut and J.H. Sloan, Illinois preprint ILL-TH-90-#12 (1990).

[16] J. Kogut, talk presented at the 1990 International Workshop on Strong Coupling Gauge Theories and Beyond, Nagoya, Japan, July 1990.

[17] E. Dagotto, talk presented at the 1990 International Workshop on Strong Coupling Gauge Theories and Beyond, Nagoya, Japan, July 1990.

NON-PERTURBATIVE EXTERNAL FIELD EFFECTS IN QED*

Janos Polonyi [+] [†]

Center for Theoretical Physics
Laboratory for Nuclear Science and Department of Physics
Massachusetts Institute of Technology
Cambridge, Massachusetts 02139 U.S.A.

INTRODUCTION

QED is the most thoroughly studied quantum field theory. The predictions up to the order $O(e^8)$ perturbation expansion have been confronted with the experiments with success[1] and the manner how ultraviolet divergences are handled is supported by the phenomenology related to the axial anomaly[2]. But QED cannot be reduced to a set of the Feynman diagrams, there are phenomena which are inherently non-perturbative in their nature. The well known examples are the bound state formation and the strong interactions at short distances. They can be handled by resumming an infinite subset of Feynman diagrams by methods such as the Bethe-Salpeter equation[3] and the renormalization group[4].

Another possibility to study the underlying rich structure of the Quantum Field Theory is the introduction of some physical external parameter. The goal of this talk is to survey the possibilities of non-perturbative phenomena related to external field.

THE RENORMALIZATION GROUP

Consider the space of relevant coupling constants for QED. Two axis are well known, they correspond to the electric charge e and the electron mass m. The rest is not entirely mapped yet, they should include four fermion couplings and constants controlling the non-linearity of the Maxwell equations. These non-orthodox coupling constants are suppressed in Fig. 1.

---

* This work is supported in part by funds provided by the U. S. Department of Energy (D.O.E.) under contract #DE-AC02-76ER03069.

[+] On leave of absence from CRIP, Budapest, HUNGARY

[†] Supported in part by the Alfred P. Sloan Foundation and the NSF-PYI grant PHY-8958079.

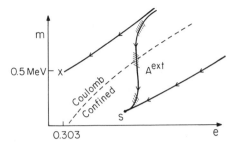

Fig. 1. The qualitative features of the renormalization group flow for QED.

The point X stands for QED as we know it. The coupling constant e is determined by the unpolarized electron-electron elastic scattering

$$d\sigma_{\mathrm{exp}}(\mathbf{q}_1, \mathbf{q}_2) = e^4 F(s, t, u, m) \tag{1}$$

Here the left hand side contains the cross section which is provided by the experiments performed with electrons having momenta $\mathbf{q}_1, \mathbf{q}_2$. The right hand side is the leading order expression[5] in perturbation expansion and defines e. The electron mass m is fixed by the equation

$$S_{\mathrm{exp}}^{-1}(\mathbf{q}) = i\mathbf{q}\gamma + i\sqrt{\mathbf{q}^2 + m^2}\gamma_0 - m \tag{2}$$

where the left hand side is the value of the inverse electron propagator at some momentum $\mathbf{q}$ extracted from Coulomb scattering experiments. The conditions (1) and (2) are called renormalization conditions imposed at scale q_{ren} when $q_1 = q_2 = q = q_{\mathrm{ren}}$ and they are used to fix the renormalized coupling constants $e(q_{\mathrm{ren}}), m(q_{\mathrm{ren}})$ corresponding to this scale. This scale dependence is described by the renormalization group which gives the appropriate choice of the renormalized coupling constants to reproduce the physics at the scale q_{ren}. Due to the non-asymptotically free behavior of QED the renormalization conditions are best to impose in the infrared, $q_{\mathrm{ren}} \sim 0$. The success of the perturbation expansion suggests the absence of any other coupling constant under the usual circumstances.

The assemble of points in the coupling constant space which belongs to the same physical theory is called the renormalized trajectory and its points are parametrized by q_{ren}. The arrows indicate the infrared direction in Fig. 1. The point denoted by X is an infrared fixed point since it is reached as $q_{\mathrm{ren}} \to 0$. It is worthwhile to mention that this differs from the usual terminology of statistical mechanics. There the bare coupling constants are used and their dimension is removed by the lattice spacing i.e. cutoff. Thus they would consider $m_0 = \frac{m_{\mathrm{bare}}}{\Lambda}$ instead of m where Λ is the cutoff. Furthermore it is the lattice spacing rather than the renormalization scale what is changed. It is important to keep in mind that the bare coupling constants can be considered as renormalized coupling constants defined at the cutoff. Thus the bare mass is interpreted as the physical mass scale when the cutoff is sent into the infrared during the blocking procedure of statistical mechanics. The fixed point where m_0 is independent from the cutoff can have either $m = 0$ or $m = \infty$.

Suppose that the theory is investigated at high energies and $q_{\mathrm{ren}} \to \infty$. As long as the renormalization conditions can be fulfilled the renormalized trajectory may tend either to a finite point or to an infinite point of the plane (e, m). This point or asymptota is the ultraviolet fixed point. It is not known whether QED has such an ultraviolet fixed point i.e. the renormalization conditions can be satisfied by a finite choice of the coupling constants as $q_{ren} \rightarrow \infty$. This is due to the non-asymptotically free behavior of QED which generates large coupling constants at short distance and invalidates perturbation expansion. If there is no ultraviolet fixed point then QED is not renormalizable. The renormalized trajectory ends in this case at a finite scale q_{max} beyond which the cutoff cannot be advanced without violating some of the low energy content of the theory. Then QED becomes a low energy effective theory which can be used only up to the momentum scale q_{max}. Fortunately the existence of the ultraviolet fixed point in QED has limited relevance for realistic applications. In fact, the scale where the naive Landau pole appears to invalidate the perturbation expansion is $q_{crit} = m e^{\frac{6\pi^2}{e^2}} \sim m 10^{280}$ which is well beyond the onset of the electro-weak theory $q_{ew} = O(100)$ GeV. It will be sufficient for our purposes to trace the renormalization group trajectory up to a high enough momentum scale which is less than q_{ew}.

Imagine a world where the physical experiments lead to the electric charge $e = O(1)$ in the infrared. By pursuing the high energy behavior we would follow another renormalized trajectory which starts at the point S in Fig. 1. It is the subject of a long series of papers[6,7] that QED may have an ultraviolet fixed point in this case, for the latest summary of the situation see Elbio Dagotto's talk at this School. Naturally the possible existence of this fixed point has nothing to do with the predictions of QED in our world. If exists, it represents the only non-asymptotically free ultraviolet fixed point known so far in Quantum Field Theory and challenges theoretical physicists.

Even if the strong coupling QED has no ultraviolet fixed point it is a non-trivial effective theory. Its properties are partly known from numerical studies in lattice gauge theory[7,8]: The potential between static charges is linear and the chiral symmetry is broken dynamically for massless electrons. Furthermore it develops additional relevant coupling constants when the ultraviolet scaling dimension of an operator becomes less or equal to four at the infrared fixed point due to the anomalous dimension contributions. This is because the naive power counting to classify the ultraviolet divergences allows such operators in the theory. As the value of the coupling constant e is lowered at the infrared fixed point the usual QED with Coulomb law is recovered at some value $e_{crit}(m)$. We can assume that all other unusual coupling constants vanish at this point. The line $e_{crit}(m)$ is the boundary of the Coulomb and the confining phases on the plane (e, m).

THE EXTERNAL FIELD PROBLEM

Suppose that we create an external field A_μ^{ext} and repeat the renormalization group analysis. The physical content of the theory is specified by the bare coupling constants at a large enough value of the cutoff and should be the same in the presence of the external field. In order to keep the physics unchanged by A_μ^{ext} we have to assume the absence of any impact of the external field on the dynamics at scales beyond the cutoff. In other words, the external field cannot change the ultraviolet fixed point. But the renormalized coupling constants which are the difference of the bare one and the coefficients of the counterterms depend on A_μ^{ext} in general. In fact, the external field induces a change in the left hand side of (1) and (2). By changing the scale of the renormalization conditions we form another renormalized trajectory with the given A_μ^{ext} which must approach the original fixed point as $q_{ren} \rightarrow \infty$.

We now compare two different QED models. One of them has a non-vanishing external field and for the other $A_\mu^{ext} = 0$. The bare coupling constants are fine tuned in such a manner that both have the same renormalized coupling constants defined at some infrared point. The agreement of the renormalized coupling constants reflects some kind of similarity of these models despite of their different physical environment.

The similarities are obvious. Some qualitative features of the dynamics close to the renormalization scale should be similar owing to the same renormalized coupling strength. If the renormalized charges belong to the confining phase in the absence of the external field then the other model with external field but the same coupling constant should be confining as well. But the external field distinguishes the two cases which can be made explicit by tracing the renormalized trajectories back to the ultraviolet regime. If the external field is weak then both model should belong to the same ultraviolet fixed point. But it may happen for strong enough external field that the model with and without the external field belongs to the Coulomb and the confining ultraviolet fixed point, respectively as depicted in Fig. 1. If this case can be realized then the model with external field is rather peculiar: It confines electrons in the infrared and exhibits Coulomb-like behavior at shorter length scales. The strength of the of the effects due to the external field is measured by the rate how the renormalized trajectories with and without external field depart as the renormalization scale is changed. In addition violent external field dependence is expected when the renormalized trajectory crosses the phase boundary $e_{\mathrm{crit}}(m)$. The shaded regions in Fig. 1 are where non-perturbative external field effects should be observed.

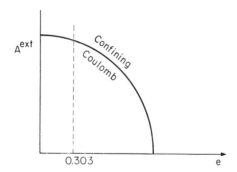

Fig. 2.   A possible phase diagram for QED.

Another way to look into this issue is to monitor the change of the renormalized coupling constant at an infrared scale as the function of the external field. Suppose that the coupling constant becomes larger with the external field. Then one expects to reach the confining phase boundary at large enough field values and to form a phase structure as shown in Fig. 2. What is important for us from such a phase diagram is the value of the critical external field at the physical coupling strength $e_{\mathrm{phys}} = \sqrt{\frac{4\pi}{137}} \sim 0.303$.

Note that this scenario can be found in any model with qualitatively different "phases". Furthermore the existence of the ultraviolet fixed point is not essential, we may use the coupling constants at a large but finite value of the ultraviolet cutoff as long as the effects of the external field are negligible at that scale.

Another system with similar behavior is scalar electrodynamics in the presence of homogenous external magnetic field. The system of charged scalar particles has two phases, with normal or superconducting ground state. In the latter case the global gauge invariance is broken spontaneously and the Higgs mechanism gives rise to the Meissner effect. As the external magnetic field is increased in the superconducting vacuum the normal ground state will be energetically favorable and transition to the normal phase occurs. The qualitative behavior shown in Fig. 3. can be obtained from Fig. 1 by identifying the normal and the superconducting ultraviolet fixed point with confining and the Coulomb-type fixed points, respectively.

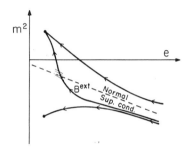

Fig. 3.   The qualitative features of the renormalization group flow for scalar QED.

We may distinguish three possibilities depending on the scale where the external field modifies the trajectories significantly or drives them through a phase boundary. Such a non-perturbative phenomenon may take place at (i) $q_{\text{ren}} = 0$, (ii) $q_{\text{ren}} = $ finite and (iii) $q_{\text{ren}} = \infty$. The scenario (ii) appears to be the most reasonable. It is realized in the case of scalar electrodynamics where the magnitude of the external field is the natural source of the non-perturbative scale. The case (i) corresponds to phase transitions triggered by the external field.

Return now to the case (ii) as depicted in Fig. 1. Suppose that we gradually change the external field so as to realize scenario (iii). Then the dashed line representing the phase boundary must approach the ultraviolet fixed point. The difficult point of this scenario is clear: The physical external field with finite gradient and magnitude is supposed to modify the dynamics at $q = \infty$. An important realization of this scenario is when the external field gives non-vanishing

$$Q_A = \frac{1}{8\pi^2} \int d^4x \, \epsilon^{\mu\nu\rho\sigma} F_{\mu\nu}^{\text{ext}} F_{\rho\sigma}^{\text{ext}} \tag{3}$$

In fact, the axial charge is not conserved for the Dirac sea in the presence of such external field and the axial anomaly is understood[9] in terms of an induced compensating axial current at the "bottom" of the Dirac sea i.e. at $q = \infty$.

An interesting conjecture was put forward in Ref. [10]. In trying to describe the unusual e^+e^- peaks observed at the GSI[11] as QED effects the main difficulty is the explanation of the relatively large energy shifts in the spectrum compared to the usual scale of Rydberg. It has been suggested that the strong field in the vicinity of compound nucleus with sufficiently high Z might alter the renormalization of QED. If the change due to the external field takes place in the ultraviolet then its effect might become large enough until we integrate the renormalization group equations down to the observed infrared region. The unusually large energy shift in the infrared should come from the unusually high momentum scale where the external field begins to modify the dynamics seriously. Thus this suggestion belongs effectively to the case (iii). It would be interesting to identify some physically acceptable external field in this class.

An alternative scenario will be suggested below. The infrared singularities in the case (i) might provide the large factors and energy scales can be understood by staying in the conventional scale region.

The supercritical vacuum expected for large Z nucleus[12] belongs to the case (ii). This is because the effects related to the supercriticality emerge mainly at the length scale of the diving bound state. One might speculate whether the volume where the interesting effects are observed can be increased as to make the non-perturbative phenomena more important. One ultimately arrives to the question whether there are physical external fields in the class (i). The straightforward extension of the supercritical vacuum phenomenon is to consider external potential which is supercritical over a larger volume. Thus we turn to constant external A_0 field which is the subject of the rest of this talk[13].

THE $eA_0 = \mu$ EXTERNAL FIELD PROBLEM

In the discussion of the constant external A_0 problem first one has to clarify whether this external field cannot be gauged away and why the mere shift of the energy can have any significant effect. The answer to the first question lies in the Gauss' Law which is the equation of motion corresponding to A_0. In eliminating A_0 in the temporal gauge formalism we have to impose Gauss' Law externally[14] which amounts to the introduction of a variable which appears formally in the same manner as a static A_0. In other words, the static component of A_0 cannot be eliminated by gauge transformation. The second objection can be answered in a less formal way. The hamiltonian of the external field problem is

$$H_\mu = H_0 - \mu Q \tag{4}$$

where Q is the charge operator. The Hilbert space of the temporal gauge formalism consists of the direct sum of superselection sectors with given charge Q. These sectors are not connected by gauge invariant operators and the physical excitations build on the ground state of (4) give one such a class. The role of μ is to select this class and to constraint the contributions to observables onto this class by the help of gauge invariance.

Since all what μ does is to select a charge sector the μ dependence of the observables is not analytical. To see this we have to place the system into a finite quantization box with size L where the energy spectrum becomes discrete and the typical level spacing is $O(\frac{1}{L})$. In order to change the selected sector μ has to produce a level crossing for the ground state which requires a shift $\Delta\mu = O(\frac{1}{L})$. Thus the μ dependence is step function-like with step size $O(\frac{1}{L})$. This non-analycity creates serious finite size effects in lattice gauge theory[15] and renders Schwinger's famous computation of the polarization of the Dirac sea by an external field[16] unreliable when chemical potential is present. This latter can be understood by recalling that the external field dependence of an expectation value $\langle 0 \rangle_A$ in the presence of the external field A is obtained by integrating the functional derivative $\langle 0 \rangle_A = \int dA \frac{\delta \langle 0 \rangle_A}{\delta A}$ in this computation. The derivative of a step function is singular and its integral requires special care[17].

PERTURBATION EXPANSION

The perturbation expansion in the presence of the external potential $eA_0 = \mu$ is organized in such a manner that the additional electrons in the ground state compared to the Dirac sea are taken into account in the free hamiltonian. The free propagator with respect of the Fermi sphere

$$
\begin{aligned}
iG_0(x-y) &= \langle 0_F|T[\psi(x)\bar\psi(y)]|0_F\rangle \\
&= \Theta(x^0 - y^0)\langle 0_F|\psi(x)\bar\psi(y)|0_F\rangle - \Theta(y^0 - x^0)\langle 0_F|\bar\psi(y)\psi(x)|0_F\rangle
\end{aligned}
\tag{5}
$$

is computed by using the expansion

$$
\psi(x) = \int_{k>k_F} u(\mathbf{k},\sigma)e^{-ikx}a_{\mathbf{k},\sigma} + \int_{0<k<k_F} u(\mathbf{k},\sigma)e^{ikx}\bar b^\dagger_{-\mathbf{k},\sigma} + \int_k v(\mathbf{k},\sigma)e^{ikx}b^\dagger_{-\mathbf{k},\sigma}
\tag{6}
$$

Here $kx = k^0x^0 - \mathbf{kx}$, $(\slashed{k}-m)u = (\slashed{k}+m)v = 0$, the Fermi sphere is defined as

$$
a_{\mathbf{k},\sigma}|0_F> = b_{\mathbf{k},\sigma}|0_F> = \bar b_{\mathbf{k},\sigma}|0_F> = 0
\tag{7}
$$

and the integration symbol stands for[18]

$$
\int_k = \int \frac{d\mathbf{k}}{(2\pi)^3}\frac{m}{\omega_k}
\tag{8}
$$

with $k^0 = \omega_k = \sqrt{m^2 + k^2}$ The non-vanishing contributions are

$$
\langle 0_F|\psi(x)\bar\psi(y)|0_F\rangle = \left\langle \int_{k>k_F} u_{\mathbf{k},\sigma}e^{-ikx}a_{\mathbf{k},\sigma} \int_{q>k_F} \bar u_{\mathbf{q},\kappa}e^{iqy}a^\dagger_{\mathbf{q},\kappa}\right\rangle
\tag{9}
$$

and

$$
\begin{aligned}
\langle 0_F|\bar\psi(y)\psi(x)|0_F\rangle &= \left\langle \int_{0<k<k_F} \bar u_{\mathbf{k},\sigma}e^{-iky}\bar b_{-\mathbf{k},\sigma} \int_{0<q<k_F} u_{\mathbf{q},\kappa}e^{iqx}\bar b^\dagger_{-\mathbf{q},\kappa}\right\rangle \\
&+ \left\langle \int_k \bar v_{\mathbf{k},\sigma}e^{-iky}b_{-\mathbf{k},\sigma} \int_q v_{\mathbf{q},\kappa}e^{iqx}b^\dagger_{-\mathbf{q},\kappa}\right\rangle
\end{aligned}
\tag{10}
$$

The projection operators

$$
\Sigma_{\sigma=1,2} u_{\mathbf{k},\sigma}\bar u_{\mathbf{k},\sigma} = \frac{m+\slashed{k}}{2m}
\tag{11}
$$

and

$$
\Sigma_{\sigma=1,2} v_{\mathbf{k},\sigma}\bar v_{\mathbf{k},\sigma} = \frac{m-\slashed{k}}{2m}
\tag{12}
$$

are used to write

$$
\begin{aligned}
iG_0(x-y) &= \Theta(x^0-y^0)\int_{k>k_F} \frac{d\mathbf{k}}{(2\pi)^3}\frac{\slashed{k}+m}{2\omega_k}e^{-ik(x-y)} \\
&-\Theta(y^0-x^0)\{\int_{0<k<k_F}\frac{d\mathbf{k}}{(2\pi)^3}\frac{\slashed{k}+m}{2\omega_k}e^{ik(x-y)} + \int \frac{d\mathbf{k}}{(2\pi)^3}\frac{m-\slashed{k}}{2\omega_k}e^{ik(x-y)}\}
\end{aligned}
\tag{13}
$$

It is a matter of few lines of algebra to check that (13) can be written as

$$
\int \frac{dk^0}{2\pi}\int \frac{d\mathbf{k}}{(2\pi)^3}\frac{m+\slashed{k}}{(k^0-\omega_k+i\epsilon_{k^0})(k^0+\omega_k-i\epsilon_{k^0})}e^{-ik(x-y)}
\tag{14}
$$

247

if the transformation $\mathbf{k} \to -\mathbf{k}$ is performed in the last integral of (13). The required pole structure of the integrand in (14) is displayed in Fig. 4. The states up to the energy μ are filled up in the Fermi sphere and the corresponding poles have positive imaginary part. The vacant states with energy larger than μ give poles on the lower plane. The comparison with the usual case with $\mu = 0$ shows that ϵ_{k^0} should be negative for $-\mu < k^0 < 0$ to flip the sign of the imaginary part and positive otherwise. We shall take the choice

$$\epsilon_{k^0} = \epsilon k^0 (k^0 + \mu) \tag{15}$$

The final expression for the propagator is[19]

$$G_0(x) = \int \frac{dk}{(2\pi)^4} \frac{\not{k} + \mu\gamma^0 + m}{(k^0 + \mu)^2 - \mathbf{k}^2 - m^2 + i\epsilon_{k^0}} e^{-ikx} \tag{16}$$

Where the energy k^0 is measured from the Fermi surface. One can see that the shift of the energy $k^0 \to k^0 + \mu$ is trivial and the actual choice of the vacuum is hidden in (15).

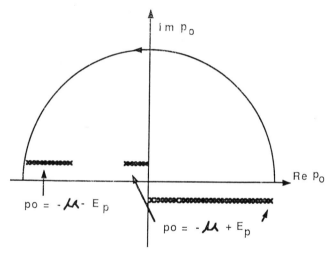

Fig. 4. The poles of the propagator on the complex energy plane for the Fermi sphere.

The usual procedure to evaluate the loop integrals is to perform Wick rotation $k^0 \to ik^0$. The Euclidean propagator is

$$G_{E0}(x) = -\int \frac{dk}{(2\pi)^4} \frac{\not{k} + i\mu\gamma^0 + m}{(k^0 + i\mu)^2 + \mathbf{k}^2 + m^2} e^{-ikx} \tag{17}$$

The important simplification in computing the loop integrals with finite μ is that integration along the line $k^0 = ik_E^0$ can be written as the integration along the line $k^0 = ik_E^0 - \mu$ plus the sum of residues between $k^0 = -\mu$ and $k^0 = 0$. The first piece is the usual contribution at $\mu = 0$ and the second one comprises the chemical potential dependent contributions. Note that chemical potential dependent pieces are ultraviolet finite and the counterterms can be chosen to be μ dependent.

One can shown that the Green functions are chemical potential independent for $-m < \mu < m$. When the Fermi surface is outside of the gap then non-perturbative effects are expected. This is because there is no gap for particle-hole excitations and the vacuum will strongly be mixed with such states. The characteristic length scale for such excitation is $k_F^{-1} = \frac{1}{\sqrt{\mu^2 - m^2}}$ which diverges as the non-relativistic chemical potential $\mu_{\mathrm{nr}} = \mu - m$ and the density tends to zero. Such a mixing can be taken into account in the free hamiltonian by computing the photon self energy and including it into the photon propagator. The result is the Thomas-Fermi screening of the electric field by generating the mass $m_{TF}^2 = e^2 \frac{k_F m}{\pi^2}$ for A_0. The screening length $\frac{1}{m_{TF}}$ diverges in the dilute limit and causes the complete breakdown of the perturbation expansion for large r_s where r_s is the average separation of the electrons in units of Bohr radius a_B and is related to the density as $\frac{4\pi}{3}(r_s a_B)^3 = \frac{1}{n}$.

The one loop radiative correction to the vertex function

$$\Gamma_\mu^{(1)}(p, p') = \int \frac{d^4 k}{(2\pi)^4}$$

$$\times (-ie\gamma_\alpha) \frac{i}{\not{p} + \not{k} + \mu\gamma_0 - m} (-ie\gamma_\mu) \frac{i}{\not{p}' + \not{k} + \mu\gamma_0 - m} (-ie\gamma_\beta) \frac{-ig^{\alpha\beta}}{k^2 - m_\alpha^2}$$

$$(18)$$

where $m_\nu^2 = m_{TF}^2 \delta_{0,\nu}$ has two interesting contributions[13]. To see them we sandwich the vertex function between the free electron spinors and consider the correction to the electric charge in Coulomb scattering

$$\frac{\delta^{(1)} e(\mathbf{q})}{e} = \frac{i}{e} \bar{u}(\mathbf{p}, +) \Gamma^{(1)}(p, p') u(\mathbf{p}', +) \tag{19}$$

where $\mathbf{q} = \mathbf{p} - \mathbf{p}'$ and $p^0 = p'^0$. The electric contribution ($\nu = 0$) to the chemical potential dependent contour integral is

$$\frac{\delta_E^{(1)} e(\mathbf{q})}{e} = e^2 \int \frac{d^3 k}{(2\pi)^3} \frac{\theta(\mu - E_\mathbf{k})}{2m}$$

$$\times \left\{ \frac{4m^2}{[\mathbf{q}^2 + 2\mathbf{k}\cdot\mathbf{q}][m_{TF}^2 + (\mathbf{k} - \mathbf{p})^2]} + \frac{4m^2}{[\mathbf{q}^2 - 2\mathbf{k}\cdot\mathbf{q}][m_{TF}^2 + (\mathbf{k} - \mathbf{p}')^2]} \right\}$$

$$(20)$$

which can be written as

$$\frac{\delta_E^{(1)} e(\mathbf{q})}{e} \sim \frac{2\pi^2}{k_F} \int \frac{d^3 k}{(2\pi)^3} \theta(\mu - E_\mathbf{k}) \left\{ \frac{1}{\mathbf{q}^2 + 2\mathbf{k}\cdot\mathbf{q}} + \frac{1}{\mathbf{q}^2 - 2\mathbf{k}\cdot\mathbf{q}} \right\} \tag{21}$$

if $k_F^2, p^2 < m_{TF}^2 = 0.66 r_s k_F^2$ which holds for $r_s > 1.5$. Note the absence of the electric charge in (21), it is cancelled by the screening mass coming from the expansion of the photon propagator. (21) is close to 0.5 for $p < 2k_F$ and causes infinitely many Feynman graphs to contribute to the vertex function in $O(e)$. The rainbow graph where there are N parallel soft photon lines connecting the electron lines gives approximately 2^{-N} close to the Fermi surface and yields $e \to 2e$ after resummation. But other graphs in the same order remain to be included making our understanding of the homogenous electron gas rather vague from $1.5 \sim r_s$ up to the region where perturbation expansion finally breaks down.

The other interesting contribution corresponds to the transverse photon propagator in (18). After some algebra one finds

$$\frac{\delta_M^{(1)} e(\mathbf{q})}{e} = e^2 \int \frac{d^3 k}{(2\pi)^3} \frac{\theta(\mu - E_\mathbf{k})}{2E_\mathbf{k}} \times$$

$$\times \left\{ \frac{2(\mathbf{p}' - \mathbf{p})^2 - (\mathbf{k} + \mathbf{p}')^2 - 2(\mathbf{k} - \mathbf{p})^2}{(\mathbf{q}^2 + 2\mathbf{k}\cdot\mathbf{q})(\mathbf{k} - \mathbf{p})^2} + \frac{2(\mathbf{p}' - \mathbf{p})^2 - (\mathbf{k} + \mathbf{p})^2 - 2(\mathbf{k} - \mathbf{p}')^2}{(\mathbf{q}^2 - 2\mathbf{k}\cdot\mathbf{q})(\mathbf{k} - \mathbf{p}')^2} \right\}$$

$$(22)$$

which can be evaluated for $\mathbf{p} \rightarrow \mathbf{p}'$

$$
\begin{aligned}
\frac{\delta_M(\mathbf{q})}{e} &= -\frac{e^2}{8\pi^2 m} \int_0^{k_F} dk \left\{ \frac{3k}{2p} \ln \left| \frac{k-p}{k+p} \right| + \frac{4kp}{k^2 - p^2} \ln \left| \frac{k-p}{k+p} \right| \right\} \\
&= \frac{m_{TF}^2}{8m^2} \frac{p}{k_F} \ln^2 \left| \frac{2k_F}{k_F - p} \right| + \frac{m_{TF}^2}{8m^2} \mathcal{O}(1)
\end{aligned} \tag{23}
$$

The relevance of this result will be discussed in the next Section.

THE KLN THEOREM

The Rayleigh-Schrödinger perturbation expansion becomes divergent term by term when the free hamiltonian has degenerate eigenstates connected by the perturbation. The problem of the degeneracy singularities can be rephrased by pointing out that the approximate diagonalization of the hamiltonian which is provided by the expansion has non-analytic coupling constant dependence in the degenerate subspaces. To avoid these singularities we have to turn to the degenerate perturbation expansion where the perturbation is diagonalized in each order within the degenerate subspaces.

When the spectrum is continuous then the dangerous subspace $D_\epsilon(E)$ consists of states whose non-perturbed energy lies in the interval $E - \epsilon < E_0 < E + \epsilon$. The usual mass-shall singularities in theories with massless particles arise from the small energy denominator when the intermediate states from D_ϵ are considered. Thus one has to diagonalize the perturbation within these subspaces in each order. This diagonalization is rather complicated and the explicit expressions are usually not available. The crucial observation of Lee and Nauenberg[20] was that this diagonalization can be avoided for certain observables. They considered the expression for cross section and pointed out that it can be written in terms of the trace of the time evolution operator as long as the theory is unitary. The trace is over the subspace of the initial and final asymptotic states. Such an expression is formally independent of the choice of the basis vectors within the degenerate subspaces as long as the traces extend over the whole degenerate subspace. Thus the inclusive cross sections where the initial and final states are summed over degenerate subspaces can formally be written in terms of the basis vectors where the perturbation is diagonal. The degeneracy singularities must cancel for such observables. This is the generalization of the results obtained earlier by Bloch and Nordsieck[21] and Kinoshita[22]. Another virtue of Lee and Nauenberg's argument that it is verified not only formally but order by order in perturbation expansion. Thus any degeneracy singularity observed in a perturbative scheme for a properly defined cross section indicates the breakdown of unitarity. The Feynman graphs which complete the cancellation are usually generated by bound state or collective mode formation.

The singularity (23) at the Fermi surface is a typical degeneracy singularity due to the absence of the mass gap in the magnetic propagator. When radiative corrections are computed for Coulomb scattering cross section in electron plasma then the singularity (23) remains uncancelled at $O(e^4)$. There must be some non-perturbative mechanism to generate new degrees of freedom whose contributions cancel this divergence for the theory be unitary. The only stable mechanism to screen the magnetic field is the supercurrent. Thus the singularity (23) is considered as an indication of the superconducting ground state of homogenous electron gas. The KLN theorem can be made shaper by discarding those couplings from the interaction hamiltonian which are not contributing to the degeneracy singularities if the hamiltonian remains Hermitian. The non-perturbative mechanism evoked before must be present in a truncated QED without Coulomb interaction and having soft photon vertices only. It is conjectured that the soft magnetic interactions form Cooper-pairs which are not destabilized by the inclusion of the screened Coulomb repulsion. The transverse photons are the analogue of phonons when the homogenous gas is compared with the conductance band of the solid states.

Unfortunately no physical systems can be used to prove our disprove our conjecture. The infrared singularities are modified in a fundamental manner in the presence of any inhomogeneities which are necessary to keep a realistic electron gas together at densities where the perturbation expansion has a chance to work.

Finally we note that (23) contains no approximation and remains valid as long as perturbation expansion is reliable. The infrared divergence due to the unscreened magnetic field makes the magnetic interactions important even for the non-relativistic regime. The usual model hamiltonian for non-relativistic electrons where the instantaneous Coulomb interaction is retained only

$$
H_{\mathrm{nr}} = \int \frac{d\mathbf{p}}{(2\pi)^3} \frac{\mathbf{p}^2}{2m} \psi^\dagger(\mathbf{p})\psi(\mathbf{p})
$$
$$
- \frac{e^2}{2} \int \frac{d\mathbf{p}}{(2\pi)^3} \frac{d\mathbf{q}}{(2\pi)^3} \frac{d\mathbf{k}}{(2\pi)^3} \frac{4\pi}{k^2} \psi^\dagger(\mathbf{p})\psi^\dagger(\mathbf{q})\psi(\mathbf{p}+\mathbf{k})\psi(\mathbf{q}-\mathbf{k})
$$

(24)

has a rather different ground state than the complete QED.

DILUTE ELECTRON GAS

The results of the previous Sections enable us to make some qualitative statements about the renormalization flow diagram in the presence of the external field $eA_0 = \mu$. The electric charge is independent of μ when it is measured at distances small compared to $a_B r_s$ and tends to zero at large distances due to the Thomas-Fermi screening. The renormalized trajectory is interpolating between the $\mu = 0$ ultraviolet regime and the infrared fixed point $e = 0$. At $q_{\mathrm{ren}} \sim q_F$ e may be around 0.6 for densities $r_s > 1.5$ but we have no evidence for any singularities.

The Thomas-Fermi screening suppresses the electric interactions. We have to use the magnetic sector to define the renormalized charge in order to have a more realistic measure of the strength of the interactions. Suppose that we replace (1) by another renormalization condition which introduces e through a spin flipping cross section dominated by the magnetic interactions. The electric charge defined in such a manner agrees with the conventional one in the ultraviolet and becomes zero in the infrared if the ground state is superconducting.

There are fundamental modifications in this picture if the density of the electron gas is lowered and $\mu \to m$. Phase transitions driven by spin correlations are expected[23] to take place at $r_s = O(10)$. The renormalized trajectories have discontinuous dependence on μ at the critical values. We shall discuss the qualitative aspects of the the low density side of this interesting transition region. The hamiltonian (24) can be written in dimensionless units as

$$
H_{\mathrm{nr}} = \frac{e^2}{a_B r_s^2} \{ \int \frac{d\mathbf{p}'}{(2\pi)^3} \frac{\mathbf{p}'^2}{2} \psi^\dagger(\mathbf{p}')\psi(\mathbf{p}')
$$
$$
- \frac{r_s}{2} \int \frac{d\mathbf{p}'}{(2\pi)^3} \frac{d\mathbf{q}'}{(2\pi)^3} \frac{d\mathbf{k}'}{(2\pi)^3} \frac{4\pi}{k'^2} \psi^\dagger(\mathbf{p}')\psi^\dagger(\mathbf{q}')\psi(\mathbf{p}'+\mathbf{k}')\psi(\mathbf{q}'-\mathbf{k}') \}
$$

(25)

where $\mathbf{p}' = a_B \mathbf{p}$, e.t.c. showing that the potential energy becomes dominant at low densities and the perturbation expansion should be organized according to the kinetic energy. This approximation would give the quantum corrections in expanding around the classical system. The "ground state" consists of localized electrons in a configuration which minimizes the potential energy and has no zero point fluctuations. This is the Wigner lattice[24] which should be formed at $r_s = O(100)$ and is a three dimensional realization of localization, a phenomenon introduced originally for 2+1 dimensional electrons[25].

An electron placed the inside of the Wigner lattice behaves fundamentally different manner than either in the vacuum or in a homogenous electron gas. In fact, the single electron states lie in bands like for the ordinary solid states since the overlap between the additional electron and the lattice is small. In the low density, large lattice spacing limit $r_s \gg 100$ the mechanism to form gaps in the single electron spectrum is different. Suppose that the physical conditions stabilizing the lattice change slowly as to allow the electrons to occupy larger volume. The characteristic phonon frequency approaches zero and the amplitude of the phonon excitations increases during this process. The Coulomb field of the lattice may be considered as a quenched field acting on the additional electron which moves with finite momentum. The electron propagator can be approximated by first computing it for a given phonon configuration and averaging over the classical phonon coordinates after that.

Such a quenching leads to cancellations in the electron propagator which can be understood in the following manner. First note that the Coulomb potential is screened by the phonons. The scattering of an electron on the Wigner lattice with a given phonon configuration produces some asymptotic states and phase shifts. As we average over the phonons the fluctuation due to the phase shift will remove some of the asymptotic states. In the rather oversimplified approximation where the potential is the sum of uniformly distributed Dirac deltas all asymptotic states with non-zero momentum are removed by this mechanism[26]. Though the actual potential is more complicated it seems reasonable to expect a large gap formed at low momentum due to such cancellations in the quenched propagator.

The appearance of the gap in the single particle spectrum effects the structure and the decay of the bound states. In fact, a positronium inside of the Wigner lattice may decay only into the allowed bands and the observed e^+e^- spectrum will be rather unusual. A possible experiment where such an effect may be observable is to collide two high luminosity ion beams as depicted in Fig. 5. The δ-electrons will be produced in large abundance and they will form a gas slowly expanding due to its Coulomb repulsion. The δ-electrons can be described by the effective hamiltonian

$$H_\delta = H_{\mathrm{nr}} + b\psi^\dagger(0) \tag{26}$$

where H_{nr} is given by (25) and b is a constant if the disturbance of the beam on the electron gas is negligible. The ground state of (26) consists of an electron gas which has a source at the origin and expands in radial direction. The constant b is proportional to the beam luminosity and describes the incoming flux of the δ-electrons. Note that the electron system (26) develops its own energy and time scales which are different than those of the colliding beams.

Suppose that a slow positron is created by a hard process during the ion collision. It captures an electron immediately and drifts away from the collision region with the electron gas. If the expansion rate of the electron gas is slow enough we can apply Thomas-Fermi type of approximation. Since the δ-electron density is proportional to $\frac{1}{R^2}$ where R is the distance from the collision zone the Wigner lattice density is reached during the expansion even for large b. According to the local description the positronium is experiencing a Wigner lattice with large phonon excitations and increasing lattice spacing in time during its flight. The positronium spectrum is deformed compared to the vacuum since the continuum states cannot start inside the gap and there should be bound states with $E > 2m$ which are stabilized by the gap.

There should be another phase transition as the density is further decreased. In fact, the sufficiently dilute electron gas is not much different from the vacuum and there should be a break-up density where the gap structure disappears. This break-up density is reached at R_b in Fig. 5. The positroniums with $E > 2m$ decay immediately at that surface and the energy excess $E - 2m$ is converted into the kinetic energy of the e^+e^- pair.

SUMMARY

The application of the renormalization group is suggested to understand how external field may generate non-perturbative effects in weak coupling theories such as QED. The usual non-perturbative phenomena when a phase transition occurs at some critical values of the external field can be understood by following the renormalized trajectory from the ultraviolet region down into the infrared. The bare parameters are independent of the external field but the renormalized parameters develop non-analytical dependence on the external field at the critical points. According to the spirit of the renormalization group such a non-perturbative effect is the result of the accumulation of the perturbative corrections during the integration of the renormalization group equations from the ultraviolet to the infrared and modifies the infrared sector of the theory.

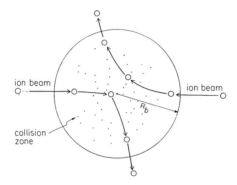

Fig. 5.   The theoretical physicist view of the colliding ion beam.

Another possibility when the renormalized coupling constants reach a crossover region at finite length scale during the integration. Then there is a rapid change from one behavior to another which appears as a non-perturbative phenomenon at that scale. If the length scale is short enough we may obtain a theory which behaves in the infrared in a similar manner than the one corresponding to different bare parameters.

Several non-perturbative effects are mentioned for one of the simplest external field in QED, $eA_0 = \mu$. It is conjectured that the ground state for $r_s < O(10)$ is superconducting due to the long range magnetic field. For $1.5 < r_s$ there are infinitely many Feynman diagrams at $O(e)$ and a partial resummation yields $e \rightarrow 2e$ for processes around the Fermi surface. Further mechanism to eliminate states and modify the bound state spectra is suggested for lower electron densities. A non-perturbative final state interaction is pointed out between the positroniums and the δ-electrons created in high yield ion beam collisions which may generate substantial changes in the e^+e^- spectra. The presented description is very superficial and a much more detailed analysis is needed to clarify this problem. It remains to be seen whether this mechanism can be observable in the GSI experiments.

REFERENCES

1. T. Kinoshita and W. B. Lindquist, 'Eight-order magnetic Moment of the Electron', Cornell Preprint, CLNS 90-980.

2. S. Adler, *Phys. Rev.* **177** (1969), 2426; J. Bell and R. Jackiw, *Nuovo Cimento* **60A** (1969) 47.

3. H. A. Bethe and E. E. Salpeter, *Phys. Rev.* **82** (1951) 309; N. Nakanishi, *Suppl. Progr. Theor. Phys.* **43** (1969) 1.

4. K. Wilson and J. Kogut, *Phys. Rep.* **12** (1974) 75; D. Amit, *Field Theory, the Renormalization Group and Critical Phenomena* (World Scientific, 1984); J. Collins, *Renormalization* (Cambridge University Press, 1984).

5. V. B. Berestetsky, E. M. Lifshitz and L. P. Pitaevsky, *Quantum Electrodynamics* (Pergamon Press, 1959), eq. (82.7).

6. V. A. Miransky and P. I. Fomin, *Sov. J. Part. Nucl.* **16** (1985) 203; P. I. Fomin, V. P. Gusynin, V. A. Miransky and Yu. A. Sitenko, *Riv. Nuovo Cimento* **6** (1983) 1; W. A. Bardeen, C. N. Leung and S. T. Love, *Phys. Rev. Lett.* **56** (1986) 1230; C. N. Leung, S. T. Love and W. A. Bardeen, *Nucl. Phys.* **B237** (1986) 649.

7. J. Kogut, E. Dagotto and A. Kocic, *Phys. Rev. Lett.* **60** (1988) 772; S. J. Hands, J. B. Kogut and E. Dagotto, *Nucl. Phys.* **B333** (1990) 551; M. Göckeler *et. al.*, *Nucl. Phys.* **B334** (1990) 527; S. P. Booth, R. D. Kenway and B. J. Pendleton, *Phys. Lett.* **228B** (1989) 115; E. Dagotto, A. Kocic and J. Kogut, *Phys. Rev. Lett.* **61** (1988) 2416; *Nucl. Phys.* **B317** (1989) 253.

8. A. Guth, *Phys. Rev.* **D21** (1980) 2291; C. B. Lang, *Nucl. Phys.* **B280** (1987) 225.

9. J. Ambjorn, J. Greensite and C. Peterson, *Nucl. Phys.* **B221** (1983) 381.

10. D. G. Caldi and A. Chodos, *Phys. Rev.* **D36** (1987) 2876.

11. J. Schweppe, *et. al.*, *Phys. Rev. Lett.* **51** (1983) 2261; M. Clemente, *et. al.*, *Phys. Lett.* **B137** (1984) 41; for a recent summary see H. Bokemeyer's and W. Koenig's talk at the School.

12. L. I. Schiff, H. Snyder and J. Weinberg, *Phys. Rev.* **57** (1940) 315; Y. B. Zeldovich and V. S. Popov, *Sov. Phys. Usp.* **14** (1972) 673; B. Müller, H. Peitz, J. Rafelski and W. Greiner, *Phys. Rev. Lett.* **28** (1972) 1235; B. Müller, J. Rafelski and W. Greiner, *Z. Phys.* **257** (1972) 62, 183; for a review see W. Greiner, B. Müller and J. Rafelski, *Quantum Electrodynamics of Strong Fields* (Springer, 1985).

13. M. C. Chu, S. Huang and J. Polonyi, "Non-Perturbative Effects in QED with Chemical Potential," MIT preprint CTP#1920, submitted to Nuclear Physics.

14. J. Polonyi, in *Quark-Gluon Plasma* (World Publisher, 1990), R. Hwa, ed.

15. M. C. Chu and J. Polonyi, unpublished.

16. J. Schwinger, *Phys. Rev.* **82** (1951) 664.

17. A. Chodos, K. Everding and D. Owen, *QED with Chemical potential: I. Case of constant magnetic field*, Yale Preprint, April, 1990.

18. C. Itzykson and J. Zuber, *Quantum Field Theory*, McGraw-Hill Book Company, 1980.

19. A. A. Abrikosov, L. P. Gorkov and I. E. Dzyaloshinski, *Methods of Quantum Field Theory in Statistical Physics*, (Dover, 1963); A. L. Fetter and J. D. Walecka, *Quantum Theory of Many-Particle Systems*, (McGraw-Hill Book Company, 1971); B. A. Freedman and L. D. McLerran, *Phys. Rev.* **D16** (1977) 1130.

20. T. D. Lee and M. Nauenberg, *Phys. Rev.* **133** (1963) B1549; T. D. Lee, *Particle Physics and Introduction to Field Theory* (Harwood Academic Publishers, 1985).

21. F. Bloch and A. Nordsieck, *Phys. Rev.* **52** (1937) 54.

22. T. Kinoshita, *J. Math. Phys.* **3** (1962) 650;

23. F. Bloch, Z. Phys. **57** (1920), 545.

24. E. P. Wigner, *Trans. Farad. Soc.* **34** (1938) 678; W. J. Carr, *Phys. Rev.* **122** (1961) 1437.

25. P. W. Anderson, *Phys. Rev.* (1958) 1492.

26. L. Lellouch and J. Polonyi, in preparation.

FUNCTIONAL INTEGRALS IN THE

STRONG COUPLING REGIME

G.V. Efimov and G. Ganbold

Laboratory of Theoretical Physics
Joint Institute for Nuclear Research, Dubna

INTRODUCTION

The path-integral approach suggested by Feynman[1] is widely used in quantum mechanics, statistical physics and quantum field theory. The standard form of these integrals is shown in (1), where the Gaussian measure with appropriate Green function describes the non-interacting free system and the nonlinear part corresponds to an interaction in this system. Various theoretical techniques have been developed to calculate this functional integral but the only method well established mathematically is the perturbative expansion at small coupling. In the strong coupling regime variational methods have a good reputation in describing physical values such as a ground state energy due to their low sensibility to errors in the choice of trial wave functions (see, for example [2]). But a variational method does not allow us to know how close is the obtained estimation to the true described value and does not give a recipe how to calculate the next correlations.

The purpose of this paper is to formulate a general method of estimation of a functional integral in the strong coupling regime. Our idea is the following. We propose that the functional integral is of the Gaussian type in the strong coupling regime but with another Green function in the Gaussian measure. The contribution of self-energy which is

Vacuum Structure in Intense Fields, Edited by
H.M. Fried and B. Muller, Plenum Press, New York, 1991

proportional to the tadpole Feynman diagrams is the main one to the formation of the new state. Thus, the mathematical problem is to take it into account correctly. It can be done by introducing the concept of the normal product according to the given Gaussian measure. We formulate the equations which make it possible to perform this program. As a result, we obtain the equivalent representation of the initial functional integral in which the main contributions of the strong interaction are concentrated in the new Green function defining the measure and in an explicit expression for the ground state energy. This representation permits us to compute small perturbation corrections.

We have applied[3] this method to the problem of polaron ground state energy. The obtained results seem promising and are in good agreements with the often-quoted results (for details see [3]).

GENERAL FORMALISM

In this section we formulate our method of calculation of functional integrals defined on the Gaussian measure. We shall consider the functional integrals of the following general type:

$$Z_\Gamma(g) = N_o \cdot \int \delta\varphi \cdot \exp\left\{-\frac{1}{2}(\varphi \cdot D_o^{-1} \cdot \varphi) + g \cdot W[\varphi]\right\} .$$
(1)

Here, we have introduced the following notation

$$(\varphi \cdot D_o^{-1} \cdot \varphi) \equiv \int_\Gamma dx \int_\Gamma dy \cdot \varphi(x) \cdot D_o^{-1}(x,y) \cdot \varphi(y) .$$
(2)

The integration in (2) is performed over a region $\Gamma \subseteq R^d$ ($d=1,2,\ldots$). Usually, the region is a box

$$\Gamma \equiv \{x: a_j \le x_j \le b_j, \ j=1,\ldots,d \},$$
(3)

where $D_o^{-1}(x,y)$ is a differential operator defined on functions $\varphi(x)$ with appropriate boundary conditions. For example

$$D_o^{-1}(x,y) = \left(-\frac{\partial^2}{\partial x^2} + m_o^2\right) \cdot \delta(x-y),$$
(4)

with periodic boundary conditions. The Green function $D_o(x,y)$ satisfies

$$\int_\Gamma dy \cdot D_o^{-1}(x,y) \cdot D_o(y,z) = \delta(x-y).$$
(5)

The normalization constant N_0 in (1) is defined by the condition

$$N_0 \cdot \int \delta\varphi \cdot \exp\left\{-\frac{1}{2}(\varphi \cdot D_0^{-1} \cdot \varphi)\right\} = 1$$

and

$$N_0 = 1/\sqrt{\det D_0^{-1}} \ . \tag{6}$$

The interaction functional $W[\varphi]$ can be written in a general form

$$W[\varphi] = \int d\mu_a e^{i(a \cdot \varphi)} \ , \tag{7}$$

where

$$(a \cdot \varphi) = \int_\Gamma dx \cdot a(x) \cdot \varphi(x)$$

and $d\mu_a$ is a measure. For example,

$$W[\varphi] = \int_\Gamma dx \cdot U(\varphi(x)) = \int_\Gamma dx \cdot \int \frac{dk}{2\pi} \tilde{U}(k) \cdot \exp\left\{i \int_\Gamma dy \cdot k \cdot \varphi(y) \cdot \delta(x-y)\right\} \ . \tag{8}$$

The parameter g is a coupling constant.

We consider that the integral (1) does exist as a functional integral and can be calculated by the perturbation method for a small coupling constant g . Our aim is to obtain a representation in which all main contributions of a strong interaction are concentrated in the Green function of the Gaussian measure.

Let us perform the following transformations in the integral (1):

$$\varphi(x) \longrightarrow \varphi(x) + b(x) \ ,$$

$$D_0^{-1}(x,y) \longrightarrow D^{-1}(x,y) \ , \tag{9}$$

where $b(x)$ and $D^{-1}(x,y)$ are arbitrary functions. The Green function $D(x,y)$ satisfies

$$\int_\Gamma dy \cdot D^{-1}(x,y) \cdot D(y,z) = \delta(x-y) \ .$$

Then, the functional integral takes the form

$$Z_\Gamma(g) = \exp\left\{\frac{1}{2} \ln\det\frac{D}{D_0} - \frac{1}{2}(b \cdot D_0^{-1} \cdot b)\right\} \cdot \int d\sigma_D \exp\left\{W_{int}[\varphi, b, D]\right\} \tag{10}$$

Here

$$d\sigma_D = N \cdot \delta\varphi \cdot \exp\left\{-\frac{1}{2}(\varphi \cdot D^{-1} \cdot \varphi)\right\}, N = 1/\sqrt{\det D^{-1}} \ ,$$

$$W_{int} = g \cdot W[\varphi + b] - (b \cdot D_0^{-1} \cdot b) - \frac{1}{2}(\varphi \cdot [D_0^{-1} - D^{-1}] \cdot \varphi) \ . \tag{11}$$

Now let us introduce the concept of the normal product according to the giving Gaussian measure $d\sigma_D$. It means that

$$e^{i(a\cdot\varphi)} =: e^{i(a\cdot\varphi)}: e^{-\frac{1}{2}(a\cdot D\cdot a)} \tag{12}$$

so that

$$\int d\sigma_D : e^{i(a\cdot\varphi)} := 1$$

or

$$\int d\sigma_D : \varphi(x_1)\cdot\ldots\cdot\varphi(x_n) := 0.$$

Then the interaction functional W_{int} (11) can be rewritten

$$W_{int} = g\cdot\int d\mu_a e^{i(a\cdot\varphi)-\frac{1}{2}(a\cdot D\cdot a)} : e^{i(a\cdot\varphi)} - 1 - i(a\cdot\varphi) + \frac{1}{2}(a\cdot\varphi)^2 :$$

$$+ \left[g\cdot\int d\mu_a e^{i(a\cdot\varphi)-\frac{1}{2}(a\cdot D\cdot a)} - \frac{1}{2}([D_o^{-1}-D^{-1}]D) \right]$$

$$+ \left[g\cdot\int d\mu_a e^{i(a\cdot\varphi)-\frac{1}{2}(a\cdot D\cdot a)} i(a\cdot\varphi) - (b\cdot D_o^{-1}\cdot\varphi) \right]$$

$$- \frac{1}{2} : \left[g\cdot\int d\mu_a e^{i(a\cdot\varphi)-\frac{1}{2}(a\cdot D\cdot a)} (a\cdot\varphi)^2 + (\varphi\cdot[D_o^{-1}-D^{-1}]\cdot\varphi) \right] : . \tag{13}$$

Our basic idea that the main contribution to a functional integral is concentrated in the Gaussian measure means that the linear and quadratic terms over the integration variables $\varphi(x)$ should be absent in the interaction part W_{int} (13). Thus, we obtain two equations

$$g\cdot\int d\mu_a ia(x)\cdot e^{i(a\cdot b)-\frac{1}{2}(a\cdot D\cdot a)} - \int_\Gamma dy\cdot b(y)\cdot D_o^{-1}(x,y) = 0,$$

$$g\cdot\int d\mu_a a(x)\cdot a(y)\cdot e^{i(a\cdot b)-\frac{1}{2}(a\cdot D\cdot a)} + D_o^{-1}(x,y) - D^{-1}(x,y) = 0 . \tag{14}$$

These equations provide the removing of above mentioned terms.

Let us introduce the functional

$$\bar{\bar{W}}[b] = \int d\mu_a \exp\left\{ i(a\cdot b) - \frac{1}{2}(a\cdot D\cdot a) \right\} . \tag{15}$$

Then, equations (14) can be rewritten in the form

$$b(x) = g\cdot\int_\Gamma dy\cdot D_o(x,y)\frac{\delta}{\delta b(y)}\bar{\bar{W}}[b],$$

$$D(x,y) = D_o(x,y) + \int_\Gamma dz\int_\Gamma dt\cdot D_o(x,z)\frac{\delta^2\,\bar{\bar{W}}[b]}{\delta b(z)\delta(t)}D(t,y) . \tag{16}$$

These equations define the functions $b(x)$ and $D(x,y)$.

Finally, we obtain

$$Z_\Gamma(g) = \exp\{W_o\}\cdot\int d\sigma_D \exp\left\{ g\cdot\bar{\bar{W}}_2[\varphi] \right\} . \tag{17}$$

Here
$$W_o = \frac{1}{2} \ln\det\frac{D}{D_o} - \frac{1}{2}(b \cdot D_o^{-1} \cdot b) - \frac{1}{2}([D_o^{-1} - D^{-1}]D) + g \cdot \bar{W}[b],$$

$$\bar{W}_2[\varphi] = g \cdot \int d\mu_a e^{i(a \cdot b) - \frac{1}{2}(a \cdot D \cdot a)} : e^{i(a \cdot \varphi)} - 1 - i(a \cdot \varphi) + \frac{1}{2}(a \cdot \varphi)^2 :$$

The functions $b(x)$ and $D(x,y)$ are the solutions of (16).

One should stress that the representations (1) and (17) are equivalent. Therefore our mathematical object $Z_\Gamma(g)$ has at least two different representations. In principle, we can get other representations if equations (16) have different solutions. We shall choose the representation in which the perturbation corrections connected with $g \cdot \bar{W}_2$ or $g \cdot W$ are at absolute minimum for the given parameters in the interaction functional $g \cdot W$.

All our transformations and equations (16) are valid for both real and complex functions in the functional integral (1).

In the case of real functional integrals equality (19) and equations (16) lead to the following conclusion. The interaction functional $\bar{W}_2[\varphi]$ in (17) satisfies

$$\int d\sigma_D \bar{W}_2[\varphi] = 0 \ . \tag{19}$$

Using Jensen's inequality we have

$$Z_\Gamma(g) \geq \exp\{W_o\}, \tag{20}$$

so that W_o defines the lower estimation for considered functional integral.

On the other hand, one can easily check that equations (16) define the minimum of the functional W_o in (18). Thus, inequality (20) is a variational estimation of the functional integral (1). Moreover, the representation (17) permits us to calculate the perturbation corrections to W_o developing the functional integral in (17) over $g \cdot \bar{W}_2[\varphi]$.

We have applied[3] this method to the problem of polaron ground state energy. The polaron is one of the simplest nontrivial models standing between quantum mechanics and quantum field theory. The results obtained by using our techniques look promising and are in good agreements with the often-quoted results on this problem. Readers are referred to [3] for details.

ACKNOWLEDGEMENTS

The authors would like to thank Profs J. Lewis and H. Fried for fruitful discussions and their hospitality at the Institute for Advance Studies, Dublin and the Cargese Summer School, Cargese.

REFERENCES

1. R. P. Feynman and A. R. Hibbs, "Quantum Mechanics and Path Integrals", McGraw Hill, N.Y., (1965).
2. Proceed. Int.Workshop on "Variational Calculations in QFT", eds. L.Polley and D.E.L.Pottinger, World Scientific, (1987).
3. G. V. Efimov and G. Ganbold, preprint of Dublin Inst., DIAS-STP-90-22, (1990).

A PHASE TRANSITION OF QED, SCHWINGER'S PROPER TIME

FORMALISM; CHEMICAL POTENTIAL AND ELECTRIC FIELD

David A. Owen

Ben Gurion University of the Negev
Department of Physics
Beer Sheva, Israel

INTRODUCTION

A part of the physics community has been intrigued for some time by the anomalous Mev e^+e^- peaks produced from heavy ion scattering. Although there have been a number of attempts to explain these strange results, none has been completely satisfying.

Briefly, these are[1] (this is only a partial list as other speakers will cover this in much more detail):

Conventional Explanations

1. Peaks are due to Nuclear Transitions

This is apparently ruled out since:

a) the position of the peaks are independent of the total charge.
b) both the electron and positron are emitted with approximately the same energies in the presence of a highly ionized nucleus
c) e^+ and e^- are emitted back to back.
d) The line width are much narrower than conventional nuclear transitions.

2. New elementary particle ϕ

a) $m_\phi \approx 1.8$ Mev does account for the Z independence of the peaks as well as the observed e^+, e^- coincidences, however, it is difficult to understand why ϕ's are predominantly produced at rest.
b) Various models of ϕ production through electromagnetic and nuclear effects give a broad spectrum for ϕ, contrary to what is observed.

Vacuum Structure in Intense Fields, Edited by
H.M. Fried and B. Muller, Plenum Press, New York, 1991

c) Natural assumptions of φ coupling to matter give too small a production rate.

d) Beam dump experiments show no evidence of a "φ" particle.

e) Multiple peaks would require the existence of several new particles.

f) Particle life time and QED constraint (see Fig. 1)

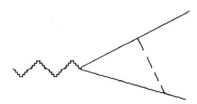

Fig. 1 This Feynman diagram represents the φ correction to the vertex diagram (where the φ propagating across the vertex is represented by the broken line)

This leads to a correction of the anomalous moment of the electron which is given by

$$a = \frac{g^2}{4\pi} < 10^{-8} \tag{1}$$

where g is the coupling constant of the φ–e vertex. The inequality is such not to upset the present agreement with experiment. However, if one calculates the half-width of the φ→e$^+$e$^-$ decay, one finds

$$\Gamma_{\phi \to e^+ e^-} = \frac{g^2}{8\pi} \sqrt{M^2 - 4m^2} \approx \frac{g^2 M}{8\pi} < 1.8 \times 10^{-8} \text{ Mev} \tag{2}$$

where M is the mass of the φ (taken to be 1.8 Mev) and m, the electron mass. Thus

$$\Delta t > \frac{\hbar}{\Gamma_{\phi \to e^+ e^-}} \approx 10^{-14} \text{ sec} \tag{3}$$

A Less Conventional Explanation

3. Non-perturbative QED

a) One of the non-perturbative approaches is to examine the magnetic interaction at short distances since this might appear to dominate the behavior of e$^+$ and e$^-$ under certain conditions. Naively one finds a potential like shown in Fig. 2 because of the short range $1/r^3$ interaction.

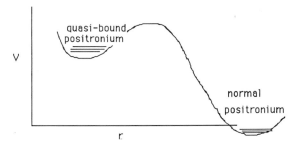

Fig. 2   A schematic   representation of the inclusion of $1/r^3$
potential.

However, the radiative corrections QED (which have not
been taken into account in the above diagram) should
ameliorate this short distance behavior so the above graph
should not be taken at face value.   b) Phase transition of
QED:It is known[2], that if we consider QED as a function of
the coupling constant, one finds that when $\alpha_c = \frac{\pi}{3}$ there is a
phase transition of QED.   The situation relevant to the
heavy ion scattering production of the exotic e^+e^- peaks,
is the question: Is it possible to induce this phase from
rapidly variation of extremely strong fields even when α
is its usual value.   This phase is a confining phase in
which chiral symmetry is broken.   The dynamical situation
can be pictured by the following schematic diagram.   In
this scenario, the false vacuum is lowered by the presence
of the external field.   When it becomes lower than the
normal vacuum, then it itself, becomes the true vacuum
(see Fig.3).

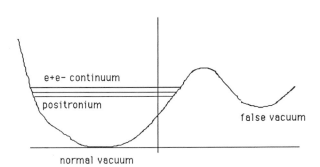

Fig. 3 This potential diagram represents a possible scenario
in which a phase transition of QED   might occur.

SCHWINGER'S FORMALISM WITH CHEMICAL POTENTIAL & NEW PHASE OF QED

Do Strong Fields Induce a QED Phase Transition?

At present the answer to this question isn't known. This is because, unfortunately, we do not have at our disposal the means to investigate non-perturbative QED in the presence of rapidly changing strong fields. One is able, however, to examine QED under more restrictive circumstances, as have the Yale group[3]. This is by using the Schwinger's proper time formalism[4] for which the non-perturbative result is known when the external field is constant. In this formalism, radiative photons are ignored.

The main thrust of Schwinger's work results from his observation that the change in the action, W is due to a vacuum current induced by an external field $A\mu$ and is given by

$$\delta W^{(1)} = ieTr\ \gamma\delta AG$$
$$= \int (dx)\ \delta A_\mu <j^\mu(x)> \qquad (4)$$

where $\delta A\mu$ is the variation in the field and $<j_\mu>=ieTr\gamma_\mu<x|G|x>$.

The two main limitations of this approach is a) it describes a static situation b) it is strictly valid only for constant fields. Another limitation, which we believe is less restrictive, is that it ignores radiative photons.

There are preciously few, non-perturbative solutions in quantum electrodynamics. Among these is the one being discussed here by Alan Chodos, the phenomenon of vacuum polarization in a constant external field and its contribution to the electron's Green function. The non-perturbative resultant was originally found by Weisskopf[5] but was rederived by Schwinger using his elegant proper-time method. The recent interest in these results and particularly with the method of Schwinger arises with the puzzlement in attempting to explaining the anomalous Mev peaks in e^+e^- spectrum resulting from heavy ion scattering[1]. Although radiative photons are neglected, the Schwinger result is non-perturbative with respect to the external field and are valid for all field strength. One of its limitation is it is restricted to essentially constant field.

The Chemical Potential with a Constant Electric Field

Nevertheless, Alan Chodos, Ken Everding and myself have attempted to extend Schwinger's formalism to include a chemical potential μ. It is our belief that the physical environment of the heavy ion collision is better represented if a chemical potential is included. This is because the ions in the collision are not completely ionized but carry a residual electron cloud. It is this cloud which the chemical potential is intended to represent.

There is an ensuing simplification if one assumes that the only external field is a constant magnetic field. Thus, our first goal was to extend Schwinger's formalism to the case in which the chemical potential is non-zero with non-zero magnetic field. This first step of our program has been completed. That is, for the case B≠0 and E=0. Thus, the cases we still wish to treat are: B=0 and E≠0; and then the general case, B≠0 and E≠0 where the non-zero values are constant.

Our second goal is to then examine the implications of the above generalized formalism and see if they indicate a phase transition of quantum electrodynamics. This part of the program cannot be implemented since we have not finished our first objectives.

To remind you what must be done, we write the Green's function with a chemical potential in the presence of an external electric field. It is

$$G = (\not\Pi + m) \frac{1}{(\Pi \cdot \gamma)^2 - m^2 + 2i\epsilon(p_0 - \mu)\, \mathrm{sgn}\, p_0} \tag{5}$$

where $\Pi_0 = (p_0 - \mu) - A_0$; $\Pi = \not\Pi = \mathbf{p} - e\mathbf{A}$.
An easy way to see how this comes about, consider the case when $A_\mu = 0$ and $\mu = 0$. Then G becomes

$$
\begin{aligned}
G &= \left\{ \frac{\Lambda^+(\mathbf{p})}{p_0 - E_p + i\epsilon} + \frac{\Lambda^-(\mathbf{p})}{p_0 + E_p - i\epsilon} \right\} \gamma_0 \\
&= \left\{ \frac{\Lambda^+(\mathbf{p})}{p_0 - E_p + i\epsilon\, \mathrm{sgn}\, p_0} + \frac{\Lambda^-(\mathbf{p})}{p_0 + E_p + i\epsilon\, \mathrm{sgn}\, p_0} \right\} \gamma_0 \\
&= (\not{p} + m) \frac{1}{(p_0 + i\epsilon\, \mathrm{sign}\, p_0)^2 - E^2(p)} \\
&\approx (\not{p} + m) \frac{1}{p_0^2 + 2ip_0\epsilon\, \mathrm{sign}\, p_0 - E^2(p)}
\end{aligned}
\tag{6}
$$

which of course is reminiscent of eq 5.

The correct boundary conditions when a chemical potential is included, as Alan Chodos has indicated in his lecture is

$$(\not\Pi + m) \frac{1}{(\Pi \cdot \gamma)^2 - m^2 + 2i\epsilon(p_0 - \mu)\, \mathrm{sign}\, p_0} \tag{7}$$

To find the Lagrangian we must evaluate

$$\langle x'(s) | x''(0) \rangle = \langle x' | e^{-i\mathcal{H}s} | x'' \rangle \tag{8}$$

This evaluation is difficult only when p and x are both present in $\mathcal{H}$. In the present case, E≠0, B=0, another difficulty can arise since the boundary condition is also a function of p_0. This is the case we now examine.

Following the route taken by Schwinger[4], we exponentiate the denominator of the propagator in eq 7. To do this we use the Feynman's calculus[6].

Let $\int_0^1 d\lambda A(\lambda) = \int_0^1 d\lambda \, [\, (\Pi_\lambda \cdot \gamma)^2 - m^2\,]$ & $\int_0^1 d\lambda B(\lambda) = \int_0^1 d\lambda \, 2(p_o - \mu) \, \text{sign } p_o$

where λ specifies the order in which operators act. The propagator in eq 7 can now be represented as

$$\frac{1}{\int_0^1 d\lambda A(\lambda) + i\varepsilon \int_0^1 d\lambda B(\lambda)} =$$

$$\frac{1}{\int_0^1 d\lambda A(\lambda) + i\varepsilon} \, \theta\!\left[\int_0^1 d\lambda B(\lambda)\right] + \frac{1}{\int_0^1 d\lambda A(\lambda) - i\varepsilon} \, \theta\!\left[-\int_0^1 d\lambda B(\lambda)\right] \qquad (9)$$

Then writing

$$\frac{1}{\int_0^1 d\lambda A(\lambda) \pm i\varepsilon} = -(\pm) i \int_0^\infty ds \, \exp\{\pm is \int_0^1 A(\lambda) \, d\lambda\} \qquad (10a)$$

$$\theta\!\left[\int_0^1 d\lambda B(\lambda)\right] = \frac{1}{2\pi i} \int_{-\infty}^\infty dq \, \frac{\exp\{iq \int_0^1 d\lambda B(\lambda)\}}{q - i\lambda} \qquad (10b)$$

Thus, we may represent the denominator of the propagator in eq 7 as

$$\frac{1}{(\Pi \cdot \gamma)^2 - m^2 + 2i\varepsilon (p_o - \mu) \, \text{sign } p_o} =$$

$$-i \int_0^\infty ds \int_{-\infty}^\infty \frac{dq}{2\pi i} \, \frac{\exp\{is[\,(\Pi \cdot \gamma)^2 - m^2 + 2i\varepsilon (p_o - \mu) \, \text{sign } p_o]\}}{q - i\lambda}$$

$$+i \int_0^\infty ds \int_{-\infty}^\infty \frac{dq}{2\pi i} \, \frac{\exp\{-is[\,(\Pi \cdot \gamma)^2 - m^2 + 2i\varepsilon (p_o - \mu) \, \text{sign } p_o]\}}{q - i\lambda} \qquad (11)$$

The "Schwingerian" procedure is to express p_o and Π_3 as functions of Δx_μ from which $<x'(s)\,|\,x''(o)>$ can be evaluated via

$$i\frac{d<x'(s)\,|\,x"(0)>}{ds} = <x'(s)\,|\,\mathcal{H}\,|\,x"(0)> \tag{12}$$

directly.

The exponential in the first term of eq 11 can be written as $\Pi^2 - \frac{1}{2}\sigma \cdot F - m^2 + 2(p_o - \mu)q\,\text{sgn}\,p_o$. Thus the Hamiltonian responsible for the proper time translation of the first term of eq 11 is

$$\mathcal{H} = -[\Pi^2 + 2(p_o - \mu)q\,\text{sgn}\,p_o] \tag{13}$$

i.e.

$$<x'\,|\frac{1}{(\Pi\cdot\gamma)^2 - m^2 + 2i\varepsilon(p_o - \mu)\,\text{sign}\,p_o}\,|\,x">=$$

$$-i\int_0^\infty ds\;e^{-im^2 s}\exp\{-\frac{1}{2}\,\sigma.F\}\int_{-\infty}^\infty dq\frac{<x'(s)\,|\,x"(0)>_1}{q - i\lambda}$$

$$+i\int_0^\infty ds\;e^{im^2 s}\exp\{\frac{1}{2}\,\sigma.F\}\int_{-\infty}^\infty dq\frac{<x'(s)\,|\,x"(0)>_2}{q - i\lambda} \tag{14}$$

where $<x'(s)\,|\,x"(0)>_1=<x'\,|\,e^{-is\mathcal{H}}\,|\,x">$ with $\mathcal{H}$ given by eq 13. A similar expression holds for the second term of eq 14.

It is interesting to look at the boundary term $2i\varepsilon(p_o - \mu)\,\text{sgn}\,p_o$ as a function of p_o (where we ignore $i\varepsilon$ for the purpose of the graph)

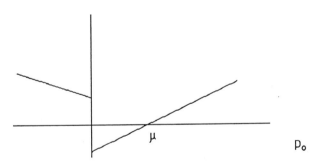

Fig. 4 A plot of $(p_o - \mu)\,\text{sgn}\,p_o$ vs p_o

It is rather discontinuous as seen in Fig. 4. Since one is only interested in the sign given by $2(p_o - \mu)\,\text{sgn}\,p_o$ one might try to replace $\text{sgn}\,p_o$ by p_o which leads to a continuous function as shown in Fig. 5.

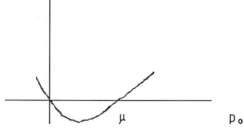

$$\text{Fig.5 A plot of } (p_0-\mu)p_0 \text{ vs } p_0$$

The Hamiltonian in eq 13 now becomes $\mathcal{H} = -[(1+q)p_0{}^2 - \pi^2 - \mu q p_0]$. Using this expression in place of eq 13 leads to integral that Alan Chodos described in his lecture. It is an integral which we are unable to evaluate. Its integrand has an infinite number of branch points and the residue at each of these cannot be expressed in closed form.

Is the problem so difficult or is there something wrong with our prescription? At present we can't answer this but we can see that there could be a problem. This can be seen returning to the Hamiltonian for which sgn $p_0 \to p_0$. We notice that its character completely changes for $q < -1$. It is no longer positive definite. This is precisely where the difficult begins with the integral over q for G. Perhaps in some way this is responsible for the observed pathology we found.

To see if we could avoid the above difficulties (i.e. the occurrence of the branch points in the q integration), we decided to use the Hamiltonian given in eq 13. To simplify matters, we chose E along the z-axis. Essentially we have to find the functions, f and g such that $p_0 = f(\Delta x_0, \Delta x_3)$ and $\Pi_3 = h(\Delta x_0, \Delta x_3)$. Then we can substitute into eq 12 and evaluate the r.h.s. as all the quantities are expressed in the x-representation . Unfortunately for $E \neq 0$, $B = 0$, $\mu \neq 0$, this program has not been fully implemented. To see the problem we have encountered we write the expressions for p_0 and Π_3 we have found.

$$p_0(s) = \{K\theta[K(s-s_0)]\cosh A + K\theta[-K(s-s_0)]\cosh B\}\sinh 2eEs$$
$$+ \{K\theta[K(s-s_0)]\sinh A + K\theta[-K(s-s_0)]\sinh B\}\cosh 2eEs$$
$$+ \theta[K(s-s_0)](\mu-q) + \theta[-K(s-s_0)](q+\mu) \tag{15}$$

$$\Pi_3 = \{K\theta[K(s-s_0)]\cosh A + K\theta[-K(s-s_0)]\cosh B\}\cosh 2eEs$$
$$+ \{K\theta[K(s-s_0)]\sinh A + K\theta[-K(s-s_0)]\sinh B\}\sinh 2eEs \tag{16}$$

with $K^2 = C - (\mu+q)^2$ and $s_0 = -\dfrac{1}{2eE}\sinh^{-1}\dfrac{q+\mu}{K} - C$

The q-dependence is much less singular that one found using the Hamiltonian in eq 15. In the usual procedure implemented by Schwinger for $\mu = 0$, one can write

$$p_o(s) = \Pi_3(0)\sinh 2eEs + p_o(0)\cosh 2eEs \qquad (17)$$

$$\Pi_3(s) = \Pi_3(0)\cosh 2eEs + p_o(0)\sinh 2eEs \qquad (18)$$

where C and C' are integration constants. Since it doesn't seem possible to write eqs 15 and 16 in the form given by eqs 17 and 18. We must therefore modify the approach followed by Schwinger.

To learn how this could be done, we have turned to the simplest case for which both $\mu=0$ and $A_\mu=0$ but maintaining the form of the boundary condition, i.e. eq 13 with $\mu=0$ and $A_\mu=0$. In a sense this is a pedagogical exercise since this is a well established result. The Hamiltonian becomes

$$\mathcal{H}\to\mathcal{H}_f = -\{p_o-\mathbf{p}^2+2p_oq \ \text{sgn} \ p_o\} \qquad (19)$$

from which it follows that

$$\frac{dx^0}{ds} = i[\mathcal{H}_f, x^0] = 2[p^0+q \ \text{sgn} \ p^0] \implies \frac{\Delta x^0}{2s} = p^0+q \ \text{sgn} \ p^0 \qquad (20)$$

$$\frac{dx^i}{ds} = 2p^i; \quad \frac{dp^i}{ds} = 0 \implies \Delta x^j = 2p^j(0)s \qquad (21)$$

Squaring $\dfrac{\Delta x^0}{2s}$ (see eq 20) we find $p_o{}^2+2p_oq \ \text{sgn} \ p_o+q^2 = \dfrac{\Delta x_0{}^2}{4s^2}$.

From which it follows (cf. eqs 19,20 & 21) that $\mathcal{H}_f = -\{\dfrac{\Delta x_0{}^2}{4s^2}$

$- \dfrac{\Delta \mathbf{x}^2}{4s^2} - q^2\}$. We still need to evaluate $(\Delta x_0)^2$ and $(\Delta \mathbf{x})^2$, i.e.

$(\Delta x_0)^2 = [x_0(s)]^2 - x_0(s)x_0(0) - x_0(0)x_0(s) + [x_0(s)]^2$ where $x_0(s)$ and $x_0(0)$ do not commute. Their commutator is found from eq 20

$$[x_0(0), x_0(s)] = -2is - 4isq\delta(p_0) \qquad (22)$$

Using $\mathcal{H}_f$ in place of $\mathcal{H}$ in eq 12 we find

$$i\frac{d\langle x'(s)|x''(0)\rangle}{ds} = -\{\frac{(x'-x'')^2}{4s^2} + \frac{2i}{s} + q^2\}\langle x'(s)|x''(0)\rangle$$

$$-i\frac{q}{s}\langle x'(s)|\delta(p_0)|x''(0)\rangle \qquad (23)$$

The last term is most readily evaluated if we introduce a complete set of momentum states. One finds

$$\langle x'(s)|\delta(p_0)|x''(0)\rangle = \frac{\exp\left\{i\dfrac{(\mathbf{x}'-\mathbf{x}'')^2}{4s}\right\}}{(2\pi)^4}\left(\frac{\pi}{s}\right)^{3/2}\exp(\frac{3\pi i}{4}) \qquad (24)$$

Equation 23 may be integrated immediately and one finds

271

$$\langle x'(s) | x''(0) \rangle = X_0 e^{-ig(s)}$$

$$-e^{-ig(s)} e^{-3\pi i/4} \int_0^s ds_1 q s_1 \frac{\exp\left\{ i \frac{(\mathbf{x'} - \mathbf{x''})^2}{4s} \right\}}{(2\pi)^4} \left(\frac{\pi}{s} \right)^{3/2} \exp(iq^2 s_1) \quad (25)$$

where X_0 is a function independent of s and g(s) is given by

$$g(s) = \frac{(x'-x'')^2}{4s} - 2i \ln s + q^2 s \quad (26)$$

I shall not pursue the precise manner in which X_0 is evaluated. It was found most efficient to go to momentum space in its evaluation. An option which might not be useful in the more general case. It is found that

$$X_0 = -\frac{i}{(2\pi)^4} \cos \left[q(x_0' - x_0'') \right] \quad (27)$$

It can be seen that with this value of X_0 the well known result for $\langle x'(s) | x''(0) \rangle$ is recovered. It is hoped that the experience involved in evaluating X_0 will help us in the more general case. Our plans are now to repeat the following cases, with this form of the boundary condition as given in eq 13: $\mu \neq 0$ with B=E=0; $\mu \neq 0$, $B \neq 0$, E=0 and then $\mu \neq 0$, B=0, $E \neq 0$. This the first two of these have been done by other means, the main object of this exercise, is really to have a way of doing the last case. This the case for which, using the other form of the boundary condition lead to a q-integral that we were unable to perform. There some indication that the form of the boundary condition may lead to more tractable integral than the one we have previously found.

REFERENCES

[1] A. Chodos, Narrow e^+e^- Peaks in Heavy-Ion Collisions; Fact and Fancy, Comments Nucl. Part. Phys. 17: 211 (1987). D. Caldi, A "New Phase of QED" on Paper, in the Lab and in the Heavens, Schwinger-Dyson Equation for a Massless Vector Theory in the Absence of a Fermion Pole, Comments Nucl. Part. Phys. 19: 137 (1989)

[2] R. Fukuda & T. Kugo, Nucl. Phys. B117:250 (1976); J. Kogut, E. Dagotto & A. Kocic, Strongly Coupled Quenched QED, Nucl. Phys. B317:253 (1989).

[3] D. Caldi, A. Chodos, K. Everding, D. Owen & Vafaeisefat, Theoretical and Phemenological Studies Concerning a Possible New Phase of QED, Phys. Rev. D39:1432 (1989)

[4] J. Schwinger, On Gauge Invariance and Vacuum Polarization, Phys. Rev. 82:664 (1951).

[5] V. Weisskopf, Kgl. Danske Videnskab Selskab, Mat.-fys. Medd. 14: 6 (1936)

[6] R. Feynman, An Operator Calculus Having Applications in Quantum Electrodynamics, Phys. Rev. 84:108 (1951)

ARE THERE BACKGROUND FIELDS THAT CAN INDUCE QED PHASE TRANSITIONS AT WEAK COUPLING?

Y. Jack Ng[1] and Y. Kikuchi[2]

[1]Institute of Field Physics
Department of Physics and Astronomy
University of North Carolina
Chapel Hill, NC 27599, USA

[2]Department of Physics
McGill University
Montreal, PQ H3A 2T8, Canada

ABSTRACT

The existence of a new, non-perturbative phase of QED is indicated by studies of Schwinger-Dyson equations and lattice calculations. The crucial question is whether the phase transition point can be driven down to $\alpha \sim 1/137$ presumably by appropriate background fields. It appears that magnetic fields potentially can induce such a phase transition. Our investigation is related to our original conjecture that the anomalous e^+e^- events at GSI are due to the decay of a new positronium system formed in the new QED phase which is induced by the electromagnetic fields of the heavy-ions.

INTRODUCTION

Several years ago we made a conjecture concerning the GSI anomalous events. We proposed that the e^+e^- peaks observed at GSI [1] are due to the decay of a bound e^+e^- system formed in a new QED strong coupling phase which is induced by the strong as well as rapidly-varying electromagnetic fields present in the heavy-ion collision experiments. [2-4] Now it appears that recent GSI experiments have put the two-body decay scenario of our conjecture in some jeopardy. But we believe the jury is still out on this question (whether our conjecture is right or wrong). In any case, our conjecture, like many other conjectures, once made, tends to take on a life of its own, in the sense that even if it fails to explain the GSI anomalous events, it has motivated us to look into a more general question. Instead of merely asking whether the electromagnetic background field present in the heavy-ion collisions can induce a QED phase transition, we now ask if there are background fields that can induce such a transition and if so, what kind of background

Vacuum Structure in Intense Fields, Edited by
H.M. Fried and B. Muller, Plenum Press, New York, 1991

fields. Of course we want the phase transition to be physically realizable i.e. at a value of the fine structure $\alpha \approx 1/137 \ll 1$, hence the qualification "at weak coupling" in the title.

THEORETICAL INTERPRETATION OF THE ANOMALOUS GSI EVENTS

Our investigation [2] was motivated by the anomalous e^+e^- events found in heavy-ion collision experiments at GSI. The multiple correlated and narrow-peak structures in electron and position spectra in the range of $\sim 3m_e$ are relatively Z-independent. At least in some experiments the energies of the e^+ and e^- are roughly equal in the center-of-mass frame and they are emitted back-to-back. Most likely these events are not due to nuclear decay or decay of new neutral elementary particles.[5] But they are, to a certain extent, consistent with two-body decays of neutral objects at rest in the center-of-mass frame.

The multiple structure in the e^+e^- spectra is clearly indicative of the formation of bound states. Now there are not too many objects that are (almost) as light as the electron or positron, and since the end products are just electrons and positrons, the simplest guess is that the constituents of the bound states are e^+ and e^-. But the bound state system cannot be the conventional positronium with a Coulomb potential since they have very different mass spectra. Hence it must be a new type of positronium. These considerations led us [2] (and two other groups [3,4]) to conjecture that the observed e^+e^- peaks at GSI are due to the decay of a bound e^+e^- system formed in a new phase of QED, which is induced by the strong and rapidly-varying electromagnetic fields of the large-Z ions. It is quite conceivable that shortly after the scattering of the heavy ions, in the absence of the strong electromagnetic fields, the new positronium system is left in a metastable vacuum and then it decays to the normal phase via quantum tunnelling into the observed e^+e^- pairs. [2,3] This theoretical scenario has the following attractive features: (1)it explains automatically why the peaks appear only in heavy-ion collisions and not elsewhere; (2) it explains the qualitative features of the experimental data;[6] (3) the proposal is a very economical one since there is no need to postulate any new particles (fields) or new types of interactions. Our proposal is also natural in that it makes use of a new, non-perturbative, "strong"-coupling, and chiral-symmetry-breaking phase of QED already known to exist from studies of the Schwinger-Dyson equations [7-11] and lattice gauge theories[11].

One may entertain the additional assumption that the new QED phase is confining so that to a fair approximation one may calculate the mass spectra of the new confined positronium system by using $V(\vec{r}) = \lambda r$ [2], the linear potential between two charged particles obtained in the (static) strong-coupling regime of lattice QED. To minimize the number of parameters in the problem one may use, for the electron mass, $m_e \approx 0.5$ MeV (although in the new chiral-symmetry breaking phase it is conceivable that the electron has an additional contribution to its mass.) From the observed e^+e^- spectra the mass splitting between the various states is typically ~ 100 keV, hence one gets $\sqrt{\lambda} \sim 200$ keV . The mean size of a typical state is ~ 1000 Fermi.[12] It is to be noted that such a large size appears to be necessary in order not to run into conflict with the existing experimental bounds on light neutral particles.[5]

Two predictions follow from our interpretation of the GSI anomalous events: (1) According to us, the strong and rapidly-varying electromagnetic fields in the heavy-ion experiments are responsible for the QED phase transition. So the anomalous e^+e^- peaks in the mass range of $\sim 1.5 - 1.8$ MeV are not expected in e^+e^- elastic scattering.[2] (Note that there is no violation of time-reversal invariance here.) (2) Since the new QED phase is characterized by spontaneous chiral symmetry breaking, we expect that the lightest new positronium state (the analogue of the pion) would be considerably lighter than the other states.[2,3] We encourage the experimentalists to search for it.

STRONG-COUPLING QED

The existence of a new, non-perturbative and strong-coupling phase of QED, the theoretical underpinning of our interpretation of the GSI events (and our on-going investigation), has been shown in two approaches using either (1) the (analytical) Schwinger-Dyson equations [7-11] or (2) the (numerical) lattice gauge calculations. Here we will go over briefly some of the ingradients in the SD equation approach.

The Schwinger-Dyson equations for QED are a set of complicated coupled integral equations. In order to make the problem tractable we will consider the SD equation for the fermion self-energy in the ladder-approximation. In this approximation the photon propagator and the vertices are replaced by the bare ones. With the choice of the Landau gauge the SD equation takes the form

$$\Sigma\,(p) = m_o + \frac{3e^2}{(2\pi)^4} \int d^4q\, \frac{1}{(p\text{-}q)^2}\, \frac{\Sigma(q)}{q^2 + \Sigma^2(q)}$$

where m_o is the bare fermion mass and we have made a Wick rotation into the Euclidean momentum space. To study the ultra-violet behaviour of the theory let us introduce an ultra-violet cut-off Λ. The intregral SD equation can be converted into a differential equation ($x \equiv p^2$, $\alpha \equiv \frac{e^2}{4\pi}$)

$$(x\,\Sigma(x))'' + \frac{3\alpha}{4\pi}\, \frac{\Sigma(x)}{x + \Sigma^2(x)} = 0$$

satisfying the boundary conditions

$$(x\,\Sigma(x))'\,|_{x=\Lambda^2} = m_o\,(\Lambda) \qquad\qquad \text{(Ultra-violet b.c.)}$$
$$x^2\,\Sigma'(x)\,|_{x=0} = 0 \qquad\qquad \text{(Infrared b.c.)}$$

By considering large x one can linearize the differential equation. The solutions take on two different forms depending on whether α is smaller or larger than a critical value $\alpha_c \equiv \pi/3$.

For weak coupling ($\alpha < \alpha_c$) one has

$$\Sigma(p) \;\overset{p\,\to\,\infty}{\underset{\sim}{}}\; p^{-1 + (1 - \frac{\alpha}{\alpha_c})^{\frac{1}{2}}}$$

with the UV boundary condition given by

$$m_o(\Lambda) \sim \Lambda^{-1 + (1 - \frac{\alpha}{\alpha_c})^{\frac{1}{2}}}$$

so that there is no spontaneous symmetry breaking solution.

For strong coupling ($\alpha > \alpha_c$) one gets

$$\Sigma(p) \sim \frac{1}{p} \frac{1}{(\frac{\alpha}{\alpha_c} - 1)^{\frac{1}{2}}} \sin [(\frac{\alpha}{\alpha_c} - 1)^{\frac{1}{2}} (\log p + \delta)]$$

with the UV boundary condition given by

$$m_o(\Lambda) \sim \frac{1}{\Lambda} \sin [\theta + \tan^{-1} (\frac{\alpha}{\alpha_c} - 1)^{\frac{1}{2}}]$$

where δ is a function of α ($\delta(\alpha) \approx 0.715$ for $\alpha \approx \alpha_c$) and

$$\theta = | \frac{\alpha}{\alpha_c} - 1 |^{\frac{1}{2}} (\log \frac{\Lambda}{\Sigma(0)} + \delta)$$

The UV boundary condition with $m_o(\Lambda) = 0$ admits infinitely many spontaneous chiral symmetry breaking solutions given by, for $\alpha \approx \alpha_c$,

$$\theta \approx n\pi - (\frac{\alpha}{\alpha_c} - 1)^{\frac{1}{2}}$$

where n - 1 counts the number of nodes in $\Sigma(p)$. For the ground state, i.e., the lowest vacuum energy, [13] n = 1 so that $0 < \theta < \pi$ yielding

$$\Sigma(0) = e^{\delta + 1} \Lambda \, e^{-\pi(\frac{\alpha}{\alpha_c} - 1)^{-\frac{1}{2}}}$$

which, for fixed α , is proportional to Λ . But, for a physically meaningful solution the fermion mass $\Sigma(0)$ must remain finite in the large Λ (i.e. continuum) limit. This is possible if α_c is the (Miransky) ultra-violet fixed point[8]:

$$\frac{\alpha(\Lambda)}{\alpha_c} = 1 + \frac{\pi^2}{\log^2 (\Lambda/\kappa)}$$

where κ is an infrared mass scale proportional to $\Sigma(0)$.

How good is the ladder approximation? The replacement of the full vertex Γ_μ by the bare vertex γ_μ turns out not to affect our linearized differential equation since we are interested in $\Sigma^2(p^2) \ll p^2$. [11] This can be seen by applying the Ward identity $q^\mu \Gamma_\mu(p, p + q) = S^{-1}(p + q) - S^{-1}(q)$. For large p, $iS^{-1}(p) \to \gamma^\mu p_\mu + $ constant, so that the Ward identity implies $\Gamma_\mu = \gamma_\mu + O(\Sigma(p)/p)$. How about the quenched approximation of replacing the full photon propagator by the bare propagator? It is harder to estimate its reliability, but numerical works [14] and lattice calculations [11] have shown that the approximation is qualitatively admissible.

The solutions to the SD equation found above give some unexpected "renormalization group" interpretations to non-perturbative QED. The anomalous dimension of the fermion mass operator defined by

$$\gamma(\alpha) = -\frac{\partial \log m_o(\Lambda)}{\partial \log \Lambda}$$

for $\alpha \approx \alpha_c$ is particularly interesting:

$$\gamma(\alpha) = 1 - (1 - \frac{\alpha}{\alpha_c})^{\frac{1}{2}} \qquad\qquad \alpha < \alpha_c$$

$$\gamma(\alpha) = 1 \qquad\qquad \alpha > \alpha_c$$

Due to this large value of the anomalous dimension the well-known Appelquist-Carazzonne decoupling theorem [15] is violated with the consequence that ultra-violet physics affects infra-red dynamics (e.g. $\Sigma(0)$). In the ladder approximation, the four-fermion operators have twice the dimension of the fermion mass operator, so that $d_{(\overline{\psi}\psi)^2} = 2(3 - \gamma)$ which approaches 4 as $\alpha \to \alpha_c$. Thus the $(\overline{\psi}\psi)^2$ interactions are relevant operators and can mix with the electromagnetic interactions.[9]

For a consistent analysis of the critical behaviors one is thus led to enlarge the pure QED Lagrangian to include all chiral invariant four-fermion interactions [9] such as those in the Nambu-Jona-Lasinio model (NJL)[16]:

$$\mathcal{L} = \overline{\psi}(i\gamma^\mu D_\mu - \mu_o) \psi + \frac{G_o}{2}[(\overline{\psi}\psi)^2 + (\overline{\psi}i\gamma_5 \psi)^2]$$

where we have now denoted the bare fermion mass by μ_o . (Note that the vector- and axial-vector four-fermion interactions are also chiral invariant. But a combination of them is, via Fierz transformation, equivalent to the scalar and pseudoscalar combination we have introduced. Also note that the pure vector $(\overline{\psi}\gamma_\mu \psi)^2$ term is irrelevant due to current conservation leading to zero anomalous dimension. On the other hand, terms like $\overline{\psi} \sigma_{\mu\nu}\psi F^{\mu\nu}$ may have to be included.[17]) In the ladder approximation the four-fermion interactions are easy to incorporate into the SD equation since the only extra contribution comes from the fermion tadpole which is independent of the external momentum. Its only effect is to change the bare fermion mass μ_o to

$$m_o = \mu_o - G_o <\overline{\psi}\psi>_o$$

without changing the differential equation. The closed fermion loop $< \overline{\psi}\psi >_o$ in the tadpole can be calculated by using the previously found solutions.

In the $\mu_o = 0$ chiral limit, the gap equation reads [9]

$$\alpha > \alpha_c : \ \tan \theta = \frac{G + 1}{G - 1}(\frac{\alpha}{\alpha_c} - 1)^{\frac{1}{2}}$$

$$\alpha < \alpha_c : \ \tanh \theta = \frac{G + 1}{G - 1}(1 - \frac{\alpha}{\alpha_c})^{\frac{1}{2}}$$

where $G \equiv G_o\Lambda^2\alpha_c/(\pi^2\alpha)$ is the renormalized four-fermion coupling. For strong coupling $(\alpha > \alpha_c)$, the vacuum solution again requires $0 < \theta < \pi$ so that the continuum limit demands $\alpha \to \alpha_c$, $G \to 1$. For weak coupling ($\alpha < \alpha_c$), (surprisingly) nontrivial solutions exist for $G > 1$: in the continuum limit θ becomes infinitely large, yielding the critical line given by [10]

$$\frac{G_0 \Lambda^2}{\pi^2} = [\, 1 + (\, 1 - \frac{\alpha}{\alpha_c})^{\frac{1}{2}}\,]^2$$

which goes smoothly to the pure NJL limit (more about this later) as $\alpha \to 0$. It may not be totally crazy to speculate that the critical behaviour for weak electromagnetic coupling at $\alpha \sim 1/137$ (and $G_0 \Lambda^2/\pi^2 \approx 4$) is reasonably approximated by that for the pure NJL model ($\alpha = 0$ and $G_0 \Lambda^2/\pi^2 = 4$, to be shown later). The pure NJL model has the distinct advantage of being amenable to simple analytical analyses involving only algebraic equations. In the next section we will take this speculation seriously and our investigation will concentrate on the NJL model.

The importance of the four-fermion interactions for physical consistency can also be glimpsed by an examination of the renormalization-group β-function. The Miransky solution for pure QED demands a dynamical running of α for $\alpha > \alpha_c$. But in the quenched approximation, the origin of the running of α is quite unclear. On the other hand, the four-fermion coupling in the QED-NJL theory is renormalized even in the ladder approximation. It is the dynamical running of the four-fermion coupling that leads to a finite dynamical fermion mass and hence an infrared meaningful theory when $G \to 1$ and $\alpha \to \alpha_c$ in the continuum limit. [9]

Needless to say, in our discussion, the dynamical degree of freedom of the photon is utterly crucial for phase transitions in QED (with or without the NJL terms.) We believe that the dynamical photon also plays an indispensible role in the critical behaviour of QED in background fields. [18]

INDUCING QED PHASE TRANSITIONS BY BACKGROUND FIELDS?

Our theoretical interpretation of the GSI events described above hinges on the speculation that there is a connection between the strong (as well as rapidly-varying) electromagnetic fields and strong-coupling QED. This connection has not been proven so far. To prove it, one will have to show that the QED critical coupling separating the two phases, which is ~1 in the absence of background fields, can be driven down to the normal weak value ~ 1/137 by the appropriate electromagnetic background fields. At present it is not known what attributes of the background fields are relevant for QED phase transitions.

The more general question we are asking now is whether there are background fields, not necessarily pure electromagnetic ones, which can induce the phase transition point to move towards *weaker* coupling.[19] So far we have examined only QED and the NJL model with some background fields that are easily amenable to analytical analysis.

As the first example let us consider QED with a (small) chemical potential μ, a non-zero value of which signifies the presence of some background e^- (or e^+). This example may have bearing on the interpretation of the GSI events since the heavy ions in the collisions are presumably surrounded by clouds of electrons. Note that the addition of a chemical potential does not break chiral symmetry by itself. For $\mu \neq 0$, the electron propagator is given by [20]

$$G_\mu (p) = \frac{\gamma \cdot p + m}{[\, p_o + i\varepsilon \, \text{sign} \, (p_o + \mu)\,]^2 - (\vec{p}^2 + m^2)}$$

where m is the electron mass and ε is infinitesimally small but positive. Accordingly the Schwinger-Dyson equation for the electron self energy in the ladder approximation with the choice of Landau gauge is [21]

$$\Sigma_\mu (p) = m_o - 3ie^2 \int \frac{d^4 q}{(2\pi)^4} \frac{1}{(p-q)^2} \frac{\Sigma_\mu (q)}{[q_o + i\varepsilon \, \text{sign} \, (q_o + \mu)]^2 - \vec{q}^2 - \Sigma_\mu^2 (q)}$$

In going to the Euclidean momentum space by a Wick rotation we have to examine more closely the pole structure of the electron propagator in the complex q_o plane. Poles are located at $q_o = \pm \sqrt{\vec{q}^2 + \Sigma_\mu^2(q)} - i\varepsilon \, \text{sign} \, (q_o + \mu)$. For $0 < |\mu| < \sqrt{\vec{q}^2 + \Sigma_\mu^2(q)}$, i.e., $\vec{q}^2 > \mu^2 - \Sigma_\mu^2(q)$, the usual two poles appear with one pole each in the upper-and lower-half planes. But for $0 < \vec{q}^2 < \mu^2 - \Sigma_\mu^2(q)$ both poles are found in either the upper (for $\mu < 0$) or lower (for $\mu > 0$) half planes. Thus in performing the Wick rotation ($q_o \to iq_o$) one picks up an additional contribution to $\Sigma_\mu(p)$ from the momentum interval $0 < \vec{q}^2 < \mu^2 - \Sigma_\mu^2(q)$ (if such an interval exists). Assuming that this momentum interval is small we can replace $\Sigma_\mu(q)$ in the integrand by $\Sigma_\mu(0)$. For large p, the extra piece of $\Sigma_\mu(p)$ is thus proportional to p^{-2}. One can solve the SD equation for large p following the same procedure as in the $\mu = 0$ case. The differential equation is not affected by the extra piece of $\Sigma_\mu(p)$. Neither is the ultra-violet boundary condition. The only effect is in the infrared region (replacing $\Sigma(0)$ by $\Sigma_\mu(0)$.) Hence for this case the critical coupling is not changed.[21]

To gain further insights let us turn to the pure NJL model which, as shown above, is relevant to our study of critical behavior of QED. In the absence of background fields the model, [16]

$$\mathcal{L}_{NJL} = \frac{G_o}{2} [\, (\overline{\psi} \, \psi)^2 + (\overline{\psi} i\gamma_5 \, \psi)^2 \,],$$

has nontrivial chiral-symmetry breaking solutions ($m \neq 0$) with the gap equation given by

$$1 = -\frac{G_o i}{4\pi^4} \int \frac{d^4 p}{p^2 + m^2 - i\varepsilon}$$

At this point, we have a choice of two different cut-off regularizations. We will make use of both of them later. If one uses a non-covariant cut-off at $|\vec{p}| = \Lambda$ to regularize the integral, the critical coupling is given by

$$\frac{2\pi^2}{G_o \Lambda^2} = 1$$

One can also adopt a covariant cut-off at $p^2_{\text{Euclidean}} = \Lambda^2$, then the critical coupling is given by

$$\frac{4\pi^2}{G_o \Lambda^2} = 1$$

which as mentioned before, is nothing other than the $\alpha = 0$ limit of the critical line for the QED-NJL model.

For the NJL model with a chemical potential μ, the gap equation

$$1 = -\frac{G_o\, i}{4\pi^4} \int d^3 p \int d\,p_o \frac{1}{-[\,p_o + i\varepsilon\, \text{sign}\,(p_o - \mu\,)]^2 + \vec{p}^2 + m^2}$$

yields the critical coupling

$$\frac{2\pi^2}{G_o \Lambda^2} = 1 - \frac{\mu^2}{\Lambda^2}$$

if we use the non-covariant cut-off. The critical coupling is thus larger than for the $\mu = 0$ case, i.e., with a chemical potential the phase transition point moves towards stronger coupling, consistent with the physical picture that a nonvanishing e^- (e^+) background density repels the e^- (e^+) but attracts the e^+ (e^-) in the e^+e^- condensate making it harder to have e^+e^- condensates the formation of which is necessary for spontaneous chiral symmetry breaking.[21]

For the NJL model in a heat bath $(T \neq 0\,)$, the gap equation is [22]

$$1 = \frac{G_o}{\pi^2} \int dp \frac{p^2}{(\,p^2 + m^2\,)^{\frac{1}{2}}} [\, 1 - \frac{2}{e^{(p^2 + m^2\,)/(kT)} + 1}\,]$$

where the non-covariant cut-off has been adopted. A nonvanishing temperature reduces the positive definite integrand, leading to a larger critical coupling. This result is physically reasonable since thermal excitations tend to dissociate the e^+e^- pairs in the condensate.

Finally let us consider the NJL model with constant electromagnetic background field. One can employ Schwinger's proper-time method [23] to cast the gap equation in the form [24]

$$1 = \frac{G_o}{2\pi^2} \int_{\Lambda^{-2}}^{\infty} \frac{ds}{s}\, e^{-m^2 s}\, \{\, (es)^2\, G \frac{\text{Re}\,[\cosh\,(esX)]}{\text{Im}\,[\cosh\,(esX)]}\, \}$$

where $G = \vec{E}\cdot\vec{H}$ and $X^2 = (\vec{H} + i\vec{E})^2$ for constant electric field $\vec{E}$ and constant magnetic field $\vec{H}$; here the covariant cut-off has been used to regularize the integral. Four limiting cases can be easily worked out. The quantity inside the braces is 1, eEscot(eEs), 1, eHs coth(eHs) respectively for the cases $\vec{E} = \vec{H} = 0$, $\vec{H} = 0$ but $\vec{E} \neq 0$, $\vec{E} \perp \vec{H}$ with E = H, and $\vec{E} = 0$ but $\vec{H} \neq 0$. (Here, e, E and H denote their respective magnitudes.) The critical coupling for the respective cases is given by [24]

$$\frac{4\pi^2}{G_0 \Lambda^2} = 1, \quad 1 - \frac{\pi e E}{2\Lambda^2}, \quad 1, \quad 1 + \frac{eH}{\Lambda^2} \ln \frac{2eH}{m^2}$$

Thus a constant electric field drives the transition point towards stronger coupling while a constant magnetic field has the opposite effect. The electric field effect is consistent with our intuition since an electric field acts on e^+ and e^- in the condensates with opposite forces. It is harder to understand the magnetic field effect. A possible explanation lies in the formation of the mixed condensate $< \overline{\psi} \sigma^{\mu\nu} \psi F_{\mu\nu} >$. While chiral symmetry guarantees that it is zero in the perturbative weak-coupling phase, a non-zero value would indicate chiral symmetry breaking. A fermion moving through this mixed condensate acquires an anomalous magnetic moment to which an existing magnetic background field can couple. So it is not unlikely that a magnetic field helps to lower the critical coupling.[25]

SUMMARY

We are not ready to give a definite answer to the question posed in the title. But if one accepts the premise that the critical behvior of QED (supplemented by appropriate relevant operators like those in the NJL model) at small electromagnetic coupling ($\alpha \sim 1/137 \ll 1$) is not too different from that of the pure NJL model, then the example considered above with a pure magnetic background field does give one a glimmer of hope that one can perhaps answer the question affirmatively. Judging from the form of the critical coupling for the pure electric and pure magnetic cases it seems even possible that in some electromagnetic field configurations a relatively small magnetic component can overcome the effect of a larger electric field. For our conjecture in connection with the GSI anomalous events to be correct, this must be the case since in heavy-ion collisions the dominant electromagnetic field is electric. A detailed study is warranted.

ACKNOWLEDGEMENTS

Y. J. Ng thanks H. Fried and B. Müller for organizing such an enjoyable summer school. He also thanks them, D. G. Caldi, M. H. Bokemeyer and especially A. Kocic and E. Dagotto for useful discussions. This work was supported in part by US DOE under Grant No. DE-FG05-85ER-40219.

REFERENCES

1. M. H. Bokemeyer, these proceedings; M. W. Koenig, these proceedings.
2. Y. J. Ng and Y. Kikuchi, Phys. Rev. D36, 2880 (1987).
3. D. G. Caldi and A. Chodos, Phys. Rev. D36, 2876 (1987). Also see Caldi, Chodos and D. A. Owen, these proceedings.

4. L. S. Celenza, V. K. Mishra, C. M. Shakin and K. F. Liu, Phys. Rev. Lett. <u>57</u>, 55 (1986). The idea in this paper is similar to that suggested in Ref. 2 and 3. (We learned of this paper near the end of our work described in Ref. 2.) See also L. S. Celenza, C. R. Ji and C. M. Shakin, Phys. Rev. <u>D36</u>, 2144 (1987).

5. See, e.g., K. Geiger et al. in Proc. of the XXIVth Recontre de Moriond, edited by O. Fackler and J. Tran Thanh Van (Editions Frontieres, France, 1989), p. 107.

6. Recent GSI data (see Ref. 1) indicate that not all e^+e^- come out back-to-back and with equal energies. Possible explanations for this (in the scenario of two-body decay) can be found in D. G. Caldi, W. Greiner, and D. A. Owen, these proceedings.

7. K. Johnson, M. Baker and R. Willey, Phys. Rev. <u>136</u>, B1111 (1964); S. Adler, Phys. Rev. <u>D5</u>, 3021 (1972); T. Maskawa and H. Nakajima, Theor. Phys. <u>52</u>, 1326 (1974); R. Fukuda and T. Kugo, Nucl. Phys. <u>B117</u>, 250 (1976).

8. V. A. Miransky, Il. Nuovo Cim. <u>90A</u>, 149 (1985); V. A. Miransky and P. I. Fomin, Sov. J. Part. Nucl. <u>16</u>, 203 (1985).

9. W. A. Bardeen, C. N. Leung and S. T. Love, Nucl. Phys. <u>B273</u>, 649 (1986); <u>B323</u>, 493 (1989).

10. K. Yamawaki, M. Bando and K. Matumoto, Phys. Rev. Lett. <u>56</u>, 1335 (1986); K. Yamawaki in Proceedings of the 1988 International Workshop on New Trends in Strong Coupling Gauge Theories, edited by M. Bando, T. Muta, K. Yamawaki (World Scientific, Singapore, 1989) p. 12, and references contained therein.

11. J. B. Kogut, E. Dagotto and A. Kocic, Phys. Rev. Lett. <u>60</u>, 772 (1988); E. Dagotto and A Kocic, these proceedings; and references contained therein. For earlier works on QED phase transitions in the framework of lattice gauge theories see K. Wilson, Phys. Rev. <u>D10</u>, 2445 (1974) and M. Creutz, Phys. Rev. Lett. <u>43</u>, 553 (1979).

12. Consult, e.g., C. Quigg and J. L. Rosner, Phys. Rep. <u>56</u>, 167 (1979).

13. The vacuum energy for the massless theory is given in J. M. Cornwall, R. Jackiw and E. Tomboulis, Phys. Rev. <u>D10</u>, 2428 (1974).

14. T. Appelquist, K. Lane and U. Mahanta, Phys. Rev. Lett. <u>61</u>, 1553 (1988).

15. T. Appelquist and J. Carazzonne, Phys. Rev. <u>D11</u>, 2856 (1975).

16. Y. Nambu and G. Jona-Lasinio, Phys. Rev. <u>122</u>, 345 (1961).

17. A. Kocic, E. Dogotto and J. B. Kogut, Phys. Lett. <u>B213</u>, 56 (1988); H. L. Yu, T. S. Lee and W. B. Yeung, Phys. Rev. <u>D39</u>, 2415 (1989), V. P. Gusynin and V. A. Kushnir, Phys. Lett. <u>B242</u>, 474 (1990).

18. If one neglects the dynamical photon in the discussion of an electron propagating in a background field one is essentially dealing with a quantum mechanical problem and not a QED quantum field problem. It is not entirely clear whether one can adequately discuss spontaneous symmetry breaking outside the framework of quantum field theory. Y. J. Ng thanks E. Dagotto and A. Kocic for an interesting discussion on this point.

19. For the case of a Coulomb field see E. Dagotto and H. W. Wyld, Phys. Lett. <u>B205</u>, 73 (1988); for the case of a weak photon-condensate background field see Y. Kikuchi and Y. J. Ng, Phys. Rev. <u>D38</u>, 3578 (1988). Also see Y. J. Ng in Proceedings of the XXIVth Moriond Conference, and R. Peccei, Nature <u>332</u>, 492 (1988).

20. See, e.g., E. V. Shuryak, "The QCD Vacuum, Hadrons and the Superdense Matter" (World Scientific, Singapore, 1988).

21. Y. Kikuchi, C. N. Leung and Y. J. Ng (unpublished).

22. J. Cleymans, A. Kocic and M. D. Scadron, Phys. Rev. <u>D39</u>, 323 (1989).

23. J. Schwinger, Phys. Rev. <u>82</u>, 664 (1951).

24. S. P. Klevansky and R. H. Lemmer, Phys. Rev. <u>D39</u>, 3478 (1989). Their calculations have been slightly simplified by us.

25. A. Kocic (private communication).

NEW QED, THE GSI PEAKS, AND BACKGROUND FIELDS

D. G. Caldi

Department of Physics and Astronomy
State University of New York at Buffalo
Buffalo, New York 14260

ABSTRACT

There is much evidence that quantum electrodynamics may have a new, non-perturbative phase, characterized by spontaneous chiral symmetry breaking and confinement. We discuss the possibility that this phase may have been seen in the heavy-ion collision experiments at GSI: the narrow peaks in the e^+e^- coincidence data are well explained as composite states of e^+e^- in this new phase of QED. We concentrate particularly on the question of whether background electromagnetic fields can induce the phase transition, and we present some new, direct, affirmative evidence that field configurations similar to those in the experiments can lead to spontaneous mass generation.

I. INTRODUCTION

Quantum Electrodynamics (QED) is the best tested of all physical theories, as well as being our best known and understood quantum field theory, serving as the paradigm for our gauge theories of the other fundamental interactions -- the electroweak theory and Quantum Chromodynamics (QCD). But the successes of QED have come in the context of perturbation theory, and not only is the perturbation series at best asymptotic, but the theory, at high energies, is trivial at best due to the well-known Landau ghost problem. However, QED alone, at least in its perturbative realization, is only an effective theory.

Another possiblitity is that QED really is a non-trivial theory all by itself, but in a different, non-perturbative phase. In the past few years there has been a revival of interest in this possibilty, with many non-perturabtive studies converging on the realization that there is another, "new" (to us, not to nature) phase of QED as one theoretically increases the coupling to of order one. This phase is very different from the one we are used to from perturbative analysis: in it chiral symmetry is spontaneously broken, and there is probably confinement as well. A number of the talks[1] at this Institute have discussed this in some detail.

Vacuum Structure in Intense Fields, Edited by
H.M. Fried and B. Muller, Plenum Press, New York, 1991

Our main purpose here is to demonstrate how far it is possible to connect[2] this new phase of QED to the intriguing collection of data which has been amassing over the past few years from experiments[3,4] done at the GSI facility in Darmstadt, involving heavy-ion collisions at energies close to the Coulomb barrier. These experiments have also been extensively reviewed[5] at this Institute, and the most striking, as well as thought-provoking, feature of the data is the appearance of at least three (so far) narrow peaks in the e^+e^- coincidence spectra in the mass range 1.6 – 1.8 MeV. It has been proposed[6] that the peaks are well-explained as bound states of e^+e^- in the new phase of QED. This proposal, though somewhat speculative in some of its aspects, at least has the advantage of not (yet) being ruled out on phenomenological grounds. The central question for the validity of this scenario is whether the strong and rapidly varying background electromagnetic fields present in the heavy-ion experiments can actually induce a phase transition to new QED, without relying on the, essentially impossible to achieve, growth of the coupling to of order one. We present some recent results[7] of computer calculations which answer this question directly and in the affirmative.

II. THE GSI PEAKS AND PHENOMENOLOGY OF NEW QED

A View of the Data

A few years ago experiments were started at the Gesellschaft für Schwerionenforschung (GSI) in Darmstadt, West Germany involving their UNILAC accelerator to look at heavy-ion collisions at energies around 6 MeV per nucleon, i.e. just around the Coulomb barrier. Two main groups have been doing these experiments: the Electron Positron Spectrometer group (EPOS) and the Orange Spectrometer group. Because the experimentalists themselves have provided reviews[5] in this volume, what will be given here is one theorist's attempt at an unbiased summary of the data.

There are now at least three peaks observed in the e^+e^- coincidence data of the EPOS group. The sum energies plus masses for these peaks are: 1.64 MeV, 1.77 MeV, and 1.83 MeV, with widths between 20 to 40 keV, but this is the experimental resolution. So the range of the lifetimes for these states is $10^{-19}s < \tau < 10^{-9}s$ (the latter from the fiducial volume). These peaks are seen by the EPOS group in U + Ta and in U + Th. The Orange group has also recently reported seeing these coincidence peaks, except that at 1.77 MeV, at nearly the same energies in U + U and U + Pb, with perhaps another peak at 1.58 MeV, and maybe others as well. For both groups, the peaks are 3 to 6 sigma effects, with the 1.83 MeV peak showing the largest signal.

What appear to be the salient features of these peaks include the following: Their positions are essentially independent of Z. The electrons and positrons often come out with equal energies, the difference being ±10 to 220 keV, with *no consistent trend* toward either positrons or electrons emerging with greater energy. The electrons and positrons often emerge back-to-back, for example in the 1.83 MeV data, the angle is consistently 180°. The possible exception to the last feature is the 1.77 MeV peak, which, both groups agree, has either the e^+ and e^- emerging at closer to 90°, or the state is much longer lived than the others. So these last two features indicate that there are sometimes aspects of a non-two-body decay.

Since the heavy-ion experiments are inherently messy, it is not surprising that experimentalists looked for another, cleaner system in which to search for these states. Reasoning from time-reversal invariance that if the objects decayed to $e^+ e^-$, then it should also be possible to produce them in very low energy electron-positron collisions, experimentalists at many different facilities around the world have been engaged in this search. The basic idea is to look for anomalous peaks in Bhabha ($e^+e^- \rightarrow e^+e^-$) scattering. So far, and this is from many different experiments, the results are all negative[8,9]. In the context of the new phase explanation, this is perhaps not surprising. It, at the least, tends to imply that the background electromagnetic fields present in the heavy-ion experiments are quite crucial for producing the states. Furthermore, there really is no problem with time-reversal invariance, since one must reverse not just the electron and positron, but also the electromagnetic fields (presumably in the form of soft photons) coming out of the heavy-ion collisions. A phase transition, by its very nature, always seems to violate time-reversal invariance: melted ice never re-crystalizes, unless you reduce the temperature below T_c again.

We should also mention one other experiment, namely, a Stanford-LBL collaboration[10] at the HILAC in which conditions similar to the EPOS group were reproduced, in this case with U on Th, but they were detecting photon pairs instead of e^+e^-. At first they reported seeing a peak decaying to $\gamma\gamma$ at 1.062 MeV. However, on further analysis they have found that the effect can be explained by known nuclear transitions.

The current experimental situation is, unfortunately, rather quiet, in that the pursuit of the e^+e^- - scattering experiments has resulted in the temporary abandonement of the heavy-ion experiments. But that will soon be remedied, with improved experiments being prepared at GSI, and the introduction of these experiments at Argonne in the near future. As an example of the many significant open questions which these new experiments will eventually, one hopes, answer, we note that it is still not known how low in Z one can go and still see the peaks.

Phenomenology of New QED

Many attempts have been made to explain the data[11,12], however, most of these have been found deficient already on phenomenological grounds. Generically the difficulties are with the following facts: the Z-independence, the narrowness, that the e^+e^- come out often with equal energies and back-to-back, and finally the appearance of many peaks.

So faced with the failure of all these more or less conventional explanations and the evidence of many peaks, some of us decided that the best hope lay in assuming that we are dealing with some sort of composite system. The question then was, a composite system of what?

A composite system, of course, gives rise to many energy levels, and the data appear to tell us that at least some of these levels decay to e^+e^-. So the simplest and most conservative assumption[6] is that the states are actually composites of e^+ and e^- (More recently, some have proposed[13] an explanation based on a composite system of totally new constituents with new forces. Although, not surprisingly, with such a specifically designed system one can match much of the data, the implementation of Occam's razor appears not inappropriate for this quite *ad hoc* approach.) To be sure, a composite system of e^+ and e^- is well known, namely positronium. But positronium has all its levels below the $2m_e$ threshold. Thus a number of groups were led to postulate[6] that these states are bound states of e^+ and e^- in a new non-perturbative phase of QED which has a somewhat higher mass scale associated with it. A familiar and rather straightforward way to generate a new mass scale is to have chiral symmetry spontaneously broken in the new phase, so that the electron has a perhaps additional contribution to its mass or a new mass, and hence the bound states can lie in the neighborhood of 1.7 MeV.

From a phenomenological viewpoint, this new-phase scenario has many attractive features. It can explain most, and perhaps eventually all, of the heretofore puzzling aspects of the data. First, in this context one can understand why the states are seen in the heavy-ion collisions but are difficult at best, and perhaps impossible, to produce in other systems which do not have the strong and rapidly varying background electromagnetic fields present when the high-Z nuclei collide. The idea is that these unusual background fields induce a phase transition to the new QED vacuum. From the size of the region where the fields are large ($E \approx 10^{16}$ V/cm ≈ 1 MeV), one can estimate the size of the region of new phase to be between 100 and 1,000 fermi. (This also agrees with an uncertainty principle argument from the narrowness of the peaks, $\Gamma \lesssim 40$ keV.) After the ions separate and the fields die off, the new vacuum becomes metastable and, since it is now false vacuum, it eventually decays, liberating the usual e^+e^- in the normal phase, plus many soft (and sometimes not so soft) photons from the decay of the vacuum bubble. Since there are no large fields now, the electron and positron can come out, as seen in the data, often with equal energies and back-to-back, or sometimes with many-body decay features if some of the vacuum-decay photons are not so soft. Because we are dealing with bound states of e^+e^-, this picture also explains why there is some preference for decay into e^+e^- rather than photons. Of course, as mentioned above, we also see why there are several states. Also, once the right electromagnetic field configurations are produced, the phase transition will take place. Thus one can understand the Z-independence of the states, since the observed levels are those in the metastable phase after the fields have dissapated, just before the phase transition back to the usual phase. We will also see below that spectrum calculations in the new phase are in very good agreement with the observed peaks. Finally, a clear prediction of this scenario is the existence of an electro-pion which should be somewhat lighter than the other states. Although the $\gamma\gamma$ state at 1.062 MeV was a good candidate for such an electro-pion, its disappearance should not discourage the search for a relatively low-lying pseudoscalar state perhaps decaying, at least part of the time, via the anomaly triangle diagram (if present in the new phase), to $\gamma\gamma$.

III. THEORETICAL EVIDENCE FOR NEW QED

Theoretical investigations of QED which have revealed a new, non-perturbative phase have been going on for close twenty years, both on the lattice[1,14-16] and in the continuum[17-19]. Until recently the phase transition has been studied mostly as a function of coupling, and as one increases the coupling to something of order one, the transition to the new phase is observed.

There are a number of versions of lattice QED. The first to be studied extensively in the context of a phase transition was the compact U(1) lattice (pure) gauge theory with Wilson's action[14,15]. Although there are no fermions in this theory, and in the naive, weak-coupling continuum limit it is a free-field theory, nevertheless, it is not a free theory on the lattice. It can be thought of as an effective theory for QED once the fermions are integrated out, leaving effective interactions among the photons. Various Monte Carlo studies[14] established that there are two phases for this theory with the phase transition taking place at lattice coupling $g \approx 1$: a confining phase in strong coupling and a Coulomb phase in weak coupling. The existence of these phases was actually rigourously proven by analytic work[15]. For some time it was thought that the confining phase could be dismissed as merely a lattice artifact. However, eventually it became clearer that, given a second-order phase transition, one could take the continuum limit from the strong-coupling side of the phase transition, and so end up with a very different, non-perturbative QED.

More recently there have been studies with fermions in various formulations[1,16] -- compact and non-compact, quenched and unquenched. In these studies one can no longer monitor confinement, but instead one looks at $\langle \overline{\psi}\psi \rangle$ to monitor spontaneous chiral-symmetry breaking. Again there are two phases with the phase transition at $g \approx 1$; for strong coupling, chiral symmetry is spontaneously broken.

In the continuum, the main approach for investigating a new phase of QED has been through the machinery of the Schwinger-Dyson equation and related Bethe-Salpeter equation studies[17-19]. These have shown that for coupling strong enough ($\alpha \gtrsim \alpha_c = \pi/3$, i. e., again order 1) there are spontaneous chiral symmetry breaking solutions. These solutions are to the linearized Schwinger-Dyson equations for the fermion self-energy $\Sigma(p)$, in the ladder (planer), quenched approximation.

One of the promising recent developments in the study of new QED has been the discovery[20] that four-fermion interactions, which in perturbative analysis are non-renormalizable, appear to become renormalizable as one approaches the phase transition to new QED. This is because the anomolous dimension of the fermion field changes so that the dimension of $\overline{\psi}\psi$ goes to 2. Hence $(\overline{\psi}\psi)^2$ is a relevant operator in the new phase and so must be included (in a chirally symmetric form) with a new coupling in the theory. This development led a number of groups, both in the continuum[21] (using the Schwinger-Dyson approach) and on the lattice[22] (using Hamiltonian strong-coupling expansions), to calculate the phase diagram for the theory now with two couplings -- the usual gauge coupling g and the four-fermion coupling A . The results are in agreement that the phase transition point shifts to weaker gauge coupling as one turns on and increases the four-fermion coupling. Indeed, in terms of α, α_{crit} can be quite weak, including $\alpha = 1/137$. One of the lessons to be learned from this is that, in general, it may be possible for the phase transition to the new phase to occur for the usual QED α of 1/137, so long as some other variable is also operating. In the context of the new phase scenario for the heavy-ion collision data, this other variable is taken to be some attribute(s) of the background electromagnetic fields, including, but not necessarily limited to, the strength of the fields. In addition, one also sees the possibility that it may be easier to establish the phase transition via electromagnetic fields if these can induce the four-fermion interactions, so that one is now considering a three-dimensional (at least) phase diagram.

One of the best pieces of indirect evidence for the new-phase scenario, besides its phenomenological successes, is the agreement of the spectrum calculated in the new phase with the observed states. We did a fourth-order $(1/g^8)$ calculation[23] of the spectrum using a strong-coupling expansion in Hamiltonian lattice QED with Kogut-Susskind fermions. A chirally invariant four-fermion interaction was included. It should be noted that the conditions in which the states were calculated, near the transition point and in the absence of background fields, actually correspond to the conditions in which the presumed physical states in the heavy-ion experiments find themselves just before decay takes place.

We now come to what has been the major challenge for the new-phase scenario: to demonstrate that the electromagnetic field configurations present in the heavy-ion collisions can themselves actually induce the phase transition. After all, the theoretically known phase transition occurs as one increases a coupling (be it gauge or fermion), but this presumably is not what is operative in the heavy-ion experiments, where there is no evidence that α is anything but its usual value. There are a number of studies[7,23-27] beginning to explore this question so central to the new-phase explanation. For example, a lattice simulation which included a static Coulomb background field showed that the critical coupling increases in this environment rather than decreases[24]. But this only tells us that static Coulomb fields are not the correct configurations, which one may have surmised both from the experiments and other considerations. At least one important attribute missing is that the fields in the experiment are rapidly varying.

Another line of attack[27] has been to incorporate background electromagnetic fields into the Schwinger-Dyson equations, again to see if α_{crit} can be driven to weaker values. So far, only rather drastic approximations have been tried, and these have been unsuccessful. There is also recent work reported at this Institute[27] incorporating a magnetic field in the Nambu-Jona-Lasinio model and seeing that the chiral symmetry breaking is enhanced by the magnetic field. In line with this is some older work of Linde[28] in the context of the Abelian Higgs model showing that $\langle \psi \rangle$ increases with an external magnetic field.

The approach[7,23] I want to concentrate on has been to look at the Schwinger proper-time effective Lagrangian with a background electromagnetic field. This Lagrangian[29] incorporates all the one-loop fermionic effects, but ignores dynamical photons (which actually might be important for driving the phase transition). I wish to report on recent evidence[7] from studies[23] looking for chiral-symmetry breaking as one tries various background field configurations in Monte Carlo simulations of the four-dimensional quantum-mechanics partition function resulting from the Schwinger formalism.

In the Schwinger formalism one writes the order parameter for chiral symmetry breaking $\langle \overline{\psi} \psi \rangle$ as a proper-time integral of a quantum mechanical transition amplitude:

$$\langle \overline{\psi} \psi \rangle = m \int_0^\infty d\tau \, \exp(-m^2\tau) \, \mathrm{tr} \, \langle \, x \, | e^{-H\tau} | \, x \, \rangle \, , \qquad (1)$$

in which H, the Hamiltonian governing evolution of $x_\mu(\tau)$, is $= -(\gamma_\mu \pi_\mu)^2$, with $\pi_\mu = p_\mu - eA_\mu^{ext}$. The signal for chiral symmetry breaking is $\langle \overline{\psi} \psi \rangle \neq 0$ as $m \to 0$, and this requires $Z(\tau) \equiv \mathrm{tr} \langle x | e^{-H\tau} | x \rangle \approx (\tau)^{-1/2}$ as $\tau \to \infty$. With no external A_μ, i.e. a free propagator, $Z(\tau)$ goes as τ^{-2} for large τ. For constant $F_{\mu\nu}$, the problem can be solved exactly as shown by Schwinger, and there is no chiral symmetry breaking for any constant $F_{\mu\nu}$ no matter how strong[23].

In the case of non-constant fields it appears that computer simulations are necessary, so it is useful to express $Z(\tau)$ as a path integral:

$$Z(x,\tau) = \mathrm{Tr} \int [Dx'_\mu] \underline{P} \exp[-S(x',\dot{x}')] , \tag{2}$$

$$S(x,\dot{x}) = \int_0^\tau d\tau' \, L[x_\mu(\tau'),\dot{x}_\mu(\tau')] , \tag{3}$$

$$L(x,\dot{x}) = \tfrac{1}{4} \dot{x}_\mu \dot{x}_\mu + ieA_\mu(x) \dot{x}_\mu - e/2 \, \sigma_{\mu\nu} F_{\mu\nu}(x) , \tag{4}$$

with $x'_\mu(0) = x'_\mu(\tau) = x_\mu$. There are some unusual problems in the numerical evaluation of (2) including the fact that we need the behaviour of Z itself and not the expectation value of some other operator, that the variables x_μ are non-compact, and that there is an imaginary term, as well as a matrix term in the Lagrangian (4). All these have contributed to rendering standard Monte Carlo techniques not very useful. Instead, we[7] have adopted and modified for our problem a method developed[30] for complex actions based on the density of states. The details will be found elsewhere[7]. Suffice it to say that we first tested our method against the previously mentioned, known, exactly analytically calculable constant-field results, and found excellent agreement so that we have confidence that the method is under control and reliable. This alone is significant since our method is a new and not straightforward extension of the spectral density method to a difficult problem not previously attempted.

We have now been looking at non-constant fields with our simulation method and can report[7] that for a particular electric field (we are in Euclidean space, so this designation is somewhat arbitrary, and the same results apply for a similar magnetic field) with spatial and temporal variation of a Gaussian type, Z *does indeed fall off as* $\tau^{-1/2}$, so that there is spontaneous chiral symmetry breaking and background electromagnetic fields rather similar to those present in the heavy-ion collision experiments can induce the phase transition to new QED. The electric field is of the form:

$$E = E_x = eE \, [\cosh^2(x/w_s)]^{-1} \exp(-t^2/2w_t{}^2) \tag{5}$$

With the coupling e times the field strength E equal to 1.0 , we searched through various values for the temporal and spatial widths, w_t and w_s , finally finding the sought-for $\tau^{-1/2}$ fall off for Z, with $w_t = w_s = 3.0$. The data and fit are shown in Fig. 1 below. It is interesting to note that these values of the parameters, with eE = 1.0 MeV setting the scale, correspond to $w_t \approx 10^{-19}$s and $w_s \approx 500$fm, in agreement with the experimental field configurations. Encouraged by these results, we are now studying other, even more realisitic configurations to gain a better understanding of this fascinating phenomenon.

Non-constant E field: eE=1.0, Wt=3.0, Ws=3.0

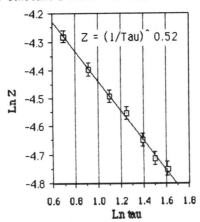

$Z = (1/Tau)^0.52$

Fig.1. Data from simulation of Z and fit.

IV. CONCLUDING REMARKS

New QED appears to be not just a theoretical learning tool, but something which nature could not resist making a reality. So far it is the only essentially complete, simple and economical explanation of the heavy-ion data. Furthermore, there is now beginning evidence that background, varying electromagnetic fields can indeed induce the phase transition. For the future, a tractable, realistic formalism for calculating things like lifetimes and decays is needed. Finally, the search is continuing for yet other systems (see, e.g., gamma-ray bursters[31]) and experiments where we may further explore new QED.

ACKNOWLEDGMENTS

Some of the work described here was done in collaboration mainly with A. Chodos and S. Vafaeisefat, as well as with F. Accetta, K. Everding, and D. Owen. It is also a pleasure to thank J. Greenberg for many enlightening discussions about the experiments. The author's work was supported in part under DOE Contract No. DE-AC02-79ER10336.

REFERENCES

1. See the contributions of E. Dagotto and A. Kocic to this volume.
2. For a recent review, see D. G. Caldi, *Comments Nucl. Part. Phys.* **19**, 137 (1989).
3. J. Schweppe et al., *Phys. Rev. Lett.* **51**, 2261 (1983); M. Clemente et al., *Phys. Lett.* **137B**, 41 (1984); T. Cowan et al., *Phys. Rev. Lett.* **54**, 1761 (1985); T. Cowan et al., *Phys. Rev. Lett.* **56**, 444 (1986); H. Tsertos et al., *Phys. Lett.* **162B**, 273 (1985); H. Tsertos et al., *Z. Phys. A* **326**, 235 (1987).
4. For a review, see T. Cowan and J. Greenberg, in *Physics of Strong Fields*, ed. W. Greiner (Plenum, New York, 1987).
5. See the contributions of H. Bokemeyer and W. Koenig to this volume.
6. D. G. Caldi and A. Chodos, *Phys. Rev.* **D36**, 2876 (1987); D. G. Caldi, A. Chodos, K. Everding, D. A. Owen, and S. Vafaeisefat, *Phys. Rev.* **D39**, 1432 (1989); L. S. Celenza, V. K. Mishra, C. M. Shakin, and K. F. Liu, *Phys. Rev. Lett.* **57**, 55 (1986); L. S. Celenza, C. R. Ji, and C. M. Shakin,

Phys. Rev. **D36**, 2144 (1987); Y. J. Ng and Y. Kikuchi, *Phys. Rev.* **D36**, 2880 (1987); Y. Kikuchi and Y. J. Ng, *Phys. Rev.* **D38**, 3578 (1988).

7. D. G. Caldi and S. Vafaeisefat, in preparation.

8. See, e.g., A. P. Mills Jr. and J. Levy, *Phys. Rev.* **D36**, 707 (1987); U. von Wimmersperg et al., *Phys. Rev. Lett.* **59**, 266 (1987); K. Maier et al., *Z. Phys. A* **326**, 527 (1987) and **330**, 173 (1988); S. H. Connell et al., *Phys. Rev. Lett.* **60**, 2242 (1988); H. Tsertos et al., *Phys. Lett.* **207B**, 273 (1988); J. van Klinken et al., *Phys. Lett.* **205B**, 223 (1988); M. Minowa et al., Univ. of Tokyo preprint (1988); E. Lorenz et al., *Phys. Lett.* **B**, to be published; J. Greenberg et al., in progress at Brookhaven.

9. For a review, see J. van Klinken, in *Tests of Fundamental Laws in Physics, Proceedings of the XXIVth Rencontre de Moriond*, eds. O. Fackler and J. Tran Thanh Van, (Editions Frontières, Gif-sur-Yvette, 1989).

10. K. Danzmann et al., *Phys. Rev. Lett.* **59**, 1885 (1987); **62**, 2353 (1989).

11. See, e.g., J. M. Cornwall and G. Tiktopoulos, *Phys. Rev.* **D39**, 334 (1989); H. M. Fried, this volume; H. Minakata, this volume.

12. For reviews, see A. Chodos, *Comments Nucl. Part. Phys.* **17**, 211 (1987) and references therein; B. Müller, in *Atomic Physics of Highly Ionized Atoms*, ed. R. Marrus (Plenum, New York, 1989).

13. S. Schramm, B. Müller, J. Reinhardt, W. Greiner, *Mod. Phys. Lett.* **A3**, 783 (1988); S. Graf, S. Schramm, J. Reinhardt, B. Müller, W. Greiner, Frankfurt preprint, UFTP 231/1989; W. Greiner, this volume.

14. See, e. g., D. G. Caldi, *Nucl. Phys.* **B220 [FS8]**, 48 (1983); and C. B. Lang, *Nucl. Phys.* **B280 [FS18]**, 225 (1987), and references therein.

15. A. Guth, *Phys. Rev.* **D21**, 2291 (1980); J. Fröhlich and T. Spencer, *Commun. Math. Phys.* **83**, 411 (1982).

16. E. Dagotto and J. Kogut, *Phys. Rev. Lett.* **59**, 617 (1987); *Nucl. Phys.* **B295 [FS21]**, 123 (1988); J. B. Kogut, E. Dagotto, and A. Kocic, *Phys. Rev. Lett.* **60**, 772 (1988); and ref. 1.

17. K. Johnson, M. Baker, and R. Willey, *Phys. Rev.* **136B**, 111 (1964); **163**, 1699 (1967); S. Adler amd W. A. Bardeen, *Phys. Rev.* **D4**, 3045 (1971); S. Adler, *Phys. Rev.* **D5**, 3021 (1972).

18. T. Maskawa and H. Nakajima, *Prog. Theor. Phys.* **52**, 1326 (1974); **54**, 860 (1975); R. Fukuda and T. Kugo, *Nucl. Phys.* **B117**, 250 (1976); V. A. Miransky, *Il Nuovo Cim.* **90A**, 149 (1985); V. A. Miransky and P. I. Fomin, *Sov. J. Part. Nucl.* **16**, 203 (1985).

19. See, e.g., R. Holdom, *Phys. Lett.* **213B**, 365 (1988); T. Appelquest, K. Lane, and U. Mahanta, *Phys. Rev. Lett.* **61**, 1553 (1988), and many talks in *Proceedings of the Nagoya Workshop 1988, and 1990* (World Scientific, Singapore, 1989 and 1991).

20. W. A. Bardeen, C. N. Leung, and S. T. Love, *Phys. Rev. Lett.* **56**, 1230 (1986); *Nucl. Phys.* **B273**, 649 (1986).

21. T. Appelquist, M. Soldate, T. Takeuchi, and L. C. R. Wijewardhana, Yale preprint,YCTP-P19-88 (August, 1988) in Proceedings of the 12th Johns Hopkins Workshop, 1988 (World Scientific, Singapore, 1989); K. Kondo, H. Mino, and K. Yamawaki, Nagoya preprint, DPNU-88-18 (June, 1988); W. A. Bardeen, C. N. Leung, and S. T. Love, in *Proceedings of the Nagoya Workshop* (World Scientific, Singapore, 1989).

22. D. G. Caldi, *Phys. Lett.* **215B**, 739 (1988).

23. D. G. Caldi, A. Chodos, K. Everding, D. A. Owen, and S. Vafaeisefat, ref. 5.

24. E. Dagotto and H. W. Wyld, *Phys. Lett.* **205B**, 73 (1988).

25. R. D. Peccei, J. Solà, and C. Wetterich, *Phys. Rev.* **D37**, 3206 (1988).

26. A. Chodos, D. Owen, and C. Sommerfield, *Phys. Lett.* **212B**, 491 (1988).

27. Y. J. Ng, this volume.

28. A. D. Linde, *Phys. Lett.* **62B**, 435 (1976).

29. See A. Chodos, this volume.

30. A. Gocksch, *Phys. Rev. Lett.* **18**, 2054 (1988); *Phys. Lett.* **206B**, 290 (1988).

31. F. S. Accetta, D. G. Caldi, and A. Chodos, *Phys. Lett.* **226B**, 175 (1989).

FUNCTIONAL APPROACH TO STRONG-COUPLING
IN (QED)₄ AND (QCD)₄

H.M. Fried

Physics Department, Brown University
Providence, RI 02912 USA

In these Lectures, the "Infrared (IR) Method" for extracting relevant low-frequency behavior is described for four-dimensional QED and QCD, using different techniques appropriate to the different theories. The subjects briefly covered are the following.

I. The IR Method in QED -

    (A) Formal, functional solutions in QED

    (B) The Fradkin representation and the IR approximation

    (C) Application to Loop Bremsstrahlung (LB)

II. An IR Approach to QCD -

    (A) Halpern's Idea (HI)

    (B) HI + Dimensional Transmutation (DT)

    (C) The IR approximation, first (qualitative) results

It will be noted that the second of these topics is on a less well-established footing than the first, even if the QED application to LB is not without approximation. The predictions of LB can nicely encompass the fourth and latest very-sharp e^+e^- "resonance", experimentally found in the collision of medium-energy heavy ions, and clear predications can be made for the next pair of back-to-back leptons expected on the basis of this model. In contrast, the work reported in the second Lecture on QCD is still at a relatively early stage, with as yet no immediate applications to confinement. In addition to the author, the people involved in different aspects[1] of this work during the past five years have been: F. Guérin, T. Grandou, and H-T Cho; in particular, the QCD material presented here was partially developed in collaboration with Dr. Cho.

Vacuum Structure in Intense Fields, Edited by
H.M. Fried and B. Muller, Plenum Press, New York, 1991

I. THE IR METHOD IN QED

(A) Functional solutions in QED (a Brief Review)

We begin with the basic "input information" of QFT, operator field equations (written for simplicity in the Feynman gauge)

$$[m + \sigma \cdot (\partial - ig[A + A^{ext}])]\psi(x) = 0,$$
$$(-\partial^2)A_\mu = ig\bar{\psi}\gamma_\mu\psi,$$

M

and Equal-Time Commutation (Anticommutation) Relations (ETC(A)Rs),

$$[A_\mu(x), A_\nu(y)]\Big|_{x_0=y_0} = 0, \quad [A_\mu(x), \partial_0 A_\nu(b)]\Big|_{x_0=y_0} = i\delta_{\mu\nu}\delta(\vec{x} - \vec{y}),$$
$$\{\psi_\alpha(x), \psi_\beta(y)\}\Big|_{y_0=y_0} = 0, \quad \{\psi_\alpha(x), \bar{\psi}_\beta(y)\}\Big|_{x_0=y_0} = \gamma_4^{\alpha\beta}\delta(\vec{x} - \vec{y}),$$

which together form an operator "shorthand" for the ∞ set of coupled n-point functions that define the theory.

In the Dirac equation above, where A_μ and ψ are operators in an appropriate Hilbert space, the quantity A^{ext} denotes a prescribed, classical field whose origin and quantum fluctuations are not relevant to the specific problem, and may be neglected; this is the "potential theory" approach, where it is often convenient to introduce an external field – such as the Coulomb field of one of the heavy, large-Z scattering ions. In contrast, a "field theory" approach would insist that there are no such entities as external fields – they are but idealizations which are occasionally useful – and that they can in any case be derived from the complete field fluctuations of relevant quanta in an appropriate, semi-classical limit. This second point of view will be adopted here, although an effective, external field will be prescribed when convenient. It is important to realize that these are, in a practical sense, complementary views: while the field theory approach may be more fundamental, it usually is not nearly as convenient as that of potential theory.

It will be useful to define the generating functional[2]

$$Z\{j, \eta, \bar{\eta}\} = < 0|(e^{i\int[jA+\bar{\eta}\psi+\bar{\psi}\eta]})_+|0 >,$$

where $j_\mu, \eta_\alpha\eta_\beta$ are "sources" whose variation give expression to the field equations and ETC(A)Rs. The formal, functional solution for $Z\{j, \eta, \bar{\eta}\}$ which contains the input information of field equations plus ETC(A)Rs, is given in terms of the causal

Green's function $G_c(x,y|A)$ satisfying

$$[m + \gamma \cdot (\partial - igA(x))] G_c(x,y|A) = \delta(x-y)$$

and the closed-loop functional L[A], defined below; here $A_\mu(z)$ is a c-no. function, or "background field" whose variations provide the difference between "second" quantization of field theory and the "first" quantization of potential theory. Finally, it should be noted that when a convenient shorthand notation is used for these quantities, such as

$$G[A] = S_c[1 - ig\gamma * AS_c]^{-1}, \; L[A] = Tr\ell n[1 - ig\gamma * AS_c],$$

the quantity A then denotes a local operator in configuration space, $< x|A_\mu|y >= A_\mu(x)\delta(x-y)$, with $< x|G_c[A]|y >= G_c(x,y|A)$.

The functional solution for $Z\{j,\eta,\bar\eta\}$ may be expressed as a functional integral over the A_μ fluctuations, or somewhat more simply in terms of the operation of a "linkage operator"

$$\exp[\mathcal{D}] = \exp[-(i/2) \int (\delta/\delta A)D_c(\delta/\delta A)],$$

whose action on all following A-dependence is equivalent to functional integration over the Gaussian measure $\exp[-(1/2)\int A(-\partial^2)A]$ of the A-dependence,

$$Z\{j,\eta,\bar\eta\} = e^{i/2 \int jD_cj} \cdot e^{\mathcal{D}} \cdot e^{i \int \bar\eta G_c[A]\eta} \cdot e^{L[A]} \Big/ < 0|S|0 >$$

Here, the c-number quantity $A_\mu(x) = \int d^4y D_{c,\mu\nu}(x-y)j_\mu(y)$, with $D_{c,\mu\nu}$ the free-photon propagator (now in an arbitrary, relativisitic gauge),

$$\tilde{D}_c(k)_{\mu\nu} = \left(\delta_{\mu\nu} - \xi\frac{k_\mu k_\nu}{k^2}\right) \cdot \frac{1}{k^2}.$$

The normalization condition $Z\{0,0,0\} = 1$ then defines the probability amplitude that the vacuum remains a vacuum for all time,

$$< 0|S|0 >= e^{\mathcal{D}} \cdot e^{L[A]}\Big|_{A_\mu \to 0},$$

while all S-matrix elements are independent of the gauge parameter ξ.

Pictorially, one may represent these quantities by sets of Feynman graphs, e.g. where $\sum$ denotes a sum over all relevant graphs.

$$G_c[A] \to \sum$$

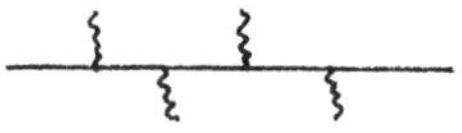

and

$$L[A] \to \sum$$

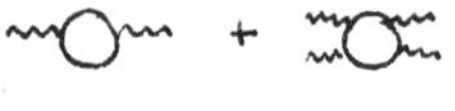

$$+ \cdots ,$$

where $\sum$ denotes a sum over all relevant graphs. The action of the linkage operator can then be expressed in a graphical way, e.g.,

$$e^{\mathcal{D}} G_c[A]\big|_{A\to 0} \Rightarrow \sum$$

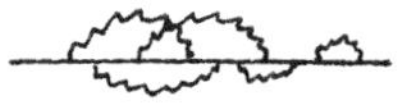

,

$$e^{\mathcal{D}} G_c^{(1)}[A] G_c^{(2)}[A]\big|_{A\to 0} \Rightarrow \sum$$

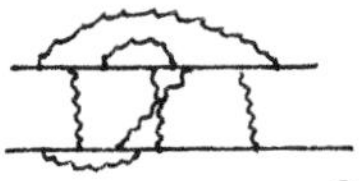

,

$$e^{\mathcal{D}} G_c^{(1)}[A] G_c^{(2)}[A] e^{L[A]}\big|_{A\to 0} \Rightarrow \sum$$

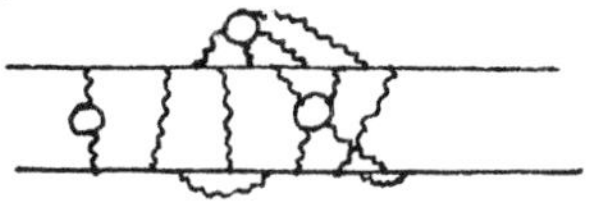

.

Another convenient representation involves the cluster decomposition property, most simply expressed for the vacuum amplitude,

$$< S >= \exp[\mathcal{D}] \exp(L[A]) = \exp\left[\sum_{n=}^{\infty} Q_n[A]/n!\right]$$

with

$$Q_n[A] = \exp[\mathcal{D}](L[A])^n|_{\text{connected}} \,.$$

This representation is most convenient for discussions of a finite number of closed fermion loops (CLF), or for the subset of "chain graphs" constructed out of such loops; here, "connected" means containing at least one virtual photon exchange between any loop and the rest of the diagram.

To avoid spurious IR divergences, we introduce a "photon mass" μ into every photon propagator, and remove it in a harmless way later on. For illustration, consider a two-particle scattering amplitude, built by "mass- shell amputation" of the relevant 4-point function,

$$T \sim e^{\mathcal{D}} G_c^{(1)}[A] G_c^{(2)}[A] e^{L[A]}\big|_{A\to 0} \,.$$

Neglecting CFLs, $\exp(L[A])- > 1$, and

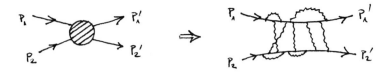

In the limit of small momentum transfer, when $p_1 - p_1' = q_1 \ll p_1$, the effect of the self-linkages along any fermion line is to renormalize the mass and wave-functions, so that those self-linkages may be dropped and renormalized parameters used,

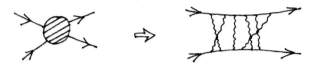

This leads to the "relativistic eikonal approximation" popular in the early '70s,

$$T(s,t) \sim is \int d^2b\, e^{i\vec{q}\cdot\vec{b}}[1 - e^{i\chi_1(s,b)}], \quad t = -q^2,$$

where $s = $ (total CM energy)^2, and the eikonal function is given by

$$\chi_1(s,b) = g^2\gamma(s)K_0(\mu b),$$

with $\gamma(s)$ a kinematical factor corresponding to the nature (scalar, vector, etc.) of the virtual photons exchanged between the scattering fermions. It turns out that the typical value of the (Euclidean) 4-momentum k of any virtual photon exchanged is $\sim q$, so that $q < p$ means $k < p$. In effect, this is a "soft" or an "IR approximation", where the momenta k of exchanged bosons is restricted; and the result, containing all powers of the coupling g^2, is non-perturbative. It is this idea which forms the essence of the IR approach to problems of strong-coupling: restrict (in a systematic manner) virtual momenta to be appropriately small, sum over all fluctuations defined in terms of such soft momenta, and one may then hope to reproduce the large-scale, strongly-coupled behavior of that nonlinear system.

In the limit of $\mu \to 0$, one finds that this eikonal amplitude differs from the <u>exact</u> Coulomb amplitude only by a divergent phase factor containing all dependence on the parameter μ, so that differential cross sections and the bound-state pole structure (obtained by continuation of the total energy below threshold) are <u>exactly</u> reproduced. It is in this sense that the sum over all relevant Feynman graphs reproduces the amplitude constructed from an explicit Coulomb potential; here, χ_1 can be expressed in terms of that effective Coulomb potential, which has been defined by the field-theory summation over Feynman graphs. Which is the more fundamental approach? Clearly, it is the field-theory construction, which provides an "equivalent potential" only in the case of small-momentum scattering, when the structure of the scattering particles is not being probed, and only the interaction between the particles is relevant.

For higher values of s, even in the small-q limit, there are physical, "multiperipheral" contributions to the scattering amplitude, constructed out of "towers" of CFLs, whose absorptive parts correspond to particles produced "from the innards" of a graph,

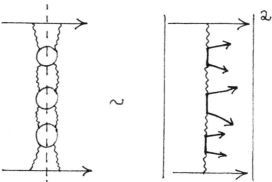

Here the eikonal becomes complex, corresponding to a "potential" which is both complex and energy-dependent. Again, from field theory one obtains a qualitatively correct physical definition of an effective potential which is relevant to that scattering process.

(B) Schwinger/Fradkin Representations

Most of the work on vacuum structure in external fields can be traced back to a seminal, 1951 paper by Schwinger,[3] in which proper-time representations for $G_c[A]$ and $L[A]$ first appeared, and in which solutions were obtained for the special case of constant $F_{\mu\nu}$ as well as the case of single-frequency, laser fields. Fifteen years later, Fradkin published[4] an independent and exact representation of $G_c[A]$ and $L[A]$, which made clear the reasons for the solubility of Schwinger's examples, and which has since been used[1] to provide the generalization of Schwinger's constant-field solutions to situations in which the $F_{\mu\nu}$ are slowly-varying. For clarity and completeness, we here write the Fradkin representations for $G_c[A]$ and $L[A]$, in terms of the fluctuations of a "proper-time-dependent four-velocity" $\phi_\mu(s)$, whose variations exactly take into account the non-community of ∂_μ and A_ν,

$$G_c(x,y|A) = i \int_0^\infty ds\, e^{-ism^2} \cdot e^{-ig \int_y^x d\xi_\mu A_\mu(\xi)} \cdot N(s).$$

$$\cdot \int d[\phi] e^{\frac{i}{4} \int_0^s ds'\phi_\mu^2} \left(m - \gamma_\mu \frac{\delta}{\delta\phi_\mu(s)} \right) \cdot < x|W(s)|y >,$$

and

$$L[A] = -\frac{1}{2} \int_0^\infty \frac{ds}{s} e^{-ism^2} \int d^4x \cdot N(s) \cdot \int d[\phi] e^{\frac{i}{4} \int_0^s ds'\phi_\mu^2}.$$

$$\cdot tr\left\{ < x|W(s)|x > - < x|W(s)|x > \big|_{g=0} \right\},$$

with

$$< x|W(s)|y > = e^{ig \int_0^1 \lambda d\lambda \int_0^s ds' \int_0^{s'} ds'' \phi_\mu(s')\phi_\nu(s'') F_{\mu\nu}(y-\lambda \int_0^{s'} \phi)}.$$

$$\cdot \left(e^{g\sigma_{\mu\nu} \int_0^s ds' F_{\mu\nu}(y-\int_0^{s'} \phi)} \right)_+ \cdot \delta\left(x - y + \int_0^s \phi \right),$$

and

$$N(s)^{-1} = \int d[\phi] e^{\frac{i}{4} \int_0^s ds'\phi_\mu^2(s')} , \quad \sigma_{\mu\nu} = \frac{1}{4} [\gamma_\mu, \gamma_\nu] ,$$

and where $\xi_\mu = \lambda x_\mu + (1 - \lambda)y_\mu$ denotes the straight-line path between x_μ and y_μ. It should be noted that all the gauge variance of $G_c(x,y|A)$ is contained in the factor $\exp[-ig \int d\xi_\mu A_\mu]$; and that $L[A]$ is gauge invariant. Very similar forms exist in other (Abelian) theories with nonlinear interactions. Incidentally, the specification $\phi_\mu(s) = dx_\mu(s)/ds$ leads immediately to the Feynman path integral representation of these quantities.

One may very well ask: Is this Progress? If the functional integrals over $\int d[\phi]$ can be evaluated only for Gaussian ϕ-dependence, which only occurs when $F_{\mu\nu}$ is constant, why is this so interesting? The answer is that the Fradkin representation provides a "natural" way of extracting low-frequency $\tilde{A}_\mu(k)$ dependence, while providing systematic corrections for the high-frequency contributions. This assumes the existence of a relevant scale, with which one can distinguish "high" and "low" frequency components: in QED, that scale is given by the (renormalized) lepton-mass; in other problems, such as quark-less QCD, real fluids or nonlinear differential equations, that scale may depend upon initial (or boundary) conditions of the problem, and can be the last thing to be (self-consistently) determined.[5]

We next give a brief description of the IR, or Eikonal Approximation in QED. This version is useful when there exits no asymptotic 4-velocities, as in low-momentum-transfer scattering or production problems, where the Bloch-Nordsieck approximation is both relevant and straightforward; the present forms were developed[1] for use in estimating $< \bar{\psi}\psi >$ in IR approximation. The Method is defined in two steps:

(1) Restrict the Fourier components k of the $\tilde{A}_\mu(k)$ to be less than or on the order of $\mu_c(\tau)$, where μ_c is a constant or proper- time-dependent parameter (with the subsequent continuation $s \rightarrow \tau$ understood), by defining "soft" and "hard" A_μ according to the simplest relation

$$\tilde{A}(k) \equiv \tilde{A}(k)e^{-k^2/\mu_c^2} + \tilde{A}(k)\left[1 - e^{-k^2/\mu_c^2}\right],$$

or

$$\tilde{A}(k) \equiv \tilde{A}^S(k) + \tilde{A}^H(k),$$

and expanding in powers of $\tilde{A}^H(k)$, so that the "lowest order" terms involve $\tilde{A}^S(k)$ only. Note that this is NOT an ordinary perturbation expansion, for the "small parameter" of this approximation is a measure of the lack of physical importance of the high-frequency components. The IR Method can never reproduce sharp transitions, or any quantity sensitive to high-frequencies; but it should be a sensible approximation for nonlinear problems whose qualitative structure is most dependent upon large-scale, or low-frequency effects.

(ii) Even with this restriction to soft frequencies, the functional integration over $\int d[\phi]$ cannot be performed. One therefore chooses μ_c so that $0(k \int ds''\phi) < 1$, and performs a multipole expansion, retaining the first non-zero term – in the Fourier transform of $F_{\mu\nu}(y - \int ds''(s''))$ – as the essential, and second approximation which defines the IR Method. From the Fradkin representation, and using the variable τ rather than s, it is clear that ϕ scales as $\tau^{-1/2}$, so that the choice $\mu_c = c/\tau^{-1/2}$, where c is a constant on the order of unity, will be sufficient to define the multipole expansion. The $\int d[\phi]$ is now Gaussian – although dependent upon τ – and can be performed without further approximation. Note that if the τ-integral converges, as expected, then $\mu_c \sim m = m_e$.

In (QED)$_4$, $L_{IR}[A]$ can be easily calculated and reproduces exactly the form of Schwinger's constant-field solution,

$$L_{IR}[F] = \frac{i}{8\pi^2} \int d^4x \int_0^\infty \frac{d\tau}{\tau^3} e^{-\tau m^2} \mathcal{F}_\tau \{F_S^2/4, \, F_S \cdot {}^*F_S/4\},$$

with

$$\mathcal{F}_\tau\{\alpha,\beta\} = g^2\tau^2\beta \coth{(g\tau X_r)} \cot{(g\tau X_i)} -$$
$$- 1 - \frac{2}{3}\alpha(g\tau)^2\,,$$

and where

$$X_r = \frac{1}{\sqrt{2}}\left\{(\alpha+i\beta)^{1/2} + (\alpha-i\beta)^{1/2}\right\},$$
$$X_i = -\frac{i}{\sqrt{2}}\left\{(\alpha+i\beta)^{1/2} - (\alpha-i\beta)^{1/2}\right\}.$$

These forms are the first generalizations away from Schwinger's $k = 0$ mode solution, containing a continuous summation over all modes with $k \lesssim m_e$. Here, for each $F_{\mu\nu}$ in $L[A]$, the (leading) IR approximation produces Schwinger's form, with his constant field components F replaced by

$$F^S(x) = \left(\frac{\mu_c^2}{4\pi}\right)^2 \int d^6 z_E F(z) \exp\left[-(x-z)_E^2\mu_c^2/4\right]\,,$$

here written in Euclidean space. Note that his $L_{IR}[A]$ contains all powers of g^2, and is therefore nonperturbative in the usual sense.

(C) Application to Loop Bemsstrahlung

The LB model is an attempt to explain the observed sharp (widths ≤ 30 keV) e^+e^- peaks[6] (at total CM energies of 1.58, 1.64, 1.76 and 1.84 MeV), in terms of an ionic, eikonal scattering model to which is adjoined a single, closed-electron-loop (CEL),

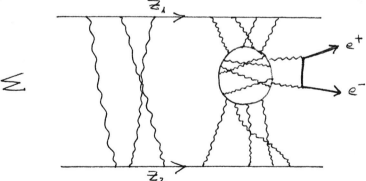

The summation here is over all soft, virtual photons exchanged between ions, between ions and the CEL, and across the CEL.

A physical picture of LB illustrates the importance of soft radiative corrections across the CEL, as follows. Imagine a pair of ions moving slowly ($v/c \sim .1$) past each other; on the natural time scale of the CEL which is to appear, $t_{\text{CEL}} \sim 1/m_e$, they are moving very slowly indeed. Suppose the ions have reached their distance of minimum approach, b; because of their like electric charge, the field strength between them is very small. Imagine that a virtual CEL appears – which can here be described in terms of charges $+\delta e$ and $-\delta e$ that exist for a time interval $\sim t_{\text{CEL}}$ in the same region – and has the effect of altering the previous, practically static, electric field between the ions. In order to do this in an interval $\lesssim t_{CEL}$, one needs a mechanism for transferring electromagnetic energy into this region of volume b^3; and that possibility is provided by the radiative corrections across the loop, or "from $+\delta e$ to $-\delta e$", which can transfer coherently a total amount of energy $\sim m_e$ in a time $\sim 1/m_e$.

Because the energy transfer is to take place electromagnetically in a time $\sim 1/m_e$, it can only occur for distances $b \lesssim 1/m_e$, which means that the effective ionic field acting on the loop, $E \sim b^{-2}$ must be sizeable: $E \lesssim m_e^2$, and no perturbative treatment in E/m_e^2 can be contemplated. If the energy transferred into this volume is to be associated with that subsequently emitted as the lepton pair of total energy $\sim 3m_e/2$, then $3m_e/2 \sim b^3 E^2$, and again $E \geq m_e^2$. The "matching" of these conditions corresponds in the model calculations to a sharp interference between those coherent radiative corrections linking ions to the loop, and those across the loop; rather than due to a "resonance", LB appears as an interference pattern of nonperturbative, low-frequency radiative corrections.

It should be emphasized that much of the kinematical details of LB are still beyond calculation – again due to the nonperturbative nature of the problem – but the qualitative features seem clear: families of narrow-width lepton peaks. One extraordinary feature of the model which unexpectedly occurs is an "enhancement factor" of form $\exp[E^2/m_e^4]$, multiplying other integrals of order unity, with the effect of compensating the extra factor of the fine-structure-constant associated with the CEL.

That family describing the LB of two virtual photons (converting into the lepton pair) which is most appropriate for CM back-to-back lepton emission, was previously denoted as the "A-family", and satisfies the Chew-Frautschi-like formula:

$$M_A^2(n, \ell) = (2m_e)^2 + \xi_A(n^2 + \ell^2)^{1/2} \,,$$

where n and ℓ are integers (which cannot simultaneously vanish), and ξ_A is a constant (which the kinematical crudeness of the model cannot provide). Before the fourth peak (at 1.58 MeV) was announced at this ASI, the 1.64 was taken to be the lowest member of an A family, $M_A(0.1)$. In this way, ξ_A was determined, and the next-higher-total mass calculated to be $M_A(1,1) = 1.84$ MeV, exactly the highest mass state seen so far. The 1.76 was assumed to be the lowest member of another family.

However, with the discovery of the lower-energy 1.58 MeV, whose pair seems to be emitted back-to-back, one may ask if another family is possible: Assume another "A'−family", with the 1.58 as its lowest state, and use this to calculate the new $\xi_{A'}$. What is the first excited state of that A' family? It comes out precisely, almost embarrassingly, to be 1.76 MeV. Is this more than chance numerology? One sees that the apparent, numerical coincidence of a correct "excited state" of the A-family is repeated for the A'-family. Two such "coincidences" would suggest that there may be some truth in LB, even if its present crudeness means that several kinematical questions remain obscure.

Perhaps the "proof" will lie in the predictions now obviously possible on the basis of these two families: the next sharp peaks to be found should have total energies of $M_A(0,2) = 1.99$ MeV, and at $M_{A'}(0,2)=2.08$ MeV. Since these peaks are so close to each other, one can expect a rather large signal at ~ 2 MeV; and this – in response to our experimental colleagues' request[6] for suggestions as to where else one should look – is the prediction which LB can make. If no peak is found at 2 MeV, then LB is wrong; but if a back-to-back peak is seen there, LB may still be a valid theoretical candidate.

One other class of predictions have been made[7] on the basis of LB, which may be worth a comment here. The Z-dependence of the peak cross sections does not seem to follow that verified for the smooth e^+ emission background, understood on the basis of supercritical potential theory[8]. One may ask what is the minimum value of Z with which one may expect LB of lepton pairs to be possible, and the results are surprisingly low, with $Z_{min} \sim 2$ for electron pairs and ~ 5 for muon pairs. A "Rutherford scattering" experiment is then, according to LB, a possible source of sharp lepton pairs, and an indirect test for the importance of closed-lepton loops in nonperturbative QED.

II. AN IR APPROACH TO QCD

Because the perturbative description of asymptotic freedom suggests that closed-gluon loops (CGL) can be more important than closed-quark loops, we consider a purely gluonic theory, with Lagrangian $\mathcal{L} = -F^2/4$, where

$$F_{\mu\nu}^a = \partial_\mu A_\nu^a - \partial_\nu A_\mu^a + g f_{abc} A_\mu^b A_\nu^c .$$

Reflection on the difficulty of defining soft-gluonic insertions (corresponding to cubic and quartic gluon couplings) into a CGL leads to the conclusion that one needs a formalism in which functional integration over all gluonic fluctuations is performed at the very beginning, a point of view originally espoused by Halpern.[9] Furthermore, $(QCD)_4$ without massive quarks is a theory with no *a priori* scale, which quantity is necessary to specify the distinction between "soft" and "hard"; and in order to introduce that scale, one is led to the method of dimensional transmutation (DT).[10] The analysis given in this Lecture is very crude, with no detailed results to be found at the end of these pages; but the method of approach may well turn out to be a useful first step towards a serious calculation of large-scale, confinement physics.

(A) Halpern's Idea

A generating functional for QCD may be defined in terms of the source $j_\mu^a(x)$ by the functional integral (FI)

$$Z\{j_\mu^a\} \sim \int d(A)\delta\left(\mathcal{F}(A)\right) \cdot \det M[A] \cdot \exp\left[-\frac{i}{4}\int F^2 + i\int j \cdot A\right],$$

where $\mathcal{F}[A] = 0$ specifies a gauge and $M[A] \sim \delta\mathcal{F}/\delta z$ defines a variation of $\mathcal{F}$ with respect to some appropriate parameter of the transformation out of that gauge; in the limit of zero source this vacuum amplitude is gauge invariant, as are quantities such as $< F^2 >, < F^2(x)F^2(y) >$, etc., which are built by appropriate functional differentiation, followed by the limit of zero source.

For simplicity, we now give up the possibility of maintaining full gauge invariance, and require that all computations be done in an axial gauge, $\mathcal{F}[A] = \eta_\mu A_\mu^a(x)$, but with η_μ unspecified. With this choice, $\det[M]$ becomes a constant, independent of A, and can be absorbed into the normalization of $\mathcal{Z}\{j\}$, while $\delta(F[A])$ can be expressed as proportional to

$$\int d[\theta_a] e^{i\int \theta_a \mu_\mu A_\mu^a},$$

so that

$$Z\{j\} \sim \int d[\theta] \int d[A] e^{i\int \theta\eta \cdot A - \frac{i}{4}\int F^2 + i\int j \cdot A}.$$

The basic difficulty here is that F is quadratic in A, and the FI cannot be performed. Following an ancient method,[11] Halpern[9] rewrote the F-dependence as

$$e^{-\frac{i}{4}\int F^2} \sim \int d[\sigma] e^{i\int \sigma^2 + i\int \sigma \cdot F},$$

with $\sigma_{\mu\nu}^a(x)$ an auxiliary, antisymmetric color field. Substitution into $\mathcal{Z}\{j\}$ allows the FI to be performed, with the result

$$Z\{j\} \sim \int d[\theta] \int d[\sigma] e^{-\frac{1}{2}TrlnK(\sigma)-\frac{i}{2}\int J\cdot K^{-1}\cdot J} \; ,$$

where

$$J_\mu^a = j_\mu^a + \theta_a \eta_\mu + 2\partial_\nu \sigma_{\mu\nu}^a$$

and

$$K_{\mu\nu}^{bc}(\sigma) = 2g(\tau^a)_{bc}\sigma_{\mu\nu}^a \; , \; (\tau^a)_{bc} = f_{abc} \; .$$

The difficulty now is not that the σ-dependence is complicated – that was to be expected – but that because $K^{-1}[\sigma]$ is a local operator in configuration space, the usual method of performing DT, *via* a gap equation, will not work.

To achieve DT, we now change the procedure slightly, rewritting QCD as a "modified" QED. Set $F_{\mu\nu}^a = f_{\mu\nu}^a + gA_\mu^b\tau_{bc}^aA_\nu^c$, where $f_{\mu\nu}^a = \partial_\mu A_\nu^a - \partial_\nu A_\mu^a$.

Then

$$e^{-\frac{i}{4}\int F^2} = e^{-\frac{i}{4}\int f^2} \cdot e^{-\frac{i}{4}\int(F^2-f^2)} \; ,$$

and the last RHS factor may be rewritten in terms of the pair of FIs over the fields $\phi_{\mu\nu}^a, \psi_{\mu\nu}^a$,

$$\sim \int d[\phi] \int d[\psi] \exp\left[i\int \psi\cdot\phi + \frac{i}{2}\int \phi\cdot(F+f) + \frac{i}{2}\int \psi\cdot(F-f) \right] \; .$$

A change of variable to $\sigma = \phi + \psi$, and the replacements

$$i\int \phi_{\mu\nu}^a(\partial_\mu A_\nu^a - \partial_\nu A_\mu^a) \rightarrow 2i\int A_\mu^a\partial_\nu\phi_{\mu\nu}^a \; ,$$

$$-\frac{i}{4}\int f^2 \rightarrow +\frac{i}{2}\int(\partial_\mu A_\mu^a)^2 - \frac{i}{2}\int A_\mu^a(-\partial^2)A_\mu^a \; ,$$

$$e^{\frac{i}{2}\int(\partial A)^2} \sim \int d[\chi_a]e^{-\frac{i}{2}\int \chi^2 - i\int A_\mu^a\partial_\mu\chi_a} \; ,$$

generate

$$Z\{j\} \sim \int d[\theta] \int d[\chi]e^{-\frac{i}{2}\int \chi^2} \cdot \int d[\sigma] \int d[\phi]e^{-i\int \phi^2+i\int \sigma\cdot\phi} \; .$$

$$\cdot \int d[A]e^{-\frac{i}{2}\int A\cdot(-\partial^2-g\sigma\cdot\tau)A+i\int J\cdot A} \; ,$$

with

$$J_\mu^a = j_\mu^a + \theta_a\eta_\mu - \partial_\mu\chi_a + 2\partial_\nu\phi_{\mu\nu}^a \; .$$

Now $K^{-1}[\sigma]$ is given by: $(-\partial^2 - g\sigma\cdot\tau)^{-1}$, a "propagator" of "mass" $g\sigma\cdot\tau$, and the conventional form of DT becomes possible. The FI over all gluonic fluctuations then yields

$$\int d[A] \to \sim \exp\left[\frac{i}{2}\int J \cdot K^{-1} \cdot J - \frac{1}{2}Tr\ln K\right],$$

and $\mathcal{Z}\{j\}$ becomes

$$Z\{j\} \sim \int d[\theta]\int d[\chi]e^{-\frac{1}{2}\int \chi^2} \cdot \int d[\sigma]e^{-\frac{1}{2}Tr\ln K(\sigma)}.$$
$$\cdot \int d[\phi]e^{\frac{1}{2}\int J \cdot K^{-1}J - i\int \phi^2 + i\int \phi \cdot \sigma}.$$

Note that the FI over ϕ is Gaussian, can be done exactly, and yields exactly Halpern's form; here, however, we shall not perform that FI until after DT is achieved.

(B) HI + Dimensional Transmutation

To define DT we shift both fields by constants, $\phi \to \phi_0 + \phi_1, \sigma \to \sigma_0 + \sigma_1$, where the fluctuating fields carry the subscript $_1$. Our choice, not unexpected, for the definition of the constant ϕ_0, σ_0, is such that

$$\frac{\delta}{\delta\phi}S[\sigma,\phi]\Big|_{\sigma_0,\phi_0} = \frac{\delta}{\delta\sigma}S[\sigma,\phi]\Big|_{\sigma_0,\phi_0} = 0,$$

where $S[\sigma,\phi]$ denotes all the exponential $\sigma-$ and $\phi-$dependence except that of the $JK^{-1}J$ term. One obtains

$$-2\phi_0 + \sigma_0 = 0,\ i\phi_0 + \frac{\delta}{\delta\sigma_0}\left(-\frac{1}{2}Tr\ln K(\sigma)\right)\Big|_{\sigma_0} = 0.$$

These generate the gap equation

$$(\sigma_0)^a_{\mu\nu} = -i\frac{\delta}{\delta\sigma^a_{\mu\nu}}Tr\ln K(\sigma)\Big|_{\sigma_0} = -iTr\left[g\tau^a\left(\frac{1}{-\partial^2 - g\sigma_0 \cdot \tau}\right)_{\nu\mu}\right],$$

which for constant σ_0 becomes

$$(\sigma_0)^a_{\mu\nu} = ig\,tr\left[\int \frac{d^4p}{(2\pi)^4}\tau^a\left(\frac{1}{p^2 - g\sigma_0 \cdot \tau}\right)_{\nu\mu}\right],\ tr = \sum_a.$$

Using

$$\left(\frac{1}{p^2 - g\sigma_0 \cdot \tau}\right)^{bc}_{\nu\mu} = \frac{\delta_{\mu\nu}\delta_{bc}}{p^2} + \frac{1}{p^2}\left(g\sigma_0 \cdot \tau\frac{1}{p^2 - g\sigma_0 \cdot \tau}\right)^{bc}_{\nu\mu},$$

and

$$tr[\tau^a] = \sum_b f_{abb} \equiv 0 ,$$

the UV divergence is only logarithmic in the expression:

$$(\sigma_0)^a_{\mu\nu} = \frac{ig^2}{(2\pi)^4} \int \frac{d^4p}{p^2} tr \left[\tau^a (\sigma_0 \cdot \tau) \frac{1}{p^2 - g\sigma_0 \cdot \tau} \right]_{\nu\mu} .$$

How is one now to choose σ_0? For simplicity a "magnetic" form is here assumed, $(\sigma_0)^a_{\mu\nu} = \epsilon_{\mu\nu34}\xi^a/\sqrt{2}$, with $\vec{\xi}$ real, and μ, ν restricted to the values 1,2. Other forms are certainly possible, but those tried seem to lead in quite similar directions.

For general $SU(N)$, $(\tau^a)_{bc}$ has $(N^2 - 1) \times (N^2 - 1)$ components, while (μ, ν) corresponds to a 2×2 structure. The, $(\sigma_0 \tau)$ can be thought of as a matrix with $[2(N^2 - 1)]^2$ elements., and when diagonalized has $2(N^2 - 1)$ real eigenvalues. Since $tr'[(\sigma_0 \tau)^{2n+1}] = 0$ for any n (where tr' denotes a sum over Lorentz and color indices), the eigenvalues must satisfy: $\sum_i(\xi_i)^{2n+1} = 0$. Half of the $e-$values are therefore the negatives of the other half; how we arrange them is quite arbitrary, and we order them so that the first $(N^2 - 1)$ are positive and the remaining $(N^2 - 1)$ are negative.

From the gap equation, one constructs

$$\sum_{a,\mu\nu} [(\sigma_0)^a_{\mu\nu}]^2 = \frac{i^2 g^2}{(2\pi)^4} \int_E \frac{d^4p}{p^2} tr' \left[(\sigma_0 \cdot \tau)^2 \frac{1}{p^2 - g\sigma_0 \cdot \tau} \right] ,$$

and, again for simplicity, this will be evaluated in $SU(2)$. One has

$$\sum_\ell \xi_\ell^2 \Rightarrow -\frac{g^2}{4\pi^2} \sum_\ell \xi_\ell^2 \int_0^\Lambda \frac{p\,dp}{p^2 - g\xi_\ell} ,$$

where we keep only the leading RHS (divergent) term. Upon rewriting in terms of the positive $e-$values $\xi_{+\ell}$, one finds

$$2\sum_{+\ell} \xi_{+\ell}^2 = -\frac{g^2}{2\pi^2} \sum_{+\ell} \xi_{+\ell}^2 \int_0^\Lambda \frac{p^3\,dp}{p^4 - g^2\xi_{+\ell}^2} ,$$

which leads to

$$\sum_{+\ell} \xi_{+\ell}^2 = -\left(\frac{g^2}{8\pi^2}\right) \sum_{+\ell} \xi_{+\ell}^2 \cdot \frac{1}{2} ln \left(\frac{\Lambda^4}{-g^2\xi_{+\ell}^2}\right) .$$

In order to proceed, it is necessary to continue to the "wrong" sign of $g^2 : g^2 \to -G^2$. This is a familiar effect,[12] and indicates that the correct ground state cannot be inferred from the (naive negative of the) Lagrangian without first "shifting", as we have done, to define a correct vacuum state. In terms of G^2, we rewrite the gap equation, define the sum-over-squares-of the positive $e-$values,

$$\xi^2 = \sum(\sigma^a_{0\mu\nu})^2 = \frac{1}{2}\sum_\ell \xi^2_\ell = \sum_{+\ell} \xi^2_{+\ell},$$

and scale out the magnitude ξ by the redefinition $\xi_{+i} = \xi\eta_{+i}$, where $\sum_i \eta^2_{+i} = 1$. Finally, we define $G\xi = \zeta M^2$, with ζ a constant given by the geometric relation

$$ln\zeta \equiv \sum_{+\ell} \eta^2_{+\ell} \, ln\left(\frac{1}{\eta_{+\ell}}\right),$$

so that the output of the gap equation reduces to the familiar statement of DT, with M constant and G vanishing as $\Lambda \to \infty$,

$$1 = \frac{G^2}{8\pi^2} ln\left(\frac{\Lambda^2}{M^2}\right).$$

Note that $K^{-1}(\sigma_0)$ can now be written as

$$K^{-1}[\sigma_0] = \left(p^2 - iG\epsilon\vec{\tau}\cdot\vec{\xi}/2\right)^{-1} = \left(p^2 - iM^2\epsilon\zeta(\vec{\tau}\cdot\hat{\xi})/\sqrt{2}\right)^{-1},$$

where $\hat{\xi} = \vec{\xi}/\xi$ and "$\varepsilon_{\mu\nu}$"$= \varepsilon_{\mu\nu34}$.

(C) The IR Approximation

After shifting, the generating functional becomes

$$Z\{j\} \sim \int d[\theta] \int d[\chi]e^{-\frac{i}{2}\int \chi^2} \cdot \int d[\sigma_1] \int d[\phi_1]\cdot$$
$$\cdot \exp\left[\frac{i}{2}\int JK^{-1}(\sigma_0 + \sigma_1)J + R(\sigma_0,\sigma_1) - i\int \phi_1^2 + i\int \sigma_1\cdot\phi_1\right],$$

with $R[\sigma_0,\sigma_1]$ representing the sum of quadratic and all higher σ_1-dependence in the expansion of $-(1/2)TrlnK[\sigma_0 + \sigma_1]$. (All the linear dependence has cancelled by virtue of the definitions of σ_0 and ϕ_0.) We now perform the Gaussian FI over ϕ_1, using $J^a_\mu = \mathcal{J}^a_\mu + 2\partial_\nu\phi^a_{1,\mu\nu}, \mathcal{J}^a_\mu = j^a_\mu + \theta_a\eta_\mu - \partial_\mu\chi^a$, with the result

$$Z\{j\} \sim \int d[\theta] \int d[\chi]e^{-\frac{i}{2}\int \chi^2} \cdot \int d[\sigma]e^{\frac{i}{2}\int \mathcal{J}\cdot K^{-1}(\sigma_0+\sigma_1)\cdot\mathcal{J}}.$$
$$\cdot \exp\left[\frac{i}{4}\int N\cdot S^{-1}N + R(\sigma_0,\sigma_1) - \frac{1}{2}TrlnS\right],$$

where

$$S^{ab}_{\mu\nu\lambda\sigma} = \delta_{ab}\delta_{\mu\nu}\delta_{\sigma\nu} - 2\vec{\partial}_\nu \left[K^{-1}(\sigma_0 + \sigma_1) \right]^{ab}_{\mu\lambda} \overleftarrow{\partial}_\sigma ,$$

and

$$N^a_{\mu\nu} = \left[\vec{\partial}_\nu (K^{-1}(\sigma_0 + \sigma_1))^{ab}_{\mu\lambda} + K^{-1}(\sigma_0 + \sigma_1)^{ba}_{\lambda\mu} \overleftarrow{\partial}_\nu \right] J^b_\lambda .$$

This is a rather complicated form; how does one proceed?

We first make an observation, then ask and answer a question. In all the above, those G–factors multiplying σ_0 are converted into the constant M; but what is $G\sigma_1$? Since G must vanish as $\Lambda \to \infty$, may one not set $G = 0$ everywhere? The answer is No: such a naive limit is improper before estimating higher-order radiative corrections, since it is quite conceivable that other (log) divergences could appear multiplying G^2, for example, and the result would be finite. In fact, there could be more singular Λ-dependence, which increases far faster than $ln(\Lambda/M)$.

However, something can be seen and can be done. Consider first the determinantal terms, $R - (1/2)TrlnS$, in particular the quadratic σ_1–dependence of this combination. One finds a result proportional to the factors

$$G^2 \cdot ln\left(\frac{\Lambda}{M} \right) \cdot \int d^4x\sigma^2(x) ,$$

so that the net quadratic σ–dependence (we henceforth drop all $_1$ subscripts) is finite, and gives

$$-\frac{i}{2} \int d^4x\sigma^2 .$$

There arises also (from $-(1/2)TrlnS$) a "tadpole" linear in σ,

$$-i\frac{M^2}{G}\epsilon_{\mu\nu34}\hat{\xi}^a \int d^4x\sigma^a(x)_{\mu\nu} ,$$

which is, in fact, a divergent phase factor. But because of the asymmetry of $\epsilon_{\mu\nu}$, this can never couple to $j_\mu \cdots j_\nu$ dependence, and one may expect its effects to cancel or be absorbed into the overall normalization. For simplicity, it is neglected it all that follows.

What are the contributions from the higher-than-quadratic σ–terms of this combination? Two arguments are possible: (i) In every term of order $(\sigma)^n, n > 2$, inspection shows no obvious UV divergence (as appeared for the quadratic terms). Hence the naive $G \rightarrow 0$ limit may be taken, and all such terms may be dropped. (ii) A better argument insists that one should calculate

$$\int d[\sigma] e^{-\frac{i}{2} \int \sigma^2}$$

over the (perturbative) expansion of these terms in powers of σ before discarding them. In fact, one finds that they <u>do</u> generate divergent contributions, containing multiples of quantities such as

$$i(VT)\Lambda^4 G^4 ln\left(\frac{\Lambda}{M}\right).$$

But these, proportional to factors of $i(VT)\Lambda^4$, correspond to purely vacuum "self-energy" graphs, and <u>should</u> be factored out of every $S-$matrix element. We may therefore neglect them, not because they vanish, but because their perturbative infinities may be identified as those of the expansion of an infinite phase factor, expected to cancel in the calculation of any physical quantity.

There remain the contributions of the non-determinantal terms, proportional to

$$\int d[\sigma] e^{-\frac{i}{2} \int \sigma^2} \cdot \exp\left[\frac{i}{2} \int \mathcal{J} \cdot K^{-1}(\sigma_0 + \sigma_1) \cdot \mathcal{J} + \frac{i}{4} \int N S^{-1} N(\sigma_0 + \sigma_1)\right].$$

Here, the UV divergences of the perturbation expansion are associated with high-frequency components of $\bar{j}(k)$. If we imagine that the original $\bar{A}(k)$ fluctuations, and associated gauge constraints, are limited to low frequencies, in particular to frequencies $k \lesssim M$, then – in such a "lowest-order IR approximation" – there are no further UV divergences, and the naive $G \rightarrow 0$ limit may be taken with assurance, yielding the pleasantly Gaussian result,

$$Z_{IR}\{j\} \sim e^{\frac{i}{2} \int j \cdot K^{-1} Y \cdot j},$$

with

$$Y = V\left[1 - \overleftarrow{\partial} \cdot \left(\vec{\partial} \cdot K^{-1} V \cdot \overleftarrow{\partial}\right)^{-1} \cdot \vec{\partial} K^{-1} V\right],$$

$$V = 1 - U \cdot \eta(\eta \cdot K^{-1} U \cdot \eta)^{-1} \eta \cdot K^{-1} U,$$

$$U = 1 - \frac{1}{4}\vec{\partial} S^{-1} W^{-1} S^{-1} \overleftarrow{\partial} K^{-1},$$

$$W = 1 + \frac{1}{2}S^{-1}.$$

What about higher-order terms in this IR, "perturbative" expansion, where the "small parameter" of the approximation is associated with the inclusion of more and more high-frequency dependence? One may expect to find the entire panoply of UV divergences expected in ordinary perturbation theory, but with one difference: Because every high-frequency correction is to be calculated in a background of an infinite number of soft, radiative corrections, one might expect that the latter could provide a measure of damping to the integrands of those momentum integrals which define the complete contribution. In many examples[13] of (at least) Abelian field theory, summing over all "soft" momenta provides a form factor which damps the "hard" four-momentum flowing through that $n-$point function. If the same type of damping occurs here, the UV behavior of the integrals over "hard" momenta so calculated could be considerably muted. Absolutely nothing is as yet known about such dependence, and we here confine the discussion to the "lowest order", IR approximation.

This Gaussian result suggests that taking all soft gluon interactions into account generates something very much like an effective "free gluon" propagator", $K^{-1}Y$, with a matrix-valued mass of order-of-magnitude M. As such, one has "confinement" of gluons to distances $\lesssim M^{-1}$, an effect that resembles some previous computer work.[14]

This result can be used to calculate quark-antiquark forces at long range, by operating with $Z[(-i)\delta/\delta A]$ upon quark wavefunctions (more properly Green's functions) dependent upon a color source A_μ^a; the only difficulty is that one must still find a way of treating the ordered exponentials[15] defining those wave functions, which suggests that such an analytic solution to quark confinement is still some distance away.

Finally, one must identify the finite mass scale M, introduced above, with some physical object, or measurement. It has recently been suggested[16] that the quantities

$$g^2 < F_{\mu\nu}^a F_{\mu\nu}^a >$$

and

$$g^3 f_{abc} < F_{\mu\nu}^a F_{\nu\sigma}^b F_{\sigma\mu}^c >$$

are measurable, and may be given by sum rules (for charmonium decay, assuming we had included quarks). If we assume these quantities are measureable (at some energy scale), then we may ask how they are related to M; in this IR approximation, g is to vanish while $< F^2 >$ is going to be divergent, and it will be interesting to see what is given by the combination. If this IR result is to be at all sensible, this quantity should be finite and independent of all UV effects.

The calculation is, in principle, quite simple: $Z[j]$ is Gaussian, and the appropriate functional derivatives are easily taken. The evaluation is made complicated only by the Y–dependence of the effective propagator, containing all the (axial) gauge structure; and hence the following simplification will be made: replace Y by 1, wherever possible, taking into account its correct form only when necessary. Without difficulty, one then obtains

$$g^2 < F^2 > \Rightarrow 4M^4 + \text{const.} \cdot \Lambda^4,$$

where the last, UV-divergent terms arise from that part of K^{-1} proportional to $(p^2)^{-1}$. Precisely these divergent terms should vanish when the K^{-1} factors defining Y are included. Without any arduous calculation, one can show that, if the result is to be independent of the axial gauge parameter η_μ – as it must – then it is also independent of this UV divergence. In this way, one can rewrite the previous equation, with some confidence, as

$$g^2 < F^2 >= 4M^4,$$

and in this fashion, a physical measurement defines the mass scale, M.

Much more work remains to be done in this approach to large-scale phenomena in QCD. For example: (i) A real calculation of the tadpole terms is necessary. (Here, $< F >$ has a tadpole divergence, one which should be removed automatically if the previously discarded σ–tadpoles are retained. The $< F^2 >$ used above are really: $< F^2 > - < F >^2$.) (ii) A better estimate of Y is necessary, and with it the explicit cancellation of gauge and UV dependence. (iii) When quarks are included, does $M \neq 0$ generate an effective quark mass, and if so, how?

When these relatively straightforward questions have been understood, one can hope to begin a realistic attack on physically interesting questions, such as an effective quark-antiquark potential. It is hoped that these non-machine IR methods will define a long overdue, semi-analytic approach to QCD.

REFERENCES

1. F. Guérin and H.M. Fried, Phys. Rev. **D33** (1986) 3039; H.M. Fried and T. Grandou, Phys. Rev. **D33** (1986) 1151; H-T Cho, T. Grandou, and H.M. Fried, Phys. Rev. **D37** (1988) 946 and 960; H.M. Fried and H-T Cho, Phys. Rev. **D41** (1990) 1489; H.M. Fried, in Proceedings of the IXth Moriond Workshop, Les Arcs, France (January 1989).

2. Pedagogical discussions of functional methods can now be found in many places, e.g. the text Quantum Field Theory, by C. Itzykson and J-B Zuber, (McGraw-Hill, 1980). To the author's best knowledge, such material first appeared in the monographs Field Theory, by J. Rzewuski, (PWN Publishers, 1972), and Functional Methods and Models in Quantum Field Theory, by H.M. Fried (MIT Press, 1972). Detailed descriptions of topics touched upon in the first Section of the present lecture may be found in the latter reference.

3. J. Schwinger, Phys. Rev. **82** (1951) 664.

4. E.S. Fradkin, Nucl. Phys. **76** (1966) 588.

5. This is discussed in Functional Methods and Eikonal Models, by H.M. Fried (Éditions Frontières, 1990).

6. H. Bokemeyer and W. Koenig, in this ASI.

7. H.M. Fried, Phys. Rev. **D42** (1990) 1857.

8. G. Soff and W. Greiner, this ASI.

9. M.B. Halpern, Phys. Rev. **D19** (1979) 517.

10. S. Coleman and E. Weinberg, Phys. Rev. **D7** (1973) 1888.

11. I. Bialynicki-Birula, J. Math. Phys. **3** (1962) 1094.

12. D.J. Gross and A. Nevue, Phys. Rev. **D10** (1974) 3235.

13. See, for example, E. Eichten and R. Jackiw, Phys. Rev. **D4** (1971) 439; H.M. Fried and T.K. Gaisser, Phys. Rev. **179** (1969) 1491; T.K. Gaisser, Phys. Rev. **D2** (1970) 1337.

14. M. Ogilivie and J.E. Mandula, Phys. Letters **B185** (1987) 127.

15. M.E. Brachet and H.M. Fried, J. Math. Phys. **28** (1987) 15; H.M. Fried, J. Math. Phys. **28** (1987) 1275; **30** (1989) 1161.

16. M.A. Shifman, A.I. Vainshtein and V.I. Zakharov, Nucl. Phys. **B147** (1979) 385, 448, 519.

The Gluon Anomaly in the Proton Spin

Jeffrey E. Mandula

Department of Energy
Division of High Energy Physics
Washington, DC 20545

In this lecture, we will review an important experiment on polarized muon-proton scattering carried out by the European Muon Collaboration (EMC)[1] and some theoretical developments that resulted from it.[2] The EMC measurement of the spin structure of deep inelastic muon scattering from protons was interpreted as indicating that essentially none of the proton's spin was carried by the spins of its quarks. A reexamination of this interpretation held that the apparent vanishing of the quark spin contribution could be due to a cancellation between an intrinsic quark contribution and a gluonic contribution induced by the triangle diagram that gives rise to the Adler-Bell-Jackiw anomaly.

Having briefly reviewed the experimental and theoretical situation, we will describe a calculation of the induced, anomalous gluonic contribution by means of a lattice QCD simulation. We will discuss some necessary theoretical and lattice preliminaries to the calculation, and the algorithm that was used. The key conclusion of the calculation is that the anomalous gluonic contribution is much too small to change the interpretation of the EMC experiment. Finally we will address some of the limitations of the calculation, and provide an estimate of its accuracy.

The principal result of the EMC experiment is that the integral of the polarized structure function g_1 is

$$\int_0^1 g_1(x, q^2 = -10.7 \text{ GeV}) \, dx = \frac{1}{2}(\frac{4}{9}\Delta u + \frac{1}{9}\Delta d + \frac{1}{9}\Delta s) = .126 \pm .010 \pm .015 \quad (1)$$

where the statistical and systematic errors are shown separately. The quoted systematic error of $\pm .015$ includes an estimate of the uncertainty due to the need to extrapolate the measured value of the structure functions to the regions above $x = .7$ and below $x = .01$.

The scaling behavior of g_1 and its connection to polarized proton quark spin fractions is known from operator product expansion analysis.[3] In particular, g_1 has at most a logarithmic dependence on q^2, and in the deep inelastic limit its integral is a sum of axial current matrix elements between proton states of polarization s:

$$4 \, m \, s_\mu \int_0^1 g_1(x) \, dx = \sum Q_i^2 \, \langle ps | \bar{q}_i \, i \, \gamma_\mu \gamma_5 \, q_i | ps \rangle \quad (2)$$

Vacuum Structure in Intense Fields, Edited by
H.M. Fried and B. Muller, Plenum Press, New York, 1991

The axial currents are the canonical spin operators for each quark, so their forward matrix elements are the fractions of the proton's spin carried by each kind of quark,

$$\langle ps | \bar{q}_i \, i \, \gamma_\mu \gamma_5 \, q_i | ps \rangle = 2 \, m \, s_\mu \, \Delta q_i \tag{3}$$

The significance of the EMC measurement stands out when combined with information from semi-leptonic axial weak decays of neutrons and hyperons. Specifically, neutron β-decay plus isospin symmetry gives the Bjorken sum rule[4]

$$\Delta u - \Delta d = g_A = 1.254 \pm .006 \tag{4}$$

while strangeness changing hyperon decay and flavor SU(3) symmetry gives[5]

$$\Delta u + \Delta d - 2 \, \Delta s = .60 \pm .12 \tag{5}$$

Putting this all together gives

$$\begin{aligned}
\Delta u &= +.74 \pm .05 \\
\Delta d &= -.51 \pm .05 \\
\Delta s &= -.19 \pm .07
\end{aligned} \tag{6}$$

and

$$\Delta u + \Delta d + \Delta s = +.04 \pm .16 \tag{7}$$

The statistical and systematic errors from the EMC result have been combined in quadrature. The errors in Eqs. (6) and (7) are all dominated by the error on the EMC result.

Eq. (7) is the source of the brouhaha created by the EMC result. The conclusion that only a small fraction, if any, of the proton's spin is carried by the spins of its constituent quarks was completely unanticipated, as was the conclusion that the strange quark's contribution was comparable to that of the down quark.

The interesting suggestion for understanding (or reinterpreting) the EMC result by separating out an anomalous gluonic contribution was made by Efremov and Teryaev,[6] Altarelli and Ross,[7] and Carlitz, Collins, and Mueller.[8] They observed that there is a contribution to each Δq_i associated with the gluonic spin distribution inside the proton which comes from the triangle diagram, shown in Figure 1, that is the source of the Adler-Bardeen-Jackiw anomaly.[9]

The evaluation of this contribution requires a choice of gauge and a careful treatment of high momentum limits. In axial gauges, at least, it comes entirely from infinite loop momenta around the triangle, so that the gluon vertices are effectively contracted to a point. In the (canonical) $A_0 = 0$ gauge, this gluonic spin contribution to each quark axial current matrix element is the matrix element of the anomalous current

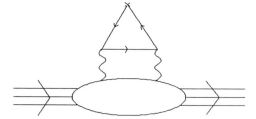

Figure 1 The triangle diagram, which induces an effective gluonic contribution to each quark spin distribution in the proton.

$$\langle ps | [j^5_\mu]_{Triangle} | ps \rangle = \frac{\alpha}{2\pi} \langle ps | K_\mu | ps \rangle = -2m \, s_\mu \, \frac{\alpha}{2\pi} \Delta g \tag{8}$$

which is the canonical gluon spin operator in that gauge.[5] We identify $-\frac{\alpha}{2\pi} \Delta g$ as the gluonic part of each quark spin fraction.

The fact that neither the isolation of this gluonic contribution nor even the final formula can be expressed in gauge invariant terms has been criticized and its physical interpretation questioned.[5,10] However, the argument that the triangle diagram meaningfully isolates a gluonic contribution hidden in a quark operator matrix element seems at least plausible, if not definitive, and we will accept it as the basis of the following calculation and analysis.

In the remainder of this lecture we describe a lattice gauge simulation of Δg by evaluating the $\langle ps | K_\mu | ps \rangle$ matrix element. We use the standard formulation: Wilson's SU(3) lattice action on a periodic Euclidean lattice to describe the gauge fields and (4-spin × 3-color component) Wilson fermions to describe the quarks. We work in the quenched approximation, which omits closed quark loop effects. Our calculation uses lattices generated by Bernard, Hockney, and Soni for a different purpose — the calculation of hadronic matrix elements that enter into weak interaction rates.[11] They made available for this purpose an ensemble of 204 $6^3 \times 10$ lattices with $\beta = 5.7$ together with quark propagators computed through 8 units of Euclidean time both forward and backwards. The quark propagators satisfy open (Neumann) boundary conditions with hopping constant $\kappa = .162$.

A key issue which arises in this calculation is the formulation of a suitable lattice approximation to the anomalous current. For the reasons discussed earlier, we work in the gauge $A_4 = 0$, which we implement on each lattice by gauge transforming all times like link variables (except at a single reference times outside the quark propagator range) to $U_4 = 1$. In the continuum, the anomalous current simplifies in this gauge to

$$K_i = 2 \, \epsilon_{ijk} \, \text{Tr} \, A_j \, F_{k4} \tag{9}$$

A convenient lattice approximation to F_{k4} is[12]

$$F_{k4}^{\text{lattice}}(n) = \frac{1}{8i} [U_k(n)U_k^\dagger(n+\hat{4}) + U_k^\dagger(n+\hat{4}-\hat{k})U_k(n-\hat{k})$$

$$+ U_k^\dagger(n-\hat{k})U_k(n-\hat{4}-\hat{k}) + U_k(n-\hat{4})U_k^\dagger(n) \tag{10}$$

$$- \text{Hermitean Conjugate}]_{\text{Traceless}}$$

It has the correct continuum limit, transforms simply under the lattice hypercubic symmetry group, and depends under gauge transformation only on the value of the gauge transformation at site n. With this expression for the field strength a convenient lattice approximation to the anomalous current is

$$K_i^{\text{lattice}}(n) = \text{Im} \, \epsilon_{ijk} \, \text{Tr} \, U_j(n-\hat{j}) \, F_{k4}^{\text{lattice}}(n) \, U_j(n) \tag{11}$$

Like $F_{k4}^{\text{lattice}}(n)$, it has the correct continuum limit and transforms simply under the hypercubic group. Under infinitesimal gauge transformations its change depends on the difference of the values of the gauge transformation at sites $n+\hat{j}$ and $n-\hat{j}$ $(j \neq i)$.

This is not a perfect approximation to the anomalous current, however, and some important properties are only recovered in the continuum. These include its response to x_4 independent gauge transformations ω,

$$\delta K_i \to \partial_4 \, \epsilon_{ijk} \, \text{Tr} \, [\partial_j \omega \, A_k] \tag{12}$$

and the implied vanishing of the proton expectation value of the gauge transformation in K_i

$$\langle ps | \delta K_i | ps \rangle \rightarrow 0 \qquad (13)$$

Other important properties of K_μ that are only recovered in the continuum are the gauge invariance of its divergence and its relation to topology changing gauge transformations.

An alternative method for evaluating Δg has the advantage that it is (almost) manifestly gauge invariant on the lattice. This algorithm also needs zero spacing limit to recover all the continuum properties. It has the disadvantage that its statistical noise is larger than that of the axial gauge algorithm. The basic idea of the algorithm is that the space components of the forward matrix elements of K_μ can be inferred from those of $\partial_\mu K_\mu$ by means of the limit

$$\langle \vec{0}s | K_i | \vec{0}s' \rangle = \lim_{\vec{p} = p\hat{i} \rightarrow \vec{0}} \frac{i}{p} \langle \vec{0}s | \partial_\mu K_\mu | \vec{p}s' \rangle \qquad (14)$$

where

$$\partial_\mu K_\mu = \frac{1}{2} \epsilon_{\mu\nu\lambda\sigma} Tr \, F_{\mu\nu} \, F_{\lambda\sigma} \qquad (15)$$

Only space components are recovered, but K_0 does not enter into the spin fraction, so this is not a problem.

To implement this on the lattice we define

$$\Delta_\mu K_\mu^{Invariant} = \frac{1}{2} \epsilon_{\mu\nu\lambda\sigma} Tr \, F_{\mu\nu}^{Latt} \, F_{\lambda\sigma}^{Latt} \qquad (16)$$

and take the momentum p to be the smallest possible on the finite lattice along each direction

$$\langle \vec{0}s | K_i | \vec{0}s' \rangle = \frac{i}{2p} (\langle \vec{0}s | \Delta_\mu K_\mu | \vec{p}s' \rangle - \langle \vec{0}s | \Delta_\mu K_\mu | -\vec{p}s' \rangle)$$
$$+ \, O(p^2) \qquad (\vec{p} \parallel \hat{i}) \qquad (17)$$

The two definitions of K_μ are not equal on the lattice, although they become equal in the continuum limit.

Of course, since K_μ is not perfectly gauge invariant, there must be some gauge dependence remaining in the covariant algorithm. That residual gauge invariance is contained in the tacit assumption that there are no unphysical singularities in matrix elements of K_μ which would interfere with the momentum limit. However, such singularities are known to occur, at least in covariant gauges. They are known as Kogut-Susskind dipoles.[13] They are not known to occur in axial gauges, and so what we have called a gauge invariant algorithm is more properly called a general axial gauge algorithm.

We project out the polarization asymmetry by tracing with $i\gamma_5\slashed{s}$, and divide by the trace of the propagator to remove the leading exponential decay, giving the convenient expression

$$\Delta g = - \frac{\text{Tr } P_+ \, i\gamma_5\slashed{s} \, \langle 0 | \Psi(x_4) \, \vec{s} \cdot \vec{K}_\mu(y_4) \, \bar{\Psi}(z_4) | 0 \rangle}{\text{Tr } \langle 0 | \Psi(x_4) \, \bar{\Psi}(z_4) | 0 \rangle} \tag{18}$$

which is independent of the normalization of the proton field. It is from this formula that we extract the value of Δg in computer simulations. To improve statistics we average over the three principal axis choices for $\vec{s}$.

In a lattice numerical simulation, the numerator of Eq. (18) measures the correlation, lattice by lattice, of two quantities which are fluctuating about 0. One is the polarization asymmetry of the proton propagator, measured by

$$\text{Tr } P_+ \, i\gamma_5\slashed{s} \, \Psi(x_4) \, \bar{\Psi}(z_4)$$

which is composed only of fermion operators, while the other is the $\vec{s}$ component of the anomalous current,

$$\vec{s} \cdot \vec{K}_\mu(y_4)$$

which is composed only of the bosonic, gauge field variables. The correlation between these fluctuations arises because, lattice by lattice, the polarization asymmetry of the proton propagator is determined by the values of the same gauge variables, through the lattice Dirac equation, that comprise the lattice anomalous current operator.

It should be noted that this aspect of the calculation is the same in both the quenched approximation and the full theory. The effect of including quark loops is only to modify the relative weighting of different gauge configurations. The core of the calculation is still the computation of gluon variable with quark variable fluctuation correlations.

We "measured" the gluon spin fraction Δg of Eq. (18) using the sample of 204 $6^3 \times 10$ lattices provided by the Bernard-Soni collaboration. With the parameters of those lattices, $\beta = 5.7$ and $\kappa = .162$, the lattice spacing is about 1.0 GeV^{-1} and the proton mass about 1.4 inverse lattice spacings.

We constructed a color singlet proton field from colored quark fields by projecting on the appropriate spin and symmetry. To extract Δg using Eq. (14), we should take x_4-y_4 and y_4-z_4 as large as possible. It is also desirable to avoid the largest values of x_4-z_4, since the quark propagator will be most affected by the boundary condition in that case. The best compromises available are $x_4-z_4 \leq 7$ with $x_4-y_4 \geq 3$ and $y_4-z_4 \geq 3$. The values of Δg are extracted from the forward and backward propagation separately, and are given along with their statistical errors in the two Tables.

Table I. Δg Using Axial Gauge Construction

Separations		Δg	
$x_4 - y_4$	$y_4 - z_4$	Forward	Backward
4	3	−.158 ± .671	+.008 ±.191
3	4	−.292 ± .367	+.009 ±.124
3	3	−.097 ± .397	+.001 ±.114

Table II. Δg Using Gauge Invariant Construction But Finite Momentum Approximation

Separations		Δg	
$x_4 - y_4$	$y_4 - z_4$	Forward	Backward
4	3	−0.83 ± 2.58	+0.07 ± .80
3	4	+2.89 ± 3.58	+0.07 ± .72
3	3	−0.15 ± 1.42	−0.08 ± .46

Several comments are in order regarding the quality of these results. One is that they are all mutually consistent, and consistent with 0. That is, we have not arrived at a value for Δg, but rather a bound, $|\Delta g| \leq .5$. Another is that the finite lattice spacing and lattice size are serious but probably not fatal limitations. The proton mass is almost 50% too large on these lattices, for example. In the zero spacing and large time limits, the value of Δg gotten from the simulation would be independent of the exact values of x_4–y_4 and y_4–z_4, so long as both were large enough. Within the statistics, there is no evidence that this is failing, and the much closer than statistics self-agreement of the forward and backward results separately indicates that the finite spacing might not be much too coarse. Similarly, quenched approximation results in other calculations, which like this one would not be structurally different in the full theory, are usually reasonable, and have not yet been seen to be qualitatively wrong.

Finally, at the fairly small values of x_4–y_4 and y_4–z_4 at which we worked (either one or both were $\Delta t = 3$), one should ask how significant is the contribution of high mass states with the same quantum numbers as the proton to the three-point Green's function from which we extracted Δg. We can estimate this by examining an extrapolation of the simulated proton propagator from the largest values of Δt, at which the single proton contribution is purest, back to $\Delta t = 3$ or 4, and seeing by how much of the simulated propagator exceeds the extrapolation. This is shown in Figure 2.

At $\Delta t = 4$ the admixture of higher states is about 14%, and at $\Delta t = 3$ it is about 40%. This is comparable to the other sources of error. The overall effect of all these considerations is to weaken the bound coming from the simulation by about a factor of two.

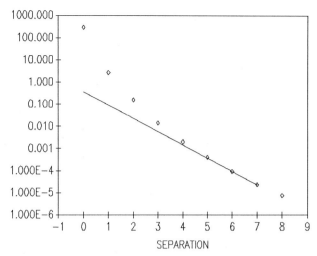

Figure 2 The proton propagator as obtained from the simulation, compared with the extrapolation back from $\Delta t = 6$ and $\Delta t = 7$.

The basic conclusion of the foregoing analysis is that the gluon spin fraction Δg is apparently less than 1 in magnitude. Since at $\beta = 6/g^2 = 5.7$ the QCD coupling is $\alpha = g^2/(4\pi) \approx .08$, this gives a bound on the correction to the total quark spin fraction of the proton's spin of $(3\alpha/2\pi)|\Delta g| \leq .05$. If, following Refs. 7-9, we modify Eq. (7) by removing the anomalous gluonic contribution to the total quark spin fraction of the proton, the result would be changed by at most .05, and each component of Eq. (6) by a third of that.

The analysis described here was undertaken to evaluate the extent to which the anomalous contribution of gluons to the spin dependent quark structure functions would modify the conclusion of the EMC experiment that the net quark contribution to the proton's spin is very small. In so far as the results using such small lattices can be trusted, the conclusion of this simulation is that the effect is an order of magnitude too small to change the thrust of the EMC conclusion.

Acknowledgements

The author wishes to express his appreciation to the Bernard-Soni collaboration for generously making available the lattices that form the basis for this analysis. The author has also greatly benefitted from valuable discussions and encouragement to all stages of this work with Carl Bender, Robert Carlitz, Robert Jaffe, Marek Karliner, James Labrenz, Alfred Mueller, Michael Ogilvie, Amarjit Soni, and Vigdor Teplitz.

References

1. The EMC Collaboration, Nucl. Phys **B328**, 1 (1989)
2. Much of the material presented in this talk is contained in J.E. Mandula, Phys. Rev. Letts. **65**, 1403 (1990), and is included here for coherence of presentation.
3. J. Kodaira, Nucl. Phys., **B165**, 129 (1979)
4. J.D. Bjorken, Phys. Rev. **148**, 1467 (1966)
5. R.L. Jaffe and A. Manohar, Nucl. Phys. **B337**, 509 (1990)

6. A.V. Efremov and O.V. Teryaev, JINR report E2-88-278 (1988)
7. G. Altarelli and G.G. Ross, Phys. Letts. **B212**, 391 (1988)
8. R.D. Carlitz, J.C. Collins, and A.H. Mueller, Phys. Letts. **214B**, 229 (1988)
9. J.S. Bell and R. Jackiw, Nuovo Cimento **A51**, 47 (1967); S.L Adler, Phys. Rev. **177**, 2426 (1969)
10. G. Bodwin and J. Qiu, Phys. Rev. D **41**, 2755 (1990)
11. See, for example, C. Bernard, T. Draper, G. Hockney, A. Rushton, and A. Soni, Phys. Rev. Letts. **55**, 2770 (1985)
12. J.E. Mandula, J. Govaerts, and G. Zweig, Nucl. Phys. **B228**, 109 (1983)
13. J. Kogut and L. Susskind, Phys. Rev. D **11**, 3594 (1975)

LECTURES ON SOLITONS

W. N. Cottingham

University of Bristol
H.H. Wills Physics Laboratory
Royal Fort
Tyndall Avenue
Bristol BS8 1TL

SOLITONS - A SIMPLE EXAMPLE IN ONE SPACE DIMENSION

The solitons of a field theory are field configurations which are twisted. A simple example is that of a metal strip with one edge fixed horizontally and hanging under gravity.

The field variable at any position x is the angle $\theta(x)$ which the strip makes with the verticle. The ground state is with $\theta(x) \equiv 0$ for all x and there are low energy configurations with $\theta(x)$ small for <u>all</u> x. A soliton can be put into the field simply by putting a twist into the strip

This soliton has a higher energy than the ground state but the ground state cannot be reached by any continuous transformation from the soliton.

In simple elasticity theory the potential energy density of a field configuration $\theta(x)$ is

$$V(\theta) = b\left[\underbrace{a^2(1 - \cos(\theta))}_{\substack{\text{gravitational} \\ \text{energy}}} + \underbrace{\left(\frac{\partial\theta}{\partial x}\right)^2}_{\substack{\text{shear} \\ \text{energy}}} \right]$$

a and b are parameters depending on the mass and shear modulous of the strip.

Vacuum Structure in Intense Fields, Edited by
H.M. Fried and B. Muller, Plenum Press, New York, 1991

The minimum energy configuration that corresponds to a static soliton located at $x \sim x_0$ is then

$$\theta(x) = \pm 4 \tan^{-1}\left[e^{a(x-x_o)}\right]$$

the + or - sign indicates a soliton or an *anti*soliton

EXERCISE   Show that the minimum energy configuration satisfies the
"SINE GORDAN EQUATION"

$$\frac{d^2\theta}{dx^2} = a^2 \sin\theta$$

and that this has solutions as above.

and clearly there are solutions with one twist (soliton) two twists (two solitons) etc.

Dynamics can be introduced into the theory by introducing kinetic energy density of motion

$$T = \frac{b}{c^2}\left(\frac{\partial\theta}{\partial t}\right)^2 \qquad \text{(with another parameter c)}$$

and a Lagrange density $L = T\text{-}V$

$$L = b\left[\frac{1}{c^2}\left(\frac{\partial\theta}{\partial t}\right)^2 - \left(\frac{\partial\theta}{\partial x}\right)^2 - a^2(1-\cos\theta)\right]$$

Near the true ground state, for small θ

$$L = b[\frac{1}{c^2}\left(\frac{\partial\theta}{\partial t}\right)^2 - \left(\frac{\partial\theta}{\partial x}\right)^2 - \frac{a^2}{2}\theta^2]$$

and there are wave like solutions $\theta = \varepsilon e^{ikx-iwt}$ with small amplitude ε and a dispersion relation

$$\frac{\omega^2}{c^2} = k^2 + \frac{a^2}{2}$$

However, there are also <u>soliton</u> solutions and the "Lorentz" invariance of L shows that a soliton (antisoliton) moving with velocity v is

$$\theta(x,t) = \pm 4 \tan^{-1}[e^{a\gamma(x-vt-x_o)}]$$

$$\gamma = \frac{1}{\sqrt{1-\dfrac{v^2}{c^2}}}$$

We do not need the detailed dynamics or kinematics to tell us that if θ changes smoothly with time then the number of twists in θ, the *soliton number*, will not change with time. The soliton number is called a TOPOLOGICAL INVARIANT; time is an inessential parameter for determining the *soliton number*. Only the space structure is important.

THE TOPOLOGICAL CURRENT

The topological current has nothing to do with a symmetry of a theory but just as the "Noether" current continuity equation expresses the local conservation law so too there exists a "topological current". In the example above

$$j^\mu = \frac{1}{2\pi} \varepsilon^{\mu\nu} \partial_\nu \theta$$

or explicitly

$$j^x(x,t) = -\frac{1}{2\pi} \frac{\partial\theta}{\partial t}$$

$$j^t(x,t) = \frac{1}{2\pi} \frac{\partial\theta}{\partial x}$$

For a topological current the continuity equation $\partial_\mu j^\mu = 0$ is a mathematical

identity; the $\dfrac{1}{2\pi}$ factor is simply to normalise .

$$\text{The Soliton number} = \int_{-\infty}^{+\infty} j^t(x)dx$$

<u>EXERCISE</u>   Show that $\theta = 4\tan^{-1}\left[\left(e^{\gamma(x-vt-x_o)}\right)\right]$
has soliton number 1 with the above convention.

A MORE COMPLICATED FORM FOR THE CURRENT

The form for the current given above can be simply guessed, but a form which is more easily generalised to two and three dimensions is the following: θ can be thought of as a point on a unit circle. That point can also be thought of as the direction of a unit vector $(n_1,n_2) = \vec{n}$ in two dimensions. (We only here still have one INDEPENDENT variable since $n_1^2 + n_2^2 = 1$).

To write the current in terms of

$\dot{\vec{n}}$ note that $d\theta = |d\vec{n}|$

$$= \vec{n}_\wedge d\vec{n} = \varepsilon_{ab} n_a dn_b$$

$\varepsilon_{12} = 1$ the antisymmetric tensor

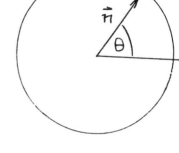

Explicitly $\qquad \vec{n} = (cos\theta, \, sin\theta)$

$$d\vec{n} = (-sin\theta, \, cos\theta)d\theta$$

$$\varepsilon_{ab} n_a dn_b = (cos^2\theta + sin^2\theta)d\theta = d\theta$$

$$\therefore \frac{\partial\theta}{\partial x} = \varepsilon_{ab} n_a \frac{\partial n_b}{\partial x} \qquad\qquad \frac{\partial\theta}{\partial t} = \varepsilon_{ab} n_a \frac{\partial n_b}{\partial t}$$

and $\quad j^\mu = \dfrac{1}{2\pi} \varepsilon^{\mu\nu} \varepsilon_{ab} n_a \partial_\nu n_b$

Yet another viewpoint is that of the topologist who can think of the line $-\infty < x < \infty$ as being parameterized as a circle S_1; for example

$$x = tan^{-1}(\alpha/2) \qquad -\pi < \alpha < \pi$$

For a topologist an N soliton field configuration is a mapping

$$\alpha \to x \to \theta$$

$$\text{or } S_1 \to S_1$$

Points around the circle "α" are mapped N times around the circle "θ". N is called the WINDING NUMBER.

The importance of this viewpoint is that mathematicians have generalised this classification to many different topological spaces.

TWO SPACE DIMENSIONS

The x,y plane can be parameterized by points on a unit sphere S_2

$$x = r \, cos\psi$$

$$y = r \, sin\psi \qquad\qquad\qquad r = tan^{-1}(1/2\alpha)$$

$$0 \le \psi \le 2\pi \qquad\qquad\qquad 0 \le \alpha \le \pi$$

Topologists tell us that any map of $S_2 \to S_1$ can be deformed smoothly onto a single point on S_1 so there is NO winding number for such maps. This tells us that there are no scaler field theories with topological solitons as in one space dimension.

The next most simple possibility is the mapping $S_2 \to S_2$ and here there are non trivial maps. Their existence can be appreciated from the fact that "you can't peel an orange without breaking the skin". For a mapping, at any fixed time t, of $S_2 \to S_2$ we need three fields $n_1(x,y;t)$ $n_2(x,y;t)$ $n_3(x,y;t)$ with the constraint.

$$n_1{}^2 + n_2{}^2 + n_3{}^2 = 1$$

The winding number (the number of coverings of the "n" unit sphere) is

$$N = \frac{1}{4\pi} \int\int \varepsilon_{abc} n_a \frac{\partial n_b}{\partial x} \frac{\partial n_c}{\partial y} \, dxdy$$

because clearly $\varepsilon_{abc} n_a \dfrac{\partial n_b}{\partial x} \dfrac{\partial n_c}{\partial y} \, dxdy = \bar{n} \cdot \left(\dfrac{\partial \bar{n}}{\partial x} \wedge \dfrac{\partial \bar{n}}{\partial y} \right) dxdy$ is a small area element on the "n" unit sphere which has total area 4π.

EXERCISE   Consider the field configuration

$$n_1 = sin\alpha cos N\Psi \qquad\qquad n_2 = sin\alpha sin N\Psi$$
$$n_3 = cos\alpha$$

Show that

$$\frac{1}{4\pi} \int \varepsilon_{abc} n_a \frac{\partial n_b}{\partial x} \frac{\partial n_c}{\partial y} \, dxdy$$

$$= \frac{1}{4\pi} \int \varepsilon_{abc} n_a \frac{\partial n_b}{\partial \alpha} \frac{\partial n_c}{\partial \psi} \, d\alpha \, d\psi = N$$

We can anticipate that field theories with Lagrange densities like

$$L = \frac{1}{2c^2} \left(\frac{\partial \bar{n}}{\partial t} \right)^2 - \frac{1}{2} \left(\frac{\partial \bar{n}}{\partial x} \right)^2 - \frac{1}{2} \left(\frac{\partial \bar{n}}{\partial y} \right)^2 - V(\bar{n}) \qquad \text{with } |\bar{n}| = 1 \text{ will have}$$

soliton solutions (see Polyakov for elaboration).

EXERCISE   Show that the continuity equation $\partial_\mu j^\mu = 0$ for the topological

current $j^\mu = \dfrac{1}{8\pi} \varepsilon^{\mu\alpha\beta} \varepsilon abc \, n_a (\partial_\alpha n_b)(\partial_\beta n_c)$ is a mathematical identity.

THREE SPACE DIMENSIONS

The generalization to three dimensions gives us the Skyrme model of the nucleon and is the first example in which the soliton ideas were applied to particle physics. The generalization is straightforward and will, I hope, look to be plausible.

Three dimensions can be parameterized by points on a three sphere S_3

(for example $x = r\sin\theta\cos\Psi$ $y = r\sin\theta\sin\Psi$ $z = r\cos\theta$ $r = \tan^{-1}(1/2\alpha)$)

Topologists tell us that there are non trivial maps $S_3 \to S_3$,So introduce the four

fields n_0 n_1 n_2 n_3 with the constraint $\sum_{a=0}^{3} n_a^2 = 1$.

Since the area of S_3 is $2\pi^2$ the winding number, or soliton number

$$N = \frac{1}{2\pi^2}\int \varepsilon_{abcd}\left(n_a \frac{\partial n_b}{\partial x}\frac{\partial n_c}{\partial y}\frac{\partial n_d}{\partial z}\right)dxdydz = \frac{1}{2\pi^2}\int \varepsilon_{abcd}\left(n_a \frac{\partial n_b}{\partial \alpha}\frac{\partial n_c}{\partial \theta}\frac{\partial n_d}{\partial \Psi}\right)d\alpha d\theta d\Psi$$

The Skyrme model is not usually written in terms of n_a but, by a simple change of notation, in terms of the "σ model" of meson physics. This model has a scaler field σ, and three pi meson like fields Φ_i i = 1,2,3 with the constraint

$\sigma^2 + \Phi_1^2 + \Phi_2^2 + \Phi_3^2 = 1$. This constraint implies only three independent fields and these are usually written in terms of the unitary matrix

$$u = \sigma + i\vec{\tau}.\vec{\Phi} = e^{i\vec{\tau}.\vec{\pi}/f_\pi}$$

$\vec{\pi} = (\pi_1, \pi_2, \pi_3)$ are the three pi meson fields f_π (= 98 MeV) is a parameter.

The most simple Lagrangian is

$$L = \frac{f_\pi^2}{4}Trace(\partial_\mu U\partial^\mu U^+) - V(\vec{\pi})$$

for $\vec{\pi}$ small $U \sim 1 + i\vec{\tau}\cdot\vec{\pi}/f_\pi$ and

$$\frac{1}{4}f_\pi^2 Trace\partial_\mu U\partial^\mu U^+ = \frac{1}{2}(\partial_\mu \pi_i)^2 = \frac{1}{2}\left[\frac{1}{c^2}\left(\frac{\partial \pi_i}{\partial t}\right)^2 - (\vec{\nabla}\pi_i)^2\right].$$

This term by itself describes three massless π mesons and has the important SU(2) x SU(2) symmetry of low energy meson physics U $\to$ AUB† with A and B arbitrary SU(2) matrices. The pion can be given a mass and the symmetry broken down to a single (isospin) SU(2) symmetry by choosing the potential

$$V(\vec{\pi}) = \frac{m_\pi^2}{2f_\pi^2}TraceU.$$

<u>EXERCISE</u>   Prove the above statements about L.

328

The Noether current associated with the symmetry $U \to AUA$ can be shown to be the axial current which couples to the W bosons and it is through this weak current that the parameter f_π is determined.

With the two parameters fixed, satisfactory predictions can be made about π-π intereactions (from higher terms in the expansion of U). The vector mesons, the ω, $\vec{\rho}$ and $\vec{a}_1$ can also be introduced to construct a phenomenological field theory of low energy meson physics (see for example ref 5). Also, and most important for this short course, this theory, which contains only meson fields, has soliton configurations. We know this from the general topology. The single static soliton centred at $\vec{r} = 0$ can be found by minimizing the energy associated with the field configuration

$$U = e^{i\vec{\tau} \cdot \hat{r} \theta(r)} = \cos\theta(r) + i\vec{\tau} \cdot \hat{r} \sin\theta(r)$$

$\hat{r} = \vec{r} / r$ is a unit vector [4].

The boundary conditions $\theta(o) = \pi$ and $\theta(\infty) = 0$ impose a single covering of the "n" sphere as $\vec{r}$ covers all space; the soliton number is one. It is important for "low energy" phenomenology that these solitons can be quite successfully interpreted as nucleons [3,4,5], soliton number is therefore BARYON NUMBER. Pressing the theory still further, many features of the nucleon-nucleon interaction can be understood as the interaction between two solitons; see for example [5].

SOLITONS IN GAUGE FIELD THEORIES, MAGNETIC MONOPOLES

We will stay in three space dimensions and I will give one example of this possibility; the Georgi-Glashow model [6].

The gauge symmetry is SU(2) with three gauge fields $X_\mu = X_\mu^b \tau^b$ b = 1,2,3,

and a field tensor $X_{\mu\nu} = \partial_\mu X_\nu - \partial_\nu X_\mu + \dfrac{ie}{2}[X_\mu, X_\nu]$. The SU(2) symmetry is

broken down to U(1) by three Higgs fields $\phi = \phi^b \tau^b$ with a potential

$$V(\phi) = (\phi^2 - a^2)^2 \qquad \phi^2 = \frac{1}{2}Trace(\phi\phi).$$ The covariant derivative of ϕ is

$$D_\mu \phi = \partial_\mu \phi + \frac{ie}{2}[X_\mu, \phi]$$

We will consider the Lagrange density

$$L = -\frac{1}{8}Trace(X^{\mu\nu}X_{\mu\nu}) - \frac{1}{4}Trace(D^\mu\phi D_\mu\phi) - V(\phi)$$

We will work in a gauge with $X_0 \equiv 0$ and be concerned with static solutions, hence

$$D_o\phi \equiv 0 \quad and \quad X_{oi} \equiv 0.$$

The potential energy density is then

$$\frac{1}{8} TraceX_{ij}X^{ij} + \frac{1}{4} TraceD^i\phi D_i\phi + V(\phi)$$

and is a sum of squares. The lowest energy density is zero when, for example $X_i \equiv 0$ and $\phi \equiv$ a constant; such that $\phi^2 = a^2$.

The soliton we now construct has a different topological interpretation to that of the Skyrmion. For the Skyrme soliton, the field at infinity had the property $n_o = 1$ $n_i = 0$: all points at ∞ where mapped onto the same point on the "n" sphere. The soliton of this model has a Higgs field configuration, as $r \to \infty$

$$\phi \to \frac{ar_i\tau_i}{r} = a\hat{r} \cdot \vec{\tau}$$

which is a mapping of the points on the ordinary sphere (S_2) of the polar angles of co-ordinate space onto the S_2 sphere in "ϕ" space

$$\phi_1^2 + \phi_2^2 + \phi_3^2 = a^2$$

This soliton is a mapping of $S_2 \to S_2$.

Can such field configurations have finite energy?? Clearly we have the

necessary property $V(\phi) \to 0$ as $r \to \infty$. However, $\partial_i\phi = \frac{a\tau_b}{r}[\delta_{ib} - \frac{r_ir_b}{r^2}]$. With no

gauge field $\int \frac{1}{4}Trace(\partial_i\phi\partial^i\phi)d\Omega = \frac{8\pi a^2}{r^2}$ and the total energy

$$8\pi \int \frac{a^2}{r^2}r^2dr \to \infty.$$

THERE MUST BE A GAUGE FIELD FOR CONVERGENCE

What must this gauge field be at large r??

$$D_i\phi = \frac{a\tau_b}{r}\left\{[\delta_{ib} - \frac{r_ir_b}{r^2}] - e\varepsilon_{bcd}X_i^c r_d\right\}$$

330

to cancel the first term the gauge field MUST BE $X_I = \frac{1}{e} \frac{\varepsilon_{ick} r_k}{r^2} \tau_b$ at large r the field tensor gives a convergent contribution to the total energy since, by substitution

$$X_{ij} = \varepsilon_{ijk} B_k$$

with $B_k = \frac{r_k r_b \tau_b}{er^4} = \frac{r_k \phi}{aer^3}$ and Trace $X_{ij} X^{ij} \sim 1/r^4$ at large r.

A finite energy field configuration of this form can be found by making the ansatz [7,8]

$$\phi = a \frac{\tau^b r_b}{r} f(r) \qquad X_i = \frac{1}{e} \frac{\varepsilon_{ick} \tau_c r_k}{r^2} g(r)$$

f(r) and g(r) are found by minimizing the energy subject to the boundary conditions (which makes the fields smooth at the origin)
f(0) = g(0) = 0 (f(∞) = g(∞) = 1 is imposed by the finite energy).

In this model the unbroken symmetry is U(1) and the coupling constant "e" identifies this as the electromagnetic symmetry. The curious property of this soliton is that it is a magentic monopole. This can be seen because the unbroken symmetry must leave φ invariant. Hence X_{ij} is invariant and the $\vec{B}$ field is the magnetic field of a magnetic monopole of stength 1/e.

REFERENCES

1.   Aspects of Symmetry, Sidney Coleman, Cambridge University Press (1985).
2.   Gauge Fields and Strings, A.M. Polyakov, Harwood Academic Publishers (1987).
3.   T.H.R. Skyrme, *Proc. Roy. Soc.*, A260:127 (1961).
4.   G.S. Adkins et al., *Nuc. Phys.*, B228:552 (1983).
5.   M. Lacombe et al., *Phys. Rev. D*, 38:1491 (1988).
6.   H. Georgi and S.L. Glashow, *Phys. Rev.*, D6:2577 (1972).
7.   G't. Hooft, *Nuc. Phys.*, B79:276 (1974).
8.   A.M. Polyakov, *J.E.T.P. Lett.*, 20:194 (1974).

RENORMALIZATION GROUP AND STABILITY

ANALYSIS OF $\lambda\phi^4$ THEORIES

M. Consoli and D. Zappalà

Dipartimento di Fisica, Università di Catania
Istituto Nazionale di Fisica Nucleare, Sezione di Catania
Catania, Italy

1. INTRODUCTION

In this paper we shall discuss the occurrence of spontaneous symmetry breaking in scalar self-interacting theories. We shall employ two different techniques which are suitable to handle non perturbative phenomena in quantum field theories: the well known canonical quantization of the Hamiltonian and the effective potential for composite operators introduced in ref. [1]. By comparing the two methods we will check our results and gain in physical insight. Differently from the perturbative analysis of the effective potential of ref. [2], spontaneous symmetry breaking does arise as a sensible phenomenon even in the absence of gauge bosons.

Before discussing field theory we shall briefly analyze in sect. 2 the quantum mechanical double well potential and the concept of Wick ordering as a necessary preliminary step for any perturbative expansion. In sect. 3 we shall present the main results for the $\lambda\phi^4$ theories based on ref. [3]. In sect. 4 the problems related to the Renormalization Group (RG) invariance of the analysis will be discussed. The cutoff independence of our results will be shown to imply the "non triviality" of a $\lambda\phi^4$ theory with a spontaneously broken phase. In sect. 5 we shall consider the extension to the $O(N)$ symmetric theory and some implications of our results about the Higgs mass.

2. QUANTUM MECHANICS. DOUBLE WELL POTENTIAL

Let us consider the Hamiltonian operator ($m^2 > 0$, $\lambda > 0$)

$$\widehat{H} = \frac{\widehat{p}^2}{2} - \frac{1}{2}m^2\widehat{x}^2 + \lambda\widehat{x}^4 \ . \tag{2.1}$$

We shall analyze this elementary system in the same approximation employed for field theory. As a first approach to the properties of the ground state, we can consider a variational procedure within the space of normalized gaussian wave functions

$$\psi(x) = \frac{1}{\pi^{1/4}\sigma^{1/2}}exp\left[-\frac{(x-x_0)^2}{2\sigma^2}\right] \tag{2.2}$$

with

$$< \psi|\widehat{x}|\psi > = x_0 \tag{2.3a}$$

$$< \psi|\widehat{x}^2|\psi > = x_0^2 + \sigma^2/2 \tag{2.3b}$$

Vacuum Structure in Intense Fields, Edited by
H.M. Fried and B. Muller, Plenum Press, New York, 1991

$$< \psi | \psi >= 1 . \tag{2.3c}$$

We therefore obtain

$$< \psi | \widehat{H} | \psi >= E(x_0, \sigma) \tag{2.4}$$

and, by extremizing E with respect to x_0 and σ, we can obtain a variational estimate of the ground state energy. At the same time we can introduce a second quantized formalism by defining ($\hbar = 1$)

$$\widehat{x} = x_0 + \widehat{\eta} \tag{2.5}$$

with

$$\widehat{\eta} = \frac{1}{\sqrt{2M}} \left(\widehat{a} + \widehat{a}^+ \right) \tag{2.6}$$

and the conjugate momentum

$$\widehat{p} = \sqrt{\frac{M}{2}} \, i \left(\widehat{a}^+ - \widehat{a} \right) \tag{2.7}$$

so that $[\widehat{x}, \widehat{p}] = i$ corresponds to $[\widehat{a}, \widehat{a}^+] = 1$.

It is well known that, by minimizing E with respect to $M = 1/\sigma^2$, we also diagonalize the quadratic part of the Hamiltonian, expressed in terms of $\widehat{a}$ and $\widehat{a}^+$. At the same time the cubic and the fourth order terms are automatically Wick ordered with all the creation operators to the left of the annihilation operators. By defining $M = M(x_0)$ the solution of the equation

$$0 = \frac{\partial E}{\partial M} = \frac{1}{4} + \frac{m^2}{4M^2} - \frac{3\lambda x_0^2}{M^2} - \frac{3\lambda}{2M^3} \tag{2.8}$$

we obtain the gaussian effective potential

$$V_g(x_0) = E(x_0, M(x_0)) \tag{2.9}$$

which represents our variational approximation to the exact effective potential defined as ($< \psi | \psi >= 1$)

$$V(x_0) = min_{\{\psi\}} < \psi | \widehat{H} | \psi > \tag{2.10}$$

with

$$< \psi | \widehat{x} | \psi >= x_0 . \tag{2.11}$$

The variational content of the gaussian approximation allows to deduce the inequality

$$V(x_0) \leq V_g(x_0) \tag{2.12}$$

which is hardly to be discovered by employing different approximations to the effective potential (e.g. the loop expansion). We shall come back to this point later on. Absolute extrema occur at those values of x_0 at which

$$0 = \frac{dV_g}{dx_0} = x_0 \left[-m^2 + 4\lambda x_0^2 + 6\frac{\lambda}{M} \right] . \tag{2.13}$$

The absolute minimum of V_g represents, in the end, the variational estimate of the ground state energy. The details concerning the simultaneous solution of eqs. (2.8) and (2.13) can be found in ref. [4]. Here we shall limit ourselves to the problem of the perturbative

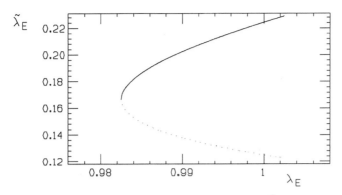

Fig. 1. The two positive roots of the cubic eq. (2.18) for $\tilde{\lambda}_E$. They both start at the value $\tilde{\lambda}_E = 1/6$.

expansion around the various extrema of the gaussian potential. We note that only when $dV_g/dx_0 = 0$ the linear terms in $\hat{a}$ and $\hat{a}^+$ vanish. Therefore only in this situation may a meaningful perturbation theory be defined. Differently, Wick ordered theories, however, may possess very different perturbative expansion parameters [5]. For instance let us consider the extremum (x_0, M_0) with $x_0 = 0$ and M_0 given by

$$M_0^2 = -m^2 + 6\,\frac{\lambda}{M_0} \tag{2.14}$$

which satisfies the bound $\lambda/M_0^3 > 1/6$; the ground state energy can, in principle, be obtained by starting from $V_g(0, M_0)$ (the relevant dimensionless coupling constant is λ/M_0^3)

$$E_0 = V_g(0, M_0) + M_0 \sum_{N=2}^{\infty} C_N \left(\frac{\lambda}{M_0^3}\right)^N \tag{2.15}$$

and the coefficients C_N can be found from the perturbative expansion around the gaussian state at $x_0 = 0$. At the same time, by combining eqs. (2.8) and (2.13), a "spontaneously broken" solution, i.e. corresponding to $x_0 \neq 0$, can be found. This class of extrema satisfy the condition

$$M^2 = 8\,\lambda\,x_0^2 \tag{2.16}$$

and the parameter M is related to M_0 by the relation

$$-2M_0^2 + 12\,\frac{\lambda}{M_0} = M^2 + 12\,\frac{\lambda}{M}\;. \tag{2.17}$$

By defining the two dimensionless coupling constants

$$\lambda_E = \frac{\lambda}{M_0^3} \tag{2.18a}$$

$$\tilde{\lambda}_E = \frac{\lambda}{M^3} \tag{2.18b}$$

we can put the relation between the perturbative expansion parameters in the form

$$\lambda_E = \frac{1}{6} + \left(\frac{\lambda_E}{\tilde{\lambda}_E}\right)^{2/3} \left(\frac{1}{12} + \tilde{\lambda}_E\right) . \tag{2.19}$$

Note that, in the weak coupling limit ($\lambda/m^3 \to 0$), $M_0 = O(\lambda)$ and therefore $\lambda/M_0^3 \to \infty$. In this situation the absolute minimum of the gaussian potential is obtained for x_0 very close to the absolute minimum of the classical potential and the coupling constant $\tilde{\lambda}_E \to 0$ since the parameter M is $O(1)$. In the strong coupling phase ($\lambda/m^3 \to \infty$) $M_0 = O(\lambda^{1/3})$ and the spontaneously broken solution disappears. The weak coupling situation is illustrated in fig. 1, where $\tilde{\lambda}_E$ turns out to be "asymptotically free" in the limit $\lambda_E \to \infty$.

3. $\lambda\phi^4$ THEORY

Very much alike the quantum mechanical example of sect. 2, a non perturbative analysis of $\lambda\phi^4$ theory can be performed by using a second quantized formalism [6] or functional techniques. A complete comparison between the two approaches is presented in ref. [4] to which we address the interested reader. Here we shall employ the functional method which allows an easier comparison with other computations as in the case of the loop expansion. At the same time, the presence of the ultraviolet divergences, makes the treatment of the field theory very different from the simple quantum mechanical example. For this reason we shall first ignore the problem of making our results cut-off independent by introducing a suitable regulator ($1/\epsilon$ in the dimensional regularization) and a bare value for the unrenormalized coupling constant which enter the stability analysis. The renormalization is then performed in sect. 4 by explicitly writing down the RG equation for the effective potential which defines, in a non perturbative way, the β-function and the anomalous dimension of the field. In order to make our treatment self-contained, we shall start by recalling the very useful formalism of the double Legendre transform[1,7] which provides the key to relate the two main approaches to the effective potential, i.e. the loop expansion and the variational method.

The starting point of our analysis is eq. (2.9a) of ref. [1], where the effective action for composite operators $\Gamma[\varphi, G]$, up to a constant, ($\hbar = 1$)

$$\Gamma[\varphi, G] = I[\varphi] + \frac{i}{2} Tr \, Ln \, DG^{-1} + \frac{i}{2} Tr \, \mathcal{D}^{-1} G + \Gamma_2[\varphi, G] \tag{3.1}$$

is explicitly given. For the $\lambda\phi^4$ case described by the Lagrangian ($m^2 > 0$, $\lambda > 0$)

$$\mathcal{L} = \frac{1}{2}(\partial_\mu \phi)^2 - \frac{1}{2} m^2 \phi^2 - \frac{\lambda}{4!}\phi^4 \tag{3.2}$$

we obtain (we restrict to the case of constant φ)

$$iD^{-1}(x, y) = -(\Box_x + m^2) \, \delta^4(x - y) \tag{3.3}$$

$$i\mathcal{D}^{-1}(x, y; \varphi) = iD^{-1}(x, y) - \frac{\lambda}{2}\varphi^2 \, \delta^4(x - y) \tag{3.4}$$

$$I[\varphi] = -\int d^4x \left(\frac{1}{2} m^2 \varphi^2 + \frac{\lambda}{4!} \, \varphi^4\right) \tag{3.5}$$

and $G(x, y)$ admits the general, space-time translationally invariant, form

$$G(x, y) = i \int \frac{d^3\vec{k}}{(2\pi)^3} \, e^{i\vec{k}\cdot(\vec{x}-\vec{y})} \int \frac{d\omega}{2\pi} \, \frac{e^{-i\omega(x_0 - y_0)}}{\omega^2 - \omega^2(k) + i\epsilon} \tag{3.6}$$

in terms of the unknown function $\omega(k)$. Finally $\Gamma_2[\varphi, G]$ is obtained by the sum of all two-particle irreducible (2PI) vacuum-vacuum graphs of the shifted theory with propagators set equal to G

$$\Gamma_2[\varphi, G] = \;\infty\; + \;\ominus\; + \;\ominus\!\!\!\!- \; + \;\bigcirc\!\!\!\!\!\bigcirc\; + \;\bigotimes\; + \;\cdots \tag{3.7}$$

where we have omitted the 2PI diagrams with more than three loops.

As shown in ref. [1] the usual effective action $\Gamma[\varphi]$ is obtained by minimizing $\Gamma[\varphi, G]$ with respect to G , i.e.

$$\Gamma[\varphi] = \Gamma[\varphi, G_0(\varphi)] \tag{3.8}$$

where $G_0(\varphi)$ is defined as the solution of

$$\frac{\delta \Gamma}{\delta G}\bigg|_{G=G_0(\varphi)} = 0 \;. \tag{3.9}$$

The stability of the underlying theory can be investigated through the relation

$$\Gamma[\varphi, G]_{static} = -\int dt \; E[\varphi, G] \tag{3.10}$$

$\Gamma[\varphi, G]_{static}$ being defined in ref. [1], and, $(<\psi|\psi> = 1)$

$$E[\varphi, G] = min_{\{\psi\}} <\psi| \,\widehat{H}\, |\psi> \tag{3.11}$$

with the conditions

$$<\psi| \,\widehat{\phi}(\vec{x})\, |\psi> = \varphi \tag{3.12}$$

$$<\psi| \,\widehat{\phi}(\vec{x})\, \widehat{\phi}(\vec{y})\, |\psi> = \varphi^2 + G(\vec{x}, \vec{y}) \tag{3.13}$$

where

$$G(\vec{x}, \vec{y}) = G(x, y)\bigg|_{x_0 = y_0} \;. \tag{3.14}$$

We stress that the above equivalence is only true for the exact quantity $\Gamma[\varphi, G]$ and that the problem of finding approximations to $E[\varphi, G]$, which enjoy stability properties, by truncating the infinite series in $\Gamma_2[\varphi, G]$, is highly non trivial. The only known case is the Hartree-Fock contribution, which is equivalent to retain the first graph in eq. (3.7), since it corresponds to a systematic variational procedure within the class[8] of the gaussian state functionals

$$\psi_g[\phi] = (DetG)^{-\frac{1}{4}} \; exp\left\{ -\frac{1}{4} \int d^3\vec{x} \int d^3\vec{y} \Big[\phi(\vec{x}) - \varphi\Big] G^{-1}(\vec{x}, \vec{y}) \Big[\phi(\vec{y}) - \varphi\Big] \right\} \;. \tag{3.15}$$

In this case we find

$$E_g[\varphi, G] = \int d^3\vec{x} \left\{ \frac{\lambda}{4!}\varphi^4 + \frac{1}{8}G^{-1}(\vec{x}, \vec{x}) + \frac{\lambda}{4}\varphi^2 G(\vec{x}, \vec{x}) \right.$$

$$\left. +\frac{1}{2}m^2\varphi^2 - \frac{1}{2}\Big(\Delta_{\vec{x}} - m^2\Big)G(\vec{x}, \vec{y})|_{\vec{x}=\vec{y}} + \frac{\lambda}{8}G(\vec{x}, \vec{x})G(\vec{x}, \vec{x}) \right\} \;. \tag{3.16}$$

For constant fields the exact effective potential is obtained by the relation

$$\int V(\varphi) \, d^3\vec{x} = E[\varphi, G_0(\varphi)] \tag{3.17}$$

and the ground state energy density $\mathcal{W}_0$ is given by

$$\mathcal{W}_0 = V(\varphi_0) \tag{3.18}$$

where φ_0 provides the absolute minimum of $V(\varphi)$. In our case we obtain analogous relations

$$\int V_g(\varphi)\, d^3\vec{x} = E_g[\varphi, \overline{G}(\varphi)]$$

(3.19)

and

$$\mathcal{W}_g = V_g(\overline{\varphi})$$

(3.20)

where $\overline{G}$ and $\overline{\varphi}$ correspond, in the gaussian approximation, to G_0 and φ_0 . The variational content of the gaussian approximation allows to deduce the fundamental inequality

$$\mathcal{W}_0 \leq \overline{\mathcal{W}}_g$$

(3.21)

which is hardly to be recovered by employing different truncations of $\Gamma_2[\varphi, G]$.

For instance, the inclusion of the second graph of eq. (3.7) in the evaluation of the ground state energy does not correspond to enlarge the gaussian subspace in a systematic way. As a consequence one can also obtain an estimate of the ground state which is below the true value. Indeed this happens in the quantum mechanical oscillator when applying the same technique (and the same truncation). Let us consider, as in ref. [9], the Hamiltonian

$$\widehat{H} = \frac{\widehat{p}^2}{2} + \frac{\lambda}{4}\widehat{x}^4 \; .$$

(3.22)

The exact ground state energy is

$$E_0 = 0.420804976...(\lambda)^{1/3}$$

(3.23)

while, by retaining the first two graphs of eq. (3.7), one obtains

$$\overline{E} = 0.419716918...(\lambda)^{1/3}$$

(3.24)

to be compared with the gaussian approximation result (the first graph in eq. (3.7))

$$E_g = 0.429267841...(\lambda)^{1/3} \; .$$

(3.25)

Better variational estimates of the ground state can be obtained by allowing for non gaussian states, as in ref. [10] where a trial wave function of the form

$$\psi_{ng} \sim exp\left(-x^3\, tghx\ \right)$$

(3.26)

was introduced. In this case the estimate of the ground state energy can be extremely good, for instance

$$E_{ng} = 0.420823654...(\lambda)^{1/3}$$

(3.27)

but, at the same time, both a perturbative improvement and an extension to quantum field theory are prohibitively difficult. This simple example confirms the unique properties of the gaussian states since they allow for consistent variational calculations and perturbative expansion.

By evaluating eq. (3.16), whose dependence on the unknown function $w(k)$ is well known to be extremized by the form

$$\omega^2(k) = \vec{k}^2 + \Omega^2$$

(3.28)

and by using dimensional regularization, we obtain ($\mathcal{V}$ = three-dimensional volume)

$$\frac{1}{V} E_g(\varphi_B, \Omega) = \frac{\lambda_B}{4!} \varphi_B{}^4 + \frac{m_B^2 \varphi_B^2}{2} + \frac{\Omega^4}{64\pi^2} \left(x - \frac{1}{2} \right)$$

$$+ \frac{1}{2} \left(m_B^2 - \Omega^2 \right) \frac{\Omega^2 x}{16\pi^2} + \frac{\lambda_B \Omega^2 \varphi_B{}^2}{64\pi^2} x + \frac{\lambda_B \Omega^4}{2048\pi^4} x^2 \qquad (3.29)$$

where ($\epsilon = n - 4$)

$$x = \frac{2}{\epsilon} + ln\Omega^2 + \gamma + ln\pi - 1 \qquad (3.30)$$

and we have, for sake of clarity added a subscript "B" to the mass, the field and the coupling constant to indicate that we are dealing with bare quantities before renormalization. From eq. (3.29) we get the coupled extremum equations

$$\frac{1}{V} \frac{\partial E_g}{\partial \varphi_B} = \varphi_B \left[\frac{\lambda_B \varphi_B{}^2}{6} + m_B^2 + \frac{\lambda_B \Omega^2}{32\pi^2} x \right] = 0 \qquad (3.31)$$

$$\frac{1}{V} \frac{\partial E_g}{\partial \Omega} = \frac{\Omega}{16\pi^2} (x + 1) \left\{ m_B^2 + \frac{\lambda_B}{2} \varphi_B{}^2 - \Omega^2 + \frac{\lambda_B \Omega^2}{32\pi^2} x \right\} = 0 \qquad (3.32)$$

whose solutions are

$$\varphi_B = 0 \qquad (3.33a)$$

$$\Omega_0^2 = m_B^2 + \frac{\lambda_B \Omega_0^2}{32\pi^2} x (\Omega_0^2) \qquad (3.33b)$$

and $\pm \overline{\varphi}_B, \overline{\Omega}$ given by

$$\overline{\Omega}^2 = \frac{\lambda_B}{3} \overline{\varphi}_B^2 \qquad (3.34)$$

$$\lambda_E - \tilde{\lambda}_E = \frac{\lambda_B}{16\pi^2} ln \frac{2 + \lambda_E}{\tilde{\lambda}_E - 1} \qquad (3.35)$$

and

$$\lambda_E = -\frac{\lambda_B}{16\pi^2} \left(\frac{2}{\epsilon} + ln\Omega_0^2 + \gamma + ln\pi - 1 \right) \qquad (3.36)$$

$$\tilde{\lambda}_E = -\frac{\lambda_B}{16\pi^2} \left(\frac{2}{\epsilon} + ln\overline{\Omega}^2 + \gamma + ln\pi - 1 \right) . \qquad (3.37)$$

Note that the continuum limit $n \to 4$ corresponds to $\epsilon \to 0^-$ as it can be deduced by comparing dimensional regularization with an ultraviolet cutoff ($2/\epsilon + \gamma + ln\pi \to -ln\Lambda^2$). This avoids the triviality bound of ref. [11] which holds for $n \geq 4$. This last remark ensures that for positive λ_B and very small and negative ϵ (i.e. very large Λ), λ_E and $\tilde{\lambda}_E$ are both positive quantities, corresponding respectively to the quantities λ/M_0^3 and λ/M^3 of the quantum mechanical example of sect. 2. In the limit $\Omega_0 \to 0$, one obtains $\lambda_E \to +\infty$ and $\tilde{\lambda}_E \to 1$, i.e. eq. (3.35) reduces to[3]

$$\lambda_B \left(\frac{2}{\epsilon} + ln\overline{\Omega}^2 + \gamma + ln\pi - 1 \right) = -16\pi^2 . \qquad (3.38)$$

We note that, differently from quantum mechanics, where $\tilde{\lambda}_E \to 0$ for $\lambda_E \to +\infty$, its quantum field theoretical analogous is 1 for $\Omega_0 \to 0$. This should not surprise since in this case $\tilde{\lambda}_E$ represents the expansion parameter in the broken phase at zero momentum which, even for very small $\lambda_B/16\pi^2$, contains a large logarithmically divergent term.

After all the situation is not different from QED where the analogous quantity $(\alpha_B/\pi) \, ln(\Lambda^2/m_e^2)$ is $O(1)$. The trend of the renormalized coupling constant at different scales will be discussed in sect. 4.

The solution $\overline{\varphi}_B$, $\overline{\Omega}$ of the spontaneously broken phase of the massless theory produces the gaussian estimate for the energy density

$$\overline{W}_g(\epsilon, \lambda_B) = - \frac{\overline{\Omega}^4}{128\pi^2} \, . \tag{3.39}$$

Since the energy density corresponding to $\varphi_B = \Omega_0 = 0$ is zero, it follows that, in the massless limit, the perturbative vacuum is unstable and spontaneous symmetry breaking is recovered in the gaussian approximation. The statement, being of variational nature, implies that, within the subspace of normalized gaussian states, the perturbative vacuum with $\varphi_B = 0$ is not the lowest, no definite conclusion being possible on the other hand, about the existence of symmetric, non gaussian states of still lower energy.

The gaussian approximation, despite of its simplicity, provides a good description of the ground state at the absolute minima of the variational procedure, lacking, however, the fundamental convexity property of the exact effective potential[12]. This problem and the explicit construction of a convex (flat in the infinite volume limit) potential have beeen solved in ref. [13], where, by the way, it has been shown that the probability of the various, degenerate quantum states enclosed by the gaussian absolute minima is exactly zero.

4. $\lambda\phi^4$ AND THE RENORMALIZATION GROUP

So far we have expressed the effective potential and the ground state using the bare coupling constant λ_B and the bare field φ_B. The renormalization procedure introduces a dependence in $V(\varphi)$, as defined in eq. (3.17), on the arbitrary subtraction point μ at which we define the renormalized coupling constant $\lambda(\mu)$ and field $\varphi(\mu)$. RG invariance implies that $V(\mu, \lambda(\mu), \varphi(\mu))$ satisfies the equation

$$\left\{ \mu \, \frac{\partial}{\partial \mu} + \beta(\lambda) \, \frac{\partial}{\partial \lambda} - \gamma(\lambda) \, \varphi \, \frac{\partial}{\partial \varphi} \right\} V = 0 \tag{4.1}$$

where we have introduced the Callan-Symanzik β-function

$$\beta(\lambda) = \mu \, \frac{\partial \lambda}{\partial \mu} \tag{4.2}$$

and the anomalous dimension of the scalar field

$$\mu \, \frac{\partial \varphi}{\partial \mu} = -\gamma(\lambda) \, \varphi \, . \tag{4.3}$$

Eqs. (4.1)÷(4.3) define the renormalization procedure beyond perturbation theory. The minimization equation

$$\frac{\partial \, V}{\partial \, \varphi} = 0 \tag{4.4}$$

fixes the boundary conditions in the (μ, λ, φ) space of the integral curve (4.2) and (4.3) at which the energy is minimized. In other words eq. (4.4), selecting special values for the RG invariant quantity

$$P = \varphi(\mu) \, exp \int^{\lambda(\mu)} dx \, \frac{\gamma(x)}{\beta(x)} \tag{4.5}$$

and thus determining its relative magnitude with respect to the basic invariant

$$Q = \mu \, exp - \int^{\lambda(\mu)} \frac{dx}{\beta(x)} \tag{4.6}$$

can be written as

$$P = \alpha \, Q \tag{4.7}$$

α being a numerical constant. The implementation of the absolute minimum condition (4.4) at some μ_0, $\varphi = f(\mu_0, \lambda_0)$, into the effective potential, provides the the ground state energy density $\mathcal{W}_0 = V(\mu_0, \lambda(\mu_0), f(\mu_0, \lambda_0))$ which, due to eq. (4.4), is a RG invariant quantity with zero anomalous dimension satisfying

$$\left\{ \mu \frac{\partial}{\partial \mu} + \beta \frac{\partial}{\partial \lambda} \right\} \mathcal{W}_0(\mu, \lambda) = 0 \; . \tag{4.8}$$

In our case we can reabsorb the polar term present in the minimum of the gaussian effective potential in a suitable definition of a $\overline{MS}$ running coupling constant $\lambda(\mu)$ (for $\mu^2 \gg \overline{\Omega}^2$)

$$\lambda(\mu) = \frac{\lambda_B}{1 + \dfrac{\lambda_B}{16\pi^2} \left(\dfrac{2}{\epsilon} + \ln \mu^2 + \gamma + \ln \pi \right)} \tag{4.9}$$

Eq. (3.35) can be read as

$$\ln \frac{\mu^2}{\overline{\Omega}^2} = \frac{16\pi^2}{\lambda(\mu)} - 1 \tag{4.10}$$

and the renormalized gaussian energy density $\mathcal{W}_g(\mu, \lambda(\mu))$ is the same quantity as in eq. (3.39) after replacing λ_B and ϵ in terms of $\lambda(\mu)$ and μ. From eqs. (4.9) and (4.10) it follows immediately that, by defining the function $\beta_g(\lambda)$

$$\beta_g(\lambda) = -\frac{\lambda^2}{8\pi^2} + O(\lambda^3) \tag{4.11}$$

$\mathcal{W}_g$ satisfies the equation

$$\left\{ \mu \frac{\partial}{\partial \mu} + \beta_g \frac{\partial}{\partial \lambda} \right\} \mathcal{W}_g(\mu, \lambda(\mu)) = 0 \; . \tag{4.12}$$

This equation means that $\mathcal{W}_g$, which has a physical meaning since it represents an upper bound to the true ground state energy density, is effectively independent from the particular choice of the scale μ. Note that, in general, eq. (4.9), which reabsorbs the polar term in the gaussian potential, does not define the exact running coupling constant. However, when considering the exact energy density $\mathcal{W}_0(\epsilon, \lambda_B)$, a suitable definition of the running coupling constant

$$\lambda^T(\mu) = \frac{\lambda_B}{1 + \dfrac{\lambda_B}{16\pi^2 \, c} \left(\dfrac{2}{\epsilon} + \ln \mu^2 + \gamma + \ln \pi \right)} \tag{4.13}$$

(the superscript T= "True") may be introduced, in terms of an unknown constant c, to renormalize $\mathcal{W}_0$, at least for small value of λ_B. It should be clear that eq. (4.13) is the only alternative left out by our variational upper bound (3.39) since, otherwise, the resulting unboundedness from below (for $n \rightarrow 4$) of the exact theory, for positive λ_B, would be very difficult to understand. As a consequence, the inequality (3.21) can be read as

$$\mathcal{W}_0(\mu, \lambda^T(\mu) \,) \leq \mathcal{W}_g(\mu, \lambda(\mu)) \tag{4.14}$$

and the RG equation

$$\left\{\mu\frac{\partial}{\partial\mu} + \beta\frac{\partial}{\partial\lambda}\right\} \mathcal{W}_0(\mu, \lambda^T(\mu)) = 0 \tag{4.15}$$

defines the "true" β-function of the theory $\beta(\lambda)$

$$\beta(\lambda) = -\frac{\lambda^2}{8\pi^2 c} + O(\lambda^3) . \tag{4.16}$$

Note that in eqs. (4.12) and (4.15) μ and λ are independent variables, as in a partial differential equation, but we are allowed to compare $\mathcal{W}_g$ and $\mathcal{W}_0$ only along integral curves $\lambda = \lambda(\mu)$ and $\lambda^T = \lambda^T(\mu)$ with the same boundary conditions, represented by a point in the μ-λ plane (μ_0, λ_0), thus corresponding to the same λ_B. If c is positive as in the gaussian approximation ($c = 1$), we can consider the same region in the μ-λ plane (large μ and small λ) as the integral curves of both the exact theory and its gaussian approximation are driven toward the ultraviolet fixed point at $\lambda = 0$. On the other hand, if one would follow the perturbative indications ($c = -2/3$) the exact theory and its gaussian approximation decouple in the continuum limit. In this case, surprising as it may be, the gaussian approximation would be, nevertheless, very appealing for consistent quantum field theoretical models of weak and electromagnetic interactions as it exhibits both spontaneous symmetry breaking and asymptotic freedom. Henceforth we shall adopt the more realistic point of view that the gaussian approximation, at least for weak coupling, being the starting point for any perturbative expansion, well reproduces the properties of the exact theory thus providing the correct sign of c.

By assuming the uniform continuity of both β and β_g, the existence and uniqueness theorem[14] states that only one integral curve (of β or β_g) can pass through any non singular point in our plane. By considering the two distinct characteristics, one of β and one of β_g, crossing a given (μ_0, λ_0) point, thus corresponding to the same λ_B, we can find a relation between the scale parameter M associated with the true ground state energy $\mathcal{W}_0$ and our variational quantity $\overline{\Omega}$. From the general solution of eq. (4.15)

$$\mathcal{W}_0 = -\mu_0^4 \; exp\left\{-4\int^{\lambda_0} \frac{dx}{\beta(x)}\right\} \tag{4.17}$$

and from eq. (4.16) one gets

$$\mathcal{W}_0 = -\frac{M^4}{128\pi^2} \tag{4.18}$$

where $\quad M = \mu_0 \; exp\left\{-(8\pi^2 c/\lambda_0)\left[1 + O(\lambda_0)\right]\right\} \quad$ is to be compared with the gaussian approximation result, eq. (4.10),

$$\overline{\Omega} = \mu_0 \; exp\left\{-\frac{8\pi^2}{\lambda_0}\left[1 + O(\lambda_0)\right]\right\} . \tag{4.19}$$

In the weak coupling limit, and by using the fundamental inequality (4.14), we deduce $0 < c \leq 1$, (or $\beta(\lambda) \leq \beta_g(\lambda)$). It should be clear that asymptotic freedom is, in general, a non perturbative statement concerning eq. (4.15). Spontaneous symmetry breaking, signaling the "essential instability" of massless self-interacting scalar theories, prevents the possibility of any consistent perturbation theory in the unbroken phase.

We note, incidentally, that from our results it follows automatically the "triviality" of the $\lambda\phi^4$ theory in the symmetric phase. Indeed by using the definition of the leading log perturbative coupling constant $\lambda^P(\mu)$ one gets the chain inequalities

$$\lambda(\mu) > \lambda_B > \lambda^P(\mu) . \tag{4.20}$$

Since (when $\epsilon \to 0^-$) $\lambda_B \to 0^+$, $\lambda^P(\mu)$ identically vanishes at any scale.

The same results about the asymptotic freedom of the theory can be obtained by using a cutoff Λ to regularize the divergences and then studying the continuum limit $\Lambda \to +\infty$ as in ref. [15].

We have not yet considered the renormalization of the field φ_B since it was not necessary to take it into account to eliminate the polar term in the gaussian ground state energy. However, since $\overline{\Omega}$, due to eq. (4.12), is RG invariant if we write the minimization condition (3.34) in terms of the scale dependent quantities $\lambda(\mu)$ and $\overline{\varphi}(\mu)$

$$\overline{\Omega}^2 = \frac{1}{3}\lambda(\mu)\,\overline{\varphi}^2(\mu) \tag{4.21}$$

$\overline{\varphi}(\mu)$ is automatically defined by eqs. (3.34), (4.9) and (4.21) and it is logarithmically increasing with μ. Eq. (4.21), in the gaussian approximation, is equivalent to eq. (4.7) with $Q^2 = \overline{\Omega}^2$ and $P^2 = \lambda\overline{\varphi}^2$.

If we impose the infinite rescaling for the field φ, as was suggested for the first time in ref. [16], we can renormalize the gaussian effective potential which has the following expression around one of its absolute minima ($\pm\overline{\varphi}$)

$$V_g(\varphi) = \frac{\lambda^2 \varphi^4}{576\pi^2} \left(ln\frac{\varphi^2}{\overline{\varphi}^2} - \frac{1}{2} \right) . \tag{4.22}$$

Then the wave function renormalization Z can be derived through the general expression

$$\left. \frac{d^2 V(\varphi)}{d\varphi^2} \right|_{\varphi=\overline{\varphi}} = \frac{m_R^2}{Z} . \tag{4.23}$$

In our case, m_R^2 is $\overline{\Omega}^2$; therefore, in the gaussian approximation,

$$Z_g = \frac{24\pi^2}{\lambda(\mu)} \tag{4.24}$$

or, by introducing explicitly the mass $\overline{\Omega}^2$, we obtain, up to a finite term in the limit $\mu \to +\infty$

$$Z_g\left(\frac{\mu^2}{\overline{\Omega}^2}\right) = \frac{3}{2} \, ln\left(\frac{\mu^2}{\overline{\Omega}^2}\right) . \tag{4.25}$$

Eq. (4.24) may be checked by means of an explicit calculation in the shifted theory[17].

Finally the renormalized coupling constant at zero momentum is defined as

$$\left. \frac{d^4 V_g(\varphi)}{d\varphi^2} \right|_{\varphi=\overline{\varphi}} = \frac{\lambda_g(0)}{Z_g^2} \tag{4.26}$$

and one obtains

$$\frac{\lambda_g(0)}{16\pi^2} = \frac{11}{2} . \tag{4.27}$$

The quantity in eq. (4.27) represents the effective coupling constant of the broken phase at low energy.

Before concluding our analysis, we shall compare our results with eq. (3.4) of ref. [2], which yields the one-loop effective potential before renormalization. By transforming their result to dimensional regularization we find

$$V_{1-loop}(\varphi) = \frac{\lambda_B \varphi_B^4}{4!} \left\{ 1 + \frac{3\lambda_B}{32\pi^2} \left[ln\left(\lambda_B \varphi_B^2\right) + \frac{2}{\epsilon} + \gamma + ln\pi + k \right] \right\} \tag{4.28}$$

where k is a numerical constant. If we would assume, following the indications of the gaussian approximation, that $\lambda_B \varphi_B^2 = \lambda(\mu)\varphi^2(\mu)$ is a RG invariant quantity, eq. (4.28) would contain all the leading logarithmic terms of the exact effective potential

$$V(\varphi) = \frac{(\lambda\varphi^2)^2}{4!} \left[\frac{1}{\lambda^T(\lambda\varphi^2)} + const \right] . \tag{4.29}$$

At the leading log level we obtain

$$\lambda^T(\mu^2) \simeq \frac{\lambda_B}{1 + \frac{3\lambda_B}{32\pi^2} \left(\frac{2}{\epsilon} + ln\,\mu^2 + \gamma + ln\,\pi \right)} \tag{4.30}$$

and, by comparing the above result with eq. (4.13), we find $c = 2/3$ in agreement with our variational bound $0 < c \leq 1$. The crucial difference with respect to ref. [2] is the RG invariance of $\lambda\varphi^2$ which is not recovered in perturbation theory, where, to the leading log accuracy, the expression

$$V_{1-loop}^{pert}(\varphi) = \frac{\lambda_B \varphi_B^4}{4!} \frac{1}{1 - \frac{3\lambda_B}{32\pi^2} \left(ln\,(\lambda_B \varphi_B^2) + \frac{2}{\epsilon} + \gamma + ln\,\pi \right)} \tag{4.31}$$

is rather obtained and the one-loop minimum disappears.

5. EXTENSION TO THE $O(N)$ THEORY AND CONCLUSIONS

The extension to the $O(N)$ symmetric massless theory is straightforward by following ref. [18]. In this case, described by the Lagrangian density ($a = 1, 2 ... N$)

$$\mathcal{L} = \frac{1}{2}(\partial_\mu \phi^a)(\partial^\mu \phi^a) - \frac{\lambda_B}{4!\,N}(\phi^a \cdot \phi^a)^2 \tag{5.1}$$

one gets the extremum conditions ($\lambda' = \lambda/N$)

$$\bar{\rho}^2(\mu) = \frac{3}{\lambda'(\mu)}\,\bar{m}_1^2 \tag{5.2}$$

$$\bar{m}_1^2 = \mu^2\,exp\left[-\frac{16\pi^2}{\lambda'(\mu)}\,\frac{3}{\sqrt{N+3}+1} + O(1) \right] \tag{5.3}$$

$$\bar{m}^2 = \frac{\bar{m}_1^2}{\sqrt{N+3}+2} \left[1 + O\left(\frac{\lambda'(\mu)}{48\pi^2}\,ln\,N \right) \right] \tag{5.4}$$

which express the relation between the mass of the Higgs particle $\bar{m}_1$, $\bar{\rho}^2 = \sum_a < \phi^a >^2$ and the mass of the Goldstone modes $\bar{m}$. Fixing $\bar{m}_1$ to set up the scale of the theory and then taking the limit $N \to \infty$, we find that $\bar{m}$ vanishes, confirming the validity of the gaussian approximation as discused in ref. [1]. Again the evaluation of the difference between the energy densities corresponding respectively to the "spontaneously broken", $(\bar{\rho}, \bar{m}_1, \bar{m})$, and the perturbative, $(0, 0, 0)$, vacuum

$$\Delta E = -\frac{1}{128\pi^2} \left[\bar{m}_1^4 + (N-1)\,\bar{m}^4 \right] \tag{5.5}$$

shows the instability of the perturbative vacuum and the presence of the spontaneous symmetry breaking. At the same time we can obtain the function β_N from eq. (5.3)

$$\beta_N(\lambda') = -\frac{\lambda'^2}{8\pi^2} \left(\frac{\sqrt{N+3}+1}{3} \right) \tag{5.6}$$

and, as in the discrete case, one can recover the presence of an ultraviolet fixed point for the exact theory as well.

A particular attention should be paid to the conditions

$$\frac{\lambda'(\mu)}{16\pi^2} \ll \frac{1}{\sqrt{N}} \ll 1 \tag{5.7}$$

ensuring that $\overline{m}_1$ is exponentially decoupled from μ in the limit $\mu \to +\infty$ and

$$\frac{\overline{m}^2}{\overline{m}_1^2} = \frac{1}{\sqrt{N}} \left[1 + O\left(\frac{lnN}{\sqrt{N}} \right) \right] \tag{5.8}$$

vanishes identically when $N \to \infty$.

Due to the non uniformity of the two limits $N \to \infty$, $\mu \to +\infty$, the conclusions of ref. [19] that an $O(N)$ invariant, self-interacting, scalar theory is trivial in the large N limit should be limited to the symmetric phase.

Finally, our analysis suggests a different approach to the bounds on the physical Higgs mass, previously based on the "triviality" of the scalar sector of the Standard Model. From our results, whatever the value of the physical mass in the broken phase, the theory, at short distances, is in a weak coupling regime. As a consequence the magnitude of the interaction constant defined by the relation (G_F=Fermi constant, M_H=Higgs mass)

$$\lambda_H = \frac{G_F}{\sqrt{2}} M_H{}^2 \tag{5.9}$$

cannot constrain the ultraviolet behavior of the theory as it represents the boundary condition, in the infrared, for a running coupling constant which, in a pure $\lambda\phi^4$ theory, would be asymptotically free. Only by exploring the stability conditions of the spontaneously broken phase may meaningful limits on the Higgs mass be obtained. However a rough estimate of its value can be obtained from the minimization condition (4.21) and the low energy effective coupling constant (4.27)

$$M_H = \left(\frac{\lambda_g(0)}{3} \overline{\varphi}^2(0) \right)^{1/2} \simeq O(1) \ TeV \tag{5.10}$$

provided that in the Standard Model the vacuum expectation value is $\overline{\varphi} \simeq 250/\sqrt{2} \ GeV$.

ACKNOWLEDGMENTS

We would like to thank V. Branchina, P. Castorina and J. M. Cornwall for many useful discussions and suggestions.

REFERENCES

1. J.M.Cornwall, R.Jackiw and E.Tomboulis, Phys. Rev., D10: 2428 (1974).
2. S.Coleman and E.Weinberg, Phys. Rev., D7: 1888 (1973).
3. V.Branchina, P.Castorina, M.Consoli and D. Zappalà, Non triviality of spontaneously broken $\lambda\phi^4$ theories, to be published on Phys. Rev. D, Brief Reports.

4. M.Consoli and A.Ciancitto, Nucl. Phys., B254: 653 (1985).

5. W.E.Caswell, Ann. Pys. (N.Y.), 123: 153 (1979).

6. P.Castorina and M.Consoli, Phys. Lett., B131: 351 (1983); P.Castorina,A.Ciancitto and M.Consoli, N. Cim., A82: 275 (1984).

7. G.Jona-Lasinio, N. Cim., 34: 1790 (1964).

8. J.Kuti, unpublished,quoted in ref. [1].

9. J.M.Cornwall, Phys. Rev., D38: 656 (1988).

10. P.Castorina, M.Consoli and D.Zappalà, N. Cim., B100: 751 (1987).

11. J.Frohlich, Nucl. Phys., B200(FS4): 281 (1982).

12. K.Symanzik, Commun. Math. Phys., 16: 48 (1970); J.Iliopulos, C.Itzykson and A.Martin, Rev. Mod. Phys., 47: 165 (1975); T.L.Curtright and C.B.Thorn, J. Math. Phys., 25: 541 (1984).

13. V. Branchina, P.Castorina and D.Zappalà, Phys. Rev., D41: 1948 (1990).

14. F.Tricomi, "Equazioni differenziali", G. Einaudi ed.,Torino (1948).

15. P.Castorina and M.Consoli, Phys. Lett., B235: 302 (1990).

16. P.M.Stevenson and R.Tarrach, Phys. Lett., B176: 436 (1986).

17. M.Consoli and G.Passarino, Phys. Lett., B165: 113 (1985).

18. Y.Brihaye and M.Consoli, N. Cim., A94: 1 (1986).

19. W.A.Bardeen and M.Moshe, Phys. Rev., D28: 1372 (1983).

BOSONIZATION AND ITS APPLICATION TO NONPERTURBATIVE

PHENOMENA IN FIELD THEORIES

Hisakazu Minakata

Department of Physics
Tokyo Metropolitan University
Setagaya, Tokyo 158
Japan

INTRODUCTION

I wish to devote my two lectures to two related but different subjects. They may be given the separate titles as follows:

I. Bosonization and its application to fermion fractionization.
II. Bosonized QED and GSI phenomena.

In the part I, I will introduce the concept of bosonization and explain how it can be useful. For this purpose the phenomenon of fermion fractionization[1] is the ideal place to do the job. Of course, it is an interesting physical phenomenon and deserves a detailed description in its own right. It is the phenomenon that a soliton receives fractional quantum numbers from fermions which couple with it. It is believed that the electric conductivity in a class of one-dimensional polymers such as polyacetylene is due to this phenomenon.[2]

I also describe the bosonized treatment of the magnetic monopole-fermion system. By examining the three-dimensional object we will learn how to bosonize the system in 3+1 dimensions. It should provide an appropriate introduction to the part II. Moreover, it is of physical interest in the context of supersymmetric grand unified theories. It is noted[3] that fermion fractionization inevitably occurs in a wide class of such theories.

Vacuum Structure in Intense Fields, Edited by
H.M. Fried and B. Muller, Plenum Press, New York, 1991

In the part II, I will apply the method of bosonization to QED around a large-Z nucleus. This problem should be of physical importance in relationship with the mysterious phenomena seen in the GSI heavy ion experiments. We will see that some interesting structures emerge from our calculation, which might explain the characteristic features of the data.

FERMION FRACTIONIZATION

Let us consider the scalar field theory described by the Lagrangian

$$L = \frac{1}{2}(\partial_\mu \phi)^2 - V(\phi) \tag{1}$$

where the potential $V(\phi)$ is assumed to have a double well structure. The simplest possibility is $V(\phi) = a(\phi^2 - v^2)^2$. The following discussions, however, will not depend on any detailed structures of the potential but only rest on the existence of the two degenerate local minima.

It is well known that the system (1) admits a soliton solution, so called the kink solution. The corresponding field configuration interpolates the two degenerate minima as shown in Fig. 1.

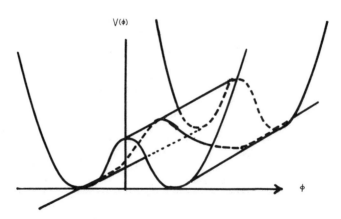

Fig. 1. A schematic illustration of the soliton configuration with double-well potential.

This feature of the solution leads to the localized energy density and most notably, to the topological stability of the solitons.

Now let us couple a fermion field with the system (1) and see what happens. For definiteness I consider the system

$$L = \frac{1}{2}(\partial_\mu \phi)^2 - V(\phi) + i\bar{\psi}\gamma^\mu \partial_\mu \psi - g\phi\bar{\psi}\psi. \tag{2}$$

One of the most distinctive feature of this system is the existence of the fermion zero mode, called the Jackiw-Rebbi zero mode in the presence of the background soliton field.[1] Its existence can be easily checked by noticing that the Dirac equation

$$-i\gamma^1 \frac{\partial}{\partial x}\psi(x) + g\phi_{cl}(x)\psi(x) = E\gamma^0\psi(x) \tag{3}$$

possesses a zero energy solution (Here we are taking a γ matrix convension $\gamma^0 = \sigma_1$, $\gamma^1 = i\sigma_3$)

$$\psi_0(x) = \text{const. exp} \left[-g \int_0^x dy\, \phi_{cl}(y) \right] \begin{bmatrix} 1 \\ 0 \end{bmatrix} \tag{4}$$

Notice that this solution is normalizable only because the background soliton configurations at $x \to \pm\infty$ differ in sign; $\phi_{cl}(\pm\infty) = \pm v$ for the simplest choice of the potential mentioned earlier.

What is the physical interpretation of the zero mode? The answer is that it implies the degeneracy of the soliton state. To see this we expand the Dirac field in terms of the complete orthonormal eigenstates as

$$\psi(x) = b_0\psi_0(x) + \sum_{n \neq 0}b_0\psi_n(x) + \sum_n d_n^+\psi_n^*(x) \tag{5}$$

where the zero energy solution we just obtained above is included (must be!) together with other massive modes. We note that fermions are massive because $|\phi| \neq 0$ in the vacuum.

We further note that the zero mode operator b_0 has to satisfy the anticommutation relation

$$\{b_0, b_0^+\}=1 \tag{6}$$

due to the canonical anticommutation relation of the $\psi(x)$ field. Then a physical soliton state has to respect the anticommutation relation (6). Namely, it must form an irreducible representation of this algebra. It is well known that it can be given by the "spin" up and down states. Thus, we have degenerate solitonic ground states $|+\rangle$ and $|-\rangle$ each connected by the zero mode operators

$$b_0^+|-\rangle = |+\rangle, \quad b_0|+\rangle = |-\rangle. \tag{7}$$

Now we observe that since $b_0(b_0^+)$ has fermion number $-1(+1)$, $|+\rangle$ and $|-\rangle$ differ by fermion number 1. If we adopt the usual definition of the fermion number[4]

$$F = \frac{1}{2}(b_0^+ b_0 - b_0 b_0^+) \tag{8}$$

which respects the charge conjugation symmetry, then the degenerate soliton states obey

$$F|\pm\rangle = \pm\frac{1}{2}|\pm\rangle. \tag{9}$$

It means that the soliton states are doubly degenerate and carry fermion number 1/2!

The similar treatment can go through in the case of fermion-coupled 't Hooft-Polyakov magnetic monopole. In this case we have two-state degeneracy for isospinor fermion and four-state degeneracy for isovector fermion. In the latter case not only the fermion number but also the angular momentum "fractionize".

BOSONIZATION

A well known thesis in field theories in 1+1 dimensions is that a fermion theory can be mapped into corresponding boson theory, and vice versa. This fact (or this procedure) is called the bosonization.[5][6] It has a long histry and may be traced back to Tomonaga' s treatment[7] of sound waves in many-fermion system. Here I present it as an operator correspondence in the massless free fermion/boson basis:

Fermi		Bose	

$$i\,\bar{\psi}\,\gamma^{\mu}\,\partial_{\mu}\,\psi \qquad\leftrightarrow\qquad \frac{1}{2}(\partial_{\mu}\phi)^2$$

$$\bar{\psi}\,\gamma^{\mu}\,\psi \qquad\leftrightarrow\qquad -\frac{1}{\sqrt{\pi}}\varepsilon^{\mu\nu}\,\partial_{\nu}\,\phi \qquad\qquad (10)$$

$$\bar{\psi}\,\psi \qquad\leftrightarrow\qquad \mu N_{\mu}\cos(2\sqrt{\pi}\,\phi)$$

In the last line of (10) the paramenter μ and the simbol N_{μ} appear. It imply to take normal ordering with an arbitrary tiny mass μ which is introduced to cure the infrared catastrophe inherent in two-dimensional massless boson theories. While it is an important but subtle problem I will not dwell on it further. For the topics of the part I it is nearly an irrelevant point. Its implication for the part II subject will be briefly mentioned there.

The correspondence between the boson and the fermion operators implies the equivalence of the whole set of Green's functions calculated via two different variables. Let us examine an instructive example to grasp how it works. I denote the right- and the left-handed components of a two-dimensional fermion field as ψ_R and ψ_L , respectively. I calculate the two point correlation function of the operator

$$\bar{\psi}_L\psi_R = \mu e^{i2\sqrt{\pi}\phi} \qquad\qquad (11)$$

as follows:

$$\begin{aligned}
\langle\,\bar{\psi}_L\psi_R,\ \bar{\psi}_R\psi_L\,\rangle &= \mu^2\langle\,e^{i2\sqrt{\pi}\phi(x)},\ e^{-i2\sqrt{\pi}\phi(y)}\,\rangle\\
&= \mu^2\ \exp[\ 4\pi\langle\phi(x),\phi(y)\rangle\ ]\\
&= \mu^2\ \exp\{\ 4\pi(-1/4\pi)\ \ln[\mu^2(x-y)^2]\}\\
&= \frac{1}{(x-y)^2}.
\end{aligned} \qquad\qquad (12)$$

which is nothing but the result of fermion one-loop computation. In (12) I used the fact that the two-point function of the boson field is a logarithm in two dimensions. I note that the arbitrary cut off μ cancells out at the last line in (12), a desired result. This corresponds to the fact that massless free fermion theory is well defined with no need of infrared cut off.

In applying the bosonization to a wider class of theories it is useful to have a direct operator correspondence relation between fermion and boson fields due to Mandelstam[6]

$$\psi(x,t) = \sqrt{\frac{\mu}{2\pi}} \begin{bmatrix} -iN_\mu \exp\{ i \sqrt{\pi} [\phi(x,t) + \tilde{\phi}(x,t)] \} \\ N_\mu \exp\{ i \sqrt{\pi} [-\phi(x,t) + \tilde{\phi}(x,t)] \} \end{bmatrix} \tag{13}$$

where

$$\tilde{\phi}(x,t) = \lim_{\varepsilon \to \infty} \int_x^\infty dy \ e^{-\varepsilon y} \phi(y,t) \tag{14}$$

Hereafter, we take the two-dimensional γ matrix convension as $\gamma^0 = \tau_2$, $\gamma^1 = i\tau_1$, and $\gamma_5 = \tau_3$. It is possible to check that the Mandelstam formula (13), after a careful point splitting procedure, reproduces the correspondence relations in (10).

Some of the students may be puzzed by the fact that the theory of fermions which possesses the fermionic charges (fermion number, electric charge, etc.) can be mapped into the theory of neutral bosons. How the neutral boson supports the fermionic charges? To find the answer we rewrite the expression of the fermionic charge, fermion number for simplicity, in terms of the Bose variable,

$$F = \int dx \ \psi^+\psi = \frac{1}{\sqrt{\pi}} \int dx \ \frac{\partial}{\partial x} \phi(x),$$

$$= \frac{1}{\sqrt{\pi}}[\phi(\infty) - \phi(-\infty)]. \tag{15}$$

Now we find the answer. The Noether charge (charge guaranteed to exist by Noether's theorem due to an invariance of the system) of the fermionic theory is the topological charge in the boson theory. It is amusing to observe that the very reason for its stability also supports the fermionic charge. Fermions are solitons in 1+1 dimensions.

As an application of the bosonization technique I treat the free massive fermions, the simplest possible example. If we bosonize the fermionic Lagrangian

$$L = i\bar{\psi}\gamma^\mu \partial_\mu \psi - m\bar{\psi}\psi \tag{16}$$

we obtain, using the operator correspondence (10),

$$L = \frac{1}{2}(\partial_\mu \phi)^2 - m\mu\cos(2\sqrt{\pi}\phi), \tag{17}$$

as an equivalent boson theory. This is nothing but the sine-Gordon theory. It is well known that this theory has solitons (kinks) in its classical and the quantum spectrum. Let us compute the fermionic charge to reveal the nature of the solitons. Recalling that a minimum energy soliton configuration interpolates two nearest neighbor degenerate vacua of (17), we obtain

$$F = \frac{1}{\sqrt{\pi}}[\phi(\infty) - \phi(-\infty)\,] = 1 \tag{18}$$

The free massive fermions in 1+1 dimensions are the sine-Gordon solitons.

Before closing this rather elementary account of the bosonization, I want to give a short remark. The free massive fermion theory is bosonized as an interacting theory using the free massless fermion basis regarding the mass term as an interaction.[8] Therefore, while the fermions live in the Hilbert space of the free theory (the transformation from the free massless fermion basis to the free massive one is trivial), the corresponding boson theory is fully interacting theory. The latter does not appear to have a smooth limit to the free theory in its solitonic sector. The fact that the bosonized theory is in the interacting Hilbert space may have important implications. The structure of the Hilbert space of the fermion theory is largely altered by adding even weak coupling interactions. But it may not be in the boson theory because it was already in the interacting Hilbert space.

Via a similar procedure one can in principle bosonize any interacting fermion theories in 1+1 dimensions.

FERMION FRACTIONIZATION VIA BOSONIZATION

I now examine how the phenomenon of fermion fractionization looks like in the light of the bosonization[9]. Let's start by bosonizing the coupled fermion-kink system described by the Lagrangian (2). The bosonic equivalent can be written as

$$L = \frac{1}{2}(\partial_\mu\phi)^2 - V(\phi) + \frac{1}{2}(\partial_\mu\chi)^2 - gm\mu\cos(2\sqrt{\pi}\chi), \qquad (19)$$

where I have denoted the bosonized fermion field as χ. Can we still find the solitons in (19)? The answer is yes, but it is a composite object consisting not only of solitonic ϕ configuration but also of χ's.[10] Namely, in order to have minimum energy χ must shift by $\frac{\sqrt{\pi}}{2}$ at the point where ϕ changes sign. Figure 2 is a rough draft of the field configurations of this ϕ-χ complex.

Now, I compute the fermion number of the solitons.

$$F = \frac{1}{\sqrt{\pi}}[\phi(\infty) - \phi(-\infty)] = \frac{1}{2}. \qquad (20)$$

This is the phenomenon of fermion fractionization reproduced by the bosonized framework. I want to emphasize that this way of looking the phenomenon clearly reveals how soliton configuration obtains the fermionic charge: The fermion fractionization is not the phenomena that a fermionic hair is put on the solitons, but the phenomena in which whole structure of the soliton configuration is disturbed by the nontrivial change in the Dirac vacuum. The soliton becomes truely boson-fermion composites through the phenomenon.

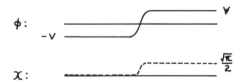

Fig. 2. A rough sketch of the field configrations of the ϕ-χ complex.

We also learned an important lesson through the above treatment. Classical computation in the bosonized theory gives a correct (=exact) induced charge on solitons. As far as I know this remains true for all known systems not only in 1+1 dimensions but also in 3+1 dimensions.

Some of you may be surprised by my statement that the classical treatment of the bosonized theory produces the exact result. In the fermionic treatment we evaluate the following diagram in order to compute the induced current $<j_\mu(x)>$ on solitons,

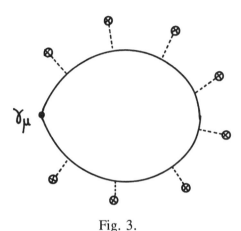

Fig. 3.

where $\otimes$ indicates the effect of the background soliton configuration. Therefore, it is the one-loop effect from the viewpoint od the fermion theory, whereas it is the tree-level result in the boson theory. The resolution of this apparent puzzle is in fact very simple in this case: A factor of ħ actually comes in into the correspondence relation:

$$j^\mu(x) = -\frac{\hbar}{\sqrt{\pi}}\epsilon^{\mu\nu}\partial_\nu\phi(x). \tag{21}$$

Thus the classical result of the Bose theory corresponds to the one-loop result of the Fermi theory.

There exist numerous investigations on this diagrammatic approach to the fermion fractionization. The close relationship with the index theorems[11] is particularly noteworthy. I only cite here the pioneering work[12] and a comprehensive review article.[13]

MAGNETIC MONOPOLES WITH FRACTIONAL CHARGES

I now turn to the fermion-soliton system in 3+1 dimensions. As I mentioned before an orthodox treatment using the Jackiw-Rebbi zero mode exists also for this system. Here I want to explain the bosonized treatment of the same system since our machinary is now well oiled and was proven to be powerful. However, I shall borrow one crucial knowledge from the orthodox analysis; the Jackiw-Rebbi zero modes exist only in the lowest partial wave components of the fermion fields.

The theory which we are to deal with is the SU(2) Georgi-Glashow model with one isodoublet of fermion fields described by the Lagrangian

$$L = -\frac{1}{4}F_{\mu\nu}{}^a F^{a\mu\nu} + \frac{1}{2}D_\mu\Phi_a D^\mu\Phi_a - \frac{g^2\lambda^2}{2v^2}\left[\Phi_a\Phi_a - \frac{v^2}{g^2}\right]^2$$

$$+\bar\psi(i(\partial + gA)\psi - g\,G\,\bar\psi\frac{\tau_a}{2}\Phi_a\psi. \tag{22}$$

The gauge fields A_μ and the Higgs fields Φ are in the adjoint representation of the gauge group SU(2), whereas the fermion fields ψ is in the fundermental representation with the representation matrix $\tau_a/2$, τ_a being the Pauli matrix. The Higgs potential in (22) is taken so that the SU(2) gauge symmetry is spontaneously broken into U(1), which will be referred to as the electromagnetism.

It is well known that the bosonic part of the theory possesses the magnetic monopole solution whose discovery is due to 't Hooft and Polyakov.[14] It is a hedgehog configuration of the isovector Higgs fields and its stability is again due to the nontrivial topology of the hedgehog. In simple terms it is a topological defect in the broken Higgs vacuum.

It is crucially important to notice that the angular momentum receives the additional contribution from the isospin[15]

$$\mathbf{J} = \mathbf{L} + \frac{\sigma}{2} + \frac{\tau}{2} \tag{23}$$

in the solitonic sector of the theory. Therefore, the fermion has spherical partial wave J=0 ! Henceforth we shall keep only the sin the solitonic sector of the theory. Therefore, the fermion has spherical partial wave J=0 ! Henceforth we shall keep only the s-wave since this is the partial wave where the Jackiw-Rebbi zero mode lives.

Correspondingly it may be legitimate to retain only the s-wave fluctuation of the gauge fields:

$$A_0(x) = \hat{r} \cdot \frac{\tau}{2} a_0(r,t),$$

$$A_i(x) = \varepsilon_{aij}\hat{r}_j \frac{\tau_a}{2} \frac{A(r)}{gr} + \hat{r}_i (\hat{r} \cdot \frac{\tau}{2}) a_1(r,t). \tag{24}$$

The Higgs field fluctuation may be ignored :

$$\Phi_a(x) = \frac{1}{g} \hat{r}_a \phi(r). \tag{25}$$

the first term of A_i in (24) and the Higgs fields (25) constitute the magnetic monopole solution; $A(r)$ and $\phi(r)$ are the functions which rapidly grow outside the monopole core to 1 and v , respectively.

The next step of bosonizing the monopole-fermion system is the partial wave expansion of the fermion fields to select out the J=0 wave. In fact it can be systematically done for an arbitrary representation fermions (representation matrix T_a) as

$$\chi_\delta(x) = \frac{1}{r} \sum_{jm\sigma\tau} v_{jm\sigma\tau}{}^{(\delta)}(r, t) \, \Psi_{jm\sigma\tau}(\theta,\varphi) \tag{26}$$

where χ_δ is the Weyl fermion of the chirality δ; $\delta = +1$ and -1 imply the right- and left-handed fermions, respectively. The functions $\Psi_{jm\sigma\tau}$ are the simultaneous eigenfunctions of J^2, J_3, $\hat{r}\cdot\sigma$, and $\hat{r}\cdot T$ with corresponding eigenvalues $j(j+1)$, m, σ, and τ. Using the orthonormality of $\Psi_{jm\sigma\tau}$ the fermionic part of the action can be written as

$$S = \int dr \, dt \sum_\delta \sum_{jm\sigma\tau} \left[v_{jm\sigma\tau}{}^{(\delta)*} [i \, \partial_0 + \tau a_0 + i\delta\sigma(\partial_r + \tau a_1)] \, v_{jm\sigma\tau}{}^{(\delta)} \right]$$

$$+\frac{i\delta\sigma}{r}\left[j(j+1)+\frac{1}{4}-\tau^2\right]^{1/2}v_{jm\,-\sigma\tau}{}^{(\delta)*}\,v_{jm\sigma\tau}{}^{(\delta)}$$

$$-\tau\,G\,\phi\,v_{jm\sigma\tau}{}^{(\delta)*}\,v_{jm\sigma\tau}{}^{(-\delta)}\Big], \tag{27}$$

where I have ignored the terms rapidly vanish outside the monopole core. It looks like a somewhat involved calculation. But in fact it isn't; there exists an elegant way of doing computation. See, for example, Appendix of Ref. 16.

Now let us confine ourselves into $j=0$ partial wave of the isospinor fermions; $m=0$ and $\tau=1/2$. Then we can construct an artificial radial two-dimensional (r,t) fermion fields u_δ as

$$u_\delta\,(r,t)=\left[\frac{1+i}{2}+\frac{1-i}{2}\tau_3\right]\begin{bmatrix}v_{\delta\,-\delta}{}^{(\delta)}\\[2mm]v_{-\delta\,\delta}{}^{(\delta)}\end{bmatrix} \tag{28}$$

where the subscripts of v indicates σ and τ. The fermionic action finally takes the form outside the monopole core as

$$S=\int dr\,dt\,\sum_\delta\,\bar{u}_\delta\left[\gamma^0\,(i\,\partial_0-\frac{\delta}{2}\,a_1)+\gamma^1(i\,\partial_1-\frac{\delta}{2}\,a_0)\right]u_\delta$$

$$-\frac{\delta}{2}\,Gv\,\bar{u}_\delta\,\gamma^5 u_{-\delta}. \tag{29}$$

Since we are discussing the topological properties of the monopole-fermion system it is entirely legitimate to disregard the details near the monopole core. I note that no explicit profile of the magnetic monopole configuration remains in (29). Only the topological property and the resultant s-wave nature of the isospinor fermions are exhibited in the action (29).

Having obtained an effectively two-dimensional fermion system we can use our powerful machinary of bosonization. I have to note the minor change in the bosonization formula[17]

$$u_\delta\,(r,t)=\sqrt{\frac{\mu}{2\pi}}\begin{bmatrix}N_\mu\,\exp\,\{\,i\,\sqrt{\pi}\,[\phi_\delta(r,t)-\tilde{\phi}_\delta(r,t)]\}\\[2mm]i\,N_\mu\,\exp\,\{\,i\,\sqrt{\pi}\,[\phi_\delta(r,t)+\tilde{\phi}_\delta(r,t)]\}\end{bmatrix} \tag{30}$$

where $\tilde{\phi}$ is defined in (14). I should also note a new important ingredient, the Rubakov-Callan boundary condition[17][18]

$$(1 - \gamma^0) u_\delta (0, t) = 0 \tag{31}$$

at the origin. I just refer the original articles for its derivation. One can easily verify that by computing two-point functions the fermionic boundary condition (31) can be translated into the bosonic one

$$\phi'_\delta (0, t) = 0 . \tag{32}$$

Using the formula (30) one can easily bosonize the system (29). Only remaining step is to make a canonical transformation to the "physical" Bose variables Φ and Q as

$$\Phi = \frac{1}{2} (\phi_+ - \phi_-) + \frac{1}{2} \int_0^r ds \left[\pi_+(s) + \pi_-(s) \right]$$

$$Q = \frac{1}{2} (\phi_+ - \phi_-) - \frac{1}{2} \int_0^r ds \left[\pi_+(s) + \pi_-(s) \right]$$

$$\Pi = \frac{1}{2} (\pi_+ - \pi_-) + \frac{1}{2}(\phi'_+ + \phi'_-)$$

$$P = \frac{1}{2} (\pi_+ - \pi_-) - \frac{1}{2}(\phi'_+ + \phi'_-) \tag{33}$$

where π, Π, and P are canonical conjugates of ϕ, Φ, and Q, respectively. Carrying out the integration over the gauge fields in $a_0 = 0$ gauge I obtain, after the canonical transformation, the bosonized Hamiltonian

$$H = \int dr \frac{1}{2} \left(\Pi^2 + P^2 + \Phi'^2 + Q'^2 \right)$$

$$- \frac{1}{4\pi} \mu G v \left[\cos(2\sqrt{\pi}\Phi) - \cos(2\sqrt{\pi}Q) \right]$$

$$+ \frac{g^2}{32\pi^2 r^2} (\Phi + Q)^2 . \tag{34}$$

For our present purpose it is much more important to pay attention to the semiclassical boundary condition $\Phi(0) + Q(0) = 0$ at the origin so as not to have divergent Coulomb energy. This, together with the obvious one $\Phi(0) = Q(0)$ which follows by definition, leads to the boundary condition

$$\Phi(0) = Q(0) = 0. \tag{35}$$

Now let us compute the fermionic charge on monopoles. The bosonic expressions of the electric charge and the fermion number read

$$Q = \int d^3x \, \bar{\psi} \, \gamma^0 \, \hat{r} \, \frac{\tau}{2} \, \psi = -\frac{1}{2\sqrt{\pi}} [Q(r) + \Phi(\rho)]_0^\infty \, ,$$

$$F = \int d^3x \, \bar{\psi} \, \gamma^0 \, \psi = \frac{1}{\sqrt{\pi}} [\, Q(r) - \Phi(r)]_0^\infty \, . \tag{36}$$

The behavior of the Bose fields at spatial infinity is governed by the Higgs-type mass term and we have, for $Gv > 0$

$$\Phi(\infty) = 0, \quad Q(\infty) = \pm \frac{\sqrt{\pi}}{2} \, , \tag{37}$$

and for $Gv < 0$,

$$\Phi(\infty) = \pm \frac{\sqrt{\pi}}{2}, \quad Q(\infty) = 0 \, . \tag{38}$$

In each case we have

$$Q = \pm \frac{1}{4}, \quad F = \pm \frac{1}{2} \, . \tag{39}$$

This means that whole fermion quantum number is fractionized.[19] (Note that the elementary fermions have $Q = \pm 1/2$ and $F = \pm 1$.)

Some of you may be curious about the electric charge fractionization. However, there exist a natural interpretation of it as due to the effect of chiral anomaly. Also one can show that the orthodox treatment a la Jackiw and Rebbi inevitably leads to the color electric charge fractionization in a class of grand unified theories. For a detailed explanation of these points we refer Ref. 20.

This completes a rather lengthy account of the bosonized treatment of the fermion fractionization in the magnetic monopole-isospinor fermion system. It can be extended to the isovector fermion case. I refer for interested readers Ref. 16 for this topics. As I have mentioned it involves the "fractionization" of angular momentum but in a way consistent with the representation of the rotation group: the nonabelian quantum numbers never fractionize.

One may wonder the consistency between this assertion and my previous statement that the color quantum numbers fractionize in grand unified theories. In fact, it involves a rather deep issue. The global SU(3) color rotation becomes ill defined[21] and only an abelian piece fractionizes.

BOSONIZED QED

I now apply the method of bosonization preceded by the partial wave decomposition to the problem of QED around a large-Z nucleus.[22] I hope that this framework will be useful in uncovering nonperturbative structure of the theory. I will describe one such possibility in the next section.

You may rise the question; "Your bosonization scheme needs to truncate the theory by discarding all partial waves but the lowest one. What kind of considerations do justify this procedure in the present problem?" This is a good question and I have to give the answer as I did for the monopole-fermion system. To do this I have to sharpen my concern to the system. What I am interested in in studying QED with highly charged nucleus is to construct a field theory on the supercritical ground state. The latter is the state that is expected to be created around a huge charge, Z ~ 170, and it can be characterrized by the driving of the bound levels into the negative energy continuum.[23] Again, as far as Z < 300, only the lowest partial wave (j=1/2) dives. Since I am interested in the structure of the ground state and the low lying excitations it may be legitimate to keep only the j=1/2 wave.

I consider QED with an external charge Ze with spherically symmetric charge distribution $\rho(r,t)$

$$L = -\frac{1}{4}F_{\mu\nu}F^{\mu\nu} + \bar\psi(i\,\partial\!\!\!/ + e\,A\!\!\!/ - m_0)\psi - Ze\rho A_0 \,. \tag{40}$$

I shall take the similar procedure as I used for the monopole-fermion system to select out the lowest partial wave. The difference here is, of course, the angular momentum is given by

$$\mathbf{J} = \mathbf{L} + \sigma/2 \tag{41}$$

as usual.

I do not want to repeat complicated calculations here again. Referring Ref. 22 for interested readers I just give the resultant expression of the radial two-dimensional QED

$$S = \int dr \; dt \; [2\pi r^2 (\partial_0 a_1 - \partial_r a_0)^2 - 4\pi Z e r^2 \rho(r,t) a_0]$$

$$+ \int dr \; dt \; \sum_{m \; \delta} \bar{u}_m^\delta [\gamma^0 (i\partial_0 + e a_0) + \gamma^1 (i\partial_r + e a_1)] u_m^\delta$$

$$+ \frac{i}{r} \bar{u}_m^\delta \gamma_5 u_m^\delta - m_0 \bar{u}_m^\delta u_m^{-\delta} . \tag{42}$$

Here we have twice as many two-dimensional spinors as the monopole case due to spin degrees of freedom which is indicated by the index m. For the same reason the centrifugal barrier term has appeared in the fermionic part of (42).

Let us bosonize the system (42). The Mandelstam bosonization (13) is appropriate in this system. The boundary condition take a form, which is different from (32),

$$\phi_m^\delta (0, t) = 0 \tag{43}$$

if expressed by the bosonic variables. After the canonical transformation similar to (33) we have the bosonized Hamiltonian

$$H = \int dr \sum_m \frac{1}{2} \left(\Pi_m^2 + P_m^2 + \Phi_m'^2 + Q_m'^2 \right)$$

$$- \sum_{m \; \delta} \frac{1}{2\pi r^2} N_\mu \cos \left[\sqrt{\pi} \left\{ \Phi_m + Q_m - \delta \int_r^\infty ds \; [\Pi_m(s) - P_m(s)] \right\} \right]$$

$$- \sum_m \frac{\mu m_0}{\pi} N_\mu \left[\cos (2\sqrt{\pi}\Phi_m) + \cos (2\sqrt{\pi}Q_m) \right]$$

$$+ \frac{e^2}{8\pi r^2} \left[\Theta(r,t) - \frac{1}{\sqrt{\pi}} \sum_m (\Phi_m + Q_m) \right]^2 , \tag{44}$$

362

where

$$\Theta (r,t) = \int_0^r dr\, \rho(r,t) . \tag{45}$$

I should note that the centrifugal barrier term (second line of (44)) now carries $1/r^2$ not $1/r$ as in (42). This change of the power is of the quantum origin. It is produced via a point-splitting procedure conspired by the boundry condition (43). Only with this change the centrifugal barrier correctly competes with the Coulomb attraction. Unless this occurs our bosonization scheme would go astray immediately. Therefore, the second term of (44) is not just centrifugal barrier but also takes care of the quantum effect. In fact, it is this term that makes the hydrogen-like atom large enough, up to the order of the Bohr radius, in spite of the classical treatment of the Bose theory.[24]

Let us analyze the theory classically as we witnessed to be successful in the foregoing analyses. First of all the theory simplifies at spatial infinity; it reduces to the two decoupled sine-Gordon theory. Therefore, the electrons are the radial sine-Gordon solitons. This interpretation can be confirmed by computing the electric charge and the angular momentum,

$$Q = -\frac{e}{\sqrt{\pi}} \sum_m [\, \Phi_m(r) + Q_m(r)\,]_0^\infty ,$$

$$J_3 = \int d^3 x\, \psi_+ (r \times \frac{1}{i} v + \frac{\sigma}{2})_3 \psi \tag{46}$$

$$= \frac{1}{\sqrt{\pi}} \sum_{m=\pm\frac{1}{2}} m[\, \Phi_m(r) + Q_m(r)\,]_0^\infty .$$

The Φ and Q degeneracy of the electron states is nothing but the parity degeneracy. One has to form superpositions of Φ- and Q-soliton states to obtain the parity eigenstates.

By increasing the external charge Z one can observe that the ground state of the system changes from the normal QED vacuum to the supercritical one at $Z\sim170$. One has to be careful about the renormal-ordering procedure with respect to the physical radius dependent masses. I prefer not to enter into the details of this procedure.[25] But I just want to mention that the supercritical ground state is characterized by the induced electric charge around a nucleus. In our bosonic language it is a solitonic ground state.

METASTABLE EXCITATIONS ON SUPERCRITICAL GROUND STATE

I discuss the possibility of having new states in QED in strong fields using the bosonized formulation of QED just constructed.[25][26] Since the ground state is the solitonic one I can formulate the problem as the small fluctuations around this ground state. Then, following the well known procedure I expand the Bose fields as

$$\Phi = \Phi_{cl} + \phi,$$

$$Q = Q_{cl} + q, \tag{47}$$

where the small letters denote the fluctuations.

I restrict myself, for simpliciy, to working with the symmetric ansatz $\Phi_{cl} = Q_{cl}$ which means the four electron diving. I introduce new variables

$$\psi_m = \frac{1}{\sqrt{2}}(\phi_m + q_m)$$

$$\chi_m = \frac{1}{\sqrt{2}}(\phi_m - q_m) \tag{48}$$

and can show that the ψ-and the χ-fluctuations decouple. I further make the combinations

$$\psi_{\pm} = \frac{1}{\sqrt{2}}\left(\psi_{\frac{1}{2}} \pm \psi_{-\frac{1}{2}}\right) \tag{49}$$

Then I can show that the $\psi_{\pm}$ satisfies the Schrodinger-like equation

$$\omega_{\pm}^2 \psi_{\pm} = \left[-\frac{d^2}{dr^2} + V_{\pm}(r)\right]\psi_{\pm} \tag{50}$$

with $\omega_{\pm}$ being the energies of the fluctuations. Since the difference between the potentials V_+ and V_- is minor I just present $V_+(r)$ in Fig 4.

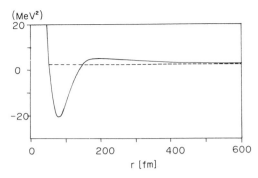

Fig 4. The potential V_+ (r) computed with the spherical uniformely charged sphere of radius R=35 fm and Z=170. The dashed line indicate ω^2, the energy squared of the metastable excitation.

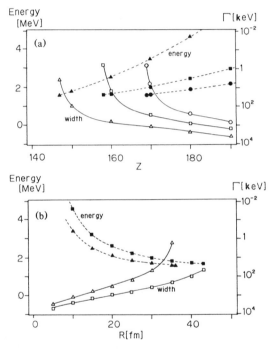

Fig 5. The energy levels and the widths of the metastable states are plotted as (a) functions of Z for R=10 fm (Δ), 22 fm ($\circ$), and 34 fm (o); (b) functions of R for Z=170 (Δ), and 180 ($\circ$). The solid (open) symbols indicate the energies (widths) of the states. The left (right) ordinate is for the energy (width) plotted in units of MeV (keV) in linear (inverse-logarithmic) scale. The solid and the dashed curves are to guide the eye.

There appears the pocket structure in the potential at the point where the ground state soliton configuration stand up. Then it is natural to ask whether there exist states trapped in this pocket. In fact there is! In Fig.5 I present the energies and the widths of this state computed by the WKB approximation.[26] Notice that the state can decay by tunneling through the barrier.

We observe in Fig. 5 that an interesting region of the parameters exists where the state is metastable with narrow width, ≤ 10 keV, and moreover the energies of the excitation are insenstive to the parameters. Such a metastable particle-like state, if created, is a good candidate for explaining the peak structures observed in the GSI experiments.[27]

One further remark: the long lifetime of the state is due to the long path length which is required to tunnel through the barrier. It is of the order of 1000 fm for narrow states. Therefore, it materializes to $e^+ e^-$ after traversing ~ 1000 fm. This implies that the Coulomb splitting between e^+ and e^- is small, ~ 200 keV. It is also a favorable feature for the explanation of the peaks.

How many metastable states? I have two at present with quantum numbers

$$\psi_+ : \quad J^{PC} = 0^{+-},$$

$$\psi_- : \quad J^{PC} = 1^{++}. \tag{51}$$

The problem is, however, that the splitting between the ψ_+ and ψ_- excitations is too tiny, ~ 10keV, in the WKB approximation. Nevertheless, I want to emphasize that our treatment based on the bosonization revealed some interesting structures in the excitation spectrum on the supercritical ground state. It remains to be seen whether their existence can be demonstrable by other independent ways.

Of course, there still remain many problems to be understood to realy explain the characteristic features of the GSI data. One of the most important is to determine in what circumstances and how the transition to the supercritical ground state occurs. This is essentially understood for the adiabatic situation[23] but what we need is that of the real heavy ion collision experiments. In a separate occasion[28] we have argued that the process of the transition may be described in close analogy to the false vacuum decay in the theory of first-order phase transition. The outcome of such treatment is the suggestion of possible dynamical enhancement of the transition which, if true, should be helpful in understanding the so called "undercritical peaks".[27]

Fortunately, there is a way to experimentally test whether the physics of the supercritical vacuum has something to do with the mysterious peaks in heavy ion collisions. It is the triple coincidence experiment[25] that measures not only the peak e^+ and e^- but also the excess e^+ due to the spontaneous positron emission. I urge our experimental colleague to seriously think about this experiment.

In these lectures I have discussed that the method of bosonization can be successfully applied to a variety of nonpertarbative questions in field theories. They include the phenomenon of fermion fractionization in 1+1 dimensions, the fractional charges on magnetic monopoles, and the QED around a highly charged external source. I emphasize that most of these problems have real physical impacts which go far beyond the academic interests. It would be very nice if we can come back to this charming place again with more extended list of successful applications of the bosonization in the foreseeable future.

ACKNOWLEDGEMENTS

I wish to thank Yumi S. Hirata for many fruitful collaborations. The works done with her constitute the main body of these lectures. I am grateful to Berndt Muller for inspiring discussions and his continuous interests in our works. The discussions with the GSI experimentalists, in particular with H. Bokemeyer, W. Koenig, and J. Greenberg, were very useful in illuminating the experimental situation. I thank Daniel Caldi, Alan Chodos, and Reijiro Fukuda for many useful conversations and encouragements.

Finally, my sincere thanks go to Herb Fried and Berndt Muller for their efforts devoted to organizing such a enjoyable meeting in such pleasant atmosphere. I thank Yamada Science Foundation for financial support.

REFERENCES

1. R. Jackiw and C. Rebbi, Phys. Rev. D13, 3398 (1976).

2. W. P. Su, J. R. Schrieffer, and A. J. Heeger,
    Phys. Rev. B22, 2099 (1980) ;
    R. Jackiw and J. R. Schrieffer, Nucl. Phys.
    B190 [FS3], 253 (1981).
3. H. Minakata, Phys. Lett. 155B, 352 (1985).
4. J. D. Bjorken and S. D. Drell, Relativistic Quantum Fields
    (McGraw Hill, New York, 1964).
5. S. Coleman, Phys. Rev. D11, 2088 (1975).
6. S. Mandelstam, Phys. Rev. D11, 3026 (1975).
7. S. Tomonaga, Prog. Theor. Phys. 5, 544 (1950).
8. M. B. Halpern, Phys. Rev. D12, 1684 (1975).
9. R. Shankar and E. Witten, Nucl. Phys. B141, 349 (1978).
10. W. A. Bardeen et al, Nucl. Phys. B218, 445 (1983).
11. C. Callias, Commun. Math. Phys. 62, 213 (1978) ;
    R. Bott and R. Seeley, ibid. 62, 235 (1978).
12. J. Goldstone and F. Wilczek, Phys. Rev. Lett. 46, 988 (1981).
13. A. J. Niemi and G. W. Semenoff, Phys. Rep. C135, 99 (1986).
14. G. 't Hooft, Nucl. Phys. B79, 276 (1974) ;
    A. M. Polyakov, JETP Lett. 20, 194 (1974).
15. R. Jackiw and C. Rebbi, Phys. Rev. Lett. 10, 1116 (1976) ;
    P. Hasenfratz and G. 't Hooft, ibid. 10, 1119 (1976) ;
    A. S. Goldhaber, ibid. 10, 1122 (1976).
16. H. Minakata, Phys. Rev. D32, 2134 (1985).
17. C. G. Callan, Jr. Phys. Rev. D26, 2058 (1982).
18. V. A. Rubakov, Nucl. Phys. B203, 311 (1982).
19. J. A. Harvey, Phys. Lett. 131B, 104 (1983).
20. Y. Hirata and H. Minakata, Phys. Rev. D34, 2519 (1986).
21. P. Nelson and A. Manohar, Phys. Rev. Lett. 50, 943 (1983) ;
    A. P. Balachandran et al., ibid. 50, 1553 (1983).
22. Y. Hirata and H. Minakata, Phys. Rev. D34, 2493 (1986).
23. W. Greiner, B. Muller, and J. Refelski, Quantum
    Electrodynamics of Strong Fields (Springer, New York, 1985).
24, Y. Hirata and H. Minakata, Phys. Rev. D35, 2619 (1987).
25. Y. S. Hirata and H. Minakata, Z. Phys. C46, 45 (1990).
26. Y. S. Hirata and H. Minakata, Phys. Rev. D39, 2813 (1989).
27. For the latest status, see the reports by Bokemeyer and
    Koenig in this volume.
28.. Y. S. Hirata and H. Minakata, Mod. Phys. Lett. A5, 1081 (1990)

REMARKS ON COSMIC STRINGS, AHARONOV-BOHM EFFECT

AND VACUUM POLARIZATION

Pawel Górnicki

Institute for Theoretical Physics
Polish Academy of Sciences
Al. Lotników 32/46, 02-668 Warsaw, Poland

The idea that intergalactic space may be populated by topological defects frozen into the physical vacuum is one of the most exciting results of so called Grand Unified Theories. The defects may be of various kind: domain walls, strings, monopoles, textures. They come into life in the very early Universe during the symmetry-breaking phase transitions. None of them has been observed.

Magnetic monopoles became the first object to be studied intensively. Their nonexistence caused disillusion and brought some interest to strings. I do not wish to review the whole topic here. An interested reader may refer to Vilenkin[1].

For simplicity I am going to discuss the case of Abelian string ie. the string that may be formed during the phase transition with spontaneous breakdown of some $\tilde{U}(1)$ gauge symmetry. The string is the tubelike region of the 'old' ($\tilde{U}(1)$ symmetric) vacuum with the 'magnetic' flux trapped inside. The word 'magnetic' refers of course to the spontaneously broken theory under discussion (not to the ordinary electromagnetism which remains unbroken). There is a strict analogy between the cosmic string (Nielsen-Olesen string[2]) and Abrikosov vortices[3] in superconductors.

STRINGS AND AHARONOV-BOHM EFFECT

I would like to investigate an interesting (and perhaps not so remote) possibility that the gauge field $\tilde{A}_\mu$ is coupled to some ordinary fermions e.g. electrons, quarks etc. It means that these fermions carry the $\tilde{U}(1)$ charge in addition to an ordinary electric one. Since the $\tilde{U}(1)$ 'photons' are presumably very massive (Grand Unification mass scale) the additional interaction between the fermions remains unnoticed being of a very short range. The same is of course true about the direct interaction of the particles and the 'magnetic' field inside the string.

Fortunately this is not the whole thing. The line of magnetic flux is surrounded by topologically nontrivial configuration of gauge field $\tilde{A}_\mu$. This field cannot be globally gauged out and (due to Stokes theorem) extends to infinity.

The classical particles moving in this field remain uneffected since

Vacuum Structure in Intense Fields, Edited by
H.M. Fried and B. Muller, Plenum Press, New York, 1991

$\tilde{A}_\mu$ is locally a pure gauge. This does not hold in quantum theory thanks to the strange phenomenon known as Aharonov-Bohm effect[4]. If the 'magnetic' flux inside the string is non-integer (as expressed in London units $F = q\Phi/2\pi$, q - relevant charge) the Aharonov-Bohm effect will present a very efficient way of coupling the matter to the string. (The flux value is connected to the charge of the relevant Higgs field, so one is not perfectly free to choose it. Nevertheless, under some fortunate circumstances values like $F = 1/2$ or $1/4$ may be available.)

The above scenario has been around for some time and already led to quite interesting results[5]. I would like to present one more development along this path.

VACUUM POLARIZATION AND AHARONOV-BOHM EFFECT

The virtual electron-positron pairs (always present in the physical vacuum) are also subjected by Aharonov-Bohm effect. This leads to various phenomena including the polarization current that flows around the string.

To visualize this phenomenon one can imagine a virtual e^+e^- pair created spontaneously near the string. If the electron passes the string from one side and the positron from the other before their annihilation they will form a loop that surrounds the flux. Although the particles move (mainly) outside the 'magnetic' field they will be influenced by the vector potential (since it is not possible to gauge it out simultaneously along the whole loop). The loop itself is associated with some current that may be clockwise or counterclockwise. The presence of the flux breaks the left-right symmetry of the problem leading to nonvanishing value of vacuum current.

To calculate this current one has to consider the Dirac fermion moving in the vicinity of a thin tube of magnetic flux (solenoid). This is particularly easy with perturbation theory but the result is applicable for $|F| \ll 1$ only. One has to begin with Schwinger's formula[6] (for general field configuration):

$$j_\mu = -\left(\frac{\alpha}{2\pi}\right)(2\pi)^{-1}\int (dk)e^{ikx}\left(\frac{\partial F_{\mu\nu}}{\partial x_\nu}\right)(k)$$
$$\int_0^1 dv(1-v^2)\int_0^\infty ds\, s^{-1}\exp\left[-m^2 s + \frac{1}{4}k^2(1-v^2)s\right]$$

(1)

It is possible easy to obtain the result in the closed form under assumption that the solenoid is infinitely thin.

Since one expects F to be a considerable fraction of 1 a more detailed (nonperturbative) analysis is necessary. The details of this calculation have been published elsewhere[7]. The final result is given bellow:

$$j_\phi(r) = \frac{-e\sin(F\pi)}{\pi^3 r^3}\int_0^\infty d\gamma\,\gamma\exp\left[-\gamma - \frac{(mr)^2}{2\gamma}\right]K_F(\gamma)$$

(2)

or after integration

$$j_\phi(r) = \frac{-e(mr)^4\sin(F\pi)}{4\pi^3 r^3}\left[\frac{1}{3}K_{2+F}(mr)K_{2-F}(mr)\right.$$
$$\left. - \frac{4}{3}K_{1+F}(mr)K_{1-F}(mr) + K_F^2(mr)\right].$$

(3)

This holds when $F \in (-1,1)$, but the extension to arbitrary values of F based on quasi-periodicity of the problem is rather straightforward. For

$F \to 0$ the result agrees with that of eq. (1). It is easy to note that the current flows only in the thin layer around the string - its thickness being of order of m_{fermion}^{-1}. The divergence at $r \to 0$ is artificial since the string is never infinitely thin.

The above formula looks complicated. In the very special case of $F = 1/2$ it takes on a simpler form:

$$j_\phi(r) = \frac{-e}{4\pi^2 r^3} \left(mr + \frac{1}{2}\right) e^{-2mr}. \tag{4}$$

Despite of its simplicity the above equation reflects basic properties of the general one (3).

It may look strange that j_ϕ does not vanish for $F = 1/2$. The current must be (and of course is) an antisymmetric function of F and the replacement $F \to F+1$ seems to be a symmetry the problem (gauge transformation). This is not so since the direct interaction of the fermions with the 'magnetic' field inside the string must not be neglected. The field acts on the 'magnetic' moment of the Dirac particle causing an additional binding for appropriate spin orientation. (There is still a possibility of some extra coupling between the string and the particles. Only the pure $\tilde{U}(1)$ gauge interaction, i.e., minimal coupling, has been assumed here.) As the result the Aharonov-Bohm effect is modified[8] and the $F \to F+1$ symmetry is broken (some of this symmetry remains however - it is what I called quasi-periodicity). For scalar[9] particles the current vanishes at $F = 1/2$.

It is interesting to note that the polarization current produces the flux itself. This flux acts upon the vacuum causing additional current that is not included in (3). Ideally one would like to perform some kind of self-consistent analysis here but this seems to be out of reach for technical reasons. Nevertheless this correction is rather small (it is higher order in coupling constant) even though the flux diverges as one goes to more and more thin strings (the divergence is only logarithmic).

Throughout the calculation the finite thickness of the string has been neglected. Indeed, numerical analysis shows that the finite size effects may be dropped if one wants to know j_ϕ at the distances equal to few times the radius of the string or greater. It is also clear that the polarization current drops to zero at the center of the string (for symmetry reasons). Unfortunately its difficult to interpolate between these points. The best one can do is to cut of the result (3) at some value of r approximately equal to the $\tilde{U}(1)$ scale.

POSSIBLE EFFECT OF POLARIZATION CURRENT

The polarization effect described here brings an additional structure to the string. The electrons interacting with the string via some hypothetical $\tilde{U}(1)$ charge carry their normal electric charges as well. This will cause the string to interact with electromagnetic fields, emit photons etc. Photon emission may be of some interest.

Normally, the $\tilde{U}(1)$ string is able to emit $\tilde{U}(1)$ 'photons' and Higgses which are very heavy. For this reason only high frequency modes of string's motion are to participate in this emission. As the result the emission rate is diminished. This is no longer true about the massless photons. On the other hand they are coupled to the polarization current which is rather weak. One may try to calculate the emission rate using the expressions given above. In this case one has to remember that they are valid for straight infinite and static string. The possible effects of curvature and accelerated motion have not be included. Nevertheless a rough estimate is certainly possible. Whether the photon emission mentioned above is of any astrophysical importance is not clear at the moment.

The electromagnetic interactions of cosmic strings were considered within the framework of so called superconducting strings[10]. In that case the string behaved as an superconducting wire. The mechanism proposed in the present paper is of course much weaker. The polarization effect mentioned here is weak by laboratory standards while the currents in superconducting strings are thought to be larger (by orders of magnitude) then those ever generated on Earth. Because of its weakness the polarization effect will be of any interest only when it proves to be the exclusive way of coupling the string to electromagnetic fields.

Acknowledgments. I would like to acknowledge the financial support from Committee of Physics (Polish Academy of Sciences).

REFERENCES

1. A. Vilenkin, *Phys. Rep.* 121 (1985), 263.
2. H. Nielsen and P. Olesen, *Nucl. Phys. B* 61 (1973), 45.
3. A. Abrikosov, *Zh. Eksp. Teor. Fiz.* 32 (1957), 1441 (*JETP Lett. (Engl. Transl.)* 5 (1957), 1173).
4. Y. Aharonov and D. Bohm, *Phys. Rev.* 115 (1959), 485; W. Ehrenberg and R. E. Siday, *Proc. Phys. Soc. London B* 62 (1949), 8.
5. M. G. Alford and F. Wilczek, *Phys. Rev. Lett.* 62 (1989), 1071; M. G. Alford, J. March-Russell and F. Wilczek *Nucl. Phys. B* 328 (1989), 140.
6. J. Schwinger, *Phys. Rev.* 82 (1951), 664.
7. P. Górnicki, *Ann. Phys. (NY)* 202 (1990).
8. C. R. Hagen, *Phys. Rev. Lett.* 64 (1990), 503.
9. E. M. Serbryanyi, *Teor. Mat. Fiz.* 64 (1985), 299 (*Theor. Math. Phys.* 64 (1985), 846).
10. E. Witten, *Nucl. Phys. B* 249 (1985), 557.

ON POWER CORRECTIONS IN FINITE TEMPERATURE QCD

A. Leonidov

P.N. Lebedev Physical Institute
117924 Leninsky pr. 53
Moscow USSR

Abstract

The temperature power corrections generated by vacuum inhomogeneous gluomagnetic fields in finite–temperature QCD are analyzed. In particular the expressions for the QCD partition function and speed of sound with corresponding nonperturbative contributions are given.

1. Nowadays the nonperturbative solution of QCD still does not exist. Therefore it is natural to try some approximate approaches. One possible philosophy is to work in the approximation where the information about the quantum numbers of particles involved is carried by the perturbative currents and the nonperturbative one is hidden in some classical background fields whose appearance reflect the complex vacuum structure of the theory. In fact this approach (originally proposed and developed by the ITEP group [1]) turned out to be extremely useful in QCD at zero temperature where the combined usage of the perturbative quark currents having the quantum numbers of some mesons and the background gluon fields (giving rise to the vacuum condensates) in which these currents propagate permitted to give a quantitative description of various properties of the considered mesons (see also [2,3]). In that case the nonperturbative corrections appeared typically as expressions like

$$\langle 0|(G^a)^2|0\rangle/Q^2, \quad \langle 0|f^{abc}\, G^a G^b G^c|0\rangle/Q^6, \ ... \qquad (1)$$

where G^a is a shorthand notation for the background gluonic field strength $G^a_{\mu\nu}$ and the trace is taken over the Lorentz indices also, f^{abc} are the structure constants of the colour group and Q_μ is a four–momentum carried by the quark current. It seems natural to exploit the similar approach in finite temperature QCD. For the meson sum rules the corresponding analysis was given in [4]. Here we will follow this line of reasoning to calculate the nonperturbative corrections to QCD partition function. This allows to calculate the nonperturbative contributions to the equation of state, speed of sound, etc. We shall discuss the simplest case of gluonic plasma in an external inhomogeneous gluomagnetic field [5]. The thermodynamics of a QCD plasma in a constant background chromomagnetic field was considered for example in [6–8]. In this case, however, the corresponding quantum computation is performed in an unstable vacuum (the gluon kinetic operator in such field has an imaginary mode) which is an obvious obstruction to the thermodynamic analysis which is essentially stationary. The more general case of inhomogeneous external fields was also discussed in [9].

2. Let us therefore calculate the QCD partition function in an external inhomogeneous chromomagnetic field. We shall be able to calculate it in a one–loop order with respect to quantum fields and up to a cubic invariant in the external field. We shall assume that the vacuum gluonic field is such that the quantum vacuum is stable (which is a possibility for an inhomogeneous field case). For simplicity let us also assume that the gluonic kinetic operator has no zero modes and that the external field obeys the classical equations of motion DG = 0.

For the partition function of a gluon plasma in an external field we have for the general case using the imaginary time formalism,

$$Z[A,T] = \int \mathcal{D}Q \ \mathcal{D}\chi \ \mathcal{D}\bar{\chi} \ \exp[-\int_0^{1/T} d\tau \int d^3x \ (\tfrac{1}{4}(F^a_{\mu\nu})^2 + \tfrac{1}{2}(D^{ext}_\mu Q^a_\mu)^2 + \bar{\chi}^a(D^{ext}) \chi^a)], \quad (2)$$

where $F^a_{\mu\nu}$ is a full gluon field strength, and the gluonic field is separated into a classical A^a_μ and quantum Q^a_μ parts (in (2) the expression is "prepared" to the subsequent one–loop computation with respect to a quantum field Q):

$$B^a_\mu = A^a_\mu + Q^a_\mu, \quad (3)$$

T is a temperature, D^{ext}_μ and D^{ext} are the covariant derivatives containing the external field A, χ^a and $\bar{\chi}^a$ — the ghost fields. We use the background field gauge. In the one – loop approximation we have for the partition function:

$$Z[A,T] = \exp[-\tfrac{1}{4}\tfrac{1}{T} \int d^3x (G^a_{\mu\nu})^2] \ (\det(K_1(A)))^{-\frac{1}{2}}(\det K_2(A))$$

$$(4)$$

where K_1 and K_2 are the gluon and ghost kinetic operators correspondingly, and $G^a_{\mu\nu}$ is a classical gluon field strength. The details of a calculation for the particular case of a chromomagnetic external field are presented in [5]. Here we give a final expression for the partition functions ratio:

$$\frac{F[A,T]}{T} = - \ln \frac{Z[A,T]}{Z[0,T]} = \tfrac{1}{4}(1 - \tfrac{g^2}{16\pi^2}\cdot 11 \ln \tfrac{M^2}{T^2}) \tfrac{1}{T} \int d^3x \ G^a_{ij} G^a_{ij}$$

$$- \tfrac{1}{4}\tfrac{g^2}{4\pi^{3/2}}\cdot 11 \tfrac{1}{\mu} \int d^3x \ G^a_{ij} G^a_{ij} + \tfrac{g^3}{64\pi^4}\cdot\tfrac{\zeta(3)}{60}\tfrac{1}{T^3}\cdot \int d^3x \ f^{abc} \ G^a_{ij} G^b_{jk} G^c_{ki}$$

$$- \tfrac{g^3}{12\pi^{3/2}}\cdot\tfrac{1}{60}\tfrac{1}{\mu^3}\cdot \int d^3x \ f^{abc} \ G^a_{ij} G^b_{jk} G^c_{ki} \quad (5)$$

where F is a free energy, M is an ultraviolet cutoff, μ − an infrared one, $G_{ij}^a(x)$ is a stationary inhomogeneous gluomagnetic field, ζ is a Riemann ζ−function, g is a charge in the initial Lagrangean. The infrared divergence comes from the terms with zero Matsubara frequency. They originate in exactly the same way as at T=0 case (see, e.g., [10]) and are hopefully absent only in an exact expression for the determinant.

Let us stress that the main advantage of using the operator method of calculations in the external field permits to keep the exact structures in the external field directly in the coordinate representation thus greately facilitating tthe analysis of renormalization issues.

3. Let us use the expression for the free energy to calculate the external field contribution to the speed of sound in quark − gluon plasma $c_s^2 = 1/(\partial E/\partial P)$, where E is an internal energy and P is a pressure calculated from (5). The easiest way of arriving at local thermodynamic quantities (i.e. of extracting an overall volume factor in the expression for the free energy) is to use a Fock − Schwinger gauge $x_\mu A_\mu^a = 0$ for the external field (this gauge commutes with the background gauge for the quantum field Q). Then considering for example the system being placed in a box with a linear scale L we have

$$\int d^3x \ G_{ij}^a(x)G_{ij}^a(x) = V \ G_{ij}^a(0) \ G_{ij}^a(0) + \frac{1}{12} \ L^2(D_k G_{ij} D_k G_{ij})|_0 = \ldots$$

where V is a volume. We shall also exploit a simplest assumption concerning the infrared − divergent terms in (5), namely that they are the artifacts of the computational method and could be omitted when computing the corresponding corrections to the pressure (these terms do not contribute to E). After renormalizing the free energy at a scale Λ we obtain for the speed of sound:

$$\frac{1}{c_s^2} = 3 \left[\frac{1 - (2b \ G_2)/(aT^4) - (6cG_3)/(aT^6)}{1 - (6b \ G_2)/(aT^4) + (6cG_3)/(aT^6)} \right] \qquad (6)$$

where by definition

$$G_2 = \frac{1}{V}\frac{1}{4} \int d^3x \ G_{ij}^a G_{ij}^a \ ; \ G_3 = \frac{1}{V} \int d^3x \ f^{abc} \ G_{ij}^a \ G_{jk}^b \ G_k^c \qquad (7)$$

and the coefficients a,b,c are equal to

$$a = \frac{32\pi^2}{15}; \ b = 11 \frac{g^2}{16\pi^2}; \ c = \frac{g^3}{64\pi^4} \cdot \frac{\zeta(3)}{60} ; \qquad (8)$$

4. In this lecture I described how to perform the first step towards the consistent description of gluon plasma in the background gluomagnetic field. To proceed further, one is to choose his (her) favorite candidate for the vacuum field and try to make contact with reality. I hope that with the help of the described formalism it will be possible to obtain consistent qualitative information on the finite − temperature QCD physics.

ACKNOWLEDGEMENTS

It is a pleasure to thank the organizers of the NATO Summer School "Vacuum Structure in Strong Fields" for the warm hospitality during the School.

REFERENCES

1. Shifman M., Vainstein A., Zakharov V.: *Nucl. Phys.* **B147** (1979) 385,448
2. Reinders L.J.,Rubinstein H.R.,Yazaki S.: *Nucl. Phys.* **B186** (1981) 109
3. Nikolaev S.N., Radyushkin A.V.: *Phys. Lett.* **B124** (1983) 243
4. Bochkarev M., Shaposhnikov M. Nuovo Cim : **33** (1983) 241
5. Leonidov A.V.: *Zeit. fur Phys.* **C47** (1990) 287
6. A.Cabo,O.K.Kalashnikov,A.E.Shabad: *Nucl. Phys.* **B185** (1981) 473
7. N.Ninomiya,N.Sakai: *Nucl.Phys.* **B190** [*FS3*] (1981) 316
8. B.Muller,J.Rafelski: *Phys. Lett.* **B101** (1981) 111
9. T.H.Hansson,I.Zahed: *Nucl. Phys.* **B292** (1987) 725
10. D.I.Dyakonov,V.Yu.Petrov,A.V.Yung: *Yadernaya Fizika* **39** (1984) 340

VACUUM STRUCTURE IN QCD

A. Patrascioiu* and E. Seiler†

*Physics Department and Center for the Study of Complex Systems
University of Arizona, Tucson, AZ 85721
†Max-Planck-Institut für Physik und Astrophysik
Werner Heisenberg Institut für Physik, P.O. Box 401212, Munich
Federal Republic of Germany

It is shown that nonlinear σ models in two dimensions and gauge theories in any dimension are characterized by the absence of an ordered state. This situation renders perturbation theory at best suspicious. The outline of a proof that the O(N) N $\geq$ 2 nonlinear σ models in 2D possess a phase characterized by algebraic decay of correlations is presented. The argument is based upon an analysis of the percolating properties of certain clusters. The result implies that the true Callan Symanzik β-function of such models is vanishing, in contradiction to the well-known predictions of asymptotic freedom obtained in perturbation theory for N > 2. The connection between topological properties and the phase structure of such models is also addressed.

Since the subject of this school is the nature of the vacuum in different theories, we would like to begin our discussion with a précis as to what is really meant by the word vacuum. In the frame work of quantum mechanics, we imagine that the system under consideration is described by a "good" Hamiltonian. Such a Hamiltonian is bounded from below and we will call vacuum its eigenstate of lowest energy. Since typically the Hamiltonian does not commute with the field operator, it follows that vacuum is not a state of sharp field, but it is characterized by some nontrivial functional. Depending upon the problem at hand, this functional may look simple or not, and, as we will now illustrate, QCD must be one of those theories with a messy vacuum.

First let us consider two solvable problems in nonrelativistic quantum mechanics: a harmonic oscillator (H = p^2 + x^2) and a particle of unity mass moving freely on the surface of the unit sphere (H = $L^2/2$). In both cases, the spectrum is discrete and in the vacuum $\vec{x}$ = 0. Yet, while in the first one $\langle x^2 \rangle$ = 0(h), in the second one $\langle \cos\theta \rangle$ = $1/_3$ = 0 (h^0). Thus, in the semiclassical limit, in the first case vacuum is practically a state of sharp position, while in the second case, since vacuum is the state with l = 0 for which the probability of finding the particle is uniform over the sphere, the fluctuations are of order 1 no matter how small is h. (That there is a dramatic difference between the two cases can also be seen by looking at the dependence of the energy gap function of h - 0(h^1) respectively 0(h^2).) The problem of the particle free on the sphere, involving large fluctuations, is a hard one, at least in the path integral approach. Indeed, although we know all classical motions - they are geodesics - we are not aware of any scheme in which that information can be used to compute vacuum expectation values.

Returning now to quantum field theory, if we wish to give it a nonperturbative definition, we may adopt the functional integral quantization approach. In the case of the nonlinear σ models in 2D, we are then very much facing the same situation as in 1D (nonrelativistic quantum mechanics). Indeed a rigorous theorem guarantees that whatever else the vacuum may do, it cannot spontaneously break the continuous $O(N)$ invariance. Thus, we are facing a hard problem of large fluctuations and a priori there is no reason to expect that perturbation theory or semiclassical approximations will yield correct results. As we will show later, in gauge theories too, another rigorous theorem guarantees again that in the vacuum the fluctuations are of order 1, no matter how small is the coupling.

Ever since Schiwinger's successful computation of the anomalous magnetic moment of the electron in 1949, quantum field theory and in particular perturbation theory have been the pillars of most developments in particle and, to some extent, condensed matter physics. The discovery of asymptotic freedom in 1973 gave the subject new impetus by promising a field theoretic explanation of strong interactions and of grand unification (GUTs). Roughly at the same time a new exciting realization took place, primarily through the work of Kosterlitz and Thouless:[1] the importance of topological properties, such as the existence of vortices, instantons, etc., in different models. It soon became commonly accepted that there is a fundamental difference between certain non-Abelian models and their abelian counterparts stemming precisely from the existence/absence of asymptotic freedom and/or of instantons.

In this lecture we sketch the main ingredients of what we believe can become a rigorous proof that, at least in two dimensions (2D), the above stated beliefs are incorrect. Specifically we prove that in 2D the $O(N)$ nonlinear σ models with standard nearest neighbor interaction undergo a transition to a phase characterized by algebraic decay of correlations if the temperature $1/\beta \to 0$ for any $N \geq 2$. Thus we show that the occurrence of such a transition, proven rigorously by Fröhlich and Spencer[2] for $O(2)$ and rederived by us here, has nothing to do with either the existence/absence of perturbative asymptotic freedom (as suggested in Refs. 3 and 4) or of instantons (as suggested in Refs. 1 and 5). Our result suggests that as conjectured by Patrascioiu,[6] in the non-Abelian cases $N \geq 3$, ordinary perturbation theory fails to produce the correct asymptotic expansion of the Green's functions in $1/\beta^n$ as the linear size of the lattice $L \to \infty$. While our results do not cover directly Yang-Mills theories in 4D, we expect that in that case too there is no fundamental difference between Abelian and non-Abelian models.

To put our findings in perspective we would like to first recall certain facts. It seems to be a true fact of nature that strong interactions are rather weak at short distances. The theoretical explanation of this experimental fact is asymptotic freedom, a property which only certain non-Abelian models are supposed to possess. However, asymptotic freedom is the result of computations performed in perturbation theory and so perhaps a first question to ask would be, what exactly is the role of perturbation theory in a theory such as QCD? Some may think that this is a pedantic question, since we know from the confrontation of the theory with experiment that perturbative QCD works. We do not share that opinion. Firstly, the experimental tests of QCD involve typically all sorts of unprovable assumptions about final state interactions, higher order corrections, etc. This is why at the best these tests are at the 10% level, not 10^{-8}, as in QED. Secondly, certain properties of strong interactions, such as the existence of a mass gap and of confinement cannot be discussed within a perturbative framework. Moreover, the general belief is that QCD is a fundamental theory of nature, not merely a successful phenomenology. In fact, the consensus is that only asymptotically free theories are nontrivial quantum field theories in 4D. Thus it appears to us that understanding the true role of perturbation theory in such models is a worthwhile enterprise.

The first task, therefore, is to give a nonperturbative definition to a theory such as QCD. Although we are ultimately interested in a continuum field theory, it seems that the lattice approach is the most natural way of defining the theory in a nonperturbative manner, consistent with gauge invariance. While a priori there may exist many other definitions, the lattice formulation has many virtues. On a finite lattice, all manipulations are well defined. The difficulty consists in controlling the infinite volume and the continuum limits. In practice the first limit can be controlled much better than the second one. Indeed, although particle physicists are used to thinking that continuum means letting the lattice spacing go to 0, a brief reflection reveals that a dimensionless way of saying that is, letting the correlation length become infinite in lattice units. Thus the first task in constructing a continuum limit is understanding the critical behavior of the lattice model. Only after finding out where the model becomes critical, can one contemplate using some sort of block spin renormalization group approach to construct a continuum theory.

In our presentation we will consider exclusively the lattice definition of quantum field theories such as QCD. We will not attempt to construct a continuum limit, but show that already before letting the cutoff go to zero, standard beliefs entertained by physicists for many years are not borne out. To be precise, we will prove that the Callan-Symanzik β-function computed in perturbation theory is incorrect in non-Abelian models. The method used allows us to prove this fact only for the nonlinear σ models in 2D, but we expect the conclusion to be equally valid in QCD in 4D. The nonlinear σ models in 2D are supposed to behave just as gauge theories in 4D for the following reasons:

a) They are strictly renormalizable in 2D (the coupling constant has dimension zero).

b) The Callan-Symanzik β-function of the O(N) nonlinear σ model is proportional to -(N-2). Thus, it vanishes for the Abelian case O(2), while its negative value for $N \geq 3$ reflects the asymptotic freedom of the non-Abelian models.[3,4]

c) The non-Abelian model O(3) possesses finite energy configurations (instantons), while the abelian one O(2) only infinite energy ones (vorteces). The former are supposed to disorder the system at all temperatures, while the latter only at sufficiently high temperatures.[1,5]

All these three properties are believed to be true for gauge theories in 4D. In addition there exists an approximate renormalization group scheme due to Migdal and Kadanoff, which relates gauge theories in 4D to σ models in 2D. Unfortunately, it is not clear just how reliable this identification is. Moreover, within the same approximation, one predicts that the non-Abelian spin models in 2D are asymptotically free, in contradiction with what we shall prove below.

Thus, at the present time, we have no hard evidence that at small coupling QCD is not asymptotically free and has no mass gap. Yet we shall argue that most likely, just as in the case of spin models in 2D, cutoff perturbation theory (on the lattice) is incorrect in non-Abelian gauge theories in 4D. We believe that the reason for this failure of perturbation theory is the absence of an ordered state in continuous-symmetry spin models in 2D and gauge theories in any D. Indeed in 2D the Mermin-Wagner theorem - a continuous symmetry cannot be broke spontaneously - guaranties that for any nonvanishing coupling g, no matter how small, the fluctuations of the spin do not become arbitrarily small on an infinite lattice. Since in practice perturbation theory is performed by expanding in the "field" (deviations from some chosen configuration), and the latter does not tend to zero as g → 0, one should not be surprised that the perturbative answer is wrong - although finite. This point can be verified in 1D, where one can use the transfer matrix formalism to compute the exact answer and compare it with the perturbative one. Carrying out this computation for the energy density ⟨s(0) · s(1)⟩ - the shortest nonzero distance available on the lattice - one finds[7] that for O(N), $N \geq 3$ the perturbative answer is finite yet wrong. Returning now to gauge theories, as we shall discussed later, Patrascioiu proved that in gauge theories in any D, even after one has completely fixed the gauge, on an infinite lattice the fluctuations do not go to zero as g → 0. Thus again perturbation theory is a priori suspect and in fact in the exactly soluable case of 2D one can verify that it fails for non-Abelian groups.

Before concluding this introduction, we would like to compare the difficulties discussed by us here with those related to the triviality of $\phi 4$ in 4D. In that case, cutoff perturbation theory (on the lattice) does provide the correct asymptotic expansion of the Green's functions. That is easy to understand since in fact in 4D the system is rather well ordered. Thus, one way to avoid triviality – assuming it is really there – is to retain a cutoff and avoid considering infinite distances (constructing a continuum limit). Our situation is different in that even with a cutoff, if one lowers too much the coupling constant, at some point the mass gap disappears; that is the interaction between two particles no longer decreases exponentially with the distance between them. Moreover, while one cannot conclude from the results presented below that cutoff perturbation theory fails at fixed (lattice) distance, it clearly leads to the incorrect conclusion regarding the Callan-Symanzik β-function.

Our analysis is based upon the representation of the O(N) models as Ising models with random couplings. This observation has been used recently[8] to propose a new scheme for nonlocal Monte Carlo updating, which has shown a remarkable reduction in critical slowing down. Consider a square lattice Λ and the following Gibbs measure associated with nearest neighbor interactions:

$$Z = \left[\prod_{i \in \Lambda} \int_{S^{N-1}} d\vec{s}_i \right] e^{\beta \sum_{\langle i,j \rangle} \vec{s}_i \cdot \vec{s}_j} \quad . \tag{1}$$

Here $\vec{s} \cdot \vec{s} = 1$ and $\vec{s} \in S^{N-1}$. Choosing some unit vector $\vec{u} \in S^{N-1}$ an alternative representation of the measure in Eq. (1) becomes possible:

$$Z = \sum_{\{\sigma\}} \left[\prod \int_{s_{\|i} > 0} ds_{\|i} \int d\vec{s}_{\perp i} \right] \exp\left\{ \rho \sum_{\langle i,j \rangle} \left[s_{\|i} \, s_{\|j} \, \sigma_i \sigma_j + \vec{s}_{\perp i} \vec{s}_{\perp j} \right] \right\} \quad , \tag{2}$$

where $s_{\|} = |\vec{s} \cdot \vec{u}|$, $\vec{s}_{\perp} \cdot \vec{u} = 0$ and $\sigma = \pm 1$. In terms of the $\{\sigma\}$, the system is that of an Ising ferromagnet, amenable to the Fortuin-Kasteleyn representation.[9] According to the latter, one is led to study the percolation properties of clusters of occupied bonds constructed as follows: a bond is placed between two spins only if they have the same sign and then only with probability

$$p_{ij} = 1 - \exp(-2\beta \, s_{\|i} s_{\|j}) \quad . \tag{3}$$

Finally, each such cluster is flipped randomly with probability 1/2. Thus the FK process involves two types of clusters: hemispherical (H) clusters of parallel Ising spins and FK clusters obtained from them by deleting bonds randomly with the probability given in Eq. (3). Fortuin and Kasteleyn showed that the mean size of the FK clusters, defined as the expectation value of the size of the cluster attached to the origin, is equal to the magnetic susceptibility of the original Ising system

$$\chi_{Is} = \frac{1}{L^2} \sum_{i,j \in \Lambda} \langle \sigma_i \sigma_j \rangle \quad . \tag{4}$$

For $L \rightarrow \infty$ the latter is finite if the 2-point function decays exponentially.

We shall argue now that for β sufficiently large, the mean FK cluster size is infinite. Let us define the following quantities: $\langle FK \rangle$ = mean FK cluster size; $\langle H \rangle$ = mean H cluster size. First we would like to show that for β sufficiently large $\langle H \rangle$ must diverge. We recall the following rigorously established facts:

1) For large β the configurations are smooth. As $\beta \rightarrow \infty$, one expects

$\langle \vec{s}(i) \cdot \vec{s}(j) \rangle = 1 - O(1/\beta)$ (proven for O(2)).[10]   Moreover for O(N) for β sufficiently large and ϵ sufficiently small Georgii[11] proved that the cluster defined by the event $|\vec{s}(i) \cdot \vec{s}(j) - 1| < \epsilon$ percolates, while Bricmont and Fontaine[12] showed that the expectation value of the event $|\vec{s}(i) \cdot \vec{s}(j) - 1| > \epsilon$ for i,j belonging to some prescribed set of neighboring sites of cardinality $|C|$ is smaller than $\exp(-\beta a|C|/\epsilon)$ for some a > 0.

2) The absence of spontaneous symmetry breaking (the Mermin-Wagner theorem).[13]

3) In 2D any Ising ferromagnet with a translational invariant Gibbs measure contains at most one percolating H cluster.[14,15]   Here connectivity is defined the ordinary way and * (diagonally) connected sites are not included in the same cluster. The theorem is valid only in 2D since only there the translational invariance of the measure guarantees that if one percolating H cluster exists at all, the origin is surrounded by closed contours of arbitrarily large size on which s ∈ H, which prevent the opposite spin clusters from percolating.

4) Russo's theorem regarding the divergence of the mean cluster size.[16]   Consider clusters defined by some event E, say s ∈ H, and with ordinary connectivity.   Their boundaries are *-connected sites belonging to the complementary event Ē, s ∉ H. Russo's theorem states that if there is neither ordinary percolation of E nor * percolation of Ē, then the mean cluster size (of both E and Ē) diverges.   This is intuitively clear since the absence of percolation of either E or Ē implies that the origin is surrounded by ever increasing contours ordinarily connected of E or *-connected of Ē, preventing each other from percolating.

To prove that for β sufficiently large $\langle H \rangle$ diverges, we argue as follows: on an infinite lattice, by ergodicity, one can calculate the original expectation values from a single typical configuration by spatial averaging. To determine what properties a typical configuration must possess, we temporarily replace the original square lattice Z^2 by R^2 and consider smooth maps (Lipschitz continuous) from R^2 onto S^{N-1} (by (1) above).   By (2) such a map has to have the property that for any $\vec{u}$ in Eq. (2) either both hemispheres percolate or * percolate or do neither.   By (3) the only allowed possibilities are that all hemispheres (only) *-percolate or no hemisphere *-percolates.   We shall prove below that only the latter possibility can occur and thus, by (4), prove that under the assumption of perfect smoothness $\langle H \rangle$ diverges.

First let us consider O(2) for which we shall prove that the typical configuration must be such that if one defines the clusters by s ∈ A where A is any connected portion of the circle, no matter how large, smoothness implies that such clusters cannot percolate.   Let A be larger than a hemisphere, Ā its complement and ĀR the portion of the circle obtained from Ā by reflection through the center of the circle.   By (3) if A percolated, Ā would form islands.   By (2), so would ĀR.   By smoothness, the islands associated with Ā and ĀR cannot touch.   The complement of Ā∪ĀR also consists of two region, say C and CR, whose clusters, by smoothness, could not touch. Consequently if A percolated, either C or CR would also have to percolate, in violation of (2) QED.   By smoothness, if no portion of the circle percolates, it also cannot *-percolate, hence by (4), the mean size of clusters associated with the spins belonging to the smallest portion of the circle (of nonvanishing measure) diverges.

The discussion of the case O(N) N ≥ 3 is slightly more complicated. First let us notice that in such situations, *-percolation is a nongeneric event, hence it should not occur in the infinite volume limit with probability 1.   Indeed if it occurred, it would require the existence of an infinity of points where both $s_{||}$ and its gradient vanish (the first condition specifies the equator, the second one *-percolation of the two hemispheres).   Generically a function of two variables $s_{||}(x,y)$ cannot obey three constraints QED.   The absence of *-percolation of the hemispheres is equivalent to the absence of the ordinary percolation of the equator.   In deriving this result we employed only smoothness (1), thus it should be valid in a wider range of models than the O(N) models, which in addition enjoy (2).   We would like to rederive the same result by specifically employing (2) and thus obtain a stronger version, namely that for O(N) even a sufficiently slightly enlarged equatorial set could not possibly percolate without violating (2). For concreteness we consider O(3) and assume that the No-So hemispheres *-percolated, hence the associated equator ordinary percolated. By (2) the same conclusion must apply to any other division of the sphere. Consider the one obtained from the original one by performing an infinitesimal rotation around the E-W axis. By (1), one can easily draw the position of the new equator, given the old one, and discover that with probability 1 the points of *-percolation have disappeared QED. ˙ In fact even slightly enlarged regions around the equator, converging to it, are as unstable under rotations, hence the stronger statement. It would be an interesting mathematical problem to determine the width of the minimal equatorial set which shows ordinary percolation without violating (2). We would venture the following guess: at $\beta = 0$ one is dealing with independent site percolation for which it is known that the critical density is 0.58, hence the width of such an equatorial set is known. Since the interactions are ferromagnetic, it seems probable that if anything, such equatorial clusters would become more robust with increasing β (for N ≥ 3 the equator is a connected set). Thus we doubt that for O(N) N ≥ 3, clusters associated with the spins pointing in the smallest region of S^{N-1} have divergent mean size, as is the case for O(2). This difference stems only from fact that the target space of the map has dimension greater than 1 and would also arise were we to consider the case of two uncoupled O(2) models.

Finally let us discuss the effect of the simplifications introduced in our discussion, namely perfect smoothness and replacing Z^2 with R^2. Since for any O(N) our arguments show that the assumption of perfect smoothness implies that clusters associated with even less than a hemisphere have divergent mean size, there is obviously no difficulty in returning from R^2 to Z^2. Similarly relaxing the perfect smoothness constraint should pose no problem. Indeed whereas defects (points where $|\vec{s}(i) \cdot \vec{s}(j) - 1| > \epsilon$) have nonzero density and hence occur with probability 1, by (1) they are very rare at large β.[11] On a square lattice the number of contours of length $|C|$ attached to a given site is of order $\exp(d|C|)$ with d < 3. Hence when β is large it is very unlikely to find chains of defects cutting across H clusters. Therefore, even when defects are allowed, we expect ⟨H⟩ to diverge for β sufficiently large. That this result is not surprising can be seen by considering the same O(N) models on a triangular lattice (by smoothness (1), one expects the details of the lattice structure to be irrelevant at β large). On such a lattice, *-and ordinary percolation are identical. By (2) and (3) one sees that for any β, no H cluster can percolate, hence by (4), ⟨H⟩ must diverge. This conclusion can easily be tested at $\beta = 0$, where the H clusters correspond to clusters of occupied sites for independent site percolation with p = 1/2: it is known that for this process $p_c = 1/2$, hence ⟨H⟩ diverges at $\beta = 0$. Moreover ⟨H⟩ is clearly nondecreasing in β QED.

Next we must show that for β sufficiently large ⟨FK⟩ must diverge. The FK clusters are obtained from the H clusters by randomly diluting bonds with probability 1-p, p given in Eq. (3). We shall prove that for $\beta(\epsilon)$ sufficiently large(small) $\epsilon > 0$ the subclusters obtained from the FK clusters by considering only those sites where $s_{||}(i) > \epsilon$ must have divergent mean size. Indeed before dilution such clusters must have divergent mean size (by (4)), since neither they nor the clusters defined by the complementary condition percolate (shown above). If we treat the dilution probability 1-p as an independent variable, it is clear that except at isolated β values, the mean size of the clusters so defined is a continuous function of p for 1-p sufficiently small. (This follows from the fact that the translational invariance of the measure requires that if say the event "the origin is surrounded by a certain contour" has nonvanishing probability, then the cluster containing that circuit must have a certain amount of robustness or else the event would not have occurred in the first place; moreover the robustness of clusters, which may be marginal at some β, is nondecreasing in β. See Ref. 16 for a similar result.) Finally the observation that for such subclusters the dilution probability (1-p) → 0 as β → ∞ completes the proof.

For O(2) our arguments give an outline of an alternative derivation of the Fröhlich and Spencer result.[2] Our approach utilizes techniques and results derived from percolation theory, while theirs relies upon the representation of the spin model as a Coulomb gas. At sufficiently large β, they exhibit this gas as a convex combination of neutral molecules of sufficiently small activities, for which Debye screening cannot occur. We believe that an alternative derivation of our results for O(N) N $\geq$ 3 could be given following Fröhlich and Spencer's methods. The proof would be by contradiction: suppose a mass gap did exist at large β. Then the clusters associated with $|s_z| > 1$-ϵ would have finite average size. They would be diluted (1-ϵ small) and by some local central limit theorem, independently distributed. As pointed out previously,[17,18] one can relate the O(N) N $\geq$ 3 to the O(2) model. For example for O(3) in spherical coordinates one has

$$
Z = \left[\prod_{i \,\in\, \Lambda} \int \sin\theta_i \, d\theta_i \, d\phi_i \right] \exp\left[\sum_{\langle i,j \rangle} \beta\cos\theta_i \cos\theta_j + \beta\sin\theta_i \sin\theta_j \cos(\phi_i - \phi_j) \right] \quad . \tag{5}
$$

Thus, with regard to the $\{\phi\}$ variables, one has an O(2) ferromagnet with random couplings. Following Richard,[17] E. Seiler we can use Ginibre's inequality and bound expectation values of ϕ-dependent observables by the values obtained by setting $\beta = 0$ in all regions where $|s_z| > 1$-ϵ and equal to $\beta\epsilon(2$-$\epsilon)$ elsewhere. Applying the usual transformations, this O(2) model can be changed into a Coulomb gas in which all regions where the effective β is 0 become perfect conductors. It is easy to verify that all of Fröhlich and Spencer's estimates go through for such a system, in which the perfectly conducting regions do not percolate and have an exponentially decaying size distribution (the existence of such regions amounts to a renormalization of the dielectric constant). Thus, at sufficiently large β one would conclude that there is no exponential decay of the $\{\phi\}$ dependent observables, contrary to the original hypothesis.

Our percolation arguments apply equally well to other ferromagnetic O(N) interactions. We find two such interactions of interest and will discuss them. In both cases we start with the ordinary nearest neighbor interaction and restrict the allowed configurations as follows: a) Prod $(\vec{s}(i) \cdot \vec{s}(j)) > 0$ for all plaquettes (i,j nearest neighbors); b) $\vec{s}(i) \cdot \vec{s}(j) > C$ for any $\langle i,j \rangle \in \Lambda$. Both modifications leave unaltered the naive continuum limit and in particular do not destroy asymptotic freedom or the existence of instantons for O(3). Model (a) is inspired by the Mack and Petkova modification of the Yang-Mills Wilson action,[19] while model (b) was analyzed numerically by us in some previous work.[20] In model (a), let $\sigma(i,j) = \pm 1$ be the sign of $\vec{s}(i) \cdot \vec{s}(j)$ and introduce site Ising variables $\tau(i)$

$$
\tau(i) = \prod_{\langle i,j \rangle \,\in\, \Gamma} \sigma(i,j) \quad . \tag{6}
$$

Because of the constraint this definition is independent of the path Γ chosen to connect the reference site 0 to site i. Following Mack and Petkova one can now prove rigorously that the disorder 2-point function associated with these new Ising variables behaves differently at small and large β (l.r.o. vs exponential decay). Also using a result of Georgii[11] one can prove that the ordinary two-point function of these Ising variables changes from exponential decay at small β to l.r.o. at large β. In gauge theories Mack and Petkova argued correctly that their result had no direct implications on the existence or absence of confinement in the modified model at large β, but only that if confinement occurred, it had to be due to the existence of thick vorteces. In the O(N) spin models, confinement translates into exponential decay of Green's functions while vortices correspond to domain walls. Since model (a) is clearly more ordered than the unmodified model, our previous results mean that at large β it has algebraic decay. Therefore, thick domain walls do not manage to produce a mass gap and by analogy, we doubt that thick vortices lead to confinement in nonabelian gauge theories in 4D.

Model (b) is interesting in that it may shed some light as to the cause of the phase transition we claim takes place in any O(N) N $\geq$ 2. According to Kosterlitz and Thouless[1] the basic difference between O(2) and O(3) is explained entirely by considerations regarding the energy of dominant configurations. For O(2) these are pairs of bound vortices and their energy diverges logarithmically with their separation; on the contrary for O(3) instantons cost a finite amount of energy, hence, they are abundant at all β, disordering the system. We studied numerically model (b) for O(2) and found that for $C \geq -0.24$ the system is exhibiting algebraic decay even at $\beta = 0$. Obviously at $\beta = 0$, the energy of a configuration is irrelevant and this critical value of C is determined entirely by entropy considerations. So what is driving the transition in an O(N) ferromagnet? Probably, as argued by Patrascioiu,[18] at low temperature the dominant effect is short range order (spin waves) and vortex-pairs or instantons are suppressed by entropy considerations. In 2D this effect materializes as algebraic decay of correlations. As the temperature increases, the density of defects increases too, triggering at some β the transition. While the fact that $C_{crt} \neq 0$ gives credence to the notion that the possibility of having vortex-like configurations (point defects) is important for the existence of a phase with exponential decay, both our theoretical results and our finding $C_{crt} = -0.24 < 0$ indicate that the original Kosterlitz-Thouless scenario failed to appreciate correctly the entropy and thus reached the false conclusion that the O(3) model has no massless phase. Nevertheless, we believe topological considerations may have a certain importance; namely the relevance of the existence of point defects suggests that for the O(N) models of type (b) in D dimensions at $\beta = 0$, $C_{crt} < 0$ only for $N \leq D$.

Finally a word about perturbation theory. The existence of a line of critical points in the O(N) $N \geq 3$ models implies that their true Callan-Symanzik β-functions are vanishing. It is well known that in ordinary perturbation theory, one finds a β-function proportional to $-(N-2)$, reflecting the famous asymptotic freedom.[3,4] The conclusion therefore is that for these nonabelian models, perturbation theory fails to produce the correct asymptotic expansion in $1/\beta^n$, even though the answers are infrared finite order by order. A natural way to understand this failure is to recall the Mermin-Wagner theorem,[13] which guarantees that for $L \to \infty$ the fluctuations of the spin (from any given orientation) do not become arbitrarily small as $\beta \to \infty$, $\beta < \infty$. A similar result regarding the size of the fluctuations was proven for gauge theories by Patrascioiu.[6] Thus we expect ordinary perturbation theory to fail also in nonabelian gauge theories in 4D.

REFERENCES

1) J. M. Kosterlitz and D. J. Thouless, J. Phys. (Paris) 32:581 (1973).

2) J. Fröhlich and T. Spencer, Comm. Math. Phys. 81:527 (1981).

3) E. Brezin and J. Zinn-Justin, Phys. Rev. B14:3110 (1976).

4) A. M. Polyakov, Phys. Lett. 59B:79 (1975).

5) A. A. Belavin and A. M. Polyakov, Zh. Eksp. Teor. Fiz. 22:503 (1975).

6) A. Patrascioiu, Phys. Rev. Lett. 54:2292 (1985).

7) A. Patrascioiu, Phys. Rev. Lett. 56:1023 (1986).

8) A. Patrascioiu, "Employing the Ising representation to implement nonlocal Monte Carlo updating," Nuovo Cim. 105B:91 (1990).

9) C. M. Fortuin and P. W. Kasteleyn, J. Phys. Soc. JPN sppl 26:86 (1969).

10) J. Bricmont, J. R. Fontaine, J. L. Lebowitz, E. H. Lieb, and T. Spencer, Comm. Math. Phys. 78:545 (1981).

11) H. O. Georgii, Comm. Math. Phys. 81:455 (1981).

12) J. Bricmont and J. R. Fontaine, Comm. Math. Phys. 87:417 (1982).

13) N. D. Mermin and H. Wagner, Phys. Rev. Lett. 17:1133 (1966).

14) M. Miyamoto, Comm. Math. Phys. 41:103 (1975).

15) A. Coniglio, C. R. Nappi, F. Peruggi, and L. Russo, Comm. Math. Phys. 51:315 (1976).

16) L. Russo, Z. Wahrsch. verw. Gebiete 42:39 (1978).

17) J.-L. Richard, Phys. Lett. B184:75 (1987).

18) A. Patrascioiu, Phys. Rev. Lett. 58:2285 (1987).

19) G. Mack and V. B. Petkova, Ann. Phys. 123:442 (1979).

20) E. Seiler, I. O. Stamatescu, A. Patrascioiu, and V. Linke, Nucl. Phys. B305:623 (1988).

QUANTUM VACUUM, CONFINEMENT, AND ACCELERATION

Carlos Villarreal

Instituto de Física
Universidad Nacional Autónoma de México
Apdo. Postal 20-364, México 1000 D.F.

INTRODUCTION

Hawking radiaton by black holes [1] can be understood in terms of the Doppler distortion undergone by the modes of a quantum vacuum field in the presence of a gravitational source [2]. A very different phenomenum, the Casimir effect [3], arises from the distortion of the vacuum produced by boundaries. Both phenomena have a common origin: they are manifestations of the restrictions that event horizons or physical boundaries impose to a quantum vacuum field. Such restrictions may confer to the vacuum state a complex structure. The vacuum structure will be analysed here for inertial and non-inertial observers in vacua confined by boundaries. Gravitational effects will be analysed elsewhere in this book by A. Sarmiento.

ENERGY DENSITY SPECTRUM FOR THE CASIMIR EFFECT

A simple approach [4] to the spectral density of the Casimir effect can be performed in terms of two scalar Hertz potentials, representing the two independent degrees of freedom of the electromagnetic vacuum between parallel conducting plates . Suppose that the plates are placed at $x = (t, 0, y, z)$ and $x = (t, a, y, z)$. The Maxwell equations $F^{\mu\nu}, \nu = 0$, and $^*F^{\mu\nu}, \nu = 0$, subject to the boundary conditions that the transverse component of the electric field F^{0i} and the normal component of the magnetic field $F^{ij}(i, j = 2, 3)$ vanish at the surfaces of the plates, can be written in terms of the vector potential A^μ, which in the Lorentz gauge satisfies the wave equation $\Box A^\mu = 0$. The planar symmetry of the system permits to impose the additional constraints $\partial_2 A^2 + \partial_3 A^3 = 0$, and $\partial_o A^o + \partial_1 A^1 = 0$, which imply the existence of scalar Hertz potentials ϕ, ψ, such that $A_\mu = (\psi_{,1}, -\psi_{,0}, \phi_{,3}, -\phi_{,2})$. The Maxwell equations become $\Box\phi = 0$, $\Box\psi = 0$, where the potentials ϕ and ψ satisfy Dirichlet $\phi = 0$ and Neumann $\psi_{,1} = 0$ boundary conditions, respectively. The Dirichlet and Neumann fields can be quantised in terms of a superposition of modes

$$\phi_{\ell,k_\perp}(x) = N_{\phi,\ell,k_\perp} \exp(i\mathbf{k}_\perp \cdot \mathbf{x}) \exp(-i\omega t) \sin\left(\frac{\ell\pi}{a}x_1\right)$$

$$\psi_{\ell,k_\perp}(x) = N_{\psi,\ell,k_\perp} \exp(i\mathbf{k}_\perp \cdot \mathbf{x}) \exp(-i\omega t) \cos\left(\frac{\ell\pi}{a}x_1\right) \tag{1}$$

Vacuum Structure in Intense Fields, Edited by
H.M. Fried and B. Muller, Plenum Press, New York, 1991

where $\omega^2 = |\mathbf{k}_\perp|^2 + \left(\frac{\ell\pi}{a}\right)^2$, $\ell \in N$, $\mathbf{k}_\perp = (0, k_2, k_3) \in R^3$ and a is the distance between plates. With the normalization condition

$$|N_{\phi\ell,k_\perp}|^2 = |N_{\psi\ell,k_\perp}|^2 = [(2\pi)^2\omega a]^{-1} \qquad , \ell \neq 0$$

$$|N_{\psi 0,k_\perp}|^2 = [(2\pi)^2\omega a]^{-1} \qquad , \ell = 0,$$

the Hertz potentials become

$$\hat{\phi}(x) = \sum_{\ell=0}^\infty \int d^2\mathbf{k}_\perp [\phi_{\ell,k_\perp}(x)\hat{a}_\ell(\mathbf{k}_\perp) + \phi^*_{\ell,k_\perp}(x)\hat{a}^\dagger_\ell(\mathbf{k}_\perp)]$$

$$\hat{\psi}(x) = \sum_{\ell=0}^\infty \int d^2\mathbf{k}_\perp [\psi_{\ell,k_\perp}(x)\hat{b}_\ell(\mathbf{k}_\perp) + \psi^*_{\ell,k_\perp}(x)\hat{b}^\dagger_\ell(\mathbf{k}_\perp)] \tag{2}$$

with commutation relations $[\hat{a}_\ell(\mathbf{k}_\perp), \hat{a}^\dagger_{\ell'}(\mathbf{k}'_\perp)] = \delta(\mathbf{k}_\perp - \mathbf{k}'_\perp)\delta_{\ell\ell'}$ and $[\hat{b}_\ell(\mathbf{k}_\perp), \hat{b}^\dagger_{\ell'}(\mathbf{k}'_\perp)] = \delta(\mathbf{k}_\perp - \mathbf{k}'_\perp)\delta_{\ell\ell'}$. The Neumann and Dirichlet correlation functions

$$C^+(x, x') \equiv N(x, x') = \langle 0|\hat{\psi}(x)\hat{\psi}(x')|0\rangle \qquad C^-(x, x') \equiv D(x, x') = \langle 0|\hat{\phi}(x)\hat{\phi}(x')|0\rangle$$

can then be written as

$$C^\mp(x, x') = \sum_{\ell=0}^\infty \frac{1}{2} \int d^2\mathbf{k}_\perp |N_{\ell,k_\perp}| \times \exp\left[i\mathbf{k}_\perp \cdot (x - x') - i\omega(t - t'k)\right]$$

$$\times \left\{ \cos\left[\left(\frac{\ell\pi}{a}\right)(x_1 - x'_1)\right] \mp \cos\left[\left(\frac{\ell\pi}{a}\right)(x_1 + x'_1)\right] \right\}, \tag{3}$$

and using the formula

$$\sum_{\ell=-\infty}^\infty \delta\left(k_1 - \frac{\ell\pi}{a}\right) = \frac{a}{\pi} \sum_{\ell=-\infty}^\infty \cos(2k_1 a\ell) = \frac{a}{\pi} \sum_{\ell=-\infty}^\infty \exp(i2k_1 a\ell), \tag{4}$$

the field correlations are expressed in a form which is usually obtained by the image method: $C^\mp(x, x') = F^-(x, x') \mp F^+(x, x')$, where

$$F^\mp(x, x') \equiv -\frac{1}{4\pi^2} \sum_{\ell=-\infty}^\infty \frac{1}{(x_1 \mp x'_1 - 2a\ell)^2 + (x_2 - x'_2)^2 + (x_3 - x'_3)^2 - (t - t')^2}. \tag{5}$$

Note that $N+D = 2F^-$ and $N-D = 2F^+$. The expressions for the vector potential as well as those of the electric and magnetic fields in terms of the Hertz potentials give the field correlations directly [4], and the energy momentum tensor can be calculated with the help of F^-:

$$T_{\mu\nu}(x) = \lim_{x \to x'} T_{\mu\nu}(x, x') = \lim_{x \to x'} 2\partial_\mu\partial_\nu F^-(x, x'). \tag{6}$$

Using these tools, we can now evaluate the spectral density of the electromagnetic field as measured by an observer at rest with respect to the plates. For such purpose, choose a fixed point at a distance x from one of the plates, and take the limit $x \to x'$ and $t - t' = \sigma$. The field correlations are: $\langle E_i B_j \rangle = 0$ and

$$\frac{1}{2}(\langle E_i E_j \rangle + \langle B_i B_j \rangle) = \frac{1}{\pi^2} \sum_{\ell=-\infty}^\infty \frac{[\sigma^2 + (2a\ell)^2]\delta_{ij} - 2(2a\ell)^2 n_i n_j}{[(\sigma - i\epsilon)^2 - (2a\ell)^2]^3} \tag{7a}$$

$$\frac{1}{2}(\langle E_i E_j \rangle - \langle B_i B_j \rangle) = \frac{1}{\pi^2} \sum_{\ell=-\infty}^{\infty} \frac{-[\sigma^2 + 4(x - a\ell)^2]\delta_{ij} - 2\sigma^2 n_i n_j}{[(\sigma - i\epsilon)^2 - 4(x - a\ell)^2]^3}. \tag{7b}$$

The corresponding spectral density is obtained by performing the Fourier transforms over σ. The final result for the energy momentum tensor is

$$\tilde{T}_{\mu\nu}(\omega) = \frac{1}{6\pi^2}\omega^3(\eta_{\mu\nu} + 4t_\mu t_\nu) + \frac{1}{8\pi^2 a^3}\{(2a\omega)^2 f(2a\omega)(-\eta_{\mu\nu} + 4\eta_\mu\eta_\nu)$$
$$+ [g(2a\omega) - 2a\omega g'(2a\omega) + (2a\omega)^2 f(2a\omega)](\eta_{\mu\nu} + t_\mu t_\nu - 3\eta_\mu\eta_\nu)\}, \tag{8}$$

where $f(x) = (\pi - x)/2$ and $g(x) = (\pi^2 x)/6 - (\pi x^2)/4 + (x^3)/12$. In particular, the energy density is

$$\tilde{T}_{00}(\omega) = \frac{1}{2}(\langle \widetilde{E_i E_j} \rangle + \langle \widetilde{B_i B_j} \rangle) = \frac{\omega^3}{6}\delta_{ij} + \frac{\omega^2}{8\pi^2 a}\left[f(2a\omega)(\delta_{ij} + n_i n_j)\right.$$
$$\left. + \left(\frac{g(2a\omega)}{2(a\omega)^2} - \frac{g'(2a\omega)}{(a\omega)} - f(2a\omega)\right)(-\delta_{ij} + 3n_i n_j)\right]. \tag{9}$$

Expression (9), when integrated over ω lead to the usual Casimir energy-momentum tensor, plus zero-point term.

The graph of the function $(2\pi^2\omega^{-2})\tilde{T}_{00}$ has a sawtooth form [5]. This simple form holds only for this case. The spectral density associated to more complicated geometries, or measured in non-inertial reference frames is quite complex [7].

The image method with imaginary times allows to obtain also the energy density spectrum at finite temperatures. The corresponding Dirichlet and Neumann functions are

$$\left.\begin{array}{c} N_\beta \\ D_\beta \end{array}\right\} = \frac{1}{4\pi^2} \sum_{\substack{\ell,m=-\infty \\ l=-\infty}}^{\infty} \left(\frac{1}{(x_1 - x_1' - 2a\ell)^2 + (x_2 - x_2')^2 + (x_3 - x_3')^2 - (t - t' - im\beta)^2} \right.$$
$$\left. \mp \frac{1}{(x_1 + x_1' - 2a\ell)^2 + (x_2 - x_2')^2 + (x_3 - x_3')^2 - (t - t' - im\beta)^2} \right), \tag{10}$$

where $\beta = (kT)^{-1}$. The finite temperature correlations thus obtained are, in obvious notation

$$\langle \widetilde{E_i B_j} \rangle_\beta = 0 \tag{11}$$

$$\left.\begin{array}{c} \langle \widetilde{E_i E_j} \rangle_\beta \\ \langle \widetilde{B_i B_j} \rangle_\beta \end{array}\right\} = \left(1 + \frac{2}{e^{\beta\omega} - 1}\right)\left\{ \begin{array}{c} \langle E_i E_j \rangle \\ \langle B_i B_j \rangle \end{array}\right. \tag{12}$$

$$\tilde{T}_{\mu\nu}^\beta(\omega) = \left(1 + \frac{2}{e^{\beta\omega} - 1}\right)\tilde{T}_{\mu\nu}(\omega) \tag{13}$$

It is important to note that the zero-point, plus the Planckian terms, are factorised out in every expression. Thus, an observer inside a cavity sees a quantum vacuum distorted by the presence of the plates, and modulated by the thermal equilibrium distribution.

The above formalism can be used to study the interaction of the confined vacuum field with matter. This coupling produces effects which differ from those predicted by quantum electrodynamics in free space, *v.gr.*, the inhibition or enhacement of spontaneous emission in the presence of mirrors [6]. First order perturbation theory permits the evaluation of the transition probability per unit time for the emission of a photon pertaining to the mode $(\ell, k_\perp)^N$ or $(\ell, k_\perp)^D$. In the dipolar approximation, known results are recovered when the electric dipole moment $\boldsymbol{\mu}_{if}$ of an atom is perpendicular or parallel to the plates [4]. The total transition probabilities are given respectively by

$$w^{\parallel} = \frac{3\pi A}{\omega_o a} \left\{ \frac{1}{2} + \sum_{\ell=1}^{[\omega_o a/\pi]} \left(1 - \frac{\ell^2 \pi^2}{\omega_o^2 a^2}\right) \sin^2\left(\frac{\ell\pi}{a}x\right) \right\} \qquad (15a)$$

and

$$w^{\perp} = \frac{3\pi A}{\omega_o a} \sum_{\ell=1}^{[\omega_o a/\pi]} \left(1 + \frac{\ell^2 \pi^2}{\omega_o^2 a^2}\right) \cos^2\left(\frac{\ell\pi}{a}x\right), \qquad (15b)$$

where x is the distance from the radiating systems to the plates, and A is the Einstein coefficient. It turns out that the minimum transition frequency of the system is π/a [6].

ACCELERATED OBSERVER INSIDE A CAVITY

A nice example involving pseudothermal and confinement effects is the spectral density of a Dirichlet scalar field as measured by a uniformly accelerated observer inside an infinite prismatic cavity.

Consider first the case of the observer at rest. The correlation functions of the cavity with boundaries at $x_2 = 0, b$ and $x_3 = 0, a$ can be directly found by the image method

$$D^{\pm}(x, x') = \frac{1}{4\pi^2} \sum_{n=-\infty}^{\infty} \sum_{m=-\infty}^{\infty} \left(\frac{1}{(x_1 - x_1')^2 + (x_2 - x_2' - 2bm)^2 + (x_3 - x_3' - 2an)^2 - (t - t' \mp i\epsilon)^2} \right.$$

$$- \frac{1}{(x_1 - x_1')^2 + (x_2 + x_2' - 2bm)^2 + (x_3 - x_3' - 2an)^2 - (t - t' \mp i\epsilon)^2}$$

$$- \frac{1}{(x_1 - x_1')^2 + (x_2 + x_2' - 2bm)^2 + (x_3 + x_3' - 2an)^2 - (t - t' \mp i\epsilon)^2}$$

$$\left. + \frac{1}{(x_1 - x_1')^2 + (x_2 + x_2' - 2bm)^2 + (x_3 + x_3' - 2an)^2 - (t - t' \mp i\epsilon)^2} \right) \qquad (16)$$

By evaluating these functions for an observer at the points x and x' such that $x = (x_0 = \tau - \sigma/2, x_1, x_2 = y, x_3 = z)$ and $x' = (x_0 = \tau + \sigma/2, x_1, x_2 = y, x_3 = z)$, the correlations take the form

$$D^{\pm}(\sigma) = -\frac{1}{4\pi^2} \sum_{n=-\infty}^{\infty} \sum_{m=-\infty}^{\infty} \left(\frac{1}{(\sigma \mp i\epsilon)^2 - 4a^2 n^2 - 4b^2 m^2} - \frac{1}{(\sigma \mp i\epsilon)^2 - 4a^2 n^2 - 4(y - bm)^2} \right.$$

$$\left. - \frac{1}{(\sigma \mp i\epsilon)^2 - 4b^2 m^2 - 4(z - an)^2} + \frac{1}{(\sigma \mp i\epsilon)^2 - 4(y - bm)^2 - 4(z - an)^2} \right), \qquad (17)$$

the last three terms do not contribute to the energy density when integrated over y and z. By Fourier transforming of the remaining term one gets for the energy density per unit length:

$$\frac{dE}{d\omega} = \frac{\omega^3}{2\pi^2} ab \left(1 + \frac{1}{\omega a} f(2\omega a) + \frac{1}{\omega b} f(2\omega b) + \frac{2}{\omega} \sum_{n=1}^{\infty} \sum_{m=1}^{\infty} \frac{\sin[2\omega(a^2 n^2 + b^2 m^2)^{\frac{1}{2}}]}{(a^2 n^2 + b^2 m^2)^{\frac{1}{2}}} \right) \qquad (18)$$

with $f(x)$ the same function as before. The first term in Eq. (18) represents the zero-point field energy. The second term corresponds to the case of an observer at rest between two plates separated by a distance a apart. Indeed, dividing (18) by b and taking the limit $b \to \infty$ but keeping a finite, the usual Casimir spectrum is obtained. The third term has an identical interpretation for an observer between plates a distance b apart. Finally, the last term is an interference term due to the presence of plates in both directions. The form of the spectral density in terms of the parameter $r = a/b$, appears in Ref. [7]. As expected, when $r = 10^4$, the graph shows the the sawtooth spectrum of two parallel plates. The case $r = 10$, shows a superposition of the spectrum of each of the plates. For $r = 1$, the behaviour is very complicated, showing some kind of vacuum turbulence; however a careful investigation of this behaviour must be performed.

For the case of an accelerated observer, the equations defining his world-line: $t = \alpha^{-1}\text{sh}[\alpha(\tau \pm \sigma/2)]$, $x = \alpha^{-1}\text{ch}[\alpha(\tau \pm \sigma/2)]$, must be substituted in the correlation functions of the vacuum field with boundaries. These functions then reduce to

$$
D^{\pm}(x, x') = \frac{1}{16\pi^2} \sum_{n=-\infty}^{\infty} \sum_{m=-\infty}^{\infty}
$$

$$
\left(\frac{1}{a^2 n^2 + b^2 m^2 - \alpha^{-2}\text{sh}^2(\alpha\sigma/2 \mp i\epsilon)} - \frac{1}{a^2 n^2 + (y - bm)^2 - \alpha^{-2}\text{sh}^2(\alpha\sigma/2 \mp i\epsilon)} \right.
$$

$$
\left. - \frac{1}{(z - an)^2 + b^2 m^2 - \alpha^{-2}\text{sh}^2(\alpha\sigma/2 \mp i\epsilon)} + \frac{1}{(z - an)^2 + (y - bm)^2 - \alpha^{-2}\text{sh}^2(\alpha\sigma/2 \mp i\epsilon)} \right), \quad (19)
$$

the last three terms do not contribute to the energy density as before; the Fourier transform of the surviving term is

$$
\tilde{D}^{\pm}(\omega) = \pm \frac{2}{\pi} \frac{1}{1 - e^{\mp 2\pi\omega/\alpha}} \sum_{n=-\infty}^{\infty} \sum_{m=-\infty}^{\infty} \frac{\sin[2\omega\alpha^{-1}\text{arcsinh}(\alpha\rho)]}{\rho[1 + (\alpha\rho)^2]^{\frac{1}{2}}}, \quad (20)
$$

with $\rho = (a^2 n^2 + b^2 m^2)^{\frac{1}{2}}$. By evaluating this integral, the energy density per unit length is obtained:

$$
\frac{dE}{d\omega} = \frac{\omega^2}{\pi^2} \left(\frac{1}{2} + \frac{1}{e^{2\pi\omega/\alpha} - 1} \right) \left(ab + \frac{b}{\omega} \sum_{n=1}^{\infty} \frac{\sin[2\omega\alpha^{-1}\text{arcsinh}(\alpha an)]}{n[1 + (\alpha an)^2]^{\frac{1}{2}}} \right.
$$

$$
\left. + \frac{a}{\omega} \sum_{m=1}^{\infty} \frac{\sin[2\omega\alpha^{-1}\text{arcsinh}(\alpha bm)]}{m[1 + (\alpha bm)^2]^{\frac{1}{2}}} + \frac{2ab}{\omega} \sum_{n=1}^{\infty} \sum_{m=1}^{\infty} \frac{\sin[2\omega\alpha^{-1}\text{arcsinh}(\alpha\rho)]}{\rho[1 + (\alpha\rho)^2]^{\frac{1}{2}}} \right). \quad (21)
$$

The first bracket can be identified with a pseudothermal spectrum due to the observer's acceleration with an effective temperature given by T_α. The interpretation of the following terms is very similar to the inertial case: two expressions associated to a single pair of plates and an interference term. The turbulent behaviour of the spectrum is even more complicated than in the inertial case.

EFFECT OF GLUON CONFINEMENT ON HADRON PRODUCTION

A very simple model involving confinement and acceleration in electron-positron anihilation with hadron production can be posed in terms of the above formalism. In such reaction, a virtual photon decays into a quark-antiquark pair. The quark and antiquark move in opposite directions with large momenta, remaining united by the action of the gluonic string. Assuming a constant tension in the string, the quarks will move with a constant deceleration within an approximately cylindric hadronic bag, with a diameter a of the order of one fermi. From the slope of the Regge trajectories one derives the string tension and, assuming a mass of 300 MeV for non-strange constituent quarks, an acceleration temperature $T_\alpha \sim 100 MeV$ is obtained [8]. This is consistent with empirical 'temperatures' found in experiments. This value implies an interference parameter $\alpha a/2 \sim 5/3$. The excited decelerating quarks will emit gluons with a frequency corresponding to the energy difference between excited levels, and in proportion to the energy density of the vacuum. This modified gluon phase space would multiply the $e^+e^- \to q\bar{q}g$ hadrons matrix element. According to Ref.[12], a Planckian distribution is expected for soft gluons. However, an oscillatory behaviour similar to the one found in the spectrum of prismatic cavities might exist for hard gluons. This effect seems to be present in the e^+e^- anihilation data at $29\ GeV$. For three-jets events, relatively large values of the quadratic transverse momenta in the reaction plane are expected, while low values of this quantity are expected in the orthogonal direction. It turns out that the $<p_T^2>_{out}$ distribution may be fitted by $40 \exp(-p_T/T_\alpha)$. On the other hand, the $<p_T^2>_{in}$ distribution seems to show small oscillations, which could arise from the gluon mode distribution. Their influence may be estimated by taking into account the contribution to the $<p_T^2>_{in}$ distribution of the particles generated by hard gluons. The resulting difference between energy levels is 1.3 GeV. This is consistent with the estimation obtained (1.2 - 2.1 GeV) for the accelerated observer [12] in a prismatic cavity with transversal sides of one fermi.

REFERENCES

1. S.W. Hawking, *Nature* **248** (1974) 30
2. S. Hacyan, A. Sarmiento, G. Cocho, and F. Soto, *Phys. Rev.* **D32** (1985) 3216
3. H.B.G. Casimir, *Proc. Kon. Ned. Akad. Wet.* **51** (1948) 793
    G.Plunien, B. Muller, W. Greiner, *Phys. Rep.* **134** (1986) 87
4. S. Hacyan, R. Jáuregui, F. Soto, and C. Villarreal, *J. of Physics* **23** (1990) 2401
5. L.H. Ford, *Phys. Rev.* **4D38** (1988) 528.
6. S. Haroche and D. Kleppner, *Phys. Today* **42** (1989) 24.
7. A. Sarmiento, S. Hacyan, G. Cocho, F. Soto, and C. Villarreal, *Phys. Lett.* **A142** (1989) 194.
8. B.J. Harrington and Ch.H. Tabb, *Phys. Lett.* **496B** (1980) 362.
9. J. Bartke *et al., Nucl. Phys.* **B117** (1976) 293; **B118** (1977) 14; **B118** (1977) 360; H. Boggild *et al., Nucl. Phys.* **B57** 1(1973) 77; I.V. Ajinenko *et al., Nucl. Phys.* **B165** (1980) 1; T.F. Hoang, C.K. Chew and K.K. Phua, *Phys. Rev.* **D19** (1979) 1468. *Phys. Lett.* **A142** (1989) 194.
10. D. Bender *et al., Phys. Rev.* **D31** 1(1985) 1.
11. EHS-RCBC Collaboration, *Phys. Lett.* **B206** (1988) 371.
12. G. Cocho, S. Hacyan, F. Soto, C. Villarreal, and A. Sarmiento, Preprint IFUNAM (1990).

The (First) Three B's of the Skyrme Model

Alec J. Schramm

Department of Physics, Duke University
Durham, NC 27706, U.S.A.

Dedicated to the memory of Yossef Dothan, my teacher and friend.

It is widely believed that Quantum Chromodynamics is the correct underlying theory of the strong interaction. Such belief exists despite our rather limited ability to investigate the long wavelength limit, where the poorly understood effects of confinement set in. In fact, the very structure of the QCD vacuum remains a mystery, and it is hoped that a better understanding of this vacuum will shed considerable light on the nature of the confinement mechanism.

Of the many possible properties of the vacuum, some may be topological in origin. Indeed, topological effects in QCD, such as instantons and the existence of θ-vacua, are known to have important physical ramifications. At large distance, however, it is very difficult to calculate instanton effects, so their influence on the QCD vacuum is unknown. At these large distances, of course, the strong interaction reduces to nuclear physics, so at low energy QCD should be equivalent to a theory of pions and nucleons. The discovery of topological effects in such a theory may help in the search for the correct QCD vacuum state. With this in mind, we discuss the Skyrme model, which explores and exploits the topological nature of a pion theory of spontaneously broken chiral symmetry—and thus serves as an example of topological influences in low energy strong interactions.

Although quark mass terms in the Lagrangian of QCD explicitly break chiral symmetry, since the two lightest quarks, u and d, are in fact *very* light ($m_u = m_d \simeq 7$ MeV), the massless quark Lagrangian

$$\mathcal{L} = i\bar{q}\gamma^\mu D_\mu q = i\bar{q}_L\gamma^\mu D_\mu q_L + i\bar{q}_R\gamma^\mu D_\mu q_R \tag{1}$$

may not be a bad approximation to two flavor strong interaction physics. In eq. (1), q is a flavor doublet, D_μ is the usual QCD covariant derivative, and $q_{L,R} = \frac{1}{2}(1\pm\gamma_5)q$ are the left- and right-handed components of q. This Lagrangian possesses a classical

$$U(2)_L \times U(2)_R \simeq SU(2)_L \times SU(2)_R \times U(1)_L \times U(1)_R \tag{2}$$

chiral symmetry. Were this symmetry the end of the story, one would expect to discover parity doubling in the physical spectrum, corresponding to the identical symmetries enjoyed by both the L and R components of q. However, the only symmetry observed in the hadron spectrum is the $SU(2)$ of isospin; chiral $SU(2)_L \times SU(2)_R$ is therefore assumed to be spontaneously broken,

$$SU(2)_L \times SU(2)_R \to SU(2)_I. \tag{3}$$

The Goldstone theorem then requires the emergence of three massless bosons in the spectrum; these Goldstone bosons are identified with the pion triplet. Although not massless, the pion is the lightest hadron in the particle spectrum, so to first approximation can be treated as massless.

The three isospin degrees of freedom of the pion field can be embodied by an $SU(2)$-valued field $U(x)$. Any parametrization of $SU(2)$ will involve three independent parameters which can be related to the π's in a representation-dependent way. However, for fields $U(x)$ close to the unit matrix (the vacuum), one has the unique expansion

$$U(x) = 1 + \frac{i}{f}\vec{\pi}(x) \cdot \vec{\tau} + \dots ; \tag{4}$$

the pions are thus seen as fluctuations around the vacuum.

The field $U(x)$ transforms under chiral $U(2)_L \times U(2)_R$ as

$$U \to LUR^\dagger, \tag{5}$$

where $(L, R) \in U(2)_L \times U(2)_R$. Now the $U(1)_A$ anomaly breaks the classical symmetry of eq. (2) down to $SU(2)_L \times SU(2)_R \times U(1)_V$; moreover, under the $U(1)_V$ subgroup the transformation matrices R and L are equal, thus leaving U invariant. (This is expected since the $U(1)_V$ subgroup is generated by baryon number, and the pions do not carry baryon number.) We can therefore restrict the discussion to the group $SU(2)_L \times SU(2)_R$.

In order to determine low energy phenomena, we need a Lagrangian describing vacuum fluctuations, i.e., the pion fields. A low energy effective theory can be developed by systematically expanding $\mathcal{L}$ in increasing number of derivatives. The simplest possibility would be an $\mathcal{L}$ with no derivatives, but this would result in a local $SU(2)_L \times SU(2)_R$ symmetry, allowing one to transform any such term in $\mathcal{L}$ to an irrelevant constant. The next simplest possibility has two derivatives, and it turns out that all such terms are equivalent to

$$\mathcal{L}_2 = \frac{f^2}{4}\mathrm{Tr}[\partial_\mu U \partial^\mu U^\dagger]. \tag{6}$$

$\mathcal{L}_2$ is called the "minimal chiral Lagrangian"; inserting the expansion (4) into (6) recovers the conventional form $\frac{1}{2}(\partial_\mu \vec{\pi})^2$, leading to the identification $f = \frac{1}{2}F_\pi$ (where $F_\pi = 186$ MeV is the pion decay constant).

Before proceeding with the systematic expansion of the Lagrangian, note that there is a physical constraint on any realistic field configuration: $U(x)$ must have finite energy. For $\mathcal{L}_2$, the energy is

$$E[U] = \frac{F_\pi^2}{16}\int d^3x \; \mathrm{Tr}[\partial_0 U \partial_0 U^\dagger + \partial_i U \partial_i U^\dagger] ; \tag{7}$$

finite energy requires $\partial_\mu U \to 0$ as $|\vec{x}| \to \infty$, i.e.,

$$\lim_{|\vec{x}|\to\infty} U(\vec{x}) = U_0 = \text{constant} \tag{8}$$

for fixed t. So all physically admissible configurations $U(\vec{x})$ must map the point at infinity to the same point U_0 in the $SU(2)$ manifold. The physical space (the domain of the maps U) is thus *compactified* to the 3-sphere, $\mathbb{R}^3 + \{\infty\} = S^3$. However, $SU(2)$ (the range) is topologically also the three sphere $SU(2) = S^3$, so the fields U map S^3 to S^3. Such maps fall into topologically distinct classes, distinguished by the number of times the internal space $SU(2)$ is covered as the physical space is swept through once; this integer is often called the winding number. Static configurations have the same winding number if they can be continuously deformed into one another. Since Hamiltonian time development of a given initial configuration is continuous, the winding number is a constant of the motion. Explicitly, one can show that there exists a conserved current

$$B^\mu[U] = \frac{1}{24\pi^2} \varepsilon^{\mu\nu\alpha\beta} \, \mathrm{Tr}[U^\dagger \partial_\nu U U^\dagger \partial_\alpha U U^\dagger \partial_\beta U]. \tag{9}$$

Conservation, $\partial_\mu B^\mu = 0$, follows without resorting to the equation of motion; in other words, this conservation law is topological in nature and not the result of Noether's theorem. Indeed, the winding number

$$B[U] = -\frac{1}{24\pi^2} \varepsilon^{ijk} \, \mathrm{Tr}[U^\dagger \partial_i U U^\dagger \partial_j U U^\dagger \partial_k U] \tag{10}$$

is found from the Jacobian of the transformation $\mathbb{R}^3 \to \mathbb{R}^3 + \{\infty\} = S^3$. Configurations with non-trivial topology often describe particle-like solutions called *solitons*, having very different properties than the mesons defining the theory. To be completely stable, however, the mass of a soliton U must be small enough to prevent decay into many solitons of total winding number $B[U]$. Thus we need to look for minimum energy configurations; we will do this for the first three B's ($B = 0, 1, 2$).

I. B=0

Due to the derivatives in expression (10) for B, the minimum energy configuration is just a constant $SU(2)$ matrix $U(\vec{x}) = U_0$, trivially satisfying the boundary condition at infinity. There is certainly nothing "topological" about this configuration: solitons do not exist in the $B = 0$ sector.

II. B=1

The minimum chiral model does not have a stable, static solution with $B > 0$. This can be shown using a simple scaling argument, which shows that the energy $E[U]$ is unbounded from below. Therefore, if we want to explore the $B > 1$ sector, we need to add to the minimal Lagrangian $\mathcal{L}_2$; thus we return to the systematic expansion in powers of momentum.

Of the possible four derivative terms, Skyrme chose[1]

$$\mathcal{L}_4 = \frac{1}{32e^2} \, \mathrm{Tr} \, \left[U^\dagger \partial_\mu U, \, U^\dagger \partial_\nu U \right]^2, \tag{11}$$

where e is some constant. The scaling argument can then be used to show that with the addition of the "Skyrme term" (11), the Lagrangian $\mathcal{L} = \mathcal{L}_2 + \mathcal{L}_4$ yields stable configurations for $B \neq 0$; specifically,

$$E[U] > 3\pi^2 (F_\pi/e) \, B[U]. \tag{12}$$

By proceeding with the derivative expansion of the Lagrangian, we have found stable, topologically non-trivial solutions.

In order to find the configuration of least energy, we examine the symmetry remaining in the $B = 1$ sector—expecting that the minimum energy configuration will possess the maximum symmetry. For $B \neq 0$, eq. (10) shows that U is not invariant under translations; nor is it invariant under spatial or isospin rotations. However, due to the fact that ours is a two flavor model, we can associate each of the three directions of space with the three directions in the isospin space $SU(2)$. So even though U is not invariant under separate spatial or isospin rotations, it is invariant for the special case in which an isorotation—given by a transformation matrix A—is exactly compensated by the spatial rotation $R(A)$:

$$U(\vec{x}) \longrightarrow AU[R^{-1}(A) \cdot \vec{x}]A^{\dagger} = U(\vec{x}). \tag{13}$$

Such a generalized symmetry is called *equivariance*. This seems to be the largest symmetry group available in the $B \neq 0$ sectors, so we expect the minimum energy $B = 1$ solution to be equivariant.

Skyrme used the manifestly equivariant "hedgehog" ansatz (Fig. 1)

$$U_N(\vec{x}) = \exp[iF(r)\hat{r} \cdot \vec{\tau}]. \tag{14}$$

The function $F(r)$ is determined by minimizing the energy functional (which gives the mass of the soliton); with boundary conditions $F(0) = N\pi$, $F(\infty) = 0$, a straightforward calculation shows that the hedgehog has $B = N$. For $N = 1$, the hedgehog configuration is what is commonly known as a "skyrmion"; Skyrme identified this solution as the nucleon, and the winding number B as baryon number.

Of course, this identification is not without its problems. For example, is it possible that a theory of mesons, i.e., bosons, gives rise to fermionic solutions? It has been known for some time[2] that the skyrmion can be quantized either as a boson or a fermion—but this result relies on having only two flavors, and does not generalize

Figure 1. The Hedgehog

to include strangeness. The solution was provided by Witten,[3] who found the remarkable result that for more than two flavors, the skyrmion is a fermion only if the number of colors is odd—just as baryons are defined in QCD! Another longstanding problem involved proving the assertion that the B of eq. (10) is in fact the baryon number carried by quarks. This puzzle was solved by Balachandran et al.,[4] who showed that the chiral field U effectively polarizes the Dirac sea of quarks—with the result that the baryon number of the single soliton state is exactly given by eq. (10). As the details of these and other questions have been given in many fine review articles, we will content ourselves here by citing a few of them.[5]

The static properties of the $B = 1$ skyrmion were calculated by Adkins et al.[6] Upon quantization, equivariance imposes the constraint $I = J$; thus solutions exist for $I = J = \frac{1}{2}$ (nucleon) and $I = J = \frac{3}{2}$ (delta). In their work, the masses of the nucleon and delta were taken as input while the values of F_π and e were varied to achieve the best fit to experiment. They found that in general, theory and experiment agree to within 30%—encouraging for a model with only two free parameters.

A further interpretation of this error has been given by Witten.[7] He notes that since baryons emerge as solitons, the Lagrangian expansion claims no predictive power for mesons; rather, it maintains that baryons and their properties *can be completely determined from meson physics*. Thus the 30% disparity quoted above may simply be reflecting the crudeness of a model which accounts for only the lightest of mesons, the pions. Indeed, by including the rho and omega mesons in $\mathcal{L}$, some improvement is found for the static properties of the nucleon.[8]

III. B=2

Encouraged by these results for the $B = 1$ soliton, it is natural to expect the $B = 2$ sector to contain the next simplest nucleus: the deuteron. The easiest approach may seem to be the hedgehog configuration with boundary condition $F(0) = 2\pi$. However the mass of this soliton is roughly three times that of the $B = 1$ skyrmion, and is therefore unstable against decay into two nucleons. Another approach involves expanding to a 3 flavor model (the Goldstone bosons being the pseudoscalar octet) and injecting the hedgehog into the $SU(2)$ isospin subgroup,

$$U(\vec{x}) = \begin{pmatrix} e^{iF(r)\hat{r}\cdot\vec{\tau}} & 0 \\ 0 & 1 \end{pmatrix}. \tag{15}$$

This, however, leads to the same $B = 1$ and unstable $B = 2$ solutions already found. Alternatively the hedgehog can be embedded into the $SO(3)$ subgroup of $SU(3)$;[9] this gives a stable dibaryon ($J = 0$, $S = 2$), not the deuteron.

The deuteron, however, has $I = 0$, $J = 1$ and so cannot emerge from an equivariant configuration. But without equivariance there is little symmetry left as a guide in the search for minimum energy solutions. It may therefore be more fruitful to model the deuteron realistically, i.e., as a weakly bound state of two nucleons. In fact, Skyrme himself introduced the "product ansatz"

$$U_2 \equiv U_1(\vec{r} - \tfrac{1}{2}R\hat{z}) \, A U_1(\vec{r} + \tfrac{1}{2}R\hat{z}) \, A^\dagger \tag{16}$$

which describes two $B = 1$ skyrmions, separated by a distance R, with relative isospin orientation A. It is not difficult to show that this product does, in fact, have $B = 2$. Jackson et al.[10] used eq. (16) to extract a nucleon-nucleon potential

$$V(R, A) = E(R, A) - 2M_{B=1}. \tag{17}$$

This approach seems reasonable at large separations, where one skyrmion will not be greatly affected by the weak asymptotic field of the other. In fact for large R, the potential can be interpreted in terms of single pion exchange between baryons.

An interpretation for small separations, however, is more difficult. With large overlap between the two skyrmions, they fuse into a single configuration which is not easily resolved into individual skyrmions. (Such distortion is not surprising since the interaction energy may be comparable to the skyrmion mass.) Furthermore, the product ansatz is not the minimum energy configuration in the $B = 2$ sector, so the deuteron is not expected to be of this form. On the other hand the potential does become attractive for $A = i\tau_2$, i.e., when one skyrmion is rotated by π about a line perpendicular to the common axis of the skyrmions. Using this result, Braaten and Carson[11] were able to extract the deuteron quantum numbers from the product ansatz. Thus eq. (16) with $A = i\tau_2$ seems a good starting point in a numerical search for the correct deuteron configuration.

Many numerical investigations have been conducted.[12] At Duke,[13] we performed a full three dimensional calculation in which two skyrmions were placed side-by-side on a 3 dimensional cubic lattice; one of these skyrmions was given a π isorotation to ensure an attractive potential. We then employed a numerical relaxation technique to find the minimum energy configuration; the potential energy curve which emerged is shown in Fig. 2. The well depth, range, mass, and quadrupole moment of the configuration were found to be in reasonable agreement with deuteron properties. Moreover, the results were found to improve when a pion mass term

$$\mathcal{L}_\pi = \frac{1}{8} m_\pi^2 \, F_\pi^2 [\mathrm{Tr} U - 2] \tag{18}$$

is included in the Lagrangian—which is consistent with similar improvement found in the $B = 1$ sector.[14] Perhaps most notable, however, is that the relaxed configuration no longer appeared to be a product of chiral fields.

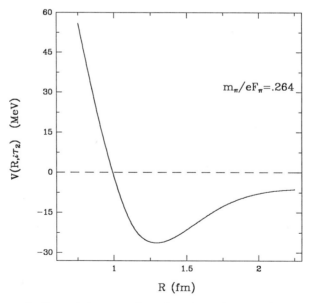

Figure 2. Potential energy for relaxed product configuration;
the values of F_π and e are taken from reference 14

Exactly what the relaxed configuration looked like, though, required more careful numerical analyses.[15] The surprising result is that the baryon number distribution of the resulting minimum energy configuration has a toroidal structure. Although this seems contrary to the usual picture of a deuteron, the calculated static properties for the toroidal configuration agree reasonably well with known deuteron properties.[16]

IV. B>2

The discovery of the unexpected toroidal structure of the deuteron solution has led to a rather unconventional approach to nuclear physics in the Skyrme model. In the deuteron search, guidance provided by either symmetry (equivariance) or intuition (the product ansatz) proved less than adequate—and these "guides" become even more difficult to use for more complex nuclei. With this in mind, Braaten and collaborators[17] have resorted to a more direct approach. They use neither an inter-nucleon potential nor the picture of nuclei as bound states of nucleons; instead, they simply identify a nucleus with the lowest quantum states of a static soliton in a given topological sector. The most unusual result of this approach is the spatial structure of these putative nuclei. A plot of the baryon number distribution for a configuration of baryon number B reveals a closed surface with $2(B-1)$ holes which meet at the center of the soliton. While this gives a rather counterintuitive model for the nucleus, they argue that one should study the properties of the soliton and see whether it successfully reproduces the physical properties of the nucleus. Certainly the results obtained for the deuteron properties lend support to this argument.

Despite the encouraging results of the Skyrme model, it is not clear that as an approach to low energy strong interactions it is any more useful than other methods, such as the quark model. Current prevailing opinion seems to be that it is not, though of course only time and further research will tell. In any event, the Skyrme model serves as an excellent paradigm in the use and application of topological methods. Moreover, the notion of the nucleon as a soliton is very intriguing—not least because this is also the description found for baryons in the large color limit of QCD.[18] This being the case, the skyrmion may yet engender insight into the complicated nature of the QCD vacuum.

Acknowledgments

I would like to thank Herb Fried and Berndt Müller for the opportunity to participate and attend the Institute, and for their efforts in making the school the great success it was. I also thank the NSF for providing travel assistance to Cargèse.

References

1. T.H.R. Skyrme, *Proc. Roy. Soc.* **A260**, 127 (1961); *Nucl. Phys.* **31**, 556 (1962).
2. D. Finkelstein and J. Rubinstein, *J. Math. Phys.* **9**, 1762 (1968).
3. E. Witten, *Nucl. Phys.* **B223**, 443 (1983).
4. A.P. Balachandran, V.P. Nair, S.G. Rajeev, and A. Stern, *Phys. Rev.* **D27**, 1153 (1983).
5. A.P. Balachandran, in *High Energy Physics 1985*, eds. M.J. Bowick and F. Gursey, World Scientific (1985); I. Zahed and G.E. Brown, *Phys. Rep.* **142**, 1 (1986); G. Holzwarth and B. Schwesinger, *Rep. Prog. Phys.* **49**, 825 (1986).
6. G. Adkins, C. Nappi, and E. Witten, *Nucl. Phys.* **B228**, 552 (1983).

7. E. Witten, in *Solitons in Nuclear and Elementary Particle Physics*, eds. A. Chodos, E. Hadjimichael, and C. Tze, World Scientific (1984).

8. G. Adkins and C. Nappi, *Phys. Lett.* **B137**, 251 (1984); G. Adkins, *Phys. Rev.* **D33**, 193 (1986).

9. A.P. Balachandran, F. Lizzi, and V.G.J. Rodgers, *Nucl. Phys.* **B256**, 525 (1985).

10. A. Jackson, A.D. Jackson, and V. Pasquier, *Nucl. Phy.* **A432**, 567 (1985).

11. E. Braaten and L. Carson, *Phys. Rev. Lett.* **56**, 1897 (1986).

12. V.B. Kopeliovich and B.E. Shtern, *Sov. Phys. JETP Lett.* **45**, 203 (1987); J.J.M. Verbaarschot, T.S. Walhout, J. Wambach, and H.W. Wyld, *Nucl. Phys.* **A468**, 520 (1987).

13. A.J. Schramm, *Phys. Rev.* **C37**, 1799 (1988); A.J. Schramm, Y. Dothan, and L.C. Biedenharn, *Phys. Lett.* **B205**, 151 (1988).

14. G. Adkins and C. Nappi *Nucl. Phys.* **B233**, 109 (1984).

15. N.S. Manton, *Phys. Lett.* **B192**, 177 (1987); J.J.M.Verbaarschot, *Phys. Lett.* **B195**, 235 (1987).

16. E. Braaten and L. Carson, *Phys. Rev.* **D38**, 3525 (1988); *Phys. Rev.* **D39**, 838 (1989).

17. E. Braaten, S. Townsend, and L. Carson, *Phys. Lett.* **B235**, 147 (1990).

18. E. Witten, *Nucl. Phys.* **B160**, 57 (1979).

PAIR PRODUCTION IN AN ELECTRIC FIELD IN A TIME-DEPENDENT GAUGE:

A TREATMENT PARALLEL TO THE COSMOLOGICAL PARTICLE PRODUCTION

I.H. Duru

Marmara Scientific and Industrial Research Centre
TUBITAK
P.K. 21, 41470 Gebze, Turkey

INTRODUCTION

In 1951 Schwinger calculated the probability for the vacuum to remain vacuum in a constant electric field from the imaginary part of the Lagrangian [1].

In 1969 Parker developed a method to study pair production in expanding universes [2]. Later path integral techniques were introduced to investigate black hole radiation and pair production in the expanding universes [3,4]. Path integral techniques are especially powerfull in the later case, in which they overcome the difficulty in the specification of the initial particle states by simply restricting the paths to lie to the future of the initial singularity [4,5].

Path integrations have also been employed to discuss pair production in a constant electric field in both the time-independent and time-dependent gauges [6,7]. In these works the Green functions are evaluated from the solutions of the wave equation, which requires a carefull discussion of the boundary conditions.

Recently the constant electric field problem has been retreated in a time dependent gauge by evaluating the Green function entirely by path integrals which has had several advantages [8]:
(1) Path integrals give a definite normalization for the Green function.
(2) The normalization of states is always implied by adiabatic WKB solution at $t \to +\infty$. (3) The use of path integrals avoids the problem of turning on and off the electromagnetic fields asymptotically. (4) The method establishes a great similarity between the pair production problems in the flat space-times with an external field and in expanding universes.

In the following section we briefly present the path integral method.

In section III we give the details of the calculation for pair production in the constant electric field.

In section IV we list some of the exactly solvable time dependent electric fields.

Vacuum Structure in Intense Fields, Edited by
H.M. Fried and B. Muller, Plenum Press, New York, 1991

GENERAL FORMULATION

The amplitude for observing a scalar particle at a space-time point x_a and an antiparticle at x_b is given by [4]

$$A= -\int d\vec{k}_a d\vec{k}_b A_0(\vec{k}_a,\vec{k}_b) d\sigma_a^\mu d\sigma_b^\nu f^*(k_a,x_a)\overset{\leftrightarrow}{\partial}_{\mu_a} G(x_a,x_b)\overset{\leftrightarrow}{\partial}_{\nu_b} f^*(k_b,x_b) \tag{1}$$

Here $f_{a,b}$ are the wave functions representing the states of the particles with momenta k_a,k_b and $d\sigma_a^\mu d\sigma_b^\nu$ are the constant time hypersurfaces. $G(x_a,x_b)$ is the propagator which carries the particle from x_b to x_a, and $A_0(\vec{k}_a,\vec{k}_b)$ is no pair production amplitude in the modes $\vec{k}_a,\vec{k}_b$.

Here we will present the explicit form of the propagator for two specific cases:
(i) For expanding universes with a spatially flat Robertson-Walker metric

$$ds^2 = -dt^2+a^2(t)d\vec{x}^2$$

we write the following path integral.

$$G(x_b,x_a)= \int_0^\infty dW\, e^{-i\mu^2 W}\int D^4 x \exp\left[\frac{i}{4}\int_0^W dw(-\dot{t}^2+\dot{a}^2(t)\dot{\vec{x}}^2)\right] \tag{2}$$

Here μ is the mass of the particle and w is the parameter time. The over dot stands for the derivative with respect to w. In general we prefer to write the Green function in terms of the phase space path integral which permits one to employ several canonical transformations:

$$G(x_b,x_a)= \int_0^\infty dW\, e^{-i\mu^2 W}\int D^4 x D^4 k\, \exp[i\int_0^W dw(k_0\dot{t}+\vec{k}.\dot{\vec{x}}+k_0^2-\vec{k}^2/a^2(t)] \tag{3}$$

Explicit graded formulations of (3) and (2) can be found in the literature[5].
(ii) For a charged particle moving in an external potential A_μ in flat space-time, the Green function is given by:

$$G(x_a,x_b)= \int_0^\infty dW\, e^{-i\mu^2 W}\int D^4 x\, \exp\frac{i}{4}\int_0^W dw(-\dot{t}^2+\dot{\vec{x}}^2+4e A.\dot{x}) \tag{4}$$

or

$$G(x_a,x_b)= \int_0^\infty dW e^{-i\mu^2 W}\int D^4 x D^4 x\, \exp[i\int_0 dw(k\,\dot{t}+\vec{k}.\dot{\vec{x}}+k_0^2-\vec{k}_0^2+e A.\dot{x})] \tag{5}$$

Explicit forms of the above path integrals are given in ref.[8].

The state functions in (1) are obtained by solving the corresponding Klein-Gordon equations, which for the Robertson-Walker space-times given by

$$[\partial_t^2+ \frac{3(da/dt)}{a}\partial_t- \frac{i}{a^2}\vec{\nabla}^2+ \frac{(da/dt)^2}{a^2} + \frac{d^2a/dt^2}{a} +\mu^2]f=0 \tag{6}$$

For a particle moving in flat space under the influence of an external electromagnetic field the wave equation takes the usual form:

$$[(i\partial_\mu -e A_\mu)^2-\mu^2]f=0 \tag{7}$$

The normalizations of the state functions are obtained by requiring them to have the WKB forms as $t \to +\infty$.

PAIR PRODUCTION IN AN ELECTRIC FIELD IN A TIME-DEPENDENT GAUGE

We choose the electromagnetic potential to be time dependent as

$$A_\mu=(0,h(t),0,0) \tag{8}$$

where $h(t)$ is an arbitrary function of time. The corresponding electric field is in the x direction:

$$\vec{E} = \frac{dh}{dt}\,\hat{x} \equiv h'(t)\hat{x} \tag{9}$$

A more familiar potential is obtained by a gauge transformation

$$A_\mu = A_\mu + \partial_\mu \Lambda = (x\, h'(t),0,0,0) \tag{10}$$

with $\Lambda = -x\, h'(t)$.

To obtain the propagator for the scalar particle moving under the influence of the field (9), we introduce (8) into (5). After translating the integration variable k_x by $k_x \to k_x - eh(t)$, we have

$$G(x_a,x_b) = \int_0^\infty dW\, e^{-i\mu^2 W} \int D^4 x D^4 k\, \exp\left[i\int_0^W dw(k\,\dot{t} + \vec{k}\cdot\dot{\vec{x}} + k_y^2 - k_z^2 - (k_x - eh(t))^2)\right] \tag{11}$$

In the above path integral the integrations over $D^3 x D^3 k$ are trivial, i.e., they are equivalent to the free case. We can then write [8]

$$G(x_a,x_b) = \int \frac{d^3k}{(2\pi)^3}\, e^{i\vec{k}\cdot(\vec{x}_b - \vec{x}_a)} G_{\vec{k}}(x_b,x_a) \tag{12}$$

Here $G_{\vec{k}}$ is the propagator for the mode $\vec{k}$:

$$G_{\vec{k}}(x_a,x_b) = \int_0^\infty dW\, e^{-i\mu^2 W} e^{-i(k_y^2 + k_z^2)W}\, K_{\vec{k}}(t_a,t_b;W) \tag{13}$$

where $K_{\vec{k}}$ is the only non-trivial part of the propagator:

$$K_{\vec{k}}(t_b,t_a;W) = \int Dt Dk_0 \exp\left[i\int_0^W dw(k_0\dot{t} + k_0^2 - (k_x - eh(t))^2)\right] \tag{14}$$

The above expression is equivalent to the non-relativistic quantum mechanical path integral for a particle of "mass"=1/2, moving under the influence of potential $(k_x - eh(t))^2$ in the "time" interval $(0,-W)$ with t playing the role of the coordinate. Note that k_x is constant fixed by $DxDk_x$ path integration.

It is possible to find exact solutions of (14) for several forms of $h(t)$. Here we will discuss the constant electric field case in detail:
For

$$h(t) = Et\ , \quad E = \text{constant} \tag{15}$$

the path integral (14) becomes of an oscillator with "frequency"=$i(2eE)$ (upside-down oscillator). It's solution is [8]

$$K_{\vec{k}}(t_a,t_b;W) = \left(\frac{ieE}{2\pi\,\text{sh}(2eEW)}\right)^{1/2} \exp\left\{\frac{ieE}{2\,\text{sh}(2eEW)}\left[(\tau_a^2 + \tau_b^2)\text{ch}(2eEW) - 2\tau_a\tau_b\right]\right\} \tag{16}$$

where

$$\tau = t - k_x/eE$$

The expression (16) can be decomposed into an expansion of parabolic cylinder functions. We then insert it into (13) and obtain after dW integration:

$$G_{\vec{k}}(x_b,x_a) = \frac{-1}{\sqrt{2eE}}\, \frac{e^{-\pi\lambda/4}}{1+e^{\pi\lambda}}\,\left[D_\gamma\left(e^{-i\pi/4}\sqrt{2eE}\tau_a\right)D_{-\gamma-1}\left(e^{i\pi/4}\sqrt{2eE}\tau_b\right)\right.$$

$$\left. + D\left(e^{3i\pi/4}\sqrt{2eE}\tau_a\right)D_{-\gamma-1}\left(e^{-3i\pi/4}\sqrt{2eE}\tau_b\right)\right] \tag{17}$$

with

$$\gamma = -\frac{1}{2}(1-i\lambda), \quad \lambda = \frac{1}{eE}(\mu^2 + k_y^2 + k_z^2) \tag{18}$$

Since in cosmological problems, as well as in the scattaring experiments one deals with the particle states defined in reasenobly free sections of space-time, we are interested in the asymptotic forms of the wave functions and the Green functions. Thus we write the late time limit of (17) by using the expansion formulae of D 's:

$$G_{\vec{k}}(x_b x_a) \cong \frac{-1}{\sqrt{2eE}} \frac{e^{-\pi\lambda/4}}{1+e^{\pi\lambda}} [e^{i\pi/4}\tau_b^\gamma \tau_a^{\gamma-1} e^{i(\tau_b^2-\tau_a^2)/4} e^{-i\pi\gamma/2}$$

$$+ e^{(3i\pi/2)\gamma} e^{3i\pi/4}\tau_a^\gamma \tau_b^{\gamma-1} e^{i(\tau_b^2-\tau_a^2)/4}$$

$$- \frac{\sqrt{2\pi}}{\Gamma(1+\gamma)} e^{i\pi(\gamma+1)} \tau_b^\gamma \tau_b^\gamma e^{i(\tau_b^2+\tau_a^2)/4}$$

$$- \frac{\sqrt{2}}{\Gamma(-\gamma)} e^{i\pi/\gamma} (\tau_a \tau_b)^{-\gamma-1} e^{-i(\tau_b^2+\tau_a^2)/4}$$

$$+ \frac{2\pi}{\Gamma(-\gamma)\Gamma(1+\gamma)} e^{i\pi/4} e^{(i\pi/4)\gamma} \tau_b^{-\gamma} \tau_a^\gamma e^{-i(\tau_a^2-\tau_a^2)/4}] \qquad (19)$$

To obtain the required state functions we solve (7) by substituting (8) with h=Et. The solution is

$$f(x,t) = \frac{1}{(2\pi)^{3/2}} e^{i\vec{k}\cdot\vec{x}} N D_{-1/2-i\lambda/2}(\sqrt{2ieE}(t-\frac{kx}{eE})) \qquad (20)$$

Following Parker, we determine the normalization by equating the $t\to\infty$ limit of (20) to the WKB solution of wave equation (7):

$$|N|^2 = \frac{e^{-\pi\lambda/8}}{(2eE)^{1/4}} \qquad (21)$$

To develop the pair production amplitude we insert (12) and (20) into (1). Integrations over $d^3x_{a,b}, d^3k_{a,b}$ fix the momenta of the particle and antipar antiparticle to be $-\vec{k}_a = \vec{k}_b = \vec{k}$.

Then, the amplitude becomes

$$A = \int d^3k \, A(\vec{k})$$

where, the amplitude for creation of a pair with momenta $\vec{k}, -\vec{k}$ is given by:

$$A(\vec{k}) = -A_0(\vec{k}) g*(\vec{k},t_a)\overset{\leftrightarrow}{\partial}_{t_a} G_{\vec{k}}(t_a,t_b)\overset{\leftrightarrow}{\partial}_{t_b} g*(-\vec{k},t_b) \qquad (22)$$

Here g is the time dependent part of the wave function (20). We are interested in detecting the pair simultaneously at very late times. Thus we insert the $t\to\infty$ forms of $G_{\vec{k}}$ and g* into (22) and develop it at $t_a=t_b=t$ limit. Only the fourth term in (19) contributes, other terms average out to zero because of rapid oscillatons. We then have:

$$A(\vec{k}) = A_0(\vec{k}) i(\sqrt{2eE})^{-i\lambda}\sqrt{2\pi} \frac{e^{-3\pi\lambda/4}}{(1+e^{-\pi\lambda})\Gamma(\frac{1}{2}-i\frac{\lambda}{2})} \qquad (23)$$

The probability for the creation of one pair with momenta $\vec{k}, -\vec{k}$ is

$$P_1(\vec{k}) = |A(\vec{k})|^2 = |A_0|^2 \frac{1}{1+e^{\pi\lambda}} \equiv |A_0|^2 w_k \qquad (24)$$

The probability for n pair creation is

$$P_n(\vec{k}) = |A_0|^2 w_k^n \qquad (25)$$

The conservation of probability condition

$$\sum_{n=0}^{\infty} P_n(\vec{k})=1$$

fixes $|A_0|^2$:

$$|A_0|^2=1-w_k \tag{26}$$

The average number of pairs with momenta $\vec{k}$ is

$$\bar{N}(\vec{k})=\sum_{n=0}^{\infty} nP_n(\vec{k})=\frac{w_k}{1-w_k}=e^{-\pi\lambda} \tag{27}$$

The average number of pairs summed over all modes is

$$N=\int dk_y \frac{L}{2\pi} \int dk_3 \frac{L}{2\pi} \int_0^{eEL} dk_0 \frac{T}{2\pi} e^{-\pi\lambda}$$

$$=\frac{(eE)^2L^3T}{(2\pi)^3} e^{-\pi\mu^2/eE} \tag{28}$$

Where L and T are the size of the space and the duration of the constant field respectively.
Note that because of the form of the exponential, the result of (28) could never be obtained by perturbative methods.

We remark that because of the Pauli principle for fermions instead of (27) we have

$$P_0(\vec{k})+P_1(\vec{k})=1 \tag{29}$$

thus $|A_0|^2$ becomes:

$$|A_0(\vec{k})|^2=\frac{1}{1+w_k}=\frac{1+e^{-\pi\lambda}}{1+2e^{-\pi\lambda}} \tag{30}$$

EXACTLY SOLVABLE TIME-DEPENDENT ELECTRIC FIELDS

The potential in the non-trivial path integral (14) is (upto the constant k_x^2):

$$V(t)=2ek_x h(t)+e^2h^2(t) \tag{31}$$

By using the known results of several non-relativistic quantum mechanical path integrals, the above potential can be solved exactly for some forms of h(t). In this section we list some cases (E_0, a_0=constants):

(i) $E(T)=E_0\delta(t)$, $h(t)=E_0\theta(t)$

$$V(t)=(2ek_x E_0+e_0^2E_0^2)\theta(t) \tag{32}$$

Path integration for the step function is solved by adopting the Hamilton-Jacobi coordinates [9]

(ii) $E(t)=E_0 \dfrac{\sin a_0 t}{\cos^2 a_0 t}$, $h(t)=\dfrac{E_0/a_0}{\cos a_0 t}$

$$V(t) = \frac{2e \frac{E_0}{a_0} - k_x}{\cos a_0 t} + \frac{e^2 E_0^2/a_0^2}{\cos^2 a_0 t} \tag{33}$$

This potential can be mapped into the Pöschl-Teller problem and then solved by using the SU(2) symmetry [10].

(iii) The previous problem is also solvable for the hyperbolic angles. This time the symmetry group is SU(1,1) [11].

(iv) $E(t) = E_0 e^{-a_0 t}$, $h(t) = -\frac{E_0}{a_0} e^{-a_0 t}$

$$V(t) = -\frac{2eE_0}{a_0} e^{-a_0 t} + \frac{e^2 E_0^2}{a_0} e^{-2a_0 t} \tag{34}$$

This problem is equivalent to the Morse potential [12].

(v) $E(t) = \frac{E_0}{t^2}$, $h(t) = -\frac{E_0}{t}$; $t > 0$

$$V(t) = -\frac{2ek_x E_0}{t} + \frac{e^2 E_0^2}{t^2}$$

The above potential is equivalent to the H-atom with an extra potential barrier which can be solved by mapping the problem into the harmonic oscillator potential [13].

REFERENCES

1. J. Schwinger, Phys.Rev. 82, 664(1951).
2. L. Parker, Phys.Rev. 183, 1057(1969).
3. J.B. Hartle and S. W. Hawking, Phys.Rev. D13, 2188(1976).
4. D. M. Chitre and J. B. Hartle, Phys.Rev. D16, 251(1977).
5. I. H. Duru and N. Unal, Phys.Rev. D34, 959(1986).
6. A. S. Lapades, Phys.Rev. D17, 2556(1978).
7. A. I. Nikishov, Zh.Eksp.Teor.Fiz. 57, 1210(1969) [Sov.Phys.JETP 30, 660(1970)]; and , N. B. Narozhnyi and A. I. Nikishov, Yad.Fiz. 11, 1072 (1970) [Sov.J.Nucl.Phys. 11, 596(1970)].
8. A. O. Barut and I. H. Duru, Phys.Rev. D41 , 1312(1990).
9. A. O. Barut and I. H. Duru, Phys.Rev. A38 , 5906(1988).
10. I. H. Duru, Phys.Rev. D30, 2121(1984).
11. M. Böhm and G. Junker, Phys.Lett. A117, 375(2986); J.Math.Phys. 28, 1978(1987).
12. I. H. Duru, Phys.Rev. D28, 2689(1983).
13. I. H. Duru and H. Kleinert, Phys.Lett. 84B, 185(1979); Fortschr.Phys. 30, 401(1982).

ZERO-POINT FIELD IN NON-INERTIAL OR CONFINED SYSTEMS

(PHYSICALLY MEASURABLE MANIFESTATIONS OF VACUUM)

Antonio Sarmiento

Instituto de Astronomía, UNAM
Apartado Postal 70-264, C.U.
México 04510, D.F., México

INTRODUCTION

The main motivation for studying the spectrum of the zero-point field in either non-inertial or confined systems is the fact that this spectrum is Lorentz invariant; that is, in order to observe it one has to go to a non-inertial system. Well known examples of phenomena associated with the zero-point field are: the Casimir effect[1], Hawking's radiation[2], accelerated mirrors radiating[3,4], and the Einstein universe[5]; in the first and the last examples there are explicit boundaries while in the other two there are horizon events: all are bounded systems and this limitation is responsible for the thermal character of the spectrum and for the modification of the states density in phase space[6,7].

Instead of following the usual interpretation of vacuum as being a sea of virtual quanta with an infinite total energy — and yet no physical reality — for which some complicated and non-rigorous techniques have been developed in order to extract physically relevant information from infinite quantities, we have constructed a formalism for the calculation and analysis of the spectrum of the zero-point field that has the following features: the formalism does not arbitrarily cut the zero-point energy off, it allows to directly derive the explicit form of the energy density spectrum, it is also applicable to stochastic classical fields, and most important, it is a simple (understandable) formalism. In what follows I shall describe in detail the formalism for the scalar field case only[8]; the generalization to arbitrary spin fields has been presented elsewhere[9], and the formalism has been applied to the calculation of the energy density spectrum of the Casimir effect[10,11], the energy density spectrum of vacuum around a cosmic string[12], the vacuum stress-energy tensor of the electromagnetic field in an arbitrarily moving frame[13,14],

Vacuum Structure in Intense Fields, Edited by
H.M. Fried and B. Muller, Plenum Press, New York, 1991

and to bounded systems and cavities[15,16]. In particular, for the case of the electromagnetic zero-point field in a uniformly rotating system[14], the energy density spectrum and the stress spectrum turn out to be non-thermal, and there is a net Poynting flux in the direction of the tangential velocity; the residual depolarization and the synchrotron radiation energy loss of electrons in storage rings at ultrarelativistic speeds can be acounted for in terms of the zero-point field[17,14]. From the study of the zero-point field in cavities we have derived a very rough 'toy model' for the confinement of quarks[18].

FORMALISM FOR SPINLESS FIELDS

Consider a field $\phi(x)$ in an arbitrary spacetime with metric $g_{\alpha\beta}$. The Lagrangean density $\mathscr{L}(x)$ depends on $\phi(x)$ and its first and second derivatives. Keeping the background metric fixed and varying ϕ, we get:

$$\delta\mathscr{L} = \left(\frac{\partial\mathscr{L}}{\partial\phi} - \partial_\mu\frac{\partial\mathscr{L}}{\partial\phi_{,\mu}} + \partial_\mu\partial_\nu\frac{\partial\mathscr{L}}{\partial\phi_{,\mu,\nu}}\right)\delta\phi + \partial_\mu\left[\left(\frac{\partial\mathscr{L}}{\partial\phi_{,\mu}} + \frac{\partial\mathscr{L}}{\partial\phi_{,\mu,\nu}}\overleftrightarrow{\partial}_\nu\right)\delta\phi\right]$$

Assuming that the background metric admits a Killing vector, then under an infinitesimal transformation of the form:

$$x^\mu \longrightarrow x'^\mu = x^\mu + \xi^\mu,$$

$g_{\mu\nu}$ remains invariant while ϕ and $\mathscr{L}$ vary according to:

$$\delta\phi = \xi^\mu\phi, \qquad \delta\mathscr{L} = \xi^\mu\mathscr{L}.$$

Inserting these variations and the fact that $\xi^\alpha_{;\alpha} = 0$ we get a conservation law for any field satisfying the Euler-Lagrange equations:

$$J^\mu_{;\mu} \equiv \frac{1}{(-g)^{1/2}}\partial_\mu[(-g)^{1/2}J^\mu] = 0,$$

where the conserved current is:

$$J^\alpha = -2(-g)^{1/2}\left[\mathscr{L}\xi^\alpha - \left(\frac{\partial\mathscr{L}}{\partial\phi_{,\alpha}} + \frac{\partial\mathscr{L}}{\partial\phi_{,\alpha,\beta}}\overleftrightarrow{\partial}_\beta\right)\left(\xi^\gamma\phi_{,\gamma}\right)\right].$$

Consider now a system of chargeless spin-zero particles with mass m which we describe by a scalar field $\phi(x)$. This noninteracting field satisfies the Klein-Gordon equation:

$$(\Box^2 + m^2 + \zeta R)\phi = 0,$$

where R is the Ricci scalar and ζ is a constant. For this case:

$$\mathscr{L} = -\frac{1}{2}(-g)^{1/2}\phi(\Box^2 + m^2 + \zeta R)\phi,$$

and the conserved current is:

$$J_\alpha = - \phi \, \overleftrightarrow{\delta}_\alpha (\xi^\beta \phi_{,\beta}).$$

Now, Minkowski spacetime admits four linearly independent and constant Killing vectors associated to the Poincaré group. Therefore, an energy-momentum tensor may be defined through:

$$J^\alpha = T^{\alpha\beta} \xi_\beta,$$

with the property that

$$T^{\alpha\beta}{}_{;\alpha} = 0,$$

due to the conservation of J^α and the linear independence of the ξ_α's.

In cartesian coordinates we explicitly have:

$$T_{\alpha\beta} = - \frac{1}{2} \phi \, \overleftrightarrow{\delta}_\alpha \, \overleftrightarrow{\delta}_\beta \, \phi.$$

We must point out that the vanishing divergence of J_α guarantees that the total energy:

$$E = \int J_\mu \, d\sigma^\mu,$$

is independent of the particular three-dimensional space (whose normal vector is $d\sigma^\mu$) over which the integration is performed.

The orbit of a timelike Killing vector can be identified with the world line $x^\alpha = x^\alpha(\tau)$ of an observer (*detector*) whose four-velocity is:

$$\frac{dx^\alpha}{d\tau} = u^\alpha = (\xi^\mu \xi_\mu)^{-1/2} \xi^\alpha,$$

where τ is the proper time; then the energy density can be unambiguously defined by:

$$e = u^\alpha J_\alpha = (\xi_\mu \xi^\mu)^{1/2} \left(-\phi \frac{d^2\phi}{d\tau^2} + \frac{d\phi}{d\tau} \frac{d\phi}{d\tau} \right),$$

where $d/d\tau = u^\beta \nabla_\beta$.

One can now perform a straightforward quantization. From the Lagrangean density and interpreting ϕ as an operator, we get:

$$J_\alpha = - \langle \phi \, \overleftrightarrow{\delta}_\alpha (\xi^\beta \phi_{,\beta}) \rangle$$

for the vacuum expectation value of the energy-momentum

four-vector, and

$$e = \frac{1}{2} (\xi^\mu \xi_\mu)^{1/2} \langle - \phi \frac{d^2\phi}{d\tau^2} - \frac{d^2\phi}{d\tau^2} \phi + 2 \frac{d\phi}{d\tau} \frac{d\phi}{d\tau} \rangle,$$

for the vacuum energy density.

It is also possible to define a divergenceless current:

$$n_\alpha = - \iota \langle \phi \overleftrightarrow{\delta}_\alpha \phi \rangle,$$

which is used to construct the scalar:

$$n \equiv u^\alpha n_\alpha = - \iota \langle \phi \frac{d\phi}{d\tau} - \frac{d\phi}{d\tau} \phi \rangle,$$

later interpreted as the particle number density.

We are now in a position to construct a formalism for the direct evaluation of the energy density or the particle number density, as measured bu a detector moving along a Killing orbit: let us simply take:

$$n = \frac{1}{\pi} \int_0^\infty \omega \left(\tilde{D}^+(\omega,\tau) - \tilde{D}^-(\omega,\tau) \right) d\omega,$$

and:

$$e = \frac{1}{\pi} (\xi^\mu \xi_\mu)^{1/2} \int_0^\infty \omega^2 \left(\tilde{D}^+(\omega,\tau) + \tilde{D}^-(\omega,\tau) \right) d\omega,$$

where ω is the frecuency measured by a detector with proper time τ and four-velocity u^μ, and $\tilde{D}^\pm$ are the Fourier transforms of the Wightman functions:

$$D^\pm(\tau+\sigma/2, \tau-\sigma/2) = \langle \phi(\tau\pm\sigma/2, \tau\mp\sigma/2) \rangle,$$

where $\phi(\tau\pm\sigma/2) \equiv \phi(x^\mu(\tau\pm\sigma/2))$. Accordingly, the particle density is:

$$f(\omega,\tau) = \frac{1}{(2\pi)^2 \omega} \left(\tilde{D}^+(\omega,\tau) - \tilde{D}^-(\omega,\tau) \right),$$

i. e., the vacuum expectation value of the field commutator or the Pauli-Jordan-Schwinger function, and the energy density per mode is:

$$de = (\xi^\mu \xi_\mu)^{1/2} \frac{\omega^2}{\pi} \left(\tilde{D}^+(\omega,\tau) + \tilde{D}^-(\omega,\tau) \right) d\omega,$$

i. e., the anticommutator or Hadamard function.

SOME EXAMPLES

1) Flat space ($\hbar = 1$)

For a massless and zero-spin field, an observer at rest measures[8] a particle number density $dn = (2\pi)^{-3}$, and an energy density $de = (\omega^3/2\pi^2)\, d\omega$; there is no special meaning in this results and the normalization is such that there is one particle in each cell of phase space. For the same case, a uniformly accelerated observer measures the same particle number density (because the same normalization has been used) but a diferent energy density[8]:

$$de = (\xi^{\mu}\xi_{\mu})^{1/2} \frac{\omega^3}{\pi^2} \left[\frac{1}{2} + \frac{1}{e^{2\pi\omega/a} - 1} \right] d\omega,$$

where there is an additional planckian term involving the acceleration a, $i.\,e.$, the accelerated observer sees the zero-point field.

In the case of a massless field with arbitrary spin ($s = 0, 1/2, 1$), an observer at rest measures $dn = (2\pi)^{-3} 4\pi\hbar\omega^2\, d\omega$ and $de = \omega/2\, dn$: the normalization used is such that there are $\hbar$ (helicity number) particles in each phase space cell of volume $(2\pi\hbar)^3$. For the same type of field, a uniformly accelerated observer measures

$$dn = (2\pi^2)^{-1} \hbar(\omega^2 + a^2 s^2)\, d\omega$$

and

$$de = \left[\pm \frac{1}{2} + \frac{1}{e^{2\pi\omega/a} \mp 1} \right] \omega\, dn,$$

where the upper sign corresponds to bosons and the lower one to fermions. In this case, the density of states is no longer proportional to $\omega^2\, d\omega$ (its value in an inertial frame) and there is a thermal energy density spectrum[19-20]. The opposite sign for bosons and fermions allows one to conjecture, in the spirit of supersymmetry, that they cancel each other in an inertial frame.

For a zero-spin field with mass m, an accelerated observer measures a particle number density which has a complicated form in terms of modified Bessel functions[21]:

$$dn = \frac{m^2\omega}{2a\pi^3} \sinh(\pi\omega/a) \left[\left| K_{1+i\omega/a}(m/a) \right|^2 - \left| K_{i\omega/a}(m/a) \right|^2 \right] d\omega,$$

possibly a consequence of the normalization being used, but the energy density spectrum still has a simple form where a Bose-Einstein function is superimposed to the zero-point energy spectrum:

$$de = 2 \left(\frac{1}{2} + \frac{1}{e^{2\pi\omega/a} - 1} \right) \omega \, dn.$$

For a field with mass m and spin $1/2$, a uniformly accelerated observer measures a particle number density:

$$dn = \frac{m^3}{2a\pi^3} \cosh(\pi\omega/a) \left[\left| K_{3/2+i\omega/a}(m/a) \right|^2 - \left| K_{1/2+i\omega/a}(m/a) \right|^2 \right] d\omega,$$

and an energy density[21]:

$$de = \left(-1 + \frac{2}{e^{2\pi\omega/a} + 1} \right) \omega \, dn,$$

i. e., the accelerated observer detects a non-zero density of 'virtual' particles with energies below mc^2 even though only states with $\hbar\omega < - mc^2$ are occupied in an inertial frame (Dirac sea). It is as if the acceleration produced a 'tide' in the Dirac sea lifting some electrons above their original energy levels and producing an overall excess of positive energy; the corresponding depletion of electrons below the level $- mc^2$ is observed as an increase in the energy associated with the positrons. From this point of view, the term $- \omega$ corresponds to the vacuum energy and the Fermi-Dirac distribution comes from the additional energy detected in the accelerated frame, the factor of two is because both, electrons and positrons, contribute to the excess energy. The acceleration needed to produce a real $e^- e^+$ pair is $a \cong mc^3/h \cong 10^{31}$ cm s^{-2}.

2) Curved spacetime - Gravitation

For a two-dimensional Schwarzschild black hole with mass M immersed in a quantized (or stochastic) massless scalar field, an observer at rest at $r = r_o$ measures $dn = (2\pi)^{-3}$ and a thermal bath[8]:

$$de = \frac{2}{\pi} \left(1 - \frac{2M}{r_o} \right)^{1/2} \left(\frac{1}{2} + \frac{1}{e^{2\pi\omega/a} - 1} \right) \omega \, d\omega$$

with a 'Hawking' temperature $kT = (8\pi M)^{-1}(1 - 2M/r_o)^{-1/2}$.

The whole energy-momentum tensor of a massless free field with spin s ($=0, 1/2, 1$) for an observer at rest in the neighbourhood of a four-dimensional Schwarzschild black hole with mass M is diagonal[22]:

$$\langle T_\mu{}^\nu \rangle = \frac{\hbar(s)}{32\pi^6} \int_0^\infty \frac{\omega(\omega^2 + 4\pi^2 s^2 k^2)}{e^{\omega/k} + (-1)^{2s+1}} \, diag(-1, 1/3, 1/3, 1/3),$$

where $\hbar(s)$ is the number of independent helicity states (1 for $s = 0$ and 2 for $s \neq 0$), and $k = (4GM)^{-1}$. There is no net flux of energy and, according to our interpretation, the energy density spectrum seen by the observer at rest is the distortion of the zero-point field created by the gravitational field of the black hole: since there is a horizon, the space is bounded and the field goes to an equilibrium state described by a thermal spectrum; this limited space also implies the abscence of long wave modes and therefore modifies the states density in phase space.

For an Einstein universe of constant radius R, an observer at rest (following a Killing orbit) detects the usual dn = $(2\pi)^{-3}$ and an energy density given by[5,8]:

$$de = \int_0^\infty \frac{\omega^3 d\omega}{2\pi^2} + \frac{1}{240\pi^2 R^3} \ .$$

SUMMARY

All the thermal effects presented in the previous lines are manifestations of the zero-point field fluctuations of the quantum field under consideration; no particle creation is necessary to understand them. These thermal effects are merely Doppler distortions of the zero-point field, their thermal character is directly linked to the discreteness originated by confining the space where the field is quantized; this confinement also implies the absence of some long wave modes which implies a modification of states density in phase space. The formalism we have constructed permits the direct calculation of the energy density spectrum observed in a non-inertial frame in terms of the vacuum expectation value of the symmetrized product of the field operator. The main difference with other formalisms is their use of the positive frequency Wightman function or Feynman Green's function only, with the justification that negative frequency contributions to the energy are eliminated. However, in the light of our analysis, this is equivalent to arbitrarily cutting the zero-point energy off.

REFERENCES

1. H. B. G. Casimir, Koninkl. Ned. Akad. Wetenschap. Proc. B51:793 (1948).
   G. Plunien, B. Müller, and W. Greiner, Phys. Rep. 134:87 (1986).

2. S. W. Hawking, Nature 248:30 (1974); Commun. Math. Phys. 43:199 (1975).

3. P. C. W. Davies, J. Phys. A8:609 (1975).

4. W. G. Unruh, Phys. Rev. D14:870 (1975).

5. L. H. Ford, Phys. Rev. D11:3370 (1975).

6. D. W. Sciama, P. Candelas, and D. Deutsch, Adv. Phys. 30:327 (1981).

7. F. Soto, G. Cocho, C. Villarreal, S. Hacyan, and A. Sarmiento, Rev. Mex. Fis. 33:389 (1987).

8. S. Hacyan, A. Sarmiento, G. Cocho and F. Soto, Phys. Rev. D32:914 (1985).

9. S. Hacyan, Phys. Rev. D32:3216 (1985).

10. A. Sarmiento, S. Hacyan, F. Soto, G. Cocho, and C. Villarreal, preprint IFUNAM (1988).

11. L. H. Ford, Phys. Rev. D38:528 (1988);
    P. C. W. Davies, Z. X. Liu, and A. C. Ottewill, Class. Quant. Gravity 6:1041 (1989)
    S. Hacyan, R. Jáuregui, F. Soto, and C. Villarreal, J. Phys. A23:2401 (1990).

12. A. Sarmiento and S. Hacyan, Phys. Rev. D38:1331 (1988).

13. S. Hacyan and A. Sarmiento, Phys. Lett. B179:287 (1986).

14. S. Hacyan and A. Sarmiento, Phys. Rev. D40:2641 (1989).

15. G. Cocho, F. Soto, S. Hacyan, and A. Sarmiento, preprint IAUNAM-186 (1986); Int. J. Theoret. Phys. 28:699 (1989)

16. A. Sarmiento, S. Hacyan, G. Cocho, F. Soto and C. Villarreal, Phys. Lett. A142:194 (1989).

17. J. S. Bell and J. M. Leinaas, Nucl. Phys. B212:131 (1983)

18. C. Villarreal, this volume.

19. P. Candelas, and D. Deutsch, Proc. Roy. Soc. London A254:79 (1977); ibid. A362:251 (1978).

20. T. Boyer, Phys. Rev. D21:2137 (1980).

21. S. Hacyan, Phys. Rev. 33:3630 (1986).

22. D. Birrel and P. C. W. Davies, "Quantum Fields in Curved Spaces (Cambridge University Press, Cambridge, 1982).

SYMPLECTIC AND LARGE-N GAUGE THEORIES

George Savvidy

Yerevan Physics Institute
Yerevan 375036, Armenia, USSR

INTRODUCTION

In our world the number of quark colors N is equal to three.Many years ago 't Hooft noticed,[1] that it is extremely useful to treat N as a free parameter and to consider the large-N limit.In the large-N limit the structure of the theory partly simplifies, that is only planar graphs survive,i.e. a graph,which does not have overlapping lines.Nonplanar graphs have factors 1/N and it is very tempting therefore to make 1/N expansion.

This result was found by the double line representation of the gluon propagator with oppositely directed arrows.The vertices always consist of the delta functions,connecting ingoing and outgoing arrows.If only gauge invariant quantities can be measured (that means that spectrum consists of color singlets only) index lines never stop and one can attach little surface to each index loop.The orientation of each of the little surfaces is defined by the corresponding arrow on its boundary.Now,glueing together these index surfaces,we shall obtain a big surface with handles attached to a sphere,so all diagrams are classified according to their numbers of handles and carries a factor N^{2-2g},where g is the number of handles.So the leading diagrams in the large-N limit are planar graphs with g=0.

't Hooft suggested,that this representation of Feynman graphs as a surface is something more than a mathematical trick.Namely,it is possible to interpret these surfaces as the world sheet of the relativistic string.Then topological structure of the perturbation series in 1/N is identical to that of the string perturbation series,so 1/N corresponds to the string coupling constant.An important physical conclusion from this analogy consists in the fact that the sum of the planar diagrams describes a free relativistic string with infinite number of stable particles.These particles are vibrational modes of a free string and 1/N expansion corresponds to an expansion with respect to a coupling strength between the hadrons.

There were many attempts to solve planar field theories.In the model with one space and one time dimension 't Hooft[2] succeeded to sum up all planar diagrams and showed,that physical mass spectrum really consists out of the straight Regge trajectory and that there is no continuum in the spectrum.In four space-time dimension Eguchi and Kawai[3] showed,that all planar diagrams can be reproduced by quenched reduced model,where the action is a function of the constant matrix fields and quenched momentum, but analytical solution of this model was not obtained. Polyakov and Migdal[4] found nonlinear functional equation, whose perturbative solution reproduces all the planar diagrams, but also were not able to solve this equation exactly.

In this lecture I would like to formulate SU(∞) gauge theories as the simplectic gauge theories, constructed on the infinite dimensional diffeomorphism group of the two dimensional surfaces Mg. Despite the fact that we still can't proof this equivalence on the quantum level, symplectic gauge theories have their own interest, as the models which, have properties very close to SU(∞) gauge theories and at the same time can be reduced to string and membrane, i.e. allow to make one more step in the direction of the t'Hooft analogy. Symplectic gauge theories naturally arise when bosonic membrane is quantized in the light cone-gauge[5,6,7].

MEMBRANE AND SYMPLECTIC GAUGE THEORIES

Let us introduce necessary notations and light-cone gauge conditions in the theory of relativistic surfaces[5,6,7]. The action, invariant under the reparametrization of three-dimensional volume swept out by a two-dimensional surface Mg of genus g, has the form:

$$S = -T \int d\tau \int_{M_g} d^2\sigma \, Det \, g^{1/2} = \int d\tau \int_{M_g} d^2\sigma \, \mathcal{L} ,$$ (1)

where

$$g_{\alpha\beta} = x^{\mu}_{\alpha} \, x_{\mu\beta} , \qquad x^{\mu}_{\alpha} = \partial x^{\mu}/\partial \xi^{\alpha}, \qquad \xi^{\alpha} = (\tau, \sigma_1, \sigma_2), \quad x(\xi)$$

is parametric representation of the swept volume, $(\sigma_1, \sigma_2) \in M_g$, $\mu = 0 \dots D-1$. The constraints have the form:

$$\omega = P^{\tau}_{\mu} P^{\mu\tau} - T^2 [x^2_{\sigma_1} x^2_{\sigma_2} - (x_{\sigma_1} \cdot x_{\sigma_2})^2] = 0,$$
$$\omega_a = P^{\tau}_{\mu} \cdot x^{\mu}_{\sigma_a} \quad , \quad a = 1, 2 ,$$ (2)

where

$$P^{\alpha}_{\mu} = -\partial\mathcal{L}/\partial x^{\mu}_{\alpha} = T\sqrt{g} \cdot g^{\alpha\beta} x_{\mu\beta} .$$

Equation of motion can be written as

$$\partial_{\alpha} P^{\alpha}_{\mu} = 0 .$$ (3)

One can choose the gauge [5,6,7]

$$x^2_{\tau} = T^{2/3} [x^2_{\sigma_1} x^2_{\sigma_2} - (x_{\sigma_1} \cdot x_{\sigma_2})^2] ,$$ (4)
$$h \cdot x = \lambda \tau , \qquad x_{\tau} \cdot x_{\sigma_a} = 0 ,$$

where

$$\lambda = (nP)/(4\pi T)^{2/3} ,$$

P_μ is total four-momentum of surface Mg. If $n_\mu = (1,1,\ldots 0,0)$, then constraints (2) and additional conditions (4) can be solved, X^- and P_-^τ can be expressed through transverse variables X^i and P_i^τ, where i=2....D-2. In this light-cone gauge the Hamiltonian has the form:

$$H = T^{-2/3} \int_{Mg} d^2\sigma \left[P_i^\tau P_i^\tau + T^2 \left((X_{\sigma_1}^i)^2 (X_{\sigma_2}^j)^2 - (X_{\sigma_1}^i X_{\sigma_2}^i)^2 \right) \right] \qquad 5)$$

and together with the Poisson bracket

$$\left[X^i(\sigma), P_j^\tau(\sigma') \right] = \delta_{ij} \cdot \delta^{(2)}(\sigma-\sigma') \qquad (6)$$

determines correct equations of motion. Variables X^i and P_i^τ are not independent, since gauge condition (4) fix parametrization incompletely. The Frobenius integrability condition for constraints ω_a has the form:

$$L(\sigma) = X_{\sigma_1}^i \cdot P_{i,\sigma_2}^\tau - X_{\sigma_2}^i \cdot P_{i,\sigma_1}^\tau = 0 . \qquad (7)$$

Expression (7) serves as a generator of residual gauge symmetry in the light-cone gauge and reduces the number of independent transverse degrees of freedom up to D-3.

One may readily be convinced that the constraint L commutes with Hamiltonian (5):

$$\left[H, L(\sigma) \right] = 0 \qquad (8)$$

and that

$$\left[L(\sigma), L(\sigma') \right] = \partial_1 \delta^{(2)}(\sigma-\sigma') \partial_2 L(\sigma') - \partial_2 \delta^{(2)}(\sigma-\sigma') \partial_1 L(\sigma') \qquad (9)$$

To arbitrary function ϵ on Mg we'll compare the quantity

$$L_\epsilon \equiv L \cdot \epsilon = \int L(\sigma) \epsilon(\sigma) d^2\sigma .$$

The action of this representation on variables X^i and P_i^τ is given by a formula

$$\delta_\epsilon X^i = \left[X^i, L_\epsilon \right] = \epsilon_{\sigma_1} X_{\sigma_2}^i - \epsilon_{\sigma_2} X_{\sigma_1}^i ,$$

$$\delta_\epsilon P_\tau^i = \left[P_\tau^i, L_\epsilon \right] = \epsilon_{\sigma_1} P_{\tau,\sigma_2}^i - \epsilon_{\sigma_2} P_{\tau,\sigma_1}^i . \qquad (10)$$

If we introduce a new bracket

417

$$\{\epsilon, \zeta\} = \partial_1 \epsilon \, \partial_2 \eta - \partial_2 \epsilon \, \partial_1 \zeta \tag{11}$$

then the Hamiltonian (5) and constrains (7) can be rewritten as

$$H = T^{-2/3} \cdot \int d^2\sigma \left[P_i^\tau P_i^\tau + T^2 \{ x^i \, x^j \}^2 \right] \tag{12a}$$

$$L = \{ x^i, P_i^\tau \} \tag{12b}$$

and the transformation (10) as

$$\delta_\epsilon x^i = \{ \epsilon, x^i \}, \quad \delta_\epsilon P_i^\tau = \{ \epsilon, P_i^\tau \}. \tag{13}$$

For arbitrary functions ϵ and ζ from (9) we'll obtain

$$[L_\epsilon, L_\zeta] = L_{\{\epsilon, \zeta\}}. \tag{14}$$

Lie algebra(9),(14) defines the Lie algebra SdiffMg of the group of area preserving diffeomorphisms of the surface Mg.So we see that area preserving diffeomorphisms SDiffMg emerged as the surviving gauge symmetry in membrane theory in the light-cone gauge.

The Hamiltonian system (12a,b) has striking similarity with the space homogeneous YM classical mechanics[8,9,10].To see that,let us consider YM field in the gauge a_o =0,restricted to the potentials depending only on time $a_i(t) \in$ G,where G is finite Lie algebra,then

$$H_{YM} = \tfrac{1}{2} tr \, e_i^2 + \tfrac{1}{4} tr [a_i a_j]^2, \tag{12c}$$

$$L_{YM} = [a_i e_i]. \tag{12d}$$

Comparing (12a,b) with(12c,d) we see that residual symmetry(12b) corresponds to time independent gauge transformation(12d) in the YM theory.The difference is that in the first case residual symmetry is infinite dimensional (SdiffMg),while in usual YM theory it is finite (commutator of the finite Lie algebra G in (12c,d) is replaced by the Poisson bracket in (12a,b).Identification between(12a,b) and(12c,d) becomes more transparent,when we establish that structure constant of SU(N),in the large-N limit,coincides with the structure constant of the SdiffMg.Remaining difference consists in the fact that membrane Hamiltonian was written in the light-cone gauge while SU(N) Hamiltonian is in the a_o =0 gauge,so that reinforcement up to Lorentz group is different in this two theories.

Now we are in a position to formulate symplectic gauge theory as a gauge theory constructed on infinite-dimensional group of area preserving diffeomorphisms SDiffMg[11,12] .Lie algebra of this group consists out of a set of functions with the bracket

$$\{ \epsilon, \zeta \} = E^{ab} \nabla_a \epsilon \, \nabla_b \zeta$$

on the two-dimensional surface Mg with metric γ and closed two-form E. Let us define symplectic gauge field as

$$\mathcal{A}_\mu^{\sigma_1 \sigma_2}(r,t) \equiv \mathcal{A}_\mu(\sigma_1,\sigma_2,r,t) \ , \qquad (15)$$

which takes its values in the Lie algebra SdiffMg, and (r,t) are coordinates in the Minkowski space M. Using the analogy

$$
\begin{array}{ll}
G & Sdiff\, Mg \ , \\
\epsilon^a(r,t) & \epsilon(\sigma,r,t) \ , \\
\mathcal{A}_\mu^a(r,t) & \mathcal{A}_\mu(\sigma,r,t) \ , \\
f^{abc}\,\mathcal{A}_\mu^b\,\mathcal{A}_\nu^c & \{\mathcal{A}_\mu,\mathcal{A}_\nu\} \ ,
\end{array}
$$

one can construct the action for this symplectic gauge theory

$$S_{SYM} = \int_{Mg}\sqrt{\gamma}\,d^2\sigma \int_M dr\,dt \left[\partial_\mu \mathcal{A}_\nu - \partial_\nu \mathcal{A}_\mu + \{\mathcal{A}_\mu, \mathcal{A}_\nu\}\right]^2 \qquad (16)$$

and Hamiltonian in the $A_0 = 0$ gauge

$$H_{SYM} = \int \sqrt{\gamma}\,d^2\sigma \int_M dr \left[E_i^2 + (\partial_i \mathcal{A}_j - \partial_j \mathcal{A}_i + \{\mathcal{A}_i, \mathcal{A}_j\})^2\right] \qquad (17)$$

together with the generator of the residual gauge transformation

$$L_{SYM} = \partial_i E_i + \{\mathcal{A}_i, E_i\} \qquad (18)$$

defines symplectic gauge theory.

In the space-homogeneous limit, when the variables A and E are independent on the coordinates r and are redefined as X and P, this gauge theory is reduced to a membrane (12-14).

This analogy makes it possible to introduce duality equation for the bosonic membrane and find topology changing solution[13,11,14]. In symplectic YM theory (16) duality equation looks like

$$E_i = \pm H_i = \pm \tfrac{1}{2}\epsilon_{ijk}\left(\partial_j \mathcal{A}_k - \partial_k \mathcal{A}_j + \{\mathcal{A}_j, \mathcal{A}_k\}\right) \qquad (19)$$

or in the space-homogeneous limit as

$$E_i = \pm \tfrac{1}{2}\epsilon_{ijk}\{\mathcal{A}_j, \mathcal{A}_k\} \ . \qquad (20)$$

Therefore one can define self-dual membranes as the embedding $X^i(\tau,\sigma,)$

419

which satisfy the equation

$$P_i{}^{\tau} = \pm \frac{1}{2} \epsilon_{ijk} \{X^j, X^k\} . \tag{21}$$

In YM theory the vacuum configurations $A = g^{-1}\partial g$ satisfy equations $F^{ij}=0$ and can be classified according to their winding number. Here, in membrane theory, all closed surfaces of given topology Mg belong to vacuum configurations (there are also string and point like configurations) and classified according to their genus g. In [13] there were found two different solutions which are, in some sense, very similar to the wormhole solution in gravity [15]. These solutions contain two different vertex' which describe transition of disk into disk with hole and into two separate disks. These saddle point solutions define interaction which exists on the classical level, and drastically differs from the string theory where interaction vertex exists only on the quantum level.

QUANTIZATION AND ANOMALIES

In quantization the classical algebras (9),(14),(18) may be modified by quantum anomalies. At present there exists a well-developed formalism which allows to describe possible quantum anomalies without perturbation theory and which is based on elements of cohomology[16], theory. This technique allows to calculate possible Schwinger terms in the commutator (14). The presence of such terms is connected with the existence of nontrivial cocycles from H^2 and H^1 of the algebra SdiffMg with the Lie bracket (11):

$$\{\epsilon, \eta\} = E^{ab}\nabla_b \epsilon \nabla_a \eta , \quad E^{ab} = \frac{\epsilon^{ab}}{\sqrt{\gamma}} \tag{22}$$

where E_{ab} is a closed 2-form which supplies Mg with symplectic structure and ∇_a is a covariant derivative. The central extension of this algebra are described by a modified Lie bracket:

$$\{(a,\epsilon); (b,\eta)\} = \left(K \cdot C^{(2)}(\epsilon,\eta), \{\epsilon,\eta\}\right) . \tag{23}$$

Where K is arbitrary real number which is called a central charge. The Jacoby identity for this commutator is equivalent to the fact that $C^{(2)}$ is a 2-cocycle:

$$dC^{(2)} = C^{(2)}(\epsilon, \{\eta, x\}) + C^{(2)}(\eta, \{x,\epsilon\}) + C^{(2)}(x, \{\epsilon,\eta\}) ,$$

where ϵ, η and x are arbitrary functions on Mg. The general cocycle has the form[17,11,18];

$$C_\xi^{(2)}(\epsilon,\eta) = \int \sqrt{\gamma} d^2\sigma \, \xi^a (\nabla_a \epsilon \eta - \nabla_a \eta \cdot \epsilon) , \tag{24}$$

if $\nabla_a \xi^a = 0$. Although locally the solution of divergenceless equation has the form $\xi^a = E^{ab}\nabla_b \xi$ this solution not always can be extended globally over the whole surface.If this is possible,then the cocycle (24)is trivial,since

$$C_{\mathcal{E}}^{(2)}(\zeta,\epsilon) = -2\int \mathcal{E}\{\epsilon,\zeta\} \sqrt{r'}\, d^2\sigma .$$

Therefore the question on cocycle nontriviality reduces to the number of divergenceless vector fields $\mathcal{E}^a$ on Mg ,which globally are not presented by means of $\mathcal{E}$ function on Mg .Comparing the divergenceless vector field $\mathcal{E}^a$ with the closed 1-form $\mathcal{E}^a E_{ab}$;we'll obtain that to vector fields representable in the form $\mathcal{E}^a = E^{ab}\nabla_b\mathcal{E}$ there will correspond an exact 1-form.Therefore the number of divergenceless vector fields non-representable in the form $E^{ab}\nabla_b\mathcal{E}$ is equal to closed 1-form factorized over exact 1-form on Mg and is equal to dimensions of the first group of cohomologies of Mg

$$\dim H'(M_g,R) = \dim R^{2g} = 2g . \tag{25}$$

Cocycle (24) determine central extension of the algebra SdiffMg(14)and as a result drive to the modification of the constraint algebra (9,14,18)of the gauge theory

$$[L_\epsilon, L_\zeta] = L_{\{\epsilon,\zeta\}} + \sum_{i=1}^{2g} K_i\, C_i^{(2)}(\epsilon,\zeta) . \tag{26}$$

This result means that in pure gauge theory without chiral fermions it is possible to have anomalies in the Gauss low if gauge group is infinite dimensional.The algebra SdiffMg has also one-dimensional right extension connected with the space H^1 .Cocycle C^1 from H^1 has the form:

$$C'(\epsilon) = (\mathcal{E}^a\nabla_a - 1)\epsilon , \tag{27}$$

where $\mathcal{E}^a$ is a vector field on Mg with constant divergence

$$\nabla_a\mathcal{E}^a = const . \tag{28}$$

The structure of extended algebra is determined by a modified bracket:

$$\{(\lambda,\epsilon); (\mu,\zeta)\} = \{\{\epsilon,\zeta\} + \mu\, C'(\epsilon) - \lambda\, C'(\zeta)\}, 0\} .$$

For this commutator the Jacoby identity is equivalent to the condition

$$dC' = C'(\{\epsilon,\zeta\}) - \{\epsilon, C'(\zeta)\} + \{\zeta, C'(\epsilon)\},$$

which coincides with the definition of 1-cocycle.Complete modification due to the right and central extension of the commutation relation for the gauge constraint algebra can be presented now as [11]

$$[L_\epsilon, L_\zeta] = L_{\{\epsilon,\zeta\} + \mu C'(\epsilon) - \lambda C'(\zeta)} + \sum_{i=1}^{2g} K_i\cdot C_i^{(2)}(\epsilon,\zeta) .) . \tag{29}$$

421

LARGE-N GAUGE THEORIES

Here I will return to the discussion of the large-N limit of the gauge fields.First of all we will see that classical symplectic gauge fields (16,17)constructed on infinite dimensional Lie group SDiffMg (22) emerges as the *specific* large-N limit of the SU(N) gauge fields and that index space of SU(N) is inverted into continuum "world-sheet" surface of SdiffMg. This can be proved in virtue of special representation of the structure constant of the SU(N).J.Hoppe was able to construct such type of representation[5] in the case of Lie algebra SdiffS[2],but it is much more transparent to do that in the case of torus T^2 and is expected to be correct for the surfaces of any genus[19].Expanding $\in$ and ℓ in terms of a complete set of periodic functions on a torus, one can obtain the Lie algebra of the area preserving diffeomorphisms (14) in Fourier space

$$\left[L_n , L_m \right] = (\bar{n} \wedge \bar{m}) L_{\bar{n}+\bar{m}} \quad , \tag{30}$$

where n $\wedge$ m $= n_1 m_2 - n_2 m_1$ .The structure constant of this algebra can be reproduced in the large-N limit out of the structure constant of SU(N). For simplicity let us consider odd N'.A basis for SU(N) may be built from two matrixes[24]

$$g = \begin{pmatrix} 1 & & & 0 \\ & \omega & & \\ & & \ddots & \\ 0 & & & \omega^{N-1} \end{pmatrix}, \quad h = \begin{pmatrix} 0 & 1 & & 0 \\ & 0 & \ddots & \\ & & \ddots & 1 \\ 1 & & & 0 \end{pmatrix}, \tag{31}$$

where ω is exp(4π i/N).They obey the identity

$$h g = \omega g h \quad , \quad g^N = h^N = 1 \quad . \tag{32}$$

The set of N*N matrices

$$J_{\bar{n}} \equiv \omega^{\frac{n_1 n_2}{2}} \cdot g^{n_1} h^{n_2}, \tag{33}$$

where

$$J_{\bar{n}}^{+} = J_{-\bar{n}} \quad , \quad \text{tr} J_{\bar{n}} = 0 \qquad \text{except for } n_1 = n_2 = 0 \quad ,$$

suffice to span an SU(N) algebra.So that the structure constant of SU(N) are

$$\left[J_{\bar{n}} , J_{\bar{m}} \right] = -2i \sin\left(\frac{2\pi}{N} \bar{n} \wedge \bar{m}\right) \cdot J_{\bar{n}+\bar{m}} \quad . \tag{34}$$

Two index representation (34) has a simple large-N limit through the identification :

$$\frac{iN}{4\pi} J_{\bar{n}} \equiv L_{\bar{n}}^{(N)} \quad .$$

Now we can discuss relation between large-N gauge fields and symplectic gauge fields (16),(17)[20],[21]. Let a_μ be an SU(N) gauge field with the matrix normalized to one

$$a_\mu \equiv a_\mu^{\bar{n}} \cdot \frac{J_{\bar{n}}}{N^{1/2}} = a_\mu^{\bar{n}} \cdot \frac{4\pi}{iN^{3/2}} L_{\bar{n}}^{(N)} \equiv \mathcal{A}_\mu^{\bar{n}} \cdot \mathcal{L}_{\bar{n}}^{(N)} , \tag{36}$$

where

$$tr(J_{\bar{n}} J_{\bar{m}}) = N \cdot \delta_{\bar{n}+\bar{m},0}; \quad tr(L_{\bar{n}}^{(N)} L_{\bar{m}}^{(N)}) = -\frac{N^3}{4\pi} \delta_{\bar{n}+\bar{m},0}$$

and

$$f_{SU(N)}^{\bar{P}\bar{n}\bar{m}} = \frac{2}{iN^{1/2}} \cdot Sin\left(\frac{2\pi}{N} \bar{n} \wedge \bar{m}\right) \delta_{\bar{P},\bar{n}+\bar{m}} .$$

Then one can find that

$$\mathcal{L}_{YM} \sim \frac{1}{\lambda^2 N^3} \cdot tr\left(\partial_\mu a_\nu - \partial_\nu a_\mu + [a_\mu, a_\nu]\right)^2$$

$$= \frac{1}{\lambda^2 N^3} \left(\partial_\mu a_\nu^{\bar{P}} - \partial_\nu a_\mu^{\bar{P}} + f_{SU(N)}^{\bar{P}\bar{n}\bar{m}} \cdot a_\mu^{\bar{n}} a_\nu^{\bar{m}}\right)^2 \tag{37}$$

$$= -\frac{1}{16\pi^2\lambda^2} \left(\partial_\mu \mathcal{A}_\nu^{\bar{P}} - \partial_\nu \mathcal{A}_\mu^{\bar{P}} + \frac{iN^{3/2}}{4\pi} f_{SU(N)}^{\bar{P}\bar{n}\bar{m}} \mathcal{A}_\mu^{\bar{n}} \mathcal{A}_\nu^{\bar{m}}\right)^2$$

and in the *special* large-N limit we have

$$-\frac{1}{16\pi^2\lambda^2} \int_{T^2} d^2\sigma \left(\partial_\mu \mathcal{A}_\nu - \partial_\nu \mathcal{A}_\mu + \{\mathcal{A}_\mu, \mathcal{A}_\nu\}\right)^2 , \tag{38}$$

where

$$\frac{a_\mu^{\bar{n}}}{N^{3/2}} \longrightarrow finite ,$$

$$\frac{iN^{3/2}}{4\pi} \cdot f_{SU(N)}^{\bar{P}\bar{n}\bar{m}} \longrightarrow (\bar{n} \wedge \bar{m}) \delta_{\bar{P},\bar{n}+\bar{m}} ,$$

$$\sum_n \mathcal{A}_\mu^{\bar{n}} exp(i\bar{n}\bar{\sigma}) \longrightarrow \mathcal{A}_\mu(\sigma) .$$

Thus, in the *special* large-N limit, when $a_\mu^{\bar{n}}/N^{3/2}$ is finite, the SU(N) group indexes inverted into surface coordinates and as the result we recover symplectic gauge fields(16,38).

For the space-time independent symplectic gauge fields, Lagrangian density (16) reduced to Schild-Eguchi density for the string, where A serves as string coordinate and surface Mg as the world-sheet

$$\int d^2\sigma \left\{ X_\mu \, X_7 \right\}^2 . \tag{39}$$

So that "naively" reduced symplectic gauge fields look like the Schild string rather than membrane.

Up to now we discussed classical gauge theories. Is it possible to extend this analyses to quantum theory?

Let us start from the usually normalized YM Lagrangian

$$\frac{N}{\lambda_o^2} \, tr \left[f_{\mu\nu}(a) \right]^2 . \tag{40}$$

Feynman diagrams with $V_3 + V_4$ vertices is associated with the factor

$$
\begin{aligned}
r &= \left(\frac{\lambda_o^2}{N} \right)^{\frac{3V_3 + 4V_4}{2}} \cdot \left(\frac{N}{\lambda_o^2} \right)^{V_3 + V_4} \cdot N^{I} \\
&= \left(\lambda_o \right)^{\frac{V_3}{2} + V_4} \cdot N^{2-2g} ,
\end{aligned}
\tag{41}
$$

where I is the number of index loops and g is the number of handles.

$$V_3 + V_4 - \frac{3V_3 + 4V_4}{2} + I = 2 - 2g . \tag{42}$$

Therefore the leading diagrams are planar ones(g=0), when λ_o is fixed. Lagrangian density (37) differs from the usual YM Lagrangian density (40) by the factor N^{-3} therefore instead of (41) we will have

$$r_s = \left(\lambda^2 N^4 \right)^{\frac{V_3}{2} + V_4} \cdot N^{2-2g} \tag{43}$$

For to have usual topological expansion like in (40) we must take the *usual* limit

$$N \to \infty , \qquad \lambda^2 N^4 = \lambda_o^2 - fixed . \tag{44}$$

That is the big difference between (37,38) and (40). The limiting procedures in this two cases have different meaning and don't coincide. In the *special* limiting procedure (38) we supposed that the gauge field fluctuations increase with N as $N^{3/2}$ and fluctuations of $tr[f(a)]^2$ as N^3, when λ is fixed,while in the ultraviolet region, where the effective interaction is small,gauge field's fluctuations is of order of N^0 and $tr[f(a)]^2$ is of order of N.So it seems reasonable to believe that such type of *specific* behavior can appear only in the strong coupling region as a result of summation of full topological series in 1/N expansion. In that sense symplectic Lagrangian (38) can appear as the effective Lagrangian.

Topological expansion for (37) or (40) has nothing to do with the topology of the surface Mg on which symplectic fields are defined. In principal the sum of all terms in the topological 1/N expansion can have as their strong coupling effective Lagrangian symplectic gauge Lagrangian (38),defined on particular surface Mg (or on finite set of such surfaces),of course,if our suggestion is correct.Self-dual solutions of the symplectic fields (20) get now completely new meaning.In the present case they determined *tunneling* transition between gauge fields defined on different infinite dimensional groups

$$\mathcal{A}_{SdiffS^2} \longleftrightarrow \mathcal{A}_{SdiffT^2} \longleftrightarrow \cdots \cdots \tag{45}$$

UNITARY REPRESENTATION OF SdiffT^2

In fact all this analyze shows that our knowledge about SdiffMg algebra representations is very obscure.In this situation,it is very important to study high dimensional diffeomophisms groups and construct their unitary representations.There are a few ways to construct unitary representation of infinite dimensional Lie algebras.The first is through the infinite-dimensional Heisenberg algebra or oscillator algebra,which serves as a basic tool for the most of the constructions of Virasoro and Kas-Moody algebras[22].and the second one is through the infinite dimensional Jacoby matrices.We use the second,more transparent,way of construction[23] and find representations of SdiffT^2 as infinite dimensional matrices.It is convenient in the beginning to do that for arbitrary diffeomorphisms of the two dimensional torus T^2 and then for area preserving diffeomorphisms[22].For that we present torus as the product of two unit circles in the complex planes   T^2=S^{1}xS^1 $\in$ CxC. Then it is easy to calculate the Lie algebra of this four dimensional transformations[22]

$$\left[L^1_{\bar{n}}, L^1_{\bar{m}} \right] = (n_1 - m_1) \cdot L^1_{\bar{n}+\bar{m}},$$
$$\left[L^2_{\bar{n}}, L^2_{\bar{m}} \right] = (n_2 - m_2) \cdot L^2_{\bar{n}+\bar{m}}, \tag{46}$$
$$\left[L^1_{\bar{n}}, L^2_{\bar{m}} \right] = -m_1 L^2_{\bar{n}+\bar{m}} + n_2 L^1_{\bar{n}+\bar{m}},$$

where

$$L^1_{\bar{n}} = - z_1^{n_1+1} z_2^{n_2} \frac{\partial}{\partial z_1},$$
$$L^2_{\bar{n}} = - z_1^{n_1} z_2^{n_2+1} \frac{\partial}{\partial z_2}, \tag{47}$$

and z$_1$,z$_2$ are coordinates in the complex planes CxC.Generators of the area preserving diffeomorphisms (30) are linear combination of L^{1}and L^2

$$L_{\bar{n}} = n_2 L^1_{\bar{n}} - n_1 L^2_{\bar{n}}. \tag{48}$$

To find matrix realization of the algebras (46) and (48) we can take appropriate basis in the vector space V$_{\alpha\beta}$[22]

$$\gamma_{\bar{K}} = z_1^{K_1+\alpha_1} z_2^{K_2+\alpha_2} \cdot (dz_1)^{\beta_1} (dz_2)^{\beta_2} \tag{49}$$

where α_1, α_2, β_1, β_2 are complex parameters. One can find that the generators L^1 and L^2 acts on that space as [22]

$$L^1_{\vec{n}} \cdot \mathcal{V}_{\vec{K}} = -(K_1 + \alpha_1 + \beta_1 n_1 + \beta_1) \cdot \mathcal{V}_{\vec{n}+\vec{K}} \ ,$$

$$L^2_{\vec{n}} \cdot \mathcal{V}_{\vec{K}} = -(K_2 + \alpha_2 + \beta_2 n_2 + \beta_2) \cdot \mathcal{V}_{\vec{n}+\vec{K}} \ . \tag{50}$$

The last formulas defines a four-parameter representation of the Lie algebra (46). Note that L^1 and L^2 are diagonal

$$L^1_{oo} \cdot \mathcal{V}_{\vec{K}} = -(K_1 + \alpha_1 + \beta_1) \mathcal{V}_{\vec{K}} \ , \quad L^2_{oo} \mathcal{V}_{\vec{K}} = -(K_2 + \alpha_2 + \beta_2) \mathcal{V}_{\vec{K}} \tag{51}$$

This representation is reducible if [22]

$$i) \ \alpha_{1,2} \in \mathbb{Z} \ , \beta_{1,2} = 0 \ , \quad ii) \ \alpha_{1,2} \in \mathbb{Z} \ , \beta_{1,2} = 1 \ . \tag{52}$$

And they are unitary if and only if

$$\alpha_1 + \beta_1 = \bar{\alpha}_1 + \bar{\beta}_1 \ , \quad \beta_1 + \bar{\beta}_1 = 1 \ ,$$

$$\alpha_2 + \beta_2 = \bar{\alpha}_2 + \bar{\beta}_2 \ , \quad \beta_2 + \bar{\beta}_2 = 1 \ . \tag{53}$$

Now we are ready to realize this elements as an infinite dimensional matrices. For that we identify $\mathcal{V}_{\vec{K}}$ with the matrix with one in the (K_1, K_2) entry and all other entries are zero and introduce four index matrices $E_{\vec{i} \vec{j}}$ with one in the $(\vec{i}, \vec{j})$ entry and all other entries are zero. Clearly that

$$E_{\vec{i} \vec{j}} \cdot \mathcal{V}_{\vec{K}} = \delta_{\vec{j} \vec{K}} \cdot \mathcal{V}_{\vec{i}} \ ,$$

$$\left[E_{\vec{i} \vec{j}}, E_{\vec{n} \vec{m}} \right] = \delta_{\vec{j} \vec{n}} \cdot E_{\vec{i} \vec{m}} - \delta_{\vec{i} \vec{m}} \cdot E_{\vec{n} \vec{j}} \ . \tag{54}$$

Using this basic elements we can express the representation (50) as an infinite dimensional matrices

$$L^1_{\vec{n}} = \sum_{\vec{K}} \left(K_1 - \alpha_1 - \beta_1 (n_1 + 1) \right) \cdot E_{\vec{K} - \vec{n}, \vec{K}} \ ,$$

$$L^2_{\vec{n}} = \sum_{\vec{K}} \left(K_2 - \alpha_2 - \beta_2 (n_2 + 1) \right) \cdot E_{\vec{K} - \vec{n}, \vec{K}} \ . \tag{55}$$

So that finally we can write two parameter matrix realization of area preserving diffeomorphisms as

$$L_{\vec{n}} = - \sum_{\vec{K}} \left(\vec{n} \wedge \vec{K} - \vec{\alpha} - \vec{\beta} \right) E_{\vec{K} - \vec{n}, \vec{K}} \tag{56}$$

426

A direct check shows that this representation satisfies commutation relation (30) and conclude that only adjoint representation of SU(N) has finite limit.

ACKNOWLEDGEMENTS

I thank the organizers of this school for providing us such marvellous place where we have had the opportunity to discuss about many fascinating topics.

REFERENCES

1. G.'t Hooft,A planar diagram theory for strong interaction, Nucl.Phys. B72:461 (1974).
2. G.'t Hooft,A two-dimensional model for mesons,Nucl.Phys.B75:461 (1974).
3. T.Eguchi and H.Kawai,Reduction of dynamical degrees of freedom in the large-N gauge theory,Phys.Rev.Lett.48:47 (1982).
4. A.M.Polyakov,"Gauge fields and strings",Harwood academic publishers, NY,(1987)
5. J.Hoppe,Quantum theory of a relativistic surface and a two-dimensional bound state problem,Ph.D. Thesis MIT (1982).
6. M.E.Laziev and G.K.Savvidy,Transversity of open relativistic membrane, Phys.Lett.198B:451 (1987)
7. G.K.Savvidy,Transversity of a massless relativistic surface.Quantization in the light-cone gauge,preprint CERN-TH 4820/87;Yad.Fiz.47:1176(1988)
8. G.Z.Baseian,S.G.Matinian and G.K.Savvidy,Nonlinear plane waves in Yang Mills theory,Pisma Zh.Eksp.Teor.Fiz.29:641(1979) (in russian).
9. S.G.Matinian,G.K.Savvidy and Ter-Arutyunyan,Classical Yang Mills mechanics.Nonlinear color oscillation,Sov.Phys.Jetp.53:421 (1981).
10. G.K.Savvidy,Classical and quantum mechanics of nonabelian gauge fields, Nucl.Phys.B246:302(1984).
11. T.A.Arakelian and G.K.Savvidy,Cocycles of area preserving diffeomorphism and anomalies in the theory of relativistic surfaces,Phys.Lett. 214B :350 (1988)
12. E.G.Floratos,J.Illiopoulos and G.Tiktopoulos,SU( ),Classical Yang Mills theories,Phys.Lett.217B:285 (1989)
13. B.Biran,E.G.Floratos and G.K.Savvidy,The self-dual closed bosonic membranes ,Phys.Lett.198B:328 (1988).
14. E.G.Floratos and G.K.Leontaris,Integrability of self-dual membranes in 4+1 dimension and toda lattice,Phys.Lett.223B:153 (1989).
15. S.V.Giddings and A.Strominger,Axion-induced topology change in quantum gravity and string theory,Nucl.Phys.B306:890 (1988)
16. D.B.Fuks,Cohomology of infinite dimensional Lie algebras,ed.R.Gamkrelidg (Steklov Institute,Moscow,1986).
17. E.G.Floratos and J.Illiopoulos,A note on the classical symmetries of the closed bosonic membranes,Phys.Lett.201B:237 (1988).
18. I.Bars,C.Pope and E.Sezgin,Central extension of area preserving membrane algebra,Phys.Lett.210B:85 (1988)
19. D.Fairlie,P.Fletcher and C.K.Zachos,Trigonometric structure constants for new infinite algebras,Phys.Lett.218B:203 (1989).
20. C.K.Zachos and D.B.Fairlie,New infinite-dimensional algebras,sine bracket,and SU( ),Phys.Lett.224B:101 (1989)
21. I.Bars,Strings from reduced large-N gauge theory via area preserving diffeomorphisms,preprint USC-90/HEP-12,February (1990).
22. V.G.Kac and A.K.Raina,Bombay lectures on"highest-weight representation of infinite dimensional Lie algebras",World Scientific 1987,Singapore.
23. G.K.Savvidy,Renormalization of infinite dimensional gauge theories, preprint TPI-MINN-90/02-T,January (1990).
24. A.A.Belavin,Hidden symmetry of the integrabel system,JETP Lett.32:169 (1980)

Summary Lecture: Vacuum Structure in Intense Fields

Berndt Müller

Department of Physics, Duke University
Durham, NC 27706, U.S.A.

Introduction

After ten very intense days of lectures on our subject of interest, little new remains to be said, but a great deal needs to be reviewed. Alluding to Pete Carruthers' inspiring lecture, one might say that, in heaven, summer schools are held at Cargèse, while in hell, one gives the summary talk at this splendid place. Despite recurring regrets to have agreed to take on this task, I have also greatly enjoyed the preparations for it, because of the very high level of the lectures presented at this school.

Probably the most exciting aspect was that the school brought together experts from two related, but quite distinct fields: those working in relativistic quantum mechanics in strong and supercritical external fields, and those doing research in strongly coupled quantum field theory. We discovered many similarities, and the discussions helped to deepen my understanding of many basic concepts in both fields. But the most fascinating aspect was the evolving possibility that strong "external" fields may induce (local) variations of the fundamental parameters of quantum electrodynamics: the electron and positron mass, and the interaction strength between charged particles. I will return to this fascinating thought later in my lecture.

The GSI Experiments

We had the good fortune to hear comprehensive reviews from H. Bokemeyer and W. Koenig of their results of the experiments conducted by their respective groups (EPOS and ORANGE) at GSI, where the emission of narrow correlated electron-positron lines was observed in collisions of very heavy ions. What is known about these so-called "anomalous" e^+e^- lines is summarized in Table 1, probably the most highly condensed table of unexplained experimental data ever assembled. I will spare you a detailed repetition of the discussion of these results, but I want to briefly remind you of the basic facts:

1. Narrow lines in the e^+e^- sum-energy spectrum have been observed in at least four different systems of heavy ions: $U + U$, $U + Th$, $U + Pb$ and $U + Ta$. The combined charge of the nuclei in these systems is 184, 182, 174, and 165, respectively. The intrinsic width of the lines appears to be less than 20 keV.

Table 1. Synopsis of the e^+e^- data from GSI.

		560 keV line	620 keV line	750 keV line	810 keV line	
U + U 184	$\langle E_\Sigma \rangle$ [keV]	554 ±6	635 ±6		815 ±8	O
	δE_Σ [keV]	12 ±6	12 ±6		15 ±6	R
	E_beam [MeV/u]	~5.8	~5.8	5.8–6.0	5.9	A
	$\theta_{e^+e^-}$	70°–150°	70°–150°		~180°	N
	$\langle E_\Delta \rangle$ [keV]	~15	+10		~0	G
	δE_Δ [keV]					E

U + U (184) not measured by EPOS

U + Th (182) not measured by ORANGE

		560 keV line	620 keV line	750 keV line	810 keV line	
U + Th 182	$\langle E_\Sigma \rangle$ [keV]		608 ±8	760 ±2	809 ±8	E
	δE_Σ [keV]		25 ±3	< 80	40 ±4	P
	E_beam [MeV/u]		5.86–5.90	~5.83	5.87–5.90	O
	$\theta_{e^+e^-}$		70°–150°		180°±40°	S
	$\langle E_\Delta \rangle$ [keV]		~−10	~+10	~+30	
	δE_Δ [keV]		~170	~140	~130	
U + Pb 174	$\langle E_\Sigma \rangle$ [keV]	574 ±6	not seen	not seen	787 ±7	O
	δE_Σ [keV]	15 ±6			15 ±6	R
	E_beam [MeV/u]	5.9	5.9	5.9	5.9	A
	$\theta_{e^+e^-}$	180°±18°			180°±18°	N
	$\langle E_\Delta \rangle$ [keV]	~0			~+10	G
	δE_Δ [keV]					E

U + Pb (174) not measured by EPOS

		560 keV line	620 keV line	750 keV line	810 keV line	
U + Ta 165	$\langle E_\Sigma \rangle$ [keV]		635 ±6	not seen	805 ±8	O
	δE_Σ [keV]		20 ±5		15 ±5	R
	E_beam [MeV/u]		6.3	6.3	6.3	A
	$\theta_{e^+e^-}$		90°–180°		90°–180°	N
	$\langle E_\Delta \rangle$ [keV]		~+100			G
	δE_Δ [keV]		~200			E
U + Ta 165	$\langle E_\Sigma \rangle$ [keV]	not seen	625 ±8	748 ±8	805 ±8	E
	δE_Σ [keV]		20 ±3	33 ±5	27 ±3	P
	E_beam [MeV/u]	5.3–6.8	6.24–6.38	5.93–6.13	6.24–6.38	O
	$\theta_{e^+e^-}$		90°–180°	60°–120°	180°±30°	S
	$\langle E_\Delta \rangle$ [keV]		+30	−150	+220	
	δE_Δ [keV]		~230	~450	~280	

2. Thanks to more systematic measurements, the two groups now agree on most essential aspects of these lines.

3. So far, four lines have been identified, with sum energies at 560, 620, 750 and 870 keV. It is possible that the locations are not identical in all measured systems.

4. Emission of the e^+e^- lines appears to be back-to-back in some cases, in some others not. The positron and the electron is sometimes emitted with equal energies, sometimes their energies differ by as much as 200 keV.

5. The yield of the line structures exhibits a pronounced beam energy dependence, the maximum occurring slightly above the Coulomb barrier. The EPOS group has found that the dependence on heavy ions differs from that of the smooth background; while the ORANGE group has shown that the signal-to-background ratio is enhanced when they trigger on nuclear inelastic collisions.

W. Greiner reported that these "anomalous" e^+e^- lines have recently also been detected by a group working at Berkeley; this should disperse the last doubts concerning their real existence. However, their phenomenology is so complex that they do not fit into any simple scheme, and I am afraid that this school has not—at least not obviously—brought us closer to the resolution of their puzzling nature. The real success of this school lies rather in the progress of basic theoretic understanding. Here I speak for myself, but I hope that most of you share my assessment. With this I turn to the theory lectures.

QED in Strong External Fields

W. Greiner and G. Soff in their lectures reviewed for us the ideas that originally motivated the GSI experiments. QED in very strong Coulomb fields of central charge $Z > \alpha^{-1} \simeq 137$ was predicted to have very special properties that were first tested in the heavy ion experiments. The most important prediction, that pairs could be created spontaneously from the vacuum, i.e. without additional energy than that stored in the static Coulomb field, for $Z > 173$ has not yet been clearly observed in the experiments. However, one must stress that this phenomenon would only be associated with line emission of positrons, if the nuclei in a heavy ion collision would stick together as a giant composite system for much more than 10^{-20}s. The absence of a clear signal for spontaneous pair production probably means that nature has not been sufficiently benevolent.

Other predictions of QED in strong Coulomb field have been verified by the experiments, such as the dramatic A^{20}-dependence of the total e^+e^- pair production cross section. The beautiful absolute agreement between theory and experiment has only been possible due to the development of analytical and numerical techniques for solving QED in *all* orders of $(Z\alpha)$, up to order α in radiative corrections, for arbitrarily strong fields. These methods have been extended by the Frankfurt theory group to time-dependent fields in order to describe e^+e^- pair production in heavy ion collisions. As Greiner and Soff showed, theory predicts neither sharp e^+ lines, nor narrow e^+e^- coincidence lines. This raises the question: what, if not QED up to order $\alpha(Z\alpha)^n$, explains the observed lines structures?

Only two serious attempts were made here to interpret the GSI data quantitatively, by W. Greiner and H. Fried. Greiner presented the results obtained in Frankfurt by showing us the virtues exhibited by a new family of light neutral, composite particles: there seems to be no contradiction to particle search experiments if the size is much larger than 1000 fm; its production almost at rest in the nuclear c.m. system can be easily understood; the presence of several states under varying conditions and the different decay characteristics of the various lines could be naturally explained. The specific model developed and studied by the Frankfurt group involves new constituent fermions $f^{\pm}$ bound together by a weak, but confining force. At short distances the f^+f^- interaction must be strongly repulsive, in order that the model does not violate QED precision data. Although the (f^+f^-) model looks quite exotic, it must be taken seriously since it successfully predicts the measured cross section of the e^+e^- lines within a factor of two!

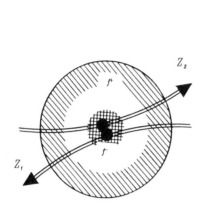

Fig. 1. The f^+f^- model.   Fig. 2. Loop bremsstrahlung.

The model of H. Fried is based on the idea that the virtual vacuum polarization modifies the local Coulomb fields, especially right between the two nuclei. Note that for $Z\alpha$ of order 1, the total absolute charge $\int d^3x|\rho_{VP}(x)|$ contained in the vacuum polarization is of order one unit of elementary charge. On the other hand, calculations show that an e^+e^- pair gets on mass shell only with probability 10^{-4}. Even a small influence of the virtual pairs on the production of real pairs could therefore modify the e^+e^- pair yield substantially. Now ρ_{VP} has Fourier frequency components $\omega \sim (2m_e)^{-1}$, which differ from those contained in the external Coulomb field. So one can easily envisage interference patterns from the two different Fourier spectra, which would be sensitive to changes in the heavy ion trajectory. Fried contends that resonance-like structures arise when the "soft" electromagnetic interaction of the vacuum polarization with itself is taken into account to all orders in α. The virtual excitations then materialize into a real pair by a bremsstrahlung-like process. Although he succeeds in fitting the observed line spectrum, many questions remain in view of the severe approximations required to get analytical results. Full-scale numerical studies of higher-order α effects are clearly needed.

A much discussed question at this school was whether the (f^+f^-) model could actually be an effective description on nonperturbative aspects of QED. In other words, could the constituents "$f^{\pm}$" be some sort of "heavy" electron, relieving the need to postulate new fundamental fermions? A. Chodos, D. Caldi, J. Ng, and D. Owen discussed this possibility in various forms, answering the question with: Yes!, if QED in a strong external field breaks chiral symmetry spontaneously (in addition to the explicit chiral symmetry breaking due to the electron mass m). The criterion for chiral symmetry breaking is:

$$\lim_{m \to 0} \langle 0|\overline{\psi}(x)\psi(x)|0\rangle \neq 0.$$

It is probably unrealistic to expect $\langle\bar\psi\psi\rangle$ to be nonzero globally, if the external field $A^\mu(x)$ is localized, but one may hope for a nonvanishing vacuum expectation value over a sufficiently extended region, say several hundred fermi. Then the electron mass would be effectively increased over the region, and the $f^\pm$ could be identified with the heavy e^+e^-. D. Caldi showed us that, making use of Schwinger's representation, we may write

$$\langle 0|\bar\psi(x)\psi(x)|0\rangle = im\int_0^\infty ds\, e^{-im^2 s} tr\langle x|e^{-i\mathcal{H}s}|x\rangle,$$

where

$$\mathcal{H} = -(\gamma\cdot\Pi)^2 = -\Pi^2 + \frac{e}{2}\sigma\cdot F,$$

and he argued that

$$U(x,x;s) = tr\,\langle x|e^{-i\mathcal{H}s}|x\rangle \xrightarrow[s\to\infty]{} Cs^{-1/2}$$

is required for spontaneous chiral symmetry breaking. Let's see again why this is so. Assume that $U(x,x;s)\to Cs^{-\alpha}$. Then by introducing the variable $x = m^2 s$, we have

$$\langle\bar\psi\psi\rangle \simeq mC\int^\infty \frac{ds}{s^\alpha}\,e^{-im^2 s} \simeq m^{2\alpha-1}C\int^\infty \frac{dx}{x^\alpha}\,e^{-x},$$

where the complete m-dependence is buried in the factor $m^{2\alpha-1}$. Obviously, a finite nonzero limit exsits only if $\alpha = \frac{1}{2}$.

In order to obtain an understanding of what type of external field can possibly produce such a behavior, let's begin with no field at all. Then we have

$$\langle x|e^{-i\mathcal{H}s}|x\rangle = \int\frac{dp_0}{2\hbar}\int\frac{d^3 p}{(2\hbar)^3}\,e^{is(p_0^2-\vec{p}^2)} = \frac{1}{(4\pi)^2 is^2}.$$

Next, consider a homogeneous magnetic field $\vec{B}$. In this case, the spatial operator $\vec\Pi^2 + e\vec\sigma\cdot\vec{B}$ is diagonalized by the Landau states

$$\varphi_n(\vec{x}_\perp)\exp(-ip_L x_L)\chi_\sigma$$

with eigenvalues

$$\epsilon_{n\sigma}(p_L) = (2n+1-\sigma)eB + p_L^2.$$

The large s behavior of $U(x,x)$ is dominated by the lowest eigenvalue, $n = 0$, $\sigma = -1$, with continuous longitudinal excitation spectrum. The asymptotic behavior is therefore

$$tr\,\langle x|e^{-i\mathcal{H}s}|x\rangle \xrightarrow[s\to\infty]{} \bar\varphi_0\varphi_0\frac{dp_0}{2\pi}\int\frac{dp_{11}}{2\pi}\,e^{is(p_0^2-p_{11}^2)} = \frac{1}{4\pi s}\bar\varphi_0(x_\perp)\varphi_0(x_\perp)$$

which is still falling too fast. However, it becomes clear that the desired $s^{-1/2}$-behavior would be obtained, if the integration over the continuous excitation spectrum of the lowest Landau level would be absent, i.e. if the lowest eigenvalue of the operator $\vec\Pi^2 + e\vec\sigma\cdot\vec{B}$ were discrete, say ϵ_0 and $\varphi_0(x)$. Then we would have

$$\text{tr}\,\langle x|e^{-i\mathcal{H}s}|x\rangle \underset{s\to\infty}{\longrightarrow} \left(\frac{1}{4\pi s}\right)^{1/2} \overline{\varphi}_0(x)\varphi_0(x)\, e^{-is\epsilon_0},$$

the desired $s^{-1/2}$ behavior needed for chiral symmetry breaking.*

Do such fields exist? I do not know. Examples of magnetic fields for which the operator $(\gamma \cdot \Pi)^2$ has negative eigenvalues were given by G. Stanciu [*J. Math. Phys.* **8**, 2043 (1967)], e.g. $H_z = H\,\text{sech}^2 ay$, but even for those the "bound" states have a continuous excitation spectrum, because charged particles are free to more along the z-direction. Maybe it would suffice to introduce an electric field that restricts this motion? Or it may be adequate to localize the particles in all directions by introducing boundary conditions. C. Martin showed in a beautiful lecture how reflecting boundary conditions can be incorporated into the Schwinger approach by means of a multiple reflection expansion. Such possibilities remain to be further explored.

Strongly Coupled "QED"

In his first lecture, H. Fried reminded us that QED is not just "Dirac particles in external fields" but also entails fluctuations of the electromagnetic field. The functional integral Z_{QED} contains an integration over *all possible* field configurations, and it is not sufficient to consider only the classical field generated by the external source. The question is then: can real QED exhibit different phases? This was discussed at length by E. Dagotto and A. Kocic, and their present answer—as I will carry it home—is: perhaps!

Let me now review their message: In order to study QED on a lattice it is necessary to include a bare four-fermion interaction of the Nambu-Jona-Lasinio type:

$$G_0[(\overline{\psi}\psi)^2 + (\overline{\psi}i\gamma_5\psi)^2].$$

One then looks for a second-order phase transition permitting to take the continuum limit and analyzes the behavior of

$$\langle\overline{\psi}\psi\rangle|_m \sim A\, m^{1/\delta}.$$

The critical line in the $\alpha_0 - G_0$ diagram obtained in quenched QED is shown in the diagram; it deviates from that obtained by an analysis of the lowest-order Schwinger Dyson (S-D) equations in that it allows a consistent continuum limit for $G_0 = 0$. This point, however, corresponds to a coupling α_0 of order 1. Even if dynamical fermions are included (work here is still in progress), it seems most likely that any continuum theory with $\alpha = \alpha_{\text{QED}} = \frac{1}{137}$ will need a bare $G_0 \neq 0$ and then exhibit strong chiral symmetry breaking even in the absence of any external fields. In other words: QED as we know it can probably not be constructed as a consistent continuum theory. [Note that this poses no real conceptual problem, since we may take the Planck

---

* At this point it may be tempting to conjecture as I did in the original summary talk that the same arguments hold for a Coulomb potential. Surely, there the lowest eigenvalues of the Dirac Hamiltonian are discrete, since the Coulomb potential has bound states; however, a more detailed inspection of the spectral properties of the operator $(\gamma \cdot \Pi)^2$ shows that it is unlikely that the desired behavior of $U(x,x;s)$ can arise for any electrostatic potential, whatever its strength is.

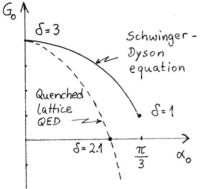

Fig. 3. Critical line for quenched QED.

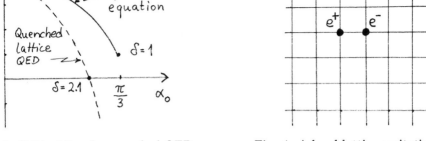

Fig. 4. A local lattice excitation.

length $\Lambda = L_p^{-1}$ as cut-off, where $\alpha(\Lambda)$ is still only about $\frac{1}{50}$, far away from the Landau pole.]

Of some relevance to the question of a possible strongly coupled continuum QED-like gauge theory is the positronium problem at strong coupling, which was discussed here by M. Bawin. He showed that within the framework of relativistic quantum constraint dynamics there is no instability in the fermion-anti-fermion system as function of the coupling constant α_0. What happens is that beyond some $\alpha_0 > \alpha_c$ there is a tendency for the bound state to collapse, but when a cut-off a is introduced, the invariant mass W of the bound state remains positive for all α_0. The asymptotic behavior

$$ W \approx 2m^2 a/\alpha_0 \qquad (\alpha_0 \gg \alpha_c) $$

indicates nonanalyticity in the coupling constant.

This result has direct bearing on the strongly coupled QED lattice theory. Consider a fermion $(e^+ e^-)$ pair excitation separated by a single lattice spacing a. For $\alpha_0 \ll 1$ the energy of this fluctuation is given by the associated kinetic energy, which must be of order a^{-1}. When the interaction becomes strong, the pair can form a bound state of size of the lattice cut-off, and its total energy will only be of order $2m^2 a/\alpha_0$, which is much less than a^{-1}. Hence at $\alpha_0 > \alpha_c$ there will be many more short-range fluctuations in the fermion vacuum, and local density terms such as $(\overline{\psi}\psi)^2$ become important.

QED at Finite Chemical Potential

This topic was discussed several times, by A. Chodos, D. Owen, J. Ng, and J. Polonyi. Before reviewing the interesting results presented here, let me briefly elaborate on an apparently trivial point: why does the addition of a constant μ to all single particle energies not correspond to a trivial gauge transformation? The answer can be given in many ways; I like to see it as follows. The partion function is given by

$$ Z = \sum_n e^{-\beta(E_n - \mu)}. $$

435

What enters in the exponent is the energy E_n of a particle state, for which a gauge invariant expression is given by $E_n = \int d^3x \langle n|T^{00}|n \rangle$, where T_{00} is the timelike component of the energy-momentum tensor. T^{00} is gauge invariant, the Hamiltonian is not; they are related by:

$$T^{00} = H - \nabla \cdot (A_{ext} \vec{E}_D) - \vec{E}_D \cdot \partial \vec{A}_{ext}/\partial t,$$

where $\vec{E}_D$ is the electrical field generated by the Dirac particle, i.e. $\nabla \cdot \vec{E}_D = e \overline{\psi} \gamma^0 \psi$. It is easy to show that this implies the relation

$$E = \int d^3x\, T^{00} = H - e\, A^0_{ext}(\infty) = \gamma^0 \vec{\gamma} \cdot \vec{\Pi} + \gamma^0 m + e[A_0(x) - A_0(\infty)],$$

i.e. the electrostatic potential at $|x| \to \infty$ does not contribute to the energy of a state. Hence explicit addition of a constant $(-\mu)$ to the energy does not correspond to a gauge transformation, but represents a nontrivial change in the physical properties.

As J. Polonyi showed in his lectures, QED at finite μ, but $A^\mu_{ext} = 0$, i.e. the relativistic electron gas, has its surprises. For example, adding up the infinite series of photon ladder diagrams to the vertex correction one finds that the long-range interaction among electrons is described by an "effective charge"

$$e + \tfrac{1}{2}e + \tfrac{1}{4}e + \ldots = 2e.$$

This leads immediately to the question whether more complicated diagrams,

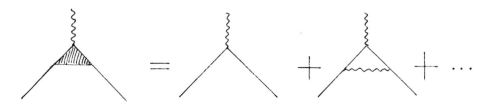

Fig. 5. Vertex corrections in QED.

or external fields, could perhaps lead to the effecting coupling

$$e \longrightarrow e_{\text{eff}} = \alpha^{-1/2} e = 12e, \quad e^2_{\text{eff}}/4\pi \simeq 1?$$

This brings us to the problem, mentioned several times during this meeting, whether a finite chemical potential is compatible with a nonvanishing electric field. I think that $\mu \neq 0$ and $\vec{E} \neq 0$ are not conceptually consistent conditions. The point is that μ can be considered as the Lagrange multiplier in the substitution

$$H \longrightarrow H - \mu J_\alpha u^\alpha$$

in the partition function $Z = \text{Tr}(e^{-\beta H})$ of the electron- positron field, where J^μ is the e^+e^- current and $u^\alpha = (1,0,0,0)$ defines the rest frame in which thermodynamic equilibrium holds. The stationarity condition of equilibrium requires that the Hamiltonian commutes with the constraint:

$$0 = i[H, J_\alpha] = \frac{d}{dt} J_\alpha = \frac{1}{m} \int d^3x\, F_{\alpha\beta} u^\beta.$$

In the rest frame, this implies $F_{i0} = E_i = 0$, i.e. the absence of an electric field. Intuitively, this condition is obvious: $\mu \neq 0$ describes the presence of a nonvanishing vacuum charge; an external electric field acts as a force dislocating this charge.

A simple consideration also explains the nonanlyticity of the limit $\mu \to 0$ in the presence of a constant electric field (if one insists on such a combination which, as I just argued, cannot correspond to thermodynamic equilibrium). This is best seen in the static gauge $A_0(x) = -\vec{E} \cdot \vec{x}$. A nonvanishing μ corresponds to the situation illustrated on the left-hand side, all stationary states filled up to $E_F = \mu$. A change $\Delta\mu$ is equivalent to a spatial translation $\Delta x = -\Delta\mu/E$. The limit $\mu = 0$ corresponds to the Fermi surface at $E_F = 0$. On the other hand, the Heisenberg-Euler-Schwinger Lagrangian describes the situation where all states moving to the left are considered as antiparticle states, all states moving to the right as particle states. Effectively, this corresponds to a Fermi energy $E_F = -\vec{E} \cdot \vec{x}$, located in the linearly changing gap between particle and antiparticle states. Obviously, this situation is not recovered by the equilibrium limit $\mu \to 0$. In many respects the dichotomy exposed here is analogous to that encountered in the resolution of Klein's paradox discussed by J. Rafelski in his second lecture.

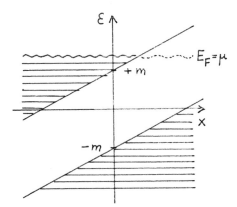

Fig. 6. Dirac sea with $E \neq 0$ and $\mu \neq 0$.

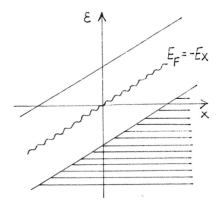

Fig. 7. Dirac sea with $E \neq 0$ and the Schwinger boundary condition.

For $E = 0$, $B \neq 0$ there arises no problem; finite μ and T are easily treated. The case $\mu = 0$, $T \neq 0$ was discussed by W. Dittrich [*Phys. Lett.* **83B**, 67 (1979)]. More interesting is the temperature- dependent effective potential $U(B,T)$ for Yang-Mills theories. There the zero temperature effective potential exhibits a minimum at finite $B \neq 0$ first discovered by G. Savvidy. The existence of unstable modes shows that this minimum does not correspond to the true ground state of the Yang-Mills field. I do not want to review here the attempts of the Copenhagen school to improve on this vacuum, but rather tell you what occurs at finite temperature. Thermal equilibrium requires the explicit exclusion of the unstable modes, which can be achieved in a gauge- and Lorentz-invariant way with the help of the proper time technique. The result for the free energy density reads:

$$-f(B,T) = \frac{T}{V} \ln Z(B,T) = -\frac{1}{4\Pi)^2} \text{tr} \int_0^\infty \frac{ds}{s^3} (gBs) J_\theta \left(0, \frac{1}{4i\pi s T^2}\right)$$
$$\cdot \left[\frac{1}{\sin igBs} - ie^{-gBs} + ie^{igBs}\frac{\Gamma(\frac{1}{2}, igBs)}{\Gamma(\frac{1}{2})}\right],$$

where J_θ is the Jacobi theta function [B. Müller, J. Rafelski, *Phys. Lett.* **101B**, 111 (1989)]. After renormalization of the $T = 0$ part, the free energy density exhibits a first-order phase transition at $T_c = 1.435\Lambda$, where Λ^4 is the non-perturbative energy density of the Savvidy vacuum:

$$\Lambda^4 = f(B_{\min}, T = 0) - f(B = 0, T = 0).$$

If this quantity is identified with the MIT-bag constant, one obtains a critical temperature $T_c \simeq 200$ MeV, which is in surprisingly good agreement with modern computer simulations of lattice Yang-Mills gauge theory.

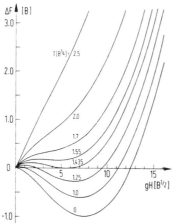

Fig. 8. Free energy density $f(B, T)$ for the Yang-Mills field.

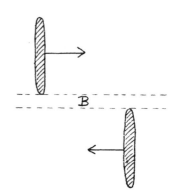

Fig. 9. Two colliding Pb nuclei.

Strong Fields in Channeling

The conceptual and phenomenological situation with respect to strong Coulomb fields in heavy ion scattering being still unclear, it is wise to consider alternative test of ideas and models. A. Sørensen described how extremely strong electromagnetic fields can arise in crystal channeling of very energetic electrons. The idea is simple: observed from the frame of the fast moving particle, the crystal appears highly Lorentz contracted and the static crystal field is enhanced by the same factor. For 100 GeV electrons we have

$$E^* = \gamma E_{\text{crystal}} \simeq 2 \cdot 10^5 E_{\text{crystal}}.$$

For appropriate materials, e.g. Ge or W, the effective field strength E^* can be many times the so-called "critical" field strength $E_c = m^2/e$. (E^*/E_c) is the expansion parameter of QED in an external electric field, hence radiative corrections become nonperturbative in E for $E^* \gg E_c$. Some of the consequences, e.g. a dramatic rise of bremsstrahlung and pair production, have been observed at CERN.

S. Graf explained that the strong fields encountered in channeling would give rise to copious production of $(f^+ f^-)$ particles, if the Frankfurt model were correct. This can be experimentally tested, and efforts are under way to do so. I think the lesson is more general: *any model that predicts peculiar effects occurring in strong fields in heavy ion collisions should also be applied to channeling fields.* A clear prediction should be made what its consequences would be in this context. If no specific signals are predicted, it should be understood why not. In many respects, e.g. due to the absence of nuclear excitation, the experimental situation may even be cleaner for channeling than for heavy ions.

Superstrong Electromagnetic Fields

The prospect of electromagnetic fields of ultimate strength emerge when the two ideas: high Z and high γ, are combined. G. Soff and M. Greiner discussed that the fields created in collisions of heavy nuclei at energies of several TeV/nucleon are sufficient to produce Higgs bosons, if the Higgs mass lies somewhere between 100 GeV and 200 GeV. The suppression caused by nuclear scattering is still debated, but it appears likely that a measurable count rate would result if Pb ions could be accelerated in the SSC. Cross section for W-bosons and possibly other yet unknown particles may be even higher.

One day after lunch I had a discussion with G. Savvidy about the question whether truly nonperturbative effects could arise on the level of electroweak theory under such conditions. E.g. it has been pointed out by Skalozub [Yad. Fiz. **41**, 1650 (1985)] that $W^\pm$ pairs would condense in the vacuum if an external magnetic field exceeds the limit $eB \geq m_w^2$. The magnetic field at the surface of a fast moving nucleus is $B = Ze\gamma/R^2$; hence the condition could be satisfied in a nuclear collision if

$$\gamma \geq (m_W R)^2 / Z\alpha \simeq 10^7.$$

With the plans for LHC and SSC, Lorentz factors of order $\gamma = 10^4$ appear within reach. This is still some way off, but not outrageously so, and I think the idea warrants some further thoughts.

Solitons, QCD, and all that

Several lectures (Cottingham, A. Schramm, H. Minakata) dealt with the intriguing aspects of the Dirac vacuum in topologically nontrivial background fields. Here the vacuum can be the agent that carries a conserved charge, e.g. baryon charge N_B in the Skyrme model, when the pion field configuration has the "hedgehog" shape

$$U(r) = \exp(i\vec{\tau} \cdot \hat{r} F(r)), \qquad N_B = \frac{F(0) - F(\infty)}{\pi};$$

another example are magnetic monopoles in unified gauge theories. H. Minakata discussed the possibility of soliton excitations of QED in strong Coulomb fields. Since the treatment is based on a bosonized formalism of spherically symmetric QED, it somewhat defies an intuitive understanding. However, if the excitations exist, they should also be seen in the equivalent fermionic formulation. This ought to be investigated. I am only aware of one ongoing study in this direction, by D. C. Ionescu at Frankfurt, who searches for collective RPA modes of the QED vacuum in Coulomb fields with $Z\alpha > 1$. It would be interesting to relate this to the bosonized formalism.

Some of the common wisdom concerning confinement in QCD was challenged by A. Patrascioiu. He did not attack QCD directly but rather dealt with the subleties of $O(N)$ lattice theories in two dimensions for $N \geq 3$. The tantalizing question of confinement here is related to the order of the continuum and infinite volume limits ($a \to 0$, $L \to \infty$). On the basis of a technically involved proof that makes use of results from percolation theory, Patrascioiu claims that the susceptibility $\chi = \sum \langle \sigma_0 \sigma_x \rangle$ diverges for small β, implying the absence of a mass gap. Apart from technical aspects, the big question is whether his arguments can be carried over to non-abelian gauge theories. Does QCD predict color confinement? Although the numerical results point to that, a rigorous proof is still outstanding.

Another simple model field theory, $\lambda\phi^4$, was the subject of M. Consoli's lectures. It has been shown to be a "trivial", i.e. noninteracting, quantum field theory in $D > 4$ dimensions, but is it trivial in four dimensions? The variational approach, based on Gaussian wave functionals

$$\Psi_G[\phi] \simeq \exp\left[-\frac{1}{2} \int dx dx' (\phi - \phi_0)_x G(x - x')(\phi - \phi_0)'_x\right]$$

indicates that a consistent nontrivial theory may exist. The basic question to be answered in this context is whether the expression $(\lambda\phi^2)$ is a renormalization group invariant. Clearly, the answer to this question has important implications for the Higgs model of spontaneous symmetry breaking which is so fundamental to most unified gauge theories.

Outlook

Let me end on a speculative note. As J. Polonyi explained in his lecture this morning, an external electromagnetic field can, and in principle will, influence the flow of renormalization group trajectories in coupling constant space (α, G, and m for QED with 4-fermion interaction). Since external fields usually have limited gradients one normally would expect that this influence is limited to the infrared, i.e. low energy, limit of the trajectory. However, it could be that the trajectories near the ultraviolet fixed point, close to which the theory is defined by a cut-off, are unstable, perhaps in the sense of a bifurcation.

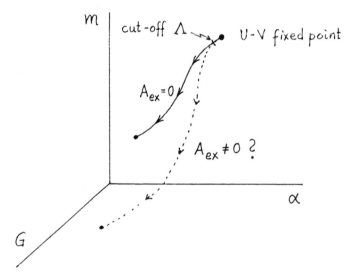

Fig. 10. RG-trajectory flow without (solid) and with (dashed) external field.

An infinitesimal change in the trajectory near this point could then result in a drastic change of the low energy end point of the trajectory. The chances that this situation is realized in nature are admittedly very slim, but it cannot be ruled out at present.

Now I remind you that most of the fundamental constants of nature, i.e. fermion and gauge boson masses, the mixing angles, and possibly even the gravitational constant, are quite likely properties of the vacuum state of some quantum field theory. If we could manipulate the vacuum state by modifying the effective coupling constants with the help of external fields, this might eventually lead to ways of controlling the fundamental constants and laws of nature! This sounds like bad science fiction, and quite likely it is. But the lectures I heard at this school and the discussions I had with many of you have made me realize that this vision might also turn out to be real science. It will be fun to continue to work in this field, and I hope we will eventually find out!

INDEX